IET TRANSPORTATION SERIES 16

ICT for Electric Vehicle Integration with the Smart Grid

Other related titles:

Volume 1	**Clean Mobility and Intelligent Transport Systems** M. Fiorini and J.-C. Lin (Editors)
Volume 2	**Energy Systems for Electric and Hybrid Vehicles** K.T. Chau (Editor)
Volume 5	**Sliding Mode Control of Vehicle Dynamics** A. Ferrara (Editor)
Volume 6	**Low Carbon Mobility for Future Cities: Principles and applications** H. Dia (Editor)
Volume 7	**Evaluation of Intelligent Road Transportation Systems: Methods and results** M. Lu (Editor)
Volume 8	**Road Pricing: Technologies, economics and acceptability** J. Walker (Editor)
Volume 9	**Autonomous Decentralized Systems and Their Applications in Transport and Infrastructure** K. Mori (Editor)
Volume 11	**Navigation and Control of Autonomous Marine Vehicles** S. Sharma and B. Subudhi (Editors)
Volume 12	**EMC and Functional Safety of Automotive Electronics** K. Borgeest
Volume 25	**Cooperative Intelligent Transport Systems: Towards high-level automated driving** M. Lu (Editor)
Volume 38	**The Electric Car** M.H. Westbrook
Volume 45	**Propulsion Systems for Hybrid Vehicles** J. Miller
Volume 79	**Vehicle-to-Grid: Linking electric vehicles to the smart grid** J. Lu and J. Hossain (Editors)

ICT for Electric Vehicle Integration with the Smart Grid

Edited by
Nand Kishor and Jesús Fraile-Ardanuy

The Institution of Engineering and Technology

Published by The Institution of Engineering and Technology, London, United Kingdom

The Institution of Engineering and Technology is registered as a Charity in England & Wales (no. 211014) and Scotland (no. SC038698).

First published 2019

The Institution of Engineering and Technology
Michael Faraday House
Six Hills Way, Stevenage
Herts, SG1 2AY, United Kingdom

www.theiet.org

British Library Cataloguing in Publication Data
A catalogue record for this product is available from the British Library

ISBN 978-1-78561-762-1 (hardback)
ISBN 978-1-78561-763-8 (PDF)

Typeset in India by MPS Limited

Contents

Preface

Smart power grid technologies pertain to the system wherein the energy resources, storage, information flow, feedback loop involve a complex decision-making in dynamic interconnected power system. Among the technologies, electric vehicles (EVs) have spurred new paradigms for smart grid operation. Recently, researchers and manufacturers in EV sector have much talked about the application of EVs in grid operation for its effective management.

EVs offer high energy efficiency and a cleaner mode of personal transportation. The EV is a new type of load being accommodated into power network. Its impact on power system will depend on its penetration level. A large-scale adoption of EVs will pose new challenges to system operators since their charging can cause technical problems, such as voltage limits violating or line congestion, mainly at the distribution level.

The fact that EV is a highly distributed electric energy storage device, its integration into the grid system gives an opportunity to support the system operation. It is possible to mitigate either partly or fully the negative impacts on integrating EVs in the grid, with maximization of positive impacts. Although EVs introduce new challenges on grid capacity, it may support the grid stability. There is a huge prospect of exploitation of EVs for smart grid applications.

Therefore, it is interesting to develop tools and strategies that allow tackling these or anticipating their consequences. In general, the so-called EV aggregators will try to maximise their benefits by allocating charging to the most favourable time periods. EVs in smart grid operation require a wide coverage with flexible and cost-effective communication networks. The communication, networking and information technologies will play a vital role in the development of smart grid by supporting two-way energy (charging & discharging of EVs) and information flow. This will enable efficient monitoring, control and optimization of power imbalance in the grid.

In general, the chapters in book contribute to address the EVs as a driving source for realizing the smart grid operation. The book includes chapters from multi-disciplinary research/industry communities, related to EVs charging schemes/technologies, and its associated communication, networking and information architectures, and ancillary services of EVs for power grid management.

Chapters 1–4 present the use of communication network/infrastructure including communication properties, cloud-based energy management service for V2F integration. Chapters 5–7 discuss power/energy management strategy applying optimization models. Chapter 8 presents peer-to-peer energy market model. Chapters 9–11 contribute in the lines of EVs as storage application for grid management. Chapters 12–15 discuss on charging/discharging strategy of EVs with integration of renewable energy resources in the distribution grid.

Nand Kishor and Jesús Fraile-Ardanuy
Editors

Chapter 1
Advanced mobility communication

Jorge Alfonso Kurano[1], David Jiménez Bermejo[1], Jesús Fraile-Ardanuy[1], Sandra Castaño[2], Julia Merino[3], and Roberto Álvaro-Hermana[4,5]

1.1 Introduction

Information technology and the massive use of data is improving people's lives by modernising infrastructures as transportation, energy and health, and promoting new citizen-centred services aiming at building the city of the future [1]. Transport is on the brink of a new era of smart mobility where infrastructure, transport means and users are increasingly interconnected to achieve optimised management models involving higher efficiency, lower costs while bringing less environmental impact. Innovation efforts are focused on applying physical internet to create improved model for holistic mobility management by means of dynamical capability to respond to a wide set of parameters, while empowering sustainability through the reduction of carbon footprint.

The urban and interurban mobility environment has been evolving in recent decades. Traditional factors that have been pushing these changes have been changes in motorisation, residential and urban sprawl, and the spatial distribution of the population, the workplaces and other societal relevant locations. Currently, more than a half of the world population is living in cities, reaching almost 72% in the European Union (EU), and it is estimated that over 80% of the European population will live in urban areas by 2020. The growth of the cities will also generate new city mobility challenges to be solved.

[1]Information Processing and Telecommunications Center (IPTC), Universidad Politécnica de Madrid, Madrid, Spain
[2]Escuela Técnica Superior de Ingeniería y Diseño Industrial (ETSIDI), Universidad Politécnica de Madrid, Madrid, Spain
[3]TECNALIA, Parque Científico y Tecnológico de Bizkaia, Derio, Spain
[4]Orkestra – Basque Institute of Competitiveness, Bilbao, Vizcaya, Spain
[5]University of Deusto, Bilbao, Vizcaya, Spain

These factors have had an impact in the key areas of transportation, efficient mobility, road safety, security and environmental sustainability. And different technological developments have followed these changes in different fields, in an attempt to mitigate the problems arising in these areas:

- Problems with daily congestion in urban areas, overload of the public transportation system, increase in the cost of mobility services (not only monetary, but also in terms of energy consumption and impact on the environment)
- An immediate and direct increase in the number of incidents, accidents, injured and fatalities related with transportation
- An immediate impact in the increase of emissions, pollutants, noise and energy consumption – basically, fossil fuels
- An increase in the risk of incidents related with the management and control of traffic and mobility in general
- Logistics and supply chain management is about providing the required products at the right place at the right time and at minimum cost. Doing so in an urban area is very challenging due to higher complexity environment. Additionally, urban logistics negatively affects the lives of the citizens living in these cities, causing traffic congestion and greenhouse gas (GHG) emissions and affecting their health due to the air pollution and noise emissions. Another important factor in urban logistics is the recent growth of e-commerce (around 14% year-over-year growth in the EU), which has increased the total number of freight movements in the city.

Intelligent Transportation Systems (ITS), understood as the systems that apply the Information and Communication Technologies (ICT) which include electronics, automatic control, computer processing and communications to the field of transport, are aimed at addressing these key areas of transportation. Thus, they can be extended to all modes of transportation and all their elements can be considered: vehicles, infrastructures and users.

In the last decades, several initiatives have had a profound impact on the way mobility is managed. Continuous development of ITS has made available the mechanisms to increase their awareness of the traffic environment to users and mobility managers. This increased awareness of the traffic and its environment conditions is an effective tool towards improving mobility efficiency, safety and environmental sustainability, but only if the parties involved actively share information. And this is after all the main reason for the ITS, Cooperative ITS and Cooperative Services concepts. The European Commission (EC) has been very active in the activities towards the improvement of aspects of road mobility efficiency, road safety and sustainability via the promotion of ITS development. At this point, transportation is on the verge of large-scale transformation through new electric vehicles, big digital platforms, governments' laws and the advances in software, data and sensing leading in new smarter integrated systems (Figure 1.1).

Interestingly, as technologies have allowed a better knowledge about the mobility environment, new trends in urban and interurban mobility have been

Figure 1.1 Cooperative systems. Area of application

identified, or have emerged as consequences of changes in other fields. The most important of these are:

- **Electric mobility.** The increase in the awareness of the limitations and dangers of the fuel-based mobility has had, as a consequence, a rapid development of the technologies related to electric mobility, from energy production and storage, to recharging points, etc.
- **Mode shifts in mobility.** Due to a combination of reasons, it is becoming increasingly less attractive to own a private vehicle. Particularly in urban areas, the increasing offer of car-sharing and car-pooling options is indicative that mobility is opening to a true shift in mobility modes.
- **Connected and automated mobility.** As a consequence of the trends towards electric and alternative mobility modes, there is more need than before of a true real-time dynamic awareness of the mobility environment. Initially conceived as awareness around a vehicle primarily for safety purposes, communication technologies and services for vehicles (ITS, C-ITS) have increasingly played a role in urban and interurban traffic management.

If we move a few years ahead with these trends, it is not difficult to picture a large proportion of purely electrical and shared/pooled vehicles, those being the core of the mobility offer in the cities. On the other hand, while the number of charging locations has increased, it is still not a completely widespread network, and definitely the number of parking spaces inside the cities is limited. It is positive though that vehicles can share their stored energy capacity, and that the private vehicle owners mostly charge their vehicles overnight at home.

It is clear that the communication needs are far from what is necessary to fulfil this scenario. We have potentially hundreds of thousand users making use of a large pool of available cars from different companies (and without counting advancing on the ideas of shared-cargo or advanced electric-based logistic urban services), with limited charging points and parking places. The need for real-time accurate information is critical, and not only about the vehicles themselves, but also about their availability, the charging levels and potential to provide energy back to the network and where and how to do this. The number of involved stakeholders in these exchanges is also relevant.

This all makes communications a central piece is pushing an all-electrical, connected/automated urban mobility scenario ahead. This chapter will present some of the technologies used today to achieve this scenario, and what seem to be the trends to address the issues that are offered Connected Vehicles and especially connected Autonomous Driving (AD) vehicles bring a whole new ecosystem with new requirements on the Cloud and the network architecture to support the new workloads and to satisfy the real-time service requirements [2]: a scenario where the emerging 5G technologies seem to be able to cover all the previously dreamed needs for a realistic and global ITS deployment (Figure 1.2).

1.2 Background

1.2.1 Cooperative ITS regulatory framework

ITS are vital for increasing safety and tackling Europe's growing emission and congestion problems. They can make transport safer, more efficient and more sustainable by applying various ICT to all modes of passenger and freight transport. Moreover, the integration of existing technologies can create new services. ITS are key to support jobs and growth in the transport sector. But in order to be effective, the roll-out of ITS needs to be coherent and properly coordinated across the EU (Figure 1.3).

The EC is working with Member States, industry and public authorities to find common solutions to the various bottlenecks for deployment. Through financial instruments, the EC supports innovative projects in ITS and, through legislative instruments, it ensures that ITS are rolled out consistently.

In the coming years, the digitalisation of transport in general and ITS in particular is expected to take a leap forward. As part of the Digital Single Market Strategy, the EC aims to make more use of ITS solutions to achieve a more efficient management of the transport network for passengers and business. ITS will be used

Figure 1.2 5G wireless technology scenario

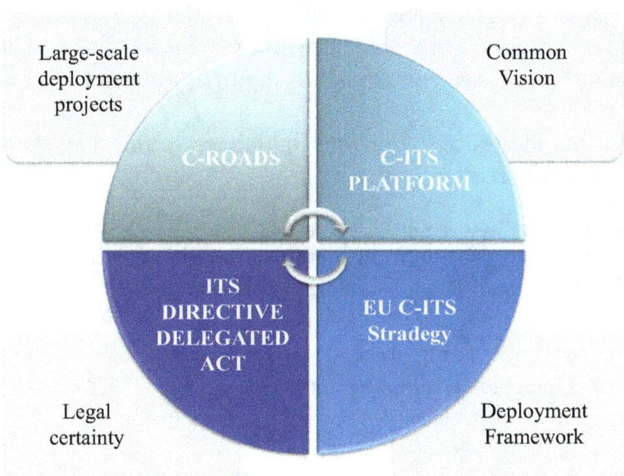

Figure 1.3 C-ITS action areas

to improve journeys and operations on specific and combined modes of transport. The EC also works to set the ground for the next generation of ITS solutions, through the deployment of Cooperative-ITS, paving the way for automation in the transport sector. C-ITS are systems that allow effective data exchange through wireless technologies so that vehicles can connect with each other, with the road infrastructure and with other road users [3].

1.2.1.1 Action plan and directive

One of the most relevant regulatory aspects in the ITS deployment was the adoption of the legal framework defined by the Directive 2010/40/EU, released in 2010 to accelerate the deployment of ITS in Europe. It aims at establishing interoperable and seamless ITS services while giving Member States the freedom to decide which systems to invest in.

It identifies a list of six priority actions encompassing:

1. The provision of EU-wide multimodal travel information services
2. The provision of EU-wide real-time traffic information services
3. Data and procedure for the provision, where possibility of road safety-related minimum universal traffic information is free of charge for users
4. The harmonised provision for an interoperable EU-wide eCall
5. The provision of information services for safe and secure parking places for trucks and commercial vehicles
6. The provision of reservation services for safe and secure parking places for trucks and commercial vehicles

It also lists priority areas in which work is to be further pursued: optimal use of road, traffic and travel data (priority area I); continuity of traffic and freight management ITS services (priority area II), ITS road safety and security applications (priority area III) and linking the vehicle with the transport infrastructure (priority area IV). These priority areas correspond to the first four priority areas of the ITS Action Plan.

There are a number of activities and initiatives in the regulatory framework derived from the ITS Directive.

- Commission Delegated Regulation (EU) No. 305/2013 with regard to the harmonised provision for an interoperable EU-wide eCall
- Commission Delegated Regulation (EU) No. 886/2013 with regard to data and procedures for the provision, where possible, of road safety-related minimum universal traffic information free of charge to users
- Commission Delegated Regulation (EU) No. 885/2013 with regard to the provision of information services for safe and secure parking places for trucks and commercial vehicles
- Commission Delegated Regulation (EU) No. 962/2015 with regard to the provision of EU-Wide real-time traffic information services
- Commission Delegated Regulation (EU) No. 2017/1926 with regard to the provision of EU-wide Multimodal Travel Information Services

Under this Directive, the EC has been adopting since then specifications (i.e., functional, technical and organisational or services provisions) to address the compatibility, interoperability and continuity of ITS solutions across the EU.

The first priorities were traffic and travel information, the eCall emergency system and intelligent truck parking.

The EC already took a major step towards the deployment and use of ITS in road transport (and interfaces to the other transport modes) in December 2008 by adopting an Action Plan. The Action Plan suggested a number of targeted measures and included the proposal for the ITS Directive. The goal was to create the momentum necessary to speed up market penetration of rather mature ITS applications and services in Europe. The initiative is supported by five co-operating Directorates-General: DG Mobility and Transport, DG Communications Networks, Content & Technology, DG Research & Innovation, DG Enterprise and Industry, and DG Climate Action.

Different reports, at national and European level, have addressed the status of implementation of the ITS Directive and Action Plan. Overall conclusions of the studies indicate that the ITS Directive has been an efficient tool for the rapid adoption of common specifications for the priority actions in the Action Plan.

Taking into account the emergence of new long-term trends (e.g., use of crowd-sourcing for transport data, partly automated driving and deployment of cooperative systems) and the possible necessity to set up new priorities, it seems necessary as a first step to prioritise the remaining actions to be addressed by the ITS Directive and the ITS Action Plan. In a second phase, it should be necessary to start preparing work for a possible revision of the ITS Directive and the supporting ITS Action Plan, taking into account the constant technical evolution of ITS and the analyses of the EC on the subject.

1.2.1.2 Cooperative, connected and automated mobility

In many respects, today's vehicles are already connected devices. However, in the very near future, they will also interact directly with each other and with the road infrastructure. This interaction is the domain of Cooperative Intelligent Transport Systems (C-ITS), which will allow road users and traffic managers to share information and use it to coordinate their actions. This cooperative element – enabled by digital connectivity between vehicles and transport infrastructure – is expected to significantly improve road safety, traffic efficiency and comfort of driving, by helping the driver to take the right decisions and adapt to the traffic situation.

Communication between vehicles, infrastructure and other road users is also crucial to increase the safety of future-automated vehicles and their full integration in the overall transport system. Cooperation, connectivity and automation are not only complementary technologies, but they also reinforce each other and will, over time, merge completely. Such an ecosystem includes the vehicles, the road infrastructure, the network infrastructure and the Cloud.

Edge Computing-based Vehicle-to-Cloud solutions enable edge cloud capabilities for different levels of AD, including Highly Autonomous Driving (HAD) and Fully Autonomous Driving (FAD) through providing different services for the driving process (e.g., High Definition real-time Maps, real-time traffic monitoring and alerts and richer passengers experience), supporting vehicles on roads to drive co-operatively and be aware of road hazards, and providing better user experience and trust to drivers and passengers.

Therefore, the EC in November 2016 adopted a European Strategy on C-ITS, a milestone initiative towards cooperative, connected and automated mobility. The objective of the C-ITS Strategy is to facilitate the convergence of investments and regulatory frameworks across the EU, in order to see deployment of mature C-ITS services in 2019 and beyond. This includes the adoption of the appropriate legal framework at EU level, the availability of EU funding for projects. The continuation of the C-ITS Platform process as well as international cooperation with other main regions of the world on all aspects related to cooperative, connected and automated vehicles. It also involves continuous coordination, in a learning-by-doing approach, with the C-ROADS platform, which gathers real-life deployment activities in Member States [4].

C-ITS platform

The EC decided early 2014 to take a more prominent role in the deployment of connected driving, by setting up a C-ITS Deployment Platform. The Platform is conceived as a cooperative framework including national authorities, C-ITS stakeholders and the Commission, in view to develop a shared vision on the interoperable deployment of C-ITS in the EU. Hence, it was expected to provide policy recommendations for the development of a roadmap and a deployment strategy for C-ITS in the EU and identify potential solutions to some critical cross-cutting issues.

In the frame of supporting the deployment of C-ITS on European roads, there are a number of C-ITS real-life pilot projects funded under TEN-T and CEF which will create new ITS services for all European road users. These projects tested vehicle-to-infrastructure and vehicle-to-vehicle interactions by using both short-range and cellular communications.

The C-ITS platform addressed the main technical (frequencies, hybrid communications, cyber-security, and access to in-vehicle data and resources) and legal issues (such as liability, data protection and privacy). Regarding access to in-vehicle data and resources, the work of the Platform was also guided by the adoption of the eCall type-approval regulation, which requested the Commission to assess the need of requirements for an interoperable, standardised, secure and open-access platform. The Platform also covered standardisation, cost–benefit analysis, business models, public acceptance, road safety and other implementation topics, international cooperation, etc [5].

Within working groups dedicated to these issues, the C-ITS Platform developed policy recommendations and proposals for action for both the Commission and also for other relevant actors along the C-ITS value chain. The progress of the development of work in the different working groups was however not completely homogeneous, due to the nature of the topics treated in them. On most issues, a shared vision emerged, including on the common solutions to address these issues;

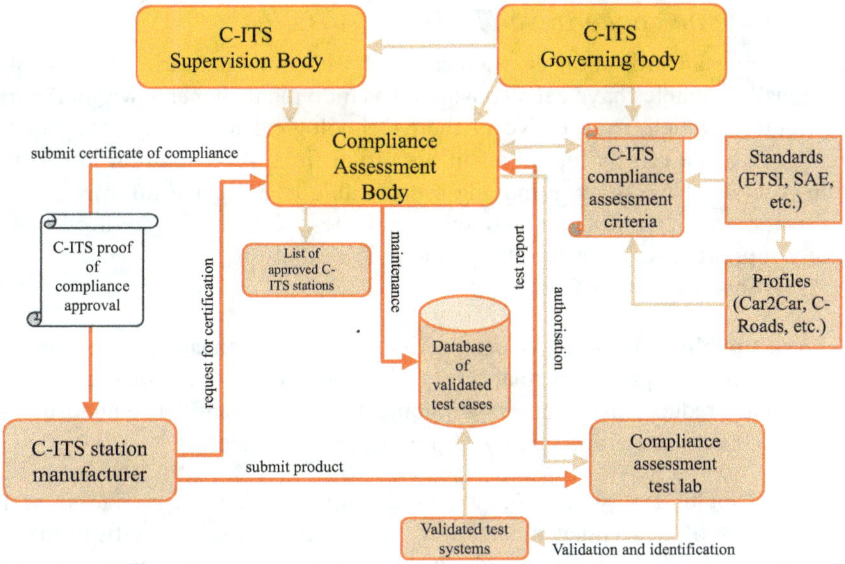

Figure 1.4 C-ITS platform

whereas on some others, the related working group succeeded on mapping the issues and possible ways to address them but did not share a common solution (Figure 1.4).

The outcomes of the C-ITS Platform address the issues related to the following aspects:

- The common technical framework necessary for the deployment of C-ITS
  - Day 1, Day 1,5 services
  - Security and Certification/Compliance assessment
  - Radio frequency and hybrid communication
  - Standardisation
  - Decentralised Congestion Control
  - Access to in-vehicle data and resources
- The legal questions related to C-ITS
  - Liability
  - Data protection and privacy issues
- The legitimacy of the deployment of C-ITS, i.e., the fact that the deployment of C-ITS can be justified and fostered at all levels
  - Road safety issues
  - Acceptance and readiness to invest
  - Costs and benefits
  - International cooperation

1.2.2 State on cooperative ITS standardisation

The topic of standardisation is technically as well as organisationally complex. Traditionally, vehicles have been considered as individual elements within the road infrastructure, which, irrespective of their technological level, rely solely on the information perceived by the driver in his visual field. As a solution to this limitation, vehicles have become moving sensors able to exchange information with the drivers, the infrastructure and other vehicles. This evolution needs to be strongly supported with a new set of standardisation effort.

As defined by project CIMEC:

> From a public road authority perspective, the use of standardised systems, interfaces and processes that are well supported by the market significantly reduce cost of buying equipment, enhance confidence and trust in reliability, safety and quality of products and services.

Standards should deal with processes and communications (interfaces) at the 'upper' layers of conceptual abstraction, gathering technical specifications that would need to be actively addressed when specifying the deployment plan for C-ITS services.

The starting point was the publication of the list of Release 1 standards for Cooperative ITS by CEN and ISO. In parallel, ECC-ITS Platform's WG7 (Standardisation) published a list with an overview of the standards being used within C-ITS deployments initiatives in Europe.

The second step consists of gathering the road operator's perspective, those standards that road operators have to actually work with in the process of deploying, operating and maintaining C-ITS services.

1.2.3 Cooperative ITS developments and initiatives

Research is often the foundation for innovation and technological advancement. In the field of Cooperative-ITS, a substantial amount of research and development has already taken place at national, European and global levels. This is crucially important for its market development and swift introduction. Increasingly, the results and findings of such research activities are needed by stakeholders to better understand the benefits of C-ITS, its overall socioeconomic impact, and increasingly, what the relationship and roles of public and private stakeholders will be to operate and manage such systems and services. Furthermore, testing and piloting are also essential to address potential weaknesses and maximise the system's performance and efficiency. However, for what concerns the testing and demonstration of C-ITS systems and services in cities, the research undertaken has been limited so far and this needs to be addressed because gaps in research activities result in development and deployment gaps at different levels later on. In particular, research of C-ITS within complex urban networks and in conjunction with a broad range of different type of vehicles including passenger and freight, and traffic management centres is lacking. The list of projects below summarises those identified by the WG that provide the best up-to-date information and useful urban specific results for local authorities.

EU Project	Related information
CIMEC	Cooperative ITS for Mobility in European Cities. http://cimec-project.eu/
Compass4D	Aims at reducing GHG and CO_2 emission https://itsforclimate.org/en/project_sheet/compass-4d/
DriveC2X	Assessment of cooperative systems through field operational tests
FOTsis	Cooperative ITS from an infrastructure perspective. http://www.fotsis.com/
FOTNet	Field Operational Test Networking and Data Sharing Support. https://fot-net.eu/
FREILOT	Looks for increasing energy efficiency in road goods transport in urban areas. http://www.ecomove-project.eu/links/freilot/
OPTICITIES	Intendsto develop and test interoperable ITS solutions to optimise urban logistics operations. http://www.opticities.com/
TEAM	Tomorrow's Elastic Adaptive Mobility. http://www.collaborative-team.eu/
VRUITS	Improving the safety and mobility of vulnerable road users
AUTOCITS	Regulation Study for Interoperability in the Adoption of Autonomous Driving in European Urban Nodes
CAPITAL	Collaborative cApacityProgramme on ITS Training-educAtion and Liaison. https://capital-project.its-elearning.eu/
C-MOBILE	Accelerating C-ITS Mobility Innovation and depLoyment in Europe. https://c-mobile-project.eu/
CODECS	COoperative ITS DEployment Coordination Support. https://www.codecs-project.eu
CO-GISTICS	Deploying of cooperative logistics. https://cogistics.eu/
C-ROADS	C-ITS services in light of cross-border harmonisation and interoperability. https://www.c-roads.eu/platform.html
C-THE-DIFFERENCE	C-ITS services to address urban mobility problems. http://c-thedifference.eu/
SPICE	Smart procurement to facilitate fast adoption of innovative sustainable transport and mobility solutions. https://spice-project.eu/
AUTOPILOT	Autonomous vehicle in a connected environment. https://autopilot-project.eu/
CO-EXIST	Towards a shared road network. https://www.h2020-coexist.eu/
CARTRE	Coordination of Automated Road Transport Deployment for Europe. https://connectedautomateddriving.eu/about-us/cartre/
MAVEN	Managing Automated Vehicles Enhances Network. http://www.maven-its.eu/

1.2.4 3GPP initiatives

1.2.4.1 1G to 5G evolution

5G is almost upon us; it is a technology that will be constructed from millions of ideas, methods, algorithms and processes. Just as 4G LTE became available when previous technologies, such as HSPA, could be further improved, 5G enters the stage when the roadmap for LTE has not been exhausted. And just as 2G coexists today with 3G and 4G, 5G will coexist with previous generations of technology.

At the end, 5G can be understood as a network of networks, providing an holistic view (and future management) of all the potential communication networks available [6].

For historical context, '1G' refers to analogue cellular technologies that became available in the 1980s. '2G' denotes initial digital systems that became available in the 1990s and that introduced services such as short messaging and lower-speed data. 3G requirements were specified by the International Telecommunication Union (ITU) as part of the International Mobile Telephone 2000 (IMT-2000) project, for which significant voice capacity improvement was a focus and digital networks had to provide 144 Kbps of throughput at mobile speeds, 384 Kbps at pedestrian speeds and 2 Mbps in indoor environments. UMTS-HSPA and CDMA2000 are the primary 3G technologies. 3G technologies began to be deployed early last decade.

In 2008, the ITU issued requirements for IMT-Advanced, which many people initially used as a definition of 4G. The focus on 4G was to improve data coverage, capacity and quality of experience. Requirements included operation in up to 40 MHz radio channels and extremely high Spectral Efficiency. The ITU required peak spectral efficiency of 15 bps/Hz and recommended operation in up to 100 MHz radio channels, resulting in a theoretical throughput rate of 1.5 Gbps. In 2009 and 2010, the term '4G' became associated with mobile broadband technologies deployed at the time, such as HSPA+, WiMAX and initial LTE deployments. Today, 4G usually refers to HSPA+ or LTE.

Although the industry is preparing for 5G, LTE capabilities will continue to improve in LTE-Advanced Pro through the rest of the decade. Many of these enhancements will come through incremental network investments. Given the scope of global wireless infrastructure, measured in hundreds of billions of dollars, offering users the most affordable service requires operators to leverage investments they have already made. 5G will eventually play an important role, but it must be timed appropriately so that the jump in capability justifies the new investment.

5G groups researching next generation wireless architecture and requirements include, among others, the ITU, the EU 5G Infrastructure Public-Private-Partnership (5G PPP), which is the framework for several projects, including METIS II (Mobile and wireless communications Enablers for the Twenty-twenty Information Society) and Next Generation Mobile Networks (NGMN). Finally, 5G Americas is actively involved in developing the vision and requirements of 5G for North, Central and South America. 5G Americas signed an MoU to collaborate with 5G-PPP.

Wireless technology has progressed to the extent that significant new capabilities are inevitable, making 5G a possible alternative to wireline broadband for many subscribers.

1.2.4.2 4G LTE advances

As competitive pressures in the mobile broadband market intensified, and as demand for capacity persistently grew, LTE became the favoured 4G solution because of its high data throughputs, low-latency, and high spectral efficiency. Specifically:

- Wider Radio Channels. LTE can be deployed in wide radio channels (for example, 10 or 20 MHz) with carrier aggregation now up to 640 MHz, although inter-band aggregation of four or five carriers (up to 100 MHz) represents a practical upper limit.
- Easiest MIMO Deployment. By using new radios and antennas, LTE facilitates MIMO Deployment, in contrast to the logistical challenges of adding antennas for MIMO to existing legacy technologies. Furthermore, MIMO gains are maximised because all user equipment supports it from the beginning.
- Best Latency Performance. For some applications, low latency (packet traversal delay) is as important as high throughput. With a low transmission time interval (TTI) of 1 millisecond (msec) and a flat architecture (fewer nodes in the core network), LTE has the lowest latency of any cellular technology.

In the same way that 3G coexists with 2G systems in integrated networks, LTE systems coexist with both 3G and 2G systems, with devices capable of 2G, 3G and 4G modes.

Beyond radio technology, the Evolved Packet Core (EPC) provides a core architecture that integrates with both legacy GSM-HSPA networks and other wireless technologies, such as CDMA2000 and Wi-Fi. The combination of EPC and LTE is referred to as the Evolved Packet System (EPS).

LTE is available in both Frequency Division Duplex (FDD) and Time Division Duplex (TDD) modes. Many deployments are based on FDD in paired spectrum. The TDD mode, however, is important for deployments in which paired spectrum is unavailable. Instances of TDD deployment include China, Europe at 2.6 GHz, U.S. Broadband Radio Service (BRS) spectrum at 2.6 GHz and the 3.5 GHz band.

The versions of LTE most widely deployed today are just the first in a series of innovations that will increase performance, efficiency and capabilities. Enhancements in the 2013 to 2016 period were the ones defined in 3GPP Releases 10, 11 and 12, and are commonly referred to as LTE-Advanced. Subsequent releases, including Releases 13 to 15, specify LTE-Advanced Pro.

Keeping in mind that different operators have varying priorities, the following list roughly ranks the most important features of LTE-Advanced and LTE-Advanced Pro for the 2017 to 2020 timeframe:

1. Carrier Aggregation. With this capability, already in use, operators can aggregate radio carriers in the same band or across disparate bands to improve throughputs (under light network load), capacity and efficiency. Carrier aggregation can also combine FDD and TDD and is the basis of LTE-U and LTE-LAA. As examples, in 2015, AT&T aggregated 700 MHz with AWS, and 700 MHz with PCS. T-Mobile aggregated 700 MHz with AWS, and AWS with PCS. Operators are now deploying three-carrier aggregation and eventually may aggregate four carriers. Release 13 introduced support for carrier aggregation of up to 32 carriers, addressing primarily the

opportunity to aggregate multiple unlicensed channels. Release 14 specifies interband carrier aggregation for up to five downlink carriers and two uplink carriers.

2. VoLTE. Initially launched in 2015 and with widespread availability in 2017, VoLTE enables operators to roll out packetized voice for LTE networks, resulting in greater voice capacity and higher voice quality.
3. Tighter Integration of LTE with Unlicensed Bands. LTE-U became available for testing in 2016, and 3GPP completed specifications for LAA in Release 13, with deployment expected around 2018. MulteFire, building on LAA, will operate without requiring a licensed carrier anchor. LTE/Wi-Fi Aggregation through LWA and LWIP are other options for operators with large Wi-Fi deployments.
4. Enhanced Support for IoT. Release 13 brings Category M1, a low-cost device option, along with Narrowband-IoT (NB-IoT), a version of the LTE radio interface specifically for IoT devices, called Category NB1.
5. Higher-Order and Full-Dimension MIMO. Deployments in 2017 use up to 4X4 MIMO. Release 14 specifies a capability called Full-Dimension MIMO, which supports configurations with as many as 32 antennas at the base station. See the section 'Smart Antennas and MIMO' and Appendix section 'LTE Smart Antennas' for further detail.
6. Dual Connectivity. Release 12 introduced the capability to combine carriers from different sectors and/or base stations (i.e. evolved Node Bs [eNBs]) through a feature called Dual Connectivity. Two architectures were defined: one that supports Packet Data Convergence Protocol aggregation between the different eNBs and one that supports separate S1 connections on the user-plane from the different eNBs to the EPC.
7. 256 QAM Downlink and 64 QAM Uplink. Defined in Release 12 and already deployed in some networks, higher-order modulation increases user throughput rates in favourable radio conditions.
8. 1 Gbps Capability. Using a combination of 256 QAM modulation, 4X4 MIMO, and aggregation of three carriers, operator networks can now reach 1 Gbps peak speeds. See below for more information.
9. V2X Communications. Release 14 specifies vehicle-to-vehicle and vehicle-to-infrastructure communications. See the section 'Cellular V2X Communications' for more information.
10. Coordinated Multi Point. CoMP (and enhanced CoMP [eCoMP]) is a process by which multiple base stations or cell sectors process a User Equipment (UE) signal simultaneously, or coordinate the transmissions to a UE, improving cell-edge performance and network efficiency. Initial usage will be on the uplink because no user device changes are required. Some networks had implemented this feature in 2017.
11. HetNet Support. HetNets integrate macro cells and small cells. A key feature is enhanced inter-cell interference coordination (eICIC), which improves the ability of a macro and a small-cell to use the same spectrum. This approach is valuable when the operator cannot dedicate spectrum to small cells. Operators

are currently evaluating eICIC, and at least one operator has deployed it. Further enhanced ICIC (feICIC) introduced in Rel-11 added advanced interference cancellation receivers into devices.

12. Ultra-Reliable and Low-Latency Communications. Being specified in Release 15, URLLC in LTE will shorten radio latency to a 1 ms range using a combination of shorter transmission time intervals and faster hybrid automatic repeat request (HARQ) error processing. See Appendix 'LTE Ultra-Reliable and Low-Latency Communications' for further details.
13. Self-Organising Networks. With SON, networks can automatically configure and optimise themselves, a capability that will be particularly important as small cells begin to proliferate. Vendor-specific methods are common for 3G networks, and trials are now occurring for 4G LTE standards-based approaches.

Other key features that will become available in the 2016–2020 timeframe include full-dimension MIMO, enhanced Multimedia Broadcast/Multicast Services (eMBMS), User-Plane Congestion Management (UPCON) and device-to-device communication (targeted initially at public-safety applications).

1.2.5 LTE-V and CV2X

The automotive industry is in the midst of a transition toward producing vehicles that are more aware of their surroundings. For many years, there has been a goal that vehicles should be able to communicate with not only other vehicles (V2V) but also with nearby infrastructure (V2I), Internet-based networks (V2N) and even pedestrians (V2P). Collectively, these use cases have become known as vehicle-to-everything (V2X) connectivity. Now, with advances in electronics, sensing technologies and computing techniques such as machine learning and computer vision, this use cases are starting to become reality. New vehicles today are capable of taking a more active role by warning drivers of potential collisions with oncoming vehicles, assisting with emergency braking and monitoring intersections, to name just a few examples. This represents a big step forward from relying on passive safety features such as seat belts and air bags [7].

In the automotive industry, this trend is viewed as the beginning of an evolution to automated and eventually fully autonomous vehicles. In an autonomous vehicle scenario, the vehicle's on-board computers will be fully capable of performing all driving operations on their own, with no human monitoring required. This is still a few years away, but today we are getting closer with the likes of Tesla providing partial automation and Google's self-driving car testing conditional automation [8].

In the United States, the NHTSA is considering using IEEE 802.11p-based Dedicated Short Range Communications (DSRC) technology for V2V communications. The technology was developed specifically for V2V applications that require critical latency of ~100ms, very high reliability and security authentication with privacy safeguards. The DSRC standard was finalised in 2009 and has been subjected to extensive testing by automakers and select large-scale trials.

Stakeholders have completed work on use of DSRC to protect vulnerable road users. The Federal Communications Commission (FCC) has allocated dedicated spectrum for transportation safety applications in 1999 in the 5.850–5.925 GHz band to ensure operation without interference that DSRC-based V2V systems plan to leverage.

However, DSRC has several weaknesses. There is no apparent path for continued evolution of the radio standard to meet changing technological and consumer needs. Additionally, as it was designed for rapid transmission of short-range basic safety messages, it is unable to meet the higher bandwidth demands of V2N applications such as AD, multimedia services. DSRC also does not have the bandwidth necessary to transmit the raw vehicle sensor data that will become increasingly common in automated vehicles. DSRC also has limited range: about 300 m.

DSRC would require the deployment of tens of thousands of roadside units (RSUs) embedded or attached to roadway infrastructure to enable an effective network along the nation's roads. This is a particular challenge in more rural areas considering the vast distances involved. State highway administrations and other roadway authorities would be responsible for deploying, managing and operating the RSUs and associated infrastructure networks, such as fibre or copper backhaul. While V2V communications do not require RSUs to perform crash-warning functions, RSUs are needed for ancillary functions such as certificate revocation list (CRL) distribution, certificate top-ups and to support other longer-range V2X use cases.

LTE and 5G can be used for these RSU functions thereby eliminating the need for highway authorities to install and maintain RSUs. That highlights another a key disadvantage for DSRC. The need for another set of radios when all new vehicles already come with embedded cellular radios. By using cellular technologies for both short- and long-range use cases, original equipment manufacturers (OEMs) can reduce vehicle bill of materials (BOM) costs while meeting or even exceeding the safety requirements.

Beyond a technology comparison, however, there are other policy considerations that will need to be resolved for LTE-based V2X to be embraced by stakeholders. These include the universal availability of V2V or other safety-related applications for vehicle owners that choose not to activate their mobile network operator SIM card for cost or privacy reasons, a revised set of liability issues and the ability of state highway authorities to interface with an LTE network that they do not operate.

Recently, attention has also been focused on cellular LTE technology which is quickly evolving to meet the needs for V2X communications. The current LTE standard in 3GPP Release 13 is not capable of meeting the low-latency and high-speed requirements of safety-critical V2V applications. Also, vehicles in areas with poor or no network coverage would be unable to communicate with each other. Despite these limitations, LTE Release 13 is capable of meeting some of the less stringent V2N use cases today. However, the completed 3GPP Release 14 LTE standard does include support for cellular-V2X (C-V2X) use cases, enabling cellular technology as an additional option for the majority of V2X applications. With LTE Release 14, direct device-to-device communication improves latency and support operation in areas without network coverage and at high relative speeds, while network broadcast capabilities can help to meet other V2X requirements. In addition, the ability to leverage existing cellular infrastructure, with its broad coverage footprint, would reduce costs and accelerate the realisation of the safety and efficiency benefits of V2X communication.

As United States regulatory agencies look towards finalising proposed legislation and the details behind V2X communication, the planning for and implementation of V2X services in Europe is progressing along a different path. In April 2015, the European Parliament passed legislation requiring all new cars to be equipped with eCall technology, which is the ability to automatically dial Europe's single emergency number in case of an accident. The law requires every new vehicle produced after April 2018 to be equipped with integrated cellular technology, thereby seeding the vehicle base with cellular-capable vehicles. To address vehicle connectivity, the EC adopted a coherent strategy in 2016. The strategy aims to promote an integrated European market that supports common priorities and would leverage both cellular communications and European Telecommunications Standards Institute – Intelligent Transport Systems – G5 (ETSI ITS-G5), a standard based on IEEE 802.11p and similar to DSRC. Spectrum resources for V2X communication in Europe have been allocated in the 5.9 GHz band, similar to the United States The strategy outlined by the EC will serve as the foundation for implementing the necessary legal framework in 2018 that will enable the commercial deployment of cooperative ITS by 2019.

A more fragmented approach is playing out in Asia. China plans to decide on unified standards for V2V and V2I communication in 2018, an important step given the country's large population and growing global economic importance. In contrast, in Japan, Toyota introduced vehicles capable of V2V and V2I communication using DSRC back in 2016 and continues to develop more advanced capabilities.

Japan's DSRC uses a different band (760 MHz) and a different standard (Association of Radio Industries and Businesses ARIB STD-109). It is still based on IEEE 802.11p but differs substantially in the physical layer. Korea has also focused significant attention on the testing of automated vehicles as of late, with the hope of deploying some automated vehicles for the 2018 Olympics in Pyeongchang. Korea has also designated spectrum in the 5.9 GHz band for ITS.

These differences around the world illustrate that there are strategic planning and deployment choices to be made. The C-V2X 3GPP standard was completed in March 2017, with products underway. LTE Release 14 C-V2X can be viewed as a necessary waypoint along the timeline for 5G development, as it supports safety-critical use cases. As 5G technology evolves, 5G-based C-V2X will be able to take advantage of the enhanced mobile broadband, ultra-reliable low-latency communication and massive-scale machine-to-machine communication, all of which will support more advanced use cases. For backward compatibility, a 5G V2X-enabled vehicle will support not only these advanced services, but also the basic safety for which LTE-based V2X was designed [9,10].

1.3 Target services

The purpose of cooperative systems is to implement a wide range of applications and services to improve safety, efficiency, comfort and reduce the environmental impact of transport through the exchange of information between vehicles infrastructure and control centres. This information allows a broadening of the horizon that it is possible to be aware of in order to make decisions and take action.

The main qualitative leap between the conventional vehicle, which includes its own systems, and the vehicle connected in a cooperative environment is that, in addition to possessing its own data and perceiving its surroundings by means of onboard sensors, it can receive information from other vehicles, infrastructure or traffic environment information providers.

Thus, in cooperative ITS, the following levels of information exchanges could be established:

- The vehicle receives information from the road or other vehicles with which it is possible to anticipate situations (retentions, work zones, adverse weather, etc.), that the driver cannot perceive.
- The infrastructure receives information from the vehicles, which allows it to know traffic variables and manage the network more efficiently.
- Information is exchanged in both directions, since vehicle and infrastructure are sources of information, and this fact increases information and its quality, benefitting both parties.

Cooperative systems (C-ITS) are considered the next qualitative leap in vehicle technology and are a priority. In order to promote the use of these C-ITS, the EC created, in November 2014, the Platform for the deployment of Cooperative ITS (C-ITS Platform) with the intention of giving a unified vision to the services derived from this technology at the European level. The Platform defined the implementation roadmap for Cooperative Systems in Europe, starting with a series of basic services, called C-ITS Day 1 services, which would be followed by subsequent releases in deployment stages of increasingly complex services, culminating with Cooperative Autonomous Driving for 2030 [5].

- C-ITS Day 1 Services.
  - Emergency electronic brake light
  - Emergency vehicle approaching
  - Slow or stationary vehicles
  - Traffic jam ahead warning
  - Hazardous location notification
  - Roadworks warning
  - Weather conditions
  - In-vehicle signage
  - In-vehicle speed limits
  - Probe vehicle data
  - Shockwave damping

- GLOSA/Time to Green (TTG)
- Signal violation/intersection safety
- Traffic signal priority request by designate vehicles

Looking at the services proposed by the Platform for Day 1 deployment, it should be noted that the stress is in the safety-critical or close to safe-critical services. Thus, the technical requirements in terms of communications coverage, network density, delay turnover times, etc., should address some of the most strict C-ITS use cases in performance.

And this can be seen in another way to identify types of services relevant for traffic and mobility. According then to the Quality of Service required, we can identify services as:

- Safety critical. Very low latencies, integrity of information critical.
- Mobility. Relevant for traffic and mobility management. Integrity and Authentication of information are very important.
- Entertainment/Infotainment. Not stringent requirements.
- Payment services. Traceability, Integrity and Authenticity of information are critical.

The big advantage of this classification is that in principle, these can be used to define families of services, in the sense that if a given infrastructure and vehicle combination of systems can provide the performance required for a certain application in a given group, it should be able to provide the performance required by other applications in the same group. Additionally, this performance-based definition of services should also help in making the services isolated or at least more isolated from the underlying technologies, and therefore enable service specifications for C-ITS infrastructures as well as 5G-based infrastructures [11].

The reality, however, is that these two visions – C-ITS versus 5G service provision – has not converged yet. And C-ITS has focused on emergency services (due to the by-design tighter control on communication and computing resources) whereas 5G has focused on the other hand on Infotainment and Mobility services (due to greater coverage and more straightforward device penetration).

1.4 Cooperative ITS current state

1.4.1 Standardisation in vehicular communications

1.4.1.1 Introduction

There are mainly two types of standards: industry standards or open standards produced by standardisation organisations such as ISO, ETSI, Internet Engineering Task Force (IETF), Institute of Electrical and Electronics Engineers (IEEE) or European Committee for Standardisation (CEN).

- ISO: Is an independent, nongovernmental organisation, the members of which are the standards organisations of the 164 member countries. It is composed of a number of Technical Committees, with the one relevant to vehicular communications being ISO/TC204. The scope of the committee is the standardisation of information, communication and control systems in the field of urban and rural surface transportation, including intermodal and multimodal aspects thereof, traveller information, traffic management, public transport, commercial transport, emergency services and commercial services in the ITS field.

- CEN: Is a public standards organisation whose mission is to foster the economy of the EU in global trading, the welfare of European citizens and the environment by providing an efficient infrastructure to interested parties for the development, maintenance and distribution of coherent sets of standards and specifications. It is composed of a number of Technical Committees, with the one relevant to vehicular communications being ISO/TC278, with a scope similar to that of the ISO/TC204.
- IEEE: The IEEE is a professional association formed in 1963, with the objectives of the educational and technical advancement of electrical and electronic engineering, telecommunications, computer engineering and allied disciplines. Not limited to standardisation activities, it includes 39 Technical Societies, acting as a major publisher of scientific journals and organiser of conferences, workshops and symposia. Its IEEE Standards Association is a leading standards development organisation for the development of industrial standards in a broad range of disciplines.
- IETF: The IETF develops and promotes voluntary Internet standards, in particular the standards that comprise the Internet Protocol (IP) suite. It is an open standards organisation, with no formal membership. The IETF started as an activity supported by the US federal government, but since 1993 it has operated as a standards development function under the auspices of the Internet Society, an international membership-based non-profit organisation.
- ETSI: The European Telecommunications Standards Institute is a non-profit organisation whose mission is to produce the telecommunications standards applicable in Europe. The Technical Committee ETSI IC ITS was established in 2007, and its work includes aspects of DSRC, CALM communications, architecture and security. It is one of the bodies usually appointed to carry out EC Mandates or Directives on standardisation.

Relevant to CEN and ISO, the Vienna Agreement signed by both institutions in 1991 aims to avoid duplication of standards between CEN and ISO, so that joint work and adoption of standards have become the norm in the past years.

Other relevant players in the standardisation arena are the governmental institutions and public authorities, which actively contribute to push standardisation bodies to specific actions towards integration and support to deployment of ITS solutions. The EC and the US department of transport have both taken throughout the years a number of actions to ensure that the standardisation efforts of the different entities were consistent and produced results that would benefit the citizens and the society. In the case of the EC, these efforts come in the form of Directives or Mandates which address a specific area of application, providing guidelines and broad objectives, together with the specific standardisation bodies which should be involved in the activities.

1.4.1.2 The ISO CALM framework

The ISO CALM communications reference architecture

An overall view of the integration effort in vehicular communications research initiatives and standardisation activities, and the resulting framework can be seen in Figure 1.5. This is known as the ISO CALM (Communications Access for Land Mobiles, in its latest definition). It is a multilayered diagram for the different technologies relevant in vehicular communications, as considered in the most recent ISO 21217 ITS Reference Architecture. The layered description follows roughly a simplified open system interconnection (OSI) stack specification, with physical (access) components at the bottom, and increasingly application-oriented components higher up in the structure.

It is important to note that the overall approach to the specification of this framework is modular. That is, components have clear specifications in terms of intended objective and scope of the technologies on which they are based. In some cases, this is more difficult; for example, those components specifying data and information flow control from the applications, or those related with security of the communications. But the goal is to try to make components modular as it will facilitate not only interoperability aspects, but also the replacement of these components as technology evolves.

ISO 21217:2014 describes the communications reference architecture of nodes called 'ITS station units' designed for deployment in ITS communication networks. The ITS station reference architecture is described in an abstract way. While ISO 21217:2014 describes a number of ITS station elements, whether a particular element is implemented in an ITS station unit depends on the specific communication requirements of the implementation. Thus, the standard defines the frame, including provisions for the most significant technologies in different levels to provide the tools for the bigger number of particular implementations possible.

ISO 21217:2014 also describes the various communication modes for peer-to-peer communications over various networks between ITS communication nodes. These nodes may be ITS station units as described in ISO 21217:2014 or any other reachable nodes.

ISO 21217:2014 specifies the minimum set of normative requirements for a physical instantiation of the ITS station based on the principles of a bounded secured managed domain.

In the following pages, some of the components of the communications framework will be highlighted, focusing on the access technologies and network technologies, as these provide the critical first link between the ITS entities, and the basic end-to-end capabilities for ITS services and applications, respectively.

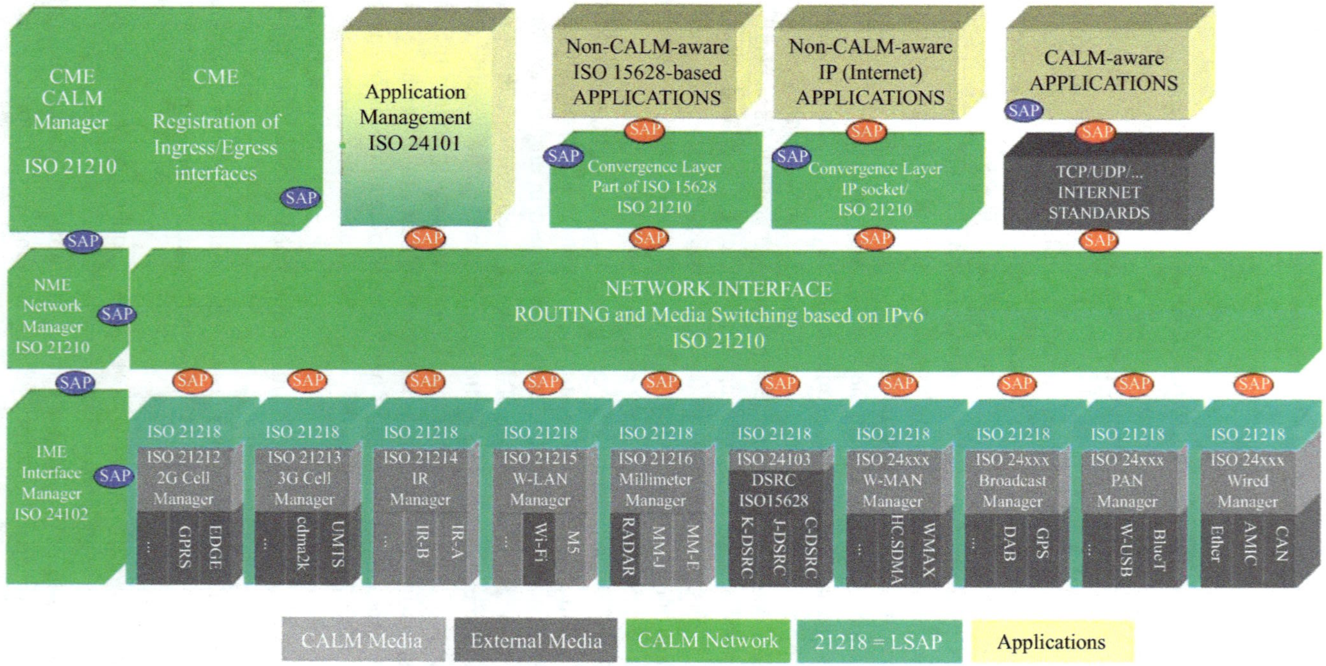

Figure 1.5 ISO CALM system architecture

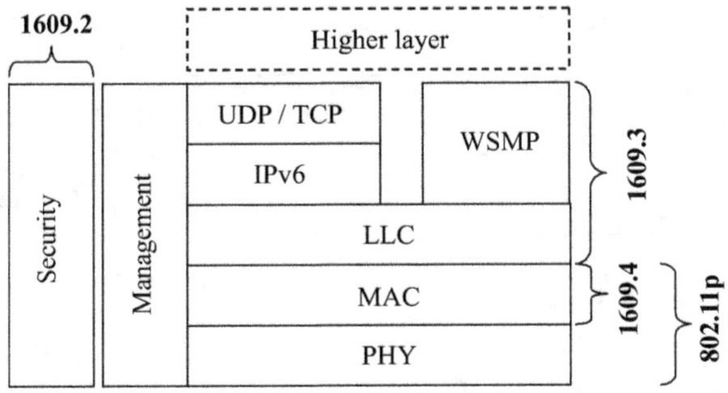

Figure 1.6 IEEE WAVE standards family and 802.11p standard scope

The ISO CALM Access media

IEEE WAVE

WAVE (Wireless Access in Vehicular Environments) is a set of standards addressed at ensuring a homogeneous access for communications between automotive manufacturers. It defines an architecture and a set of services and interfaces that collectively enable secure V2V and V2I wireless communications. As a complete architecture, it covers the complete communications stack and addresses additionally overall issues such as security and management (Figure 1.6).

The set of standards which specify the WAVE architecture are developed by the IEEE standardisation body under the title of IEEE 1609 family. In relation to the access media, IEEE 1609 WAVE relies on the IEEE 802.11p standard, which defines the physical and medium access layers of the communications stack (the two lowest levels of the OSI stack).

CEN DSRC

CEN DSRC is the European implementation of a short-range wireless communications system, and is the one traditionally used for electronic fee collection applications. In a similar way to the American WAVE, CEN DSRC is a family of standards which specify the whole DSRC stack, including PHYsical layer (PHY), dynamic link library (DLL) and application layers, and additionally considering management issues. Compliance with CEN DSRC allows multiple simultaneous RTTT (Road Transport and Traffic Telematics) applications in parallel, but the particular constraints of the application scenarios in which DSRC is likely to be used, and particularly regarding the low latency requirements, lead CEN to assume a simplified architecture for DSRC (Figure 1.7).

ETSI ITS-G5

ETSI ITS-G5 includes the specification on functionalities providing communications focused on the ETSI definition of ITS services and architecture, and based in the 5 GHz band. As the ISO CALM M5 developments, it is based on the work done in IEEE 802.11p, but adopts a reduced set of the services (at different layers) described in IEEE 802.11p (Figure 1.8).

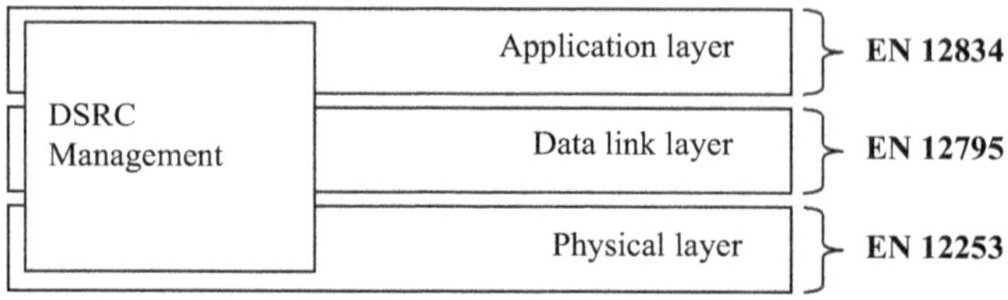

Figure 1.7 CEN DSR protocol stack

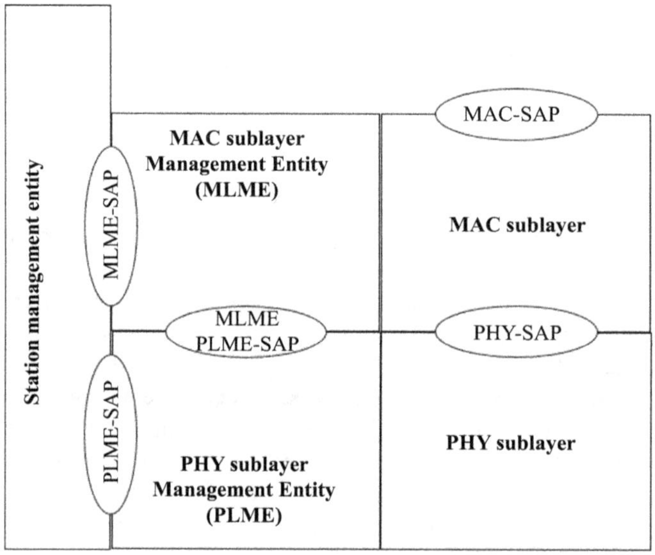

Figure 1.8 Best trade-off between robustness and computational cost

The ITS-G5 standard includes also specific requirements for different spectrum bands, in this case explicitly addressing those sub-bands in the document, as a direct response to the EC Directive on harmonised spectrum allocation in the 5.9 GHz.

The ETSI ITS-G5 standard specifies several aspects of the ITS Station reference architecture, as well as the corresponding SAPs, (Service Access Points) and particularly, the modifications made to the base IEEE 802.11p equivalent parts.

ISO CALM M5

Specifies the CALM architecture access in the 5 GHz microwave range. Its development was done in parallel with IEEE 1609 WAVE and is in fact also based at this level on the work done in IEEE 802.11p.

A CALM M5 communication interface can be integrated with CEN DSRC as a way to ensure currently deployed payment solutions (this integration expects the CEN DSRC equipment to be compliant with the CEN EN 12253:2004, CEN EN 12795:2003 and CEN EN 12834:2003 standards which specify several DSRC layers).

The specification of M5 relies in the concept of CI (Communication Interface), which in a similar way of that of the ETSI ITS-G5 describes an entity which includes aspects of the PHY, MAC and management layers of the overall ISO CALM structure and defines the corresponding access points to this entity and the compliance requirements for these SAPs (Figure 1.9).

The compliance requirements of the M5 CI are extensive, with other CALM standards and with IEEE 802.11, among others. The global concept of M5 CI is a direct implementation of the overall CALM concept of achieving interoperability at the network and transport level through the use of the Communication Adaptation Layers which isolate particular access technologies from upper layers for data transmission and CM and CI management entities to make the overall CALM system aware of the capabilities of a given CI.

IEEE 802.11

The IEEE 802.11 family of standards in general specifies the characteristics of the lower layers of the wireless LAN (Local Area Network) communications in local and metropolitan area networks. The first 802.11 standard was released in 1999, and already considered two different implementations, 802.11a and 802.11b, operating at the 5 and 2.4 GHz bands, respectively. The original 802.11 standard has been under development since then, with a number of amendments being published by IEEE, although in several cases there have not been commercial implementation of the standard documents, being 802.11a, 802.11b, 802.11g, 802.11n and 802.11p the only ones which have been widely implemented. Originally detailed in IEEE annexes to the initial 802.11 standard, these have since been integrated in the main document, so nowadays there is a single reference IEEE

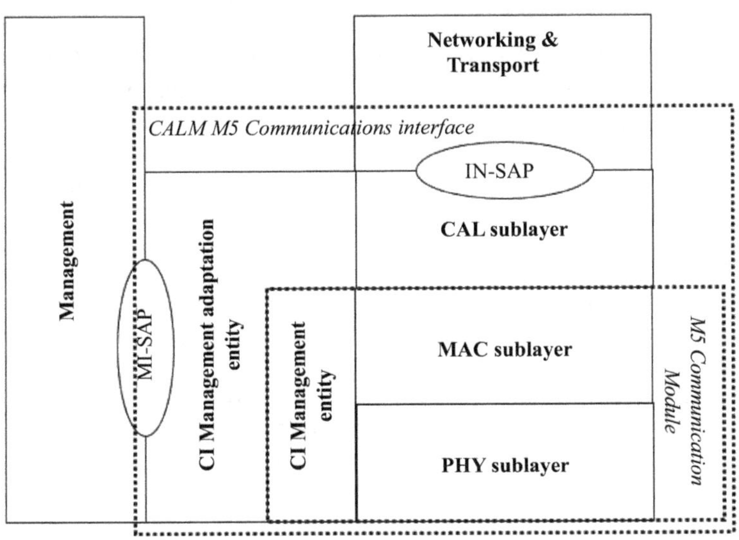

Figure 1.9 ISO CALM M5 CI architecture (adapted from ISO 21215:2010)

802.11 specification. For the purposes of clarity, and giving support to existing documentation, the suffixes of the annexes will be maintained in this document.

IEEE 802.11p

IEEE 802.11p access standard specifies the technical characteristics of the radio link of WAVE, and can be seen in turn as an IEEE validation of the ASTM E2213-03 specification for telecommunications and information exchange between road-side and vehicle systems using the 5 GHz band. It is to be noted, though, that the 11p amendment applies to a band starting also at 5 GHz, but specifies particular radio requirements (maximum transmit power, spectrum masks, etc...) for the 5.85~5.925 GHz band, probably as a result of a combined development effort with European standardisation to ensure interoperability at this level and the US FCC allocation of that band to ITS services,

> *such as traffic light control, traffic monitoring, travellers' alerts, automatic toll collection, traffic congestion detection, emergency vehicle signal preemption of traffic lights, and electronic inspection of moving trucks through data transmissions with roadside inspection facilities.*

From the standard's specification and deployment, the primary use of IEEE 802.11p compliant 5.9 GHz equipment in the WAVE environment is the provision of ITS services, with the main characteristics of high speed (27 Mbps), short range (up to 1000 m) and low latency. Integration with existing CEN DSRC PHY 5.8 GHz applications is possible by implementing the corresponding spectrum masks specified in the IEEE 802.11p standard to avoid interferences with these applications. Current development of commercial IEEE 802.11p compliant devices is in fact slowed down by the difficulty in correctly implementing these power masks, and it might be a factor to be considered in relation to the equipment to be used.

As of September 2009, IEEE 1609 approach to EFC (Electronic Fee Collection) was that it considered a part of a broader EPS concept (Electronic Payment Service), and that whenever possible, standards already specified (at higher levels, such as the CEN/ISO 14906 DSRC-based EFC application interface definition) should be used whenever possible and in agreement with the established IEEE 1609 WAVE standards structure (maybe in the shape of application profiles, for example) [12].

IEEE 802.16 WiMAX

WiMAX (Worldwide Interoperability for Microwave Access) is a communication technology trying to fill the gap between 3G and WLAN standards, thus fitting in the concept of MAN (Metropolitan Area Network). WiMAX is actually the commercial name of the IEEE 802.16 set of wireless broadband standards.

As it is the case with IEEE 802.11 Wi-Fi standards, 802.16 implementations have also evolved with time. The first release of the standard was published in 2001, specifying a point-to-multipoint broadband wireless transmission in the

10–66 GHz band, with only direct line-of-sight capability. The IEEE 802.16a evolution, released in 2003 extended the specifications for the 2–11 GHz band. However, arguably the most relevant evolution of the standard in terms of applicability to the mobile environment is the 802.16e iteration.

The ISO CALM Network layer

IETF IPv4

IPv4 is the fourth revision in the development of the IP and the first version of the protocol to be widely deployed. Together with IPv6, it is at the core of standards-based internetworking methods of the Internet. IPv4 is described in the IETF RFC 791 from 1981.

IPv4 is a connectionless protocol for use on packet-switched Link Layer networks (e.g., Ethernet). It operates on a best effort delivery model; in that it does not guarantee delivery, nor does it assure proper sequencing or avoidance of duplicate delivery. These aspects, including data integrity, are addressed by an upper layer transport protocol, such as the Transmission Control Protocol (TCP).

As one of the key aspects of the protocol, IPv4 uses 32-bit addresses, with some address blocks reserved for special purposes such as private networks and multicast addresses. As addresses were assigned to end users, an IPv4 address shortage was developing, prompting both network addressing mechanisms and the development of IPv6 to delay and eventually overcome the address exhaustion.

IPv4 addresses may be written in any notation expressing a 32-bit integer value; but for convenience, they are most often written in dot-decimal notation, which consists of four octets of the address expressed individually in decimal and separated by periods.

```
192.0.2.235
```

IETF/ISO IPv6 Networking and Mobility

IP version 6 (IPv6) is the current version of the IP, and is currently in the process of being deployed together with existing IPv4, to eventually reach an IPv6-only network scenario.

Arguably, the most important reason for the development of the IPv6 was the issue of address exhaustion caused by the relatively small addressing space of IPv4, but with the increased knowledge of the networking scenario to be faced by IPv6, its design and specification includes some significant specific improvements.

- Expanded addressing capabilities
  IPv6 increases the IP address size from 32 to 128 bits, to support more levels of addressing hierarchy, a much greater number of addressable nodes and simpler auto-configuration of addresses. The scalability of multicast routing is improved by adding a 'scope' field to multicast addresses. And a new type of address called an 'anycast address' is defined, used to send a packet to any one of a group of nodes.

  A full IPv6 address can be, for example:

  ```
  2001:0DB8:C003:0001:0000:0000:0000:F00D
  ```

- Header format simplification
  Some IPv4 headers have been dropped or made optional, to reduce the common-case processing cost of packet handling and limit the bandwidth cost of the IPv6 header.
- Improved support for extensions and options
  Changes in the way IP header options are encoded allows for more efficient forwarding, less stringent limits on the length of options and greater flexibility for introducing new options in the future.
- Flow labelling capability
  A new capability is added to enable the labelling of packets belonging to particular 'traffic' flows for which the sender requests special handling, such as non-default quality of service or 'real-time' service.
- Authentication and Privacy capabilities
  Extensions to support authentication, data integrity and optionally data confidentiality are specified for IPv6.

Even though the IPv6 specification was released (officially) in 1998, the truth is that its implementation is far from extended. There may be several reasons for this slow deployment, the assessment of which is out of the scope of this document. The fact is that very likely service deployment scenarios will include both IPv4 and IPv6 devices and therefore the introduction of some ideas about the coexistence of IPv4 and IPv6 is necessary.

The differences between IPv4 and IPv6 go beyond the extended addressing space of IPv6, and include improved and added capabilities related to self-configuration of nodes (both fixed and mobile), multicasting and both network level security and mobility facilities. Some of the solutions developed for the coexistence of IPv4 and IPv6 reduce the impact of these changes in order to maintain compatibility with the older IPv4 protocol.

Mobility in IPv6 networks

Mobility for IP networks is the implementation of a concept by means of which nodes remain reachable while moving around in an IP-based network. Each mobile node is identified by its home address, regardless of its current point of attachment to the Internet. While away from its home network, a mobile node is also associated with a CoA (Care-of Address), which provides information about the mobile node's current location. The mobility extensions implement the necessary protocols so that IP nodes cache the binding of mobile node's home address with their CoA, sending packets destined to the mobile node to this CoA, with the sender only knowing the home address of the mobile node.

It is to be noted that Mobile IP protocols solve specifically the problems related to the mobility at the network layer protocol, trying to provide upper layers with a single destination IP address independently of the particular network location of the destination device. Mobile IP does not solve mobility issues at lower layers, such as possible access technology changes or cell-to-cell handovers.

However, there is also a need to support the movement of a complete network that changes its point of attachment to the fixed infrastructure, maintaining the sessions of every device of the network. This is basically the NEMO IP Network Mobility concept, developed by the IETF.

A Mobile Network is a network segment or subnet that can move and attach to arbitrary point in the routing infrastructure. A Mobile Network can only be accessed via specific gateways called Mobile Routers that manage its movement. Mobile Networks have at least one Mobile Router serving them. A Mobile Router does not distribute the Mobile Network routes to the infrastructure at its point of attachment (i.e., in the visited network); instead, it maintains a bidirectional tunnel to a Home Agent that advertises an aggregation of Mobile Networks to the infrastructure. The Mobile Router is also the default gateway for the Mobile Network (Figure 1.10).

IEEE 1609.3 WAVE WSMP

One of the main goals of the WAVE development was to provide an implementation framework optimised for the particular requirements of vehicular environments, thus focusing mainly in air interface efficiency and low latency. While a WAVE compliant implementation of Network Services could be based on the IPv6/TCP/UDP stack, the potential of WAVE arguably lies on its WAVE-specific WAVE Short Message Protocol.

Air interface efficiency is tightly related to the signal transmission parameters, and although it can be said that these are issues which belong to lower layers of the stack, WAVE exploits the particular operating environment of the communication links and specifies a number of physical parameters in the transmission of

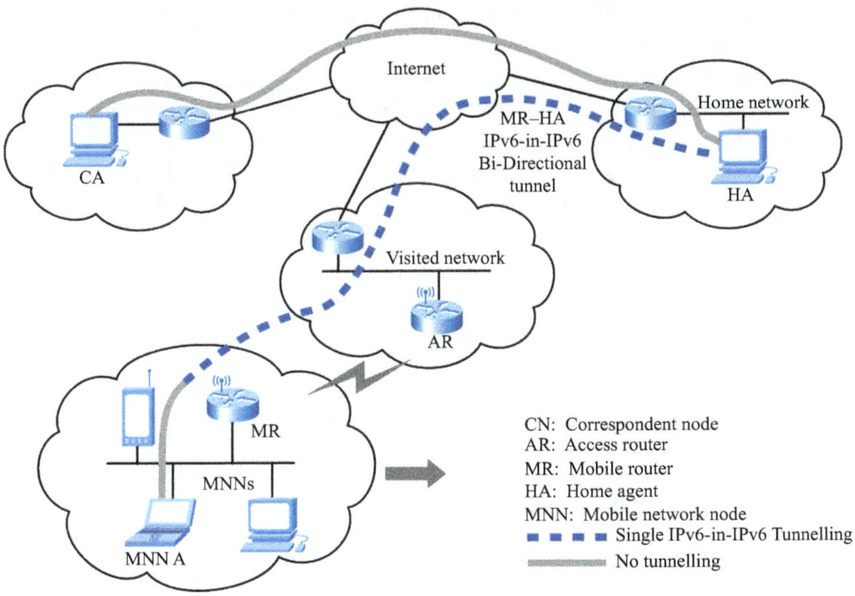

Figure 1.10 Example of NEMO basic support protocol operation

network-related data packets. Even in an IP-based stack, WAVE uses a transmitter profile containing these data. The WSMP is based on a number of primitives which allow higher layer entities to send and receive WSMs.

GeoNetworking
GeoNetworking is a practical development of a concept in which network nodes can be reached not only by means of their identified or known address, but also by means of geographic information. With GeoNetworking, it would be possible to request or send information to 'the vehicles 200 m behind me in this lane and direction', 'one vehicle approaching one intersection in front of me' or 'all the HGVs in a given geographical area' [13].

The GeoNetworking communication modes considered are:

- Unicast. Communication between a single entity and an identified destination entity at a given geographical position.
- Anycast. Communication between a single entity and a single arbitrary entity from a predefined group within a geographical area.
- Broadcast. Communication between a single entity and all entities within a given geographical area.

In combination with IPv6 Unicast, Multicast and Anycast modes, GeoNetworking + IPv6 can integrate any communication link mode for fixed or mobile scenarios.

The development of the GeoNetworking concept started at the C2C-CC consortium, after which the main effort was carried out by the GeoNet project, with the final results being incorporated into ETSI and ISO as a series of standards defining the integration of the GeoNetworking concept into the existing ITS framework.

In its more general specification, GeoNetworking is achieved by means of a C2CNet network sub-layer sitting on top of the access layer and servicing basic transport protocols. In practice, however, GeoNet and the standardisation efforts have been addressed to exploit the advanced addressing capabilities of IPv6 and the foreseeable future widespread support to IPv6-based applications to push the integration of GeoNetworking concepts with the IPv6/TCP-UDP stack communications.

Within this integrated framework, geonetworks can be seen as ad hoc subnets of a larger IP-based environment, and GeoNetworking capabilities are called upon only at certain links of the communication (C2CNet domain) when necessary depending on what the application wants to do.

1.4.2 List of related standards for cooperative ITS

- CEN EN 12253:2004: 'Road transport and traffic telematics; Dedicated Short Range Communication; Physical layer using microwave at 5.8 GHz'.
- CEN EN 12795:2003: 'Road transport and traffic telematics; Dedicated Short Range Communication (DSRC); DSRC data link layer; medium access and logical link control'.

- CEN EN 12834:2003: 'Road transport and traffic telematics; Dedicated Short Range Communication (DSRC); DSRC application layer'.
- ETSI TS 102 636-3 (V1.1.1): 'Intelligent Transport Systems (ITS); Vehicular Communications; GeoNetworking; Part 3: Network architecture', 2010–03.
- ETSI TS 102 637-1 (V1.1.1): 'Intelligent Transport Systems (ITS); Vehicular Communications; Basic Set of Applications; Part 1: Functional Requirements', 2010–09.
- ETSI ES 202 663 (V1.1.0): 'Intelligent Transport Systems (ITS); European profile standard for the physical and medium access control layer of Intelligent Transport Systems operating in the 5 GHz frequency band', 2009–11.
- ETSI EN 302 571 (V1.1.1): 'Intelligent Transport Systems (ITS); Radiocommunications equipment operating in the 5 855 to 5 925 MHz frequency band; Harmonized EN covering the essential requirements of article 3.2 of the R&TTE Directive', 2008–09.
- ETSI EN 302 665 (V1.1.1): 'Intelligent Transport Systems (ITS); Communications Architecture', 2010–09.
- IEEE Std 802.11-2007 (Revision of IEEE Std 802.11-1999): 'IEEE Standard for Information technology; Telecommunications and information exchange between systems; Local and metropolitan area networks; Specific requirements; Part 11: Wireless LAN Medium Access Control (MAC) and Physical Layer (PHY) Specifications', 2007–06.
- IEEE Std 802.11p-2010: 'IEEE Standard for Information technology; Telecommunications and information exchange between systems; Local and metropolitan area networks; Specific requirements; Part 11: Wireless LAN Medium Access Control (MAC) and Physical Layer (PHY) Specifications; Amendment 6: Wireless Access in Vehicular Environments', 2010–07.
- IEEE Std 1609.2-2006: 'IEEE Trial-Use Standard for Wireless Access in Vehicular Environments; Security Services for Applications and Management Messages', 2006–07.
- IEEE Std 1609.3-2010 (Revision of IEEE Std 1609.3-2007): 'IEEE Standard for Wireless Access in Vehicular Environments (WAVE); Networking Services', 2010–12.
- IEEE Std 1609.4-2010 (Revision of IEEE Std 1609.4-2006): 'IEEE Standard for Wireless Access in Vehicular Environments (WAVE); Multichannel Operation', 2011–02.
- IETF RFC 791: 'Internet Protocol DARPA Internet Program Protocol Specification', Information Sciences Institute, University of Southern California, 1981–09.
- IETF RFC 2460: 'Internet Protocol, Version 6 (IPv6) Specification', IETF Network Working Group, 1998–12.
- IETF RFC 2766: 'Network Address Translation – Protocol Translation (NAT-PT)', IETF Network Working Group, 2000–02.
- IETF RFC 2784: 'Generic Routing Encapsulation (GRE)', IETF Network Working Group, 2000–03.

- IETF RFC 3024: 'Reverse Tunneling for Mobile IP, revised', IETF Network Working Group, 2001–01.
- IETF RFC 3315: 'Dynamic Host Configuration Protocol for IPv6 (DHCPv6)', IETF Network Working Group, 2003–07.
- IETF RFC 3596: 'DNS Extensions to Support IP Version 6', IETF Network Working Group, 2003–10.
- IETF RFC 3775bis: 'Mobility Support in IPv6', IETF Mobile IP Working Group, 2011–09 (Expiring date for approval).
- IETF RFC 3963: 'Network Mobility (NEMO) Basic Support Protocol', IETF Network Working Group, 2005–01.
- IETF RFC 4213: 'Basic Transition Mechanisms for IPv6 Hosts and Routers', IETF Network Working Group, 2005–10.
- IETF RFC 5177: 'Network Mobility (NEMO) Extensions for Mobile IPv4', IETF Network Working Group, 2008–04.
- IETF RFC 5454: 'Dual-Stack Mobile IPv4', IETF Network Working Group, 2009–03.
- IETF RFC 5555: 'Mobile IPv6 Support for Dual Stack Hosts and Routers', IETF Networking Group, 2009–06.
- IETF RFC 5944: 'IP Mobility Support for IPv4, Revised', IETF Standards Track, 2010–11.
- ISO/DIS 21210.2: 'Intelligent Transport Systems; Communications access for land mobiles (CALM); IPv6 Networking', 2011–03 (Final Draft to be formally voted and approved).
- ISO 21215:2010: 'Intelligent Transport Systems; Communications access for land mobiles (CALM); M5', 2010–11.
- ISO/DIS 21217: 'Intelligent Transport Systems; Communications access for land mobiles (CALM); Architecture' (2010, currently under revision).
- ISO 21218:2008: 'Intelligent Transport Systems; Communications access for land mobiles (CALM); Medium service access points', 2008–08.
- ISO 24102:2010: 'Intelligent Transport Systems; Communications access for land mobiles (CALM); Management', 2010–11.
- ISO 29281:2011: 'Intelligent Transport Systems; Communications access for land mobiles (CALM); Non-IP networking', 2011–03.

1.5 5G and future mobility

5G is expected to extend vehicle automation, both inside vehicles, between vehicles and with the surrounding infrastructure. 5G brings improvements in terms of latency, bandwidth, network capacity, and further availability and reliability. Any roadmap in the future of transportation envisages the adoption of 5G. The roadmap of 5G represents a significant advance in mobile network features. Thus, the mobile industry is working with transportation vertical sector to prepare requirements and specifications for next generation communications and their impact on the automotive sector. 5G technology will support a wide range of connected devices

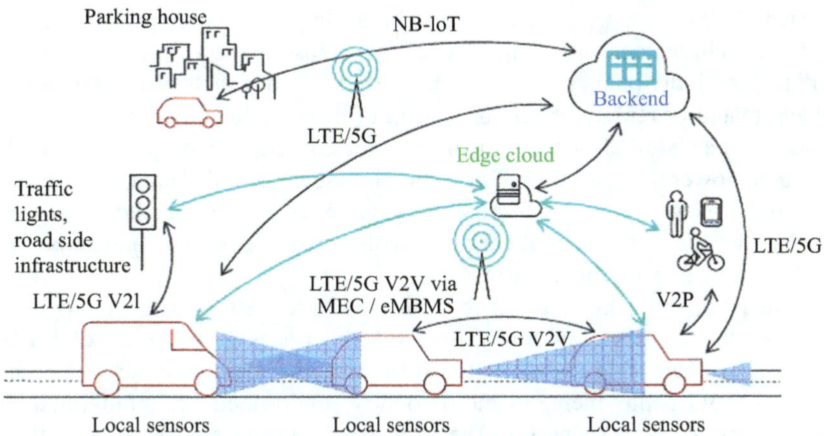

Figure 1.11 V2X communications and 5G

through ultra-wide band, high speed, high reliable connectivity and ultra-high responsivity for real-time applications. Network virtualisation and edge computing, among others, enable new services taking advantage of mobile cloud technologies. Edge computing provides compute/storage/networking capabilities at the network edge for multiple services for connected AD vehicles [14,15].

The EC has developed a 5G Action Plan for the EU, estimating 113.000 M€ of benefits from 5G deployment by 2025. The CAD vehicles market is leading the evolution of Vehicular IoT and is growing at a five-year compound annual growth rate of 45% (10 times faster than the overall car market).

From the transportation side, it is expected that 5G facilitate the delivery of connected and automated driving solutions, enabling real-time communication between cars, and exploiting the designated ITS spectrum. Connecting private vehicles and public transport with 5G will completely revolutionise the way transport will work (Figure 1.11).

With the rise of automated vehicles, vehicle-to-vehicle communication has become one of the most highly demanded developments for 5G. 5G key improvements for building transportation services are increasing bandwidth for mobile networks, and optimising massive IoT communications [16].

To conclude, 5G has a huge potential to transform the current transport systems into intelligent systems through three key drivers: standard networks, optimised communication for vehicles and interconnected digital identity systems.

1.6 Conclusions

Nowadays, both the telecom and the automotive industries are going through deep transformations. The automotive industry is evolving towards a vision where cars aim at becoming autonomous and wirelessly connected to be ready to cooperate with each other for a more efficient and safer driving. This topic opens questions on how such technologies will impact evolving transportation systems, our social world, and the individuals who live within it and whether such systems ought to be fully automated or remain under some form of direct human control.

5G is expected to close the gap and become cornerstone as the key technology that will empower this transition. For a complete success in this area, it is necessary that both the telecom and the automotive industry cooperate together to shape the future by addressing all the challenges that connected, cooperative and autonomous mobility brings into the innovation arena.

Right now, the achievements to be reached are clear and the focus rely on finding out the successful path towards CCAM services with the use of 5G technologies. The main challenges should overcome the main weakness, the uncertainties of a 5G deployment, in multi-country, multi-operator, multi-vendor and multi-car-manufacturer scenarios. The validation via trials and the optimisation of 5G technologies for CCAM services are indispensable topics to solve for the sake of commercial deployments.

Autonomy in future transportation is undoubtedly the subject of debate for a long time. All three levels – public communication, human–machine interaction and technical feasibility – co-act to sculpt the coming forms of transportation. The resulting system promises to be strikingly different from its traditional form. Some of the most evident proximal impacts will be on jobs and associated commuting patterns.

We are now witnessing an epoch of significant transition in which active control of the vehicle is being taken from the human driver and placed within the charge of the on-board computer systems themselves [17].

References

[1] President's Council of Advisors on Science and Technology. "Technology and the future of cities". February 2016.

[2] 5G Automotive Association. "Towards fully connected vehicles: edge computing for advanced automotive communications".

[3] A European strategy on cooperative, intelligent transport systems, a milestone towards cooperative, connected and automated mobility Available from http://ec.europa.eu/energy/sites/ener/files/documents/1_en_act_part1_v5.pdf.

[4] In Europe: COM(2016)588: 5G for Europe: An Action Plan and accompanying Staff Working Document SWD(2016)306. Available from http://eur-lex.europa.eu/legal-content/EN/TXT/?qid51479301654220&uri5CELEX:52016DC0588.

[5] Final Report of the C-ITS Platform, January 2016. Available from http://ec.europa.eu/transport/themes/its/c-its_en.

[6] 5G Americas, 'LTE to 5G: Cellular and Broadband Innovation', White paper, 2017.

[7] 5G Americas, 'Cellular V2X Communications Towards 5G', White Paper, 2018.
[8] OCDE, 2015. Automated and Autonomous Driving Regulation under uncertainty, International Transport Forum, Corporate Partnership Board Report.
[9] 5GPPP Automotive Working Group, 'A study on 5G V2X Deployment', version 1.0, Feb 2018.
[10] 5GAA Automotive Association, 'Timeline for deployment of C-V2X – Update', 2019.
[11] Alfonso, J., Sánchez, N., Menéndez, J. M., and Cacheiro, E., 2014. Cooperative ITS architecture - the FOTsis Project and beyond. IET Intell. Transport Syst. http://dx.doi.org/10.1049/iet-its.2014.0205, Online ISSN 1751–9578.
[12] ERTICO, 2015. Guide about technologies for future C-ITS service scenarios.
[13] Festag, A., Baldessari, R., Zhang, W., Le, L., Sarma, A., and Fukukawa, M., 2008. CAR-2-X communication for safety and infotainment in Europe. NEC Tech. J. 3 (1), 21–26.
[14] Kljaic, Z., Skorput, P., and Amin, N., 2016. The challenge of cellular cooperative ITS services based on 5G communications technology, 39th International Convention on Information and Communication Technology, Electronics and Microelectronics (MIPRO), Opatija, 2016, 587–594.
[15] 5GPPP Architecture Working Group, 'View on 5G Architecture', version 2.0, Dec 2017.
[16] 5GPPP Automotive Working Group, '5G Automotive Vision', White Paper, 2015.
[17] Hancock, P. A., 1997. Essays on the Future of Human-Machine Systems, Banta, Eden Prairie, MN.

Chapter 2
Grid integration and management of EVs through machine-to-machine communication

Sayidul Morsalin[1]*, Khizir Mahmud*[1]*, Bao Toan Phung*[1]*, and Jayashri Ravishankar*[1]

The development of vehicle-to-grid (V2G) technology for electric vehicles (EVs) enables prosumers to incorporate other energy storages and participate in large-scale bidirectional energy trading. However, the mobility of EVs makes it significantly different from any other conventional storage. For a mass-penetration of EVs, uncoordinated planning and random distribution of V2G may result in power grid instability and power quality degradation. Therefore, a real-time coordinated and automated V2G system in a distributed way is essential for effective energy management. Machine-to-machine (M2M) communication enabling bidirectional information flow between EVs and other power system components is a key element to mitigate challenges associated with their grid integration and manage all parties autonomously. M2M technology assists in improving energy efficiency and reducing any potential risk of instability in power systems. In this chapter, an in-depth study of an M2M communication-based coordinated management of EVs is presented. It includes a step-by-step description and practical implementation process of the data logging system, data transmission, and its automatic processing mechanism at the server. Additionally, the scalability issue for EV M2M communication under a 4G Long Term Evolution transceiver base station is extensively examined. Various numerical simulations with and without radio network scheduling are also presented to provide a detailed understanding of this scalability issue, taken into consideration communication delays and blocking rate.

List of abbreviations

3GPP	Third-generation partnership project
4G	Fourth-generation
AMC	Adaptive modulation and coding
AMI	Advanced metering infrastructure

[1]School of Electrical Engineering and Telecommunications, University of New South Wales, Australia

API	Application programming interfaces
CP	Contention period
DL	Downlink
DLS	Data logging system
DMS	Distribution management systems
EMS	Energy management system
EPC	Evolved Packet Core
ETSI	European Telecommunications Standards Institute
E-UTRAN	Evolved-UTRAN
EVs	Electric vehicles
G2V	Grid-to-vehicle
GPIO	General-purpose input-output
GUI	Graphical user interfaces
HAREQ	Hybrid automatic repeat request
HMI	Human–machine interface
IoE	Internet of Energy
IoT	Internet of Things
IP	Internet protocol
LTE	Long Term Evolution
M2H	Machine-to-human
M2M	Machine-to-machine
M-LWDF	Modified largest weighted delay first
MTC	Machine type communication
OFDMA	Orthogonal Frequency Division Multiple Access
PDCP	Packet data convergence protocol
PF	Proportional fair
PHY	Physical layer
PLC	Power line communication
PLR	Packet-loss ratio
QoS	Quality of service
RAN	Random access network
RB	Resource Block
RLC	Radio link control
SCADA	Supervisory control and data acquisition
SCFDMA	Single Carrier Frequency Division Multiple Access
SoC	State-of-charge
TA	Timing alignment
TCP	Transmission control protocol

TP	Transmission period
UL	Uplink
V2G	Vehicle-to-grid
WAMS	Wide area monitoring system

2.1 Introduction

Vehicle tracing and tracking by wireless machine-to-machine (M2M) communication over cellular networks will be an indispensable part of future intelligent transport. Principally, an M2M embedded system aims to interact autonomously with other electronic systems without any human intervention. Presently, the technology is gaining momentum, and facilitates a new paradigm to Internet of Things (IoT) and Internet of Energy (IOE) [1,2]. As the M2M integrated system is uprising, the technology is assumed to play promising roles in efficiently utilising the existing transport infrastructure, minimising the traffic congestion and improving the fuel efficiency [1,2].

For wireless M2M communication, Long Term Evolution (LTE), a Fourth-generation (4G) network, has already occupied a considerable share (around 2.5%) of the total M2M market in 2014 [3,4]. Moreover, LTE system is also emerging as a potential carrier of M2M systems in advanced transportation systems [5]. This is because LTE maintains a standard of quality of service (QoS) in transmitting various trip information (e.g. position and speed), which are usually fewer in bytes [6,7]. Moreover, the bidirectional flow of M2M information from vehicles to autonomous information processing centres can predict a vehicle's location and mobility in real time. Based on the received information about road conditions, the prediction can be effectively used to update the navigation systems and suggest vehicle's driver to choose a better route. Thus, M2M communication has influential roles in controlling traffic congestion as a part of the fleet management [8,9,10].

Electric vehicles (EV), both hybrid and fully electric, are a promising mode of transport for reasons of cost-effectiveness, convenience, and their environmentally friendly characteristics. A report by the International Energy Agency predicts that, in the next 4 years, global sales of EVs will treble, to six million per annum [11]. Furthermore, sales of EVs are currently 'booming' in the UK [12] and are also growing strongly in parts of the USA [13], especially when supported by government incentives. It has also been predicted that EVs will comprise more than 40% of new vehicle sales in Australia by 2030 [14,15]. These growth trends are likely to be supported, if not accelerated, by rapidly reducing battery prices (15% per annum [16]) and the increasing viability of home-storage and charging options.

Studies have reported that most vehicles are parked for 95% of their time [17,18]. If they are connected to the electricity grid while parked, then they may charge, or alternatively they can contribute any excess stored energy from their batteries to the grid, i.e. in Vehicle-to-Grid (V2G) systems [18,19,20]. In either case, the uncoordinated flow of energy can cause problems for the distribution system, such as voltage fluctuations, transformer overloading, harmonic injection, etc. [21,22].

Consequently, an energy management system (EMS) is required to minimise potential problems while maximising potential benefits for participants in distributed electricity supply systems incorporating EVs [23]. Therefore, to establish an EMS in a distributed way, an appropriately designed M2M system has been identified as a key component.

The chapter is organised as follows. Sections 2.2 and 2.3 provide a brief review of distributed EMS of EV and the important role of M2M communication.

Since data logging of EVs is a key component in an EMS, in Section 2.4, the hardware and software operation of a M2M-based data logging system (DLS) is described. The in-vehicle logging system was based on a Raspberry Pi platform with 4G modem and socket programming in Python language, and was designed primarily for automated wireless monitoring of the position and state of charge of EVs, but could also be used to log other information. The DLS sends this trip information to a remote server at regular intervals via the TCP/IP protocol.

The information rate-density that can be managed in LTE networks is limited. Therefore, to manage energy in a coordinated way, it is a concern to gain a clear perception about the scalability of EVs; in other words, how many M2M EVs can be supported under an LTE network. Section 2.5 reports both numerical simulations and analytical results; showing that up to 250 vehicles can be supported per base station before communication delays and blocking disrupt system operation.

Communication scheduling assigns radio resources in such a way that it always maintains a certain level of fairness while satisfying the best QoS among all M2M-connected EVs. Via numerical simulations, Section 2.6 presents different communication performance (e.g. delay, packet-loss) for various recognised scheduling algorithms.

2.2 M2M in distributed energy management systems

In traditional power systems, electricity is generated in a few bulky power plants and distributed to the customers through high-voltage and medium-voltage transmission lines. A limited amount of communication technologies was utilised to monitor and control the power flow to the high-voltage and medium-voltage substations. With the progress of technologies, various renewable energy sources such as wind turbine, photovoltaic, hydro-power are integrated with different scales and levels of the grid. Therefore, the complexity of the power flow increases, resulting in a need for various control, monitoring and automation technologies. The evolution of the power grid is illustrated in Figure 2.1.

One of the earliest initiatives for the control and automation of power grid is the power line communication (PLC) and supervisory control and data acquisition (SCADA) technology. The PLC technology uses the power line cables to transmit a modulated carrier signal for communication. There are two types of PLC, i.e. broadband PLC and narrowband PLC. The data rate of broadband PLC is up to

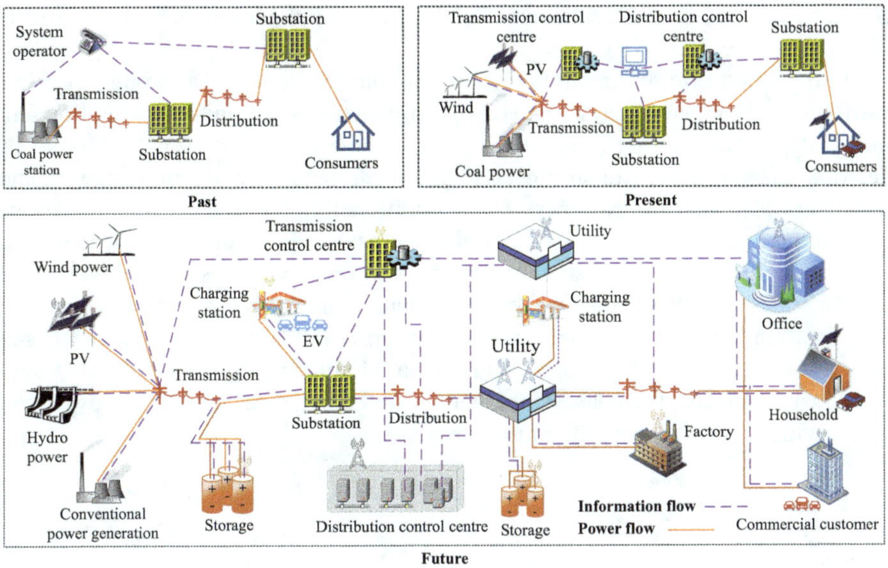

Figure 2.1 Evolution of the power grid

several hundred Mbps, with an operating frequency range of 2–250 MHz. The operating range of narrowband PLC is 3–500 kHz including both Low Data Rate Narrowband PLC and High Data Rate narrowband PLC. The Low Date Rate Narrowband PLC uses a single carrier-based approach with a data rate up to 10 kbps, whereas the High Data Rate Narrowband PLC uses a multi-carrier-based method with a data rate less than 1 Mbps.

The key applications of the PLC include fault location identification, isolation and system restoration of a medium-voltage substation. Additionally, PLC is also used for grid islanding process due to any faults in a renewable-energy-based distributed power generation system. On the other hand, the SCADA is a computer-based automation system which consists of data acquisition, data management and transmission, and graphical user interfaces (GUI) for advanced supervisory management. It also includes human–machine interface (HMI) software to facilitate centralised monitoring and control and process input and output data. Its network collects field information from various sensors and displays the data textually or graphically. Therefore, users can monitor and control the entire network under SCADA in real time from a remote location [24].

The proliferation of renewable energy sources and the integration of various energy sources, e.g. EV, fuel cell, battery storage, necessitate the need for advanced monitoring and control technologies. Moreover, some energy sources like energy storage and EVs contribute to the bidirectional energy transfer acting as a source and load at the same time. In this scenario, the usage of M2M enhances the scale of power system automation in which every device can exchange and share data for advanced management. A schematic of the M2M applications in a smart grid is shown in Figure 2.2.

In the smart grid, bidirectional M2M communication is used through SCADA distribution management systems (DMS) to control and monitor power flows in high-voltage transmission lines and the low-voltage customers' points. Moreover, electricity is generated much closer to the customers in a distributed way and consumers also act as prosumers [25]. Additionally, the penetration of various intermittent and unpredictable sources also increases significantly. Thus, to ensure power quality and system stability, M2M communication is used in multiple electric nodes to gather decision-taking information and acts such as remote power-flow data reading, remote device connection-disconnection and load control, and early outage detection [26]. The Wide Area Monitoring System (WAMS) synchronously measures the phase of the voltage and current vectors of the power grid and provides holistic visibility on the power grid.

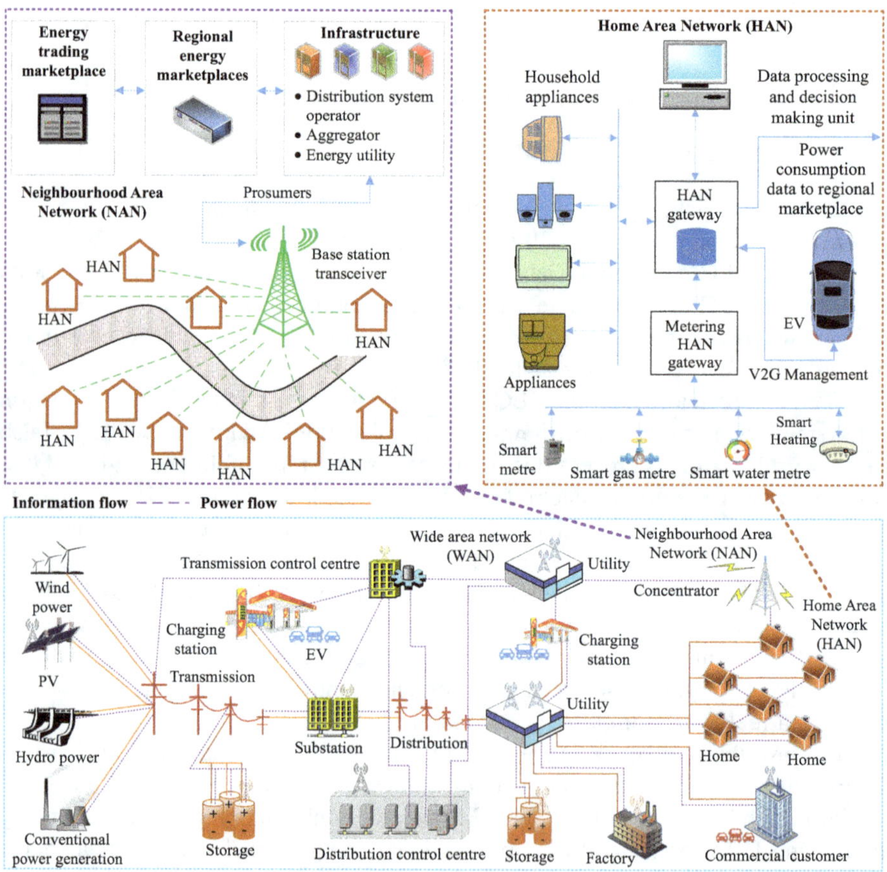

Figure 2.2 M2M applications in a smart grid

Based on the European Telecommunications Standards Institute (ETSI), an M2M network consists of the following elements:

- M2M component: it is integrated into electrical devices to transmit and execute data
- M2M gateway: it enables connectivity between M2M components and communication networks
- M2M server: it acts as a middleware to process and transmit data through various application services
- M2M area network: it provides a connection between M2M gateways and M2M components
- M2M communication network: it includes a connection between the M2M gateway and M2M server

All these elements consist of three domains, i.e. device domain for M2M components, network domain for M2M area network and gateway, and application domain for M2M server and communication network, as shown in Figure 2.3.

Various sensors and actuators are connected to the access network, either wireless (Zigbee, Wi-Fi, IEEE 802.15p, GSM, 3G, LTE, 5G, etc.) or wired (optical, xDSL, cable, etc.). The M2M area network, access network and core network suggested by ETSI are as follows:

- M2M area network: Wi-Fi, Zig-bee, PLC, M-Bus and IEEE 802.15
- M2M access network: Wi-Max, WLAN, x-DSL, Satellite, PLC and E-UTRAN
- M2M core network: 3GPP, TISPAN and ATTM

Although M2M communication helps to improve energy management, the packet delays, decreasing success ratios and the coverage areas are a few challenging factors. Figure 2.4 shows various M2M communication types, their bandwidth, speed and coverage distance.

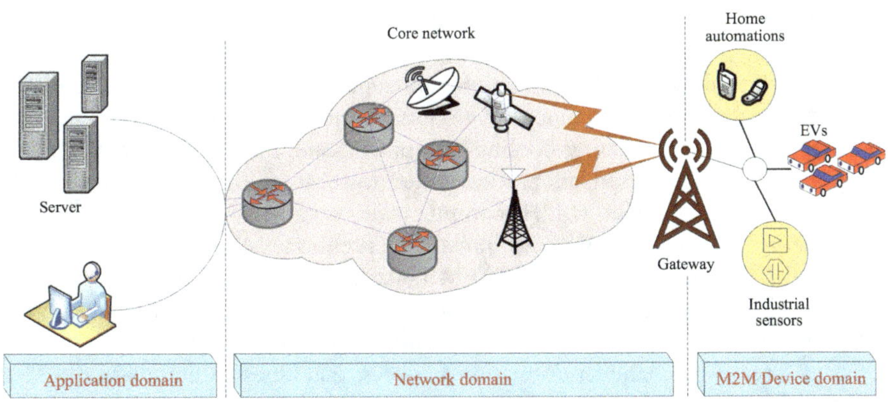

Figure 2.3 ETSI suggested M2M architecture

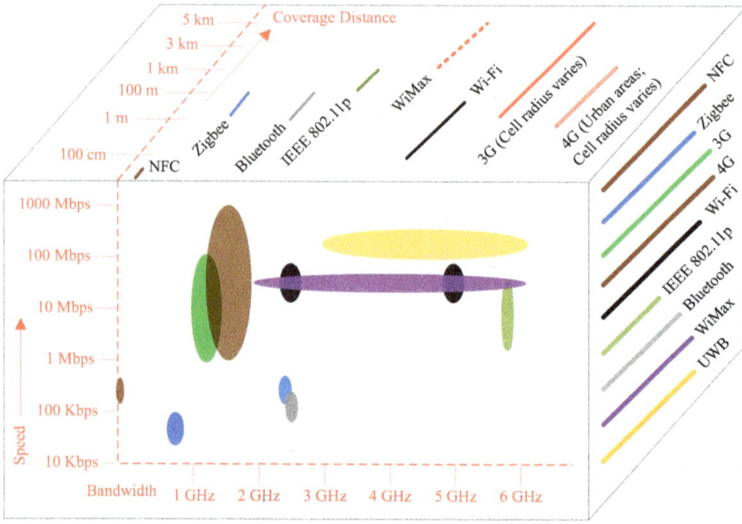

Figure 2.4 Various M2M communication types, their bandwidth, speed and coverage distance

M2M communication in various levels includes Home Area Network (HAN), Neighbourhood Area Network (NAN) or Field Area Network (FAN), Wide Area Network (WAN), Data Concentrator Unit (DCU) or Gateway, and an application or data centre, as shown in Figure 2.2. Multiple sensors in HAN, NAN and WAN would communicate through DCU to the data centre. The gateway to HAN transfers data from various domestic appliances/devices (sensor attached) through HAN DCU for domestic energy management. One such device at the household point is an advanced metering infrastructure (AMI), which acts as a home gateway or sensor. The gateways from various HANs establish secure communication with the central data aggregator, which sometimes acts as a controller to provide a decision.

Figure 2.5 shows M2M communication applications in HAN. Recently, a range of diverse embedded devices, e.g. laptops, mobile phones, digital TV, electronic appliances and heating-cooling equipment, are integrated into the home network. Various smart sensors are integrated to these devices and equipment to shift the machine-to-human (M2H) communications to M2M communications for smart energy management without human intervention. Most of the sensors are wireless and remotely read data from EMS elements such as lighting control systems, surveillance systems, grid power monitoring systems, healthcare monitoring system and home automation system. The wireless sensor networks (WSN) of these elements directly exchange information between machines to automate the total system.

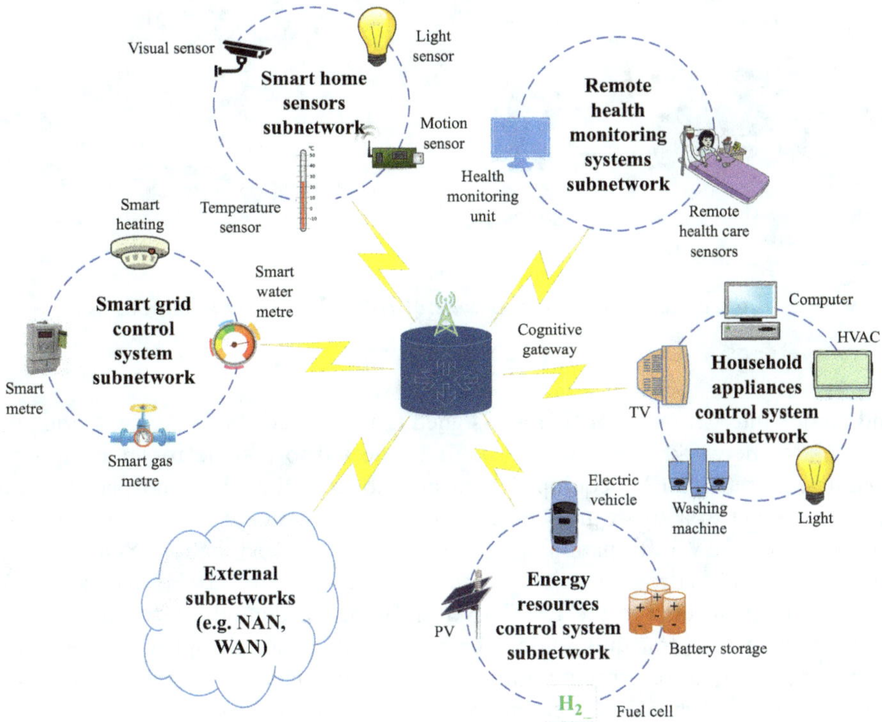

Figure 2.5 M2M communication applications in domestic energy management

2.3 M2M communication for EVs

M2M communication is used for data transfer between embedded devices to minimise human intervention and make an automated EMS. For EV, M2M is used for a coordinated charging-discharging system, battery health monitoring and traffic management. All these applications can be categorised into three basic functions:

- Real-time sensing: battery health (i.e. state-of-charge [SOC], temperature), battery charging-discharging, EV location, etc.
- Data communication: transfer sensing data to the server for monitoring or decision-making
- Data processing: process the sensor data at the server and execute decisions

The energy flow between EV and grid can be coordinated using M2M communication. From the various gateways, packets (sensing data) are sent to the central data processing unit for processing, as shown in Figure 2.6. In this process, the central processing unit is known as server and EVs are known as clients. A static

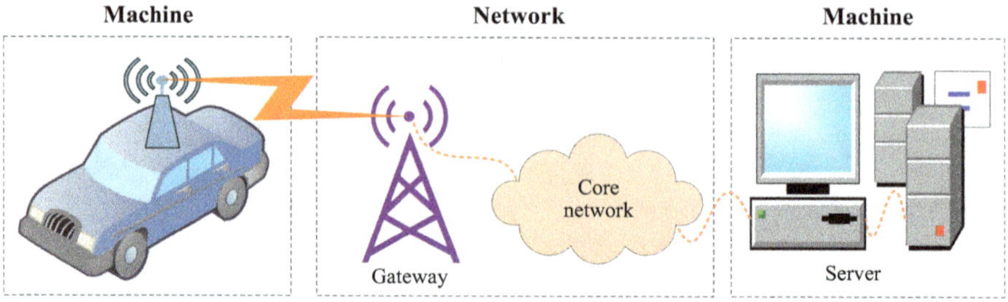

Figure 2.6 Basic components of EV M2M communication

and unique internet protocol (IP) is assigned to the server. It can also work on any private-area network, where the server is connected to a static router using port forwarding method. EVs equipped with an onboard DLS can send packets at a regular interval to a smart grid server using TCP/IP protocol. These packets may contain various EV information such as location, EV speed, battery SOC, battery temperature and battery capacity. These data (packet) can be read using mobile applications connected to the EV CAN bus through Bluetooth. Alternatively, the data reading and its transmission can be implemented using Raspberry Pi, internet modem through socket programming in Python language. Although WiMax and IEEE 802.11p is available for communication, LTE is better for its wide coverage area. However, achieving maximum information rate-density with a limited communication overhead is still a challenging issue in LTE. For large-scale EV penetration and their management using M2M, 5G communication is recommended.

2.3.1 M2M communication architecture (3GPP)

Based on the third-generation partnership project (3GPP), the functions of the M2M communications can be categorised into following parts:

- Tracking: it includes the controllable data sensing
- Logging: it includes the data transfer
- Notifications: it includes the data reception with a decision (after processing of sensing data)

The M2M communication configuration based on 3GPP for EV management in a smart grid is shown in Figure 2.7. For the configuration in the first part of Figure 2.7, communications among EVs are done using a machine type communication (MTC) which is mounted inside an operator domain. The application programming interfaces (APIs) on this MTC are provided by the operator. However, for the second configuration in Figure 2.7, MTC server is placed outside the operator domain and is not controlled by the operator.

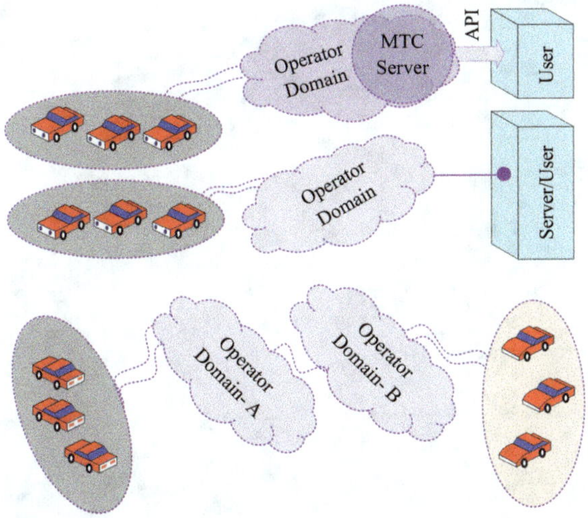

Figure 2.7 3GPP-based EV M2M communication architecture

2.4 Electric vehicle data logging systems

Data logging is an initial stage of data processing and decision execution process of M2M communication, which is used for battery management, traffic congestion and accident management, and coordinated V2G management. In this section, a step-by-step DLS modelling process is discussed. It is developed using a Raspberry Pi operating under a commercial 4G wireless network. In this DLS, EV battery SOC and EV location are read to manage the energy flows from/to EV and grid.

2.4.1 Data logging system

The radio access through cellular network is established through the DLS radio module. However, the service contract with the network authority should remain activated with the DLS. In this DLS modelling, a commercial LTE network is used. The LTE operator provides the following two types of identification to the M2M terminal:

- International Mobile Subscriber Identity (IMSI)
- Mobile Station International Subscriber Directory Number (MSISDN)

Using this IMSI and MSISDN number, EVs identification is done. EVs (alternatively known as clients) send data to a unique static IP under a private-area network where the server is placed. In this modelling process, the server was placed in PC with Linux (Ubuntu) operating system. A schematic of the DLS and automated EV monitoring system is shown in Figure 2.8.

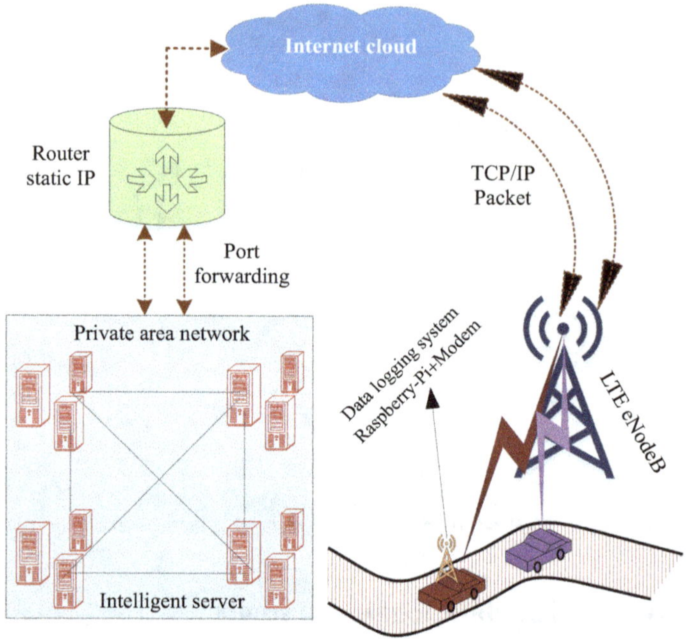

Figure 2.8 M2M-communication-based EV monitoring system

2.4.2 Hardware description

The DLS shown in Figure 2.9 functions as a client in the socket communication. The Raspberry Pi platform is interfaced with the global positioning system (GPS) and the 4G modem. The socket programming in the communication is performed using Python Language to process the battery SOC and GPS signal.

2.4.2.1 Interfacing GPS with Raspberry Pi

To track the EV location, a MicrostackTM GPS is integrated with the Raspberry Pi. A pulse per second (PPS) signal for precise time reading and a high-sensitivity internal patch antenna is embedded with this GPS. The connection of the general-purpose input-output (GPIO) pins for Raspberry with the GPS is shown in Table 2.1.

2.4.3 DLS operation

The DLS operation is shown in Figure 2.10. The latitude and longitude of the DLS setup location are identified using GPS, and it is sent to the server. DLS also sends the battery SOC and time as a payload. Figure 2.10(a) shows the client window and (b) shows the server window.

The server of the M2M communication is connected to the client with IP 49.182.28.134, and a payload of 30%. The server window shows that the SOC of the battery is 30%, and it notifies the client that the 'EV battery is running out

Figure 2.9 A practical implementation of data logging systems for M2M communication, (a) DLS setup, (b) DLS is sending data to server and processing

Table 2.1 GPS connection with Raspberry Pi

GPS pins	Raspberry Pi pins
TX	10-RX
RX	8-TX
VIN/VCC	2–5 V
GND	6-GND

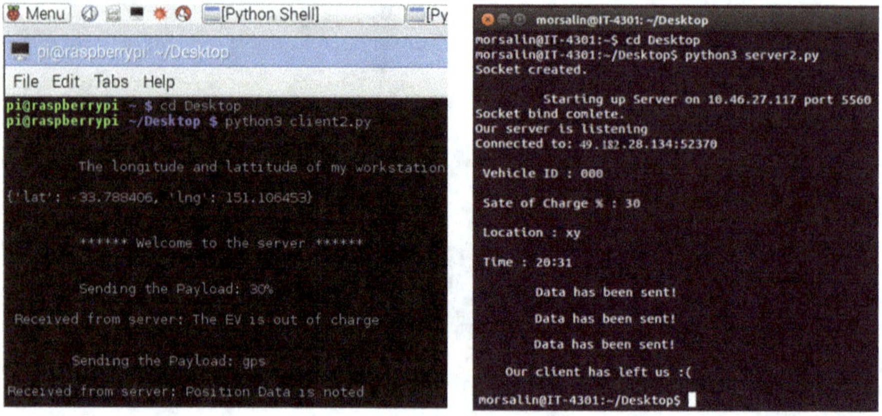

(a) (b)

Figure 2.10 DLS operation, (a). Client window, (b) Server window

of charge'. It also shows the EV position and EV id. From this 'GPS' and 'SOC' information, the server will notify drivers about the nearest charging station to charge the battery. The drivers can also send charging reservation request to the nearest EV charging station by using this DLS. Clients will always remain connected to the server to monitor position and battery SOC for better energy management. However, if the client-server connection is disconnected for more than 120 s, the server will display 'client has left' for the disconnected EV.

The designed DLS and the M2M communication can efficiently regulate EV battery charging-discharging and help to manage vehicle congestion. However, the EV penetration to the market is increasing significantly. Thus, the number of DLS systems or EVs that can be accommodated under a certain BTS needs to be investigated. Section 2.5 provides an overview of this scalability issue.

2.5 Scalability of electric vehicles

Cellular-based M2M communication is likely to become a major source of future traffic in communication networks. The M2M technology can be applied to the automated monitoring of smart grid infrastructure. For obvious reasons, LTE or 4G network has been introduced in wireless M2M systems to enhance the system capacity. However, the information rate-density accommodated by the base-station or transmitted by an M2M user is still an issue for further investigation. This section reports the scalability limits of automated electric vehicular monitoring systems in terms of blocking rates and delay versus vehicle density. The simulation outcomes from numerical modelling were found to support the approximated analytical results.

2.5.1 Radio resource and IP connection of LTE

Long Term Evolution (LTE) is a wireless communication standard which provides some extra benefits over 3.5 G Communication (e.g. HSDPA). The access network of LTE involves Evolved-UTRAN (E-UTRAN). The E-UTRAN mainly consists of two nodes: the enhanced NodeB, eNodeB (i.e. Base station) and a user equipment, UE (e.g. mobile or M2M devices). For multiple accessing, LTE uses Orthogonal Frequency Division Multiple Access (OFDMA) for the downlink (DL) and Single Carrier Frequency Division Multiple Access (SC-FDMA) for the uplink (UL) in the physical layer. The OFDMA technique allocates users in both the time and frequency domains. In the frequency domain, the LTE physical layer supports a scalable bandwidth from 1.4 to 20 MHz, divided into steps of 180 kHz, known as the Resource Block (RB). The resource allocation in both time and frequency domains is shown in Figure 2.11.

On the other hand, the largest unit in the time domain is the 10 ms radio frame, subdivided into ten sub-frames with duration of 1 ms. Later, one sub-frame splits into two slots, each of 0.5 ms. Each slot comprises seven OFDMA symbols. The resource of 12 subcarriers for duration of one transmission time interval (TTI i.e. sub-frame) is termed a Physical Resource Block (PRB). The smallest unit PRB is the Resource Element (RE). On the other hand, for uplink, SC-FDMA has similar structure and performance to OFDMA. The only exception is that SC-FDMA utilises single carrier modulation techniques. Therefore, each subcarrier contains information of all transmitted symbols whereas OFDMA subcarrier carries information related to one specific symbol.

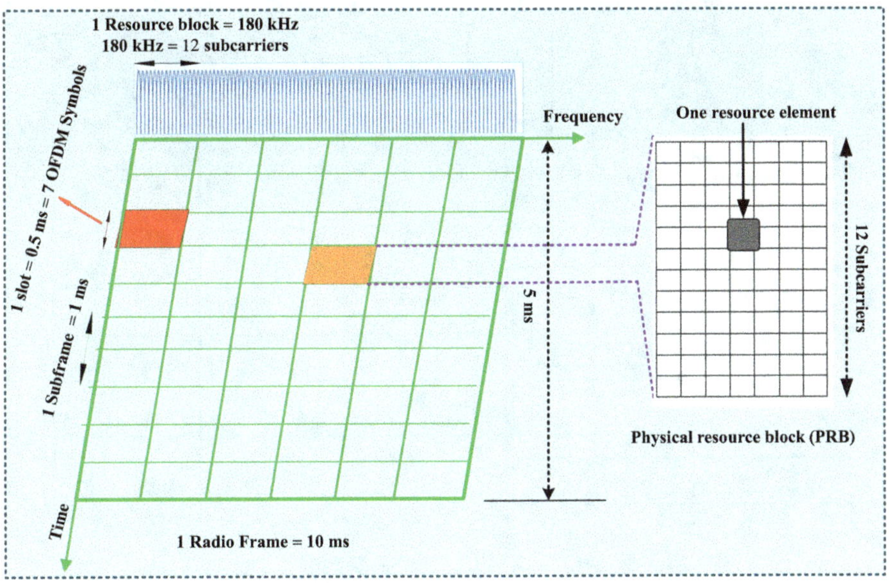

Figure 2.11 Radio resource of LTE network

For IP connection of M2M devices, as illustrated in Figure 2.12, LTE manages non-radio services by Evolved Packet Core (EPC) system. EPC mainly consists of a number of logical nodes: the Packet Data Network (PDN) Gateway (PGW), the Serving Gateway (SGW) and the Mobility Management Entity (MME). To route the IP traffic with the reasonable QoS, an EPC connection is established between the gateway of PDN (PDN-GW) and the M2M devices.

2.5.2 Radio access of M2M

The automated processes of radio access between EVs and base-station are popularly known as random access network (RAN). For M2M connection, EVs randomly follow the contention-based procedures, as illustrated in Figure 2.13.

In LTE network, EVs initially send some preamble message (msg1) to eNodeB via some radio channels at physical layers. The transmitted preamble message is a 6-bit unique id and contains mainly the information of sub-carriers. When two EVs or more select the same preamble sequence, a collision may arise. After decoding the message at the base-station, as an acknowledgement, a random-access response (RAR) message (msg-2) is ubiquitously broadcasted to all M2M users. The broadcasted message contains radio-related information such as timing alignment (TA), initial uplink grant, preamble sequence identifier and so on. Based on the received information about TA, in the next step, EVs send a request message (msg3) to the base-station to occupy its allocated uplink resource. Simultaneously, EVs start counting contention time by using a counter to avoid any possible

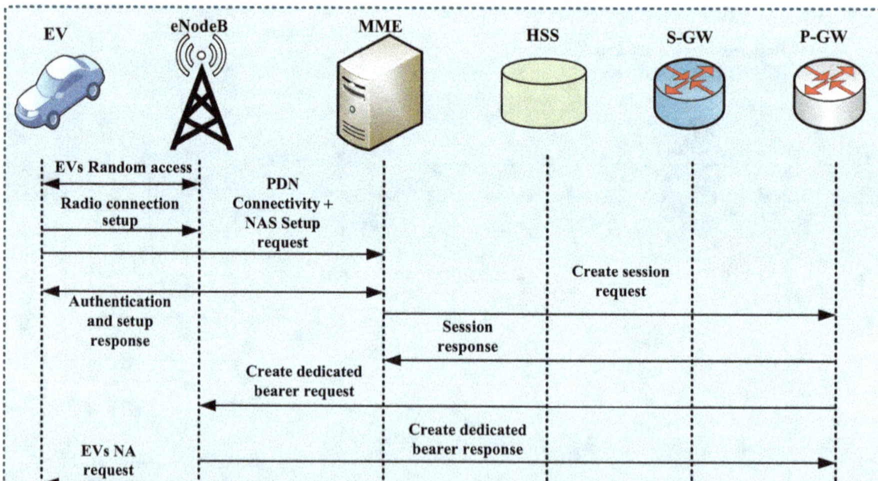

Figure 2.12 IP connection of LTE network

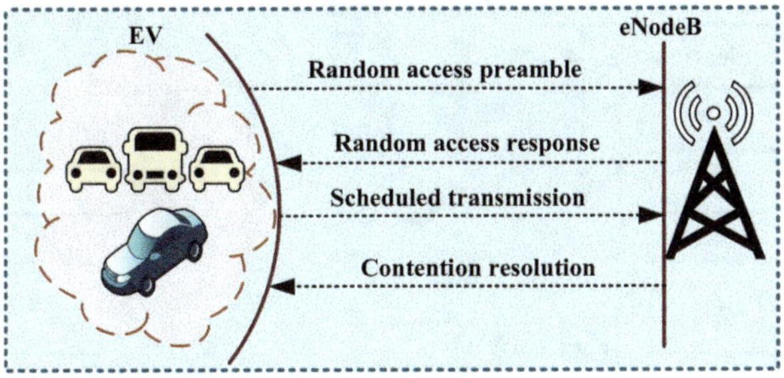

Figure 2.13 Radio access of EV

collision. It is also worthwhile to mention that if any collision occurs in the first step, still the collided message (msg1) can be detected by eNodeB.

In this circumstance, the base station is not exclusively aware of the collision and responds quickly with its RAR message. Hence, each EV again transmits msg3, which in turn results in more collision at the base-station. The colliding device will continually send msg1 until the contention timer window expires, or msg3 is entirely detected by eNodeB. If eNodeB successfully decodes one of the collided messages (msg3), it replies to the EV with a unique identifier in the fourth step. Only the EV which finds the match between the received identifier and the identity sent in the msg3 can send a feedback of Hybrid Automatic Repeat Request (HARQ) acknowledgement to eNodeB. Other M2M devices discard the received message. As a response of acknowledgement, eNodeB finally transmits a contention-resolution message (msg4) to the successful EV.

2.5.3 LTE user-plane protocol

The LTE uses a set of protocols in the user-plane. The protocols are dominantly used to transfer IP packet to eNodeB through radio channel as shown in Figure 2.14. Usually, these protocols are stacked at both M2M device of EV and eNodeB. For radio interfacing, the main protocols are the Packet Data Convergence Protocol (PDCP), the Radio Link Control (RLC), the Medium Access Control (MAC) and the Physical Layer (PHY). Every layer has its own identical role in packet transmission. For example, the PDCP layer performs compression/decompression of the IP header and thus improves the efficiency of radio access network. The RLC layer performs the concatenation and segmentation of PDCP packet while the MAC layer allocates the radio resources and performs associated scheduling at both uplink and downlink. Finally, the PHY layer accomplishes the modulation and demodulation function at both ends' terminals and sometimes performs coding/decoding. All application layer information captured by DLS is encapsulated by a Service Data Unit (SDU) and then transferred to a lower layer by using a Protocol Data Unit (PDU).

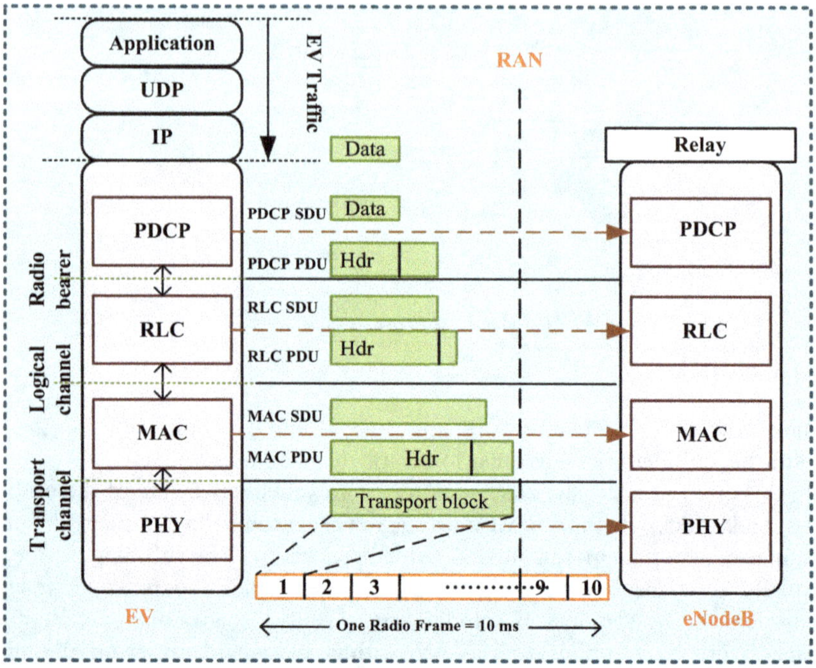

Figure 2.14 Delay vs number of EVs

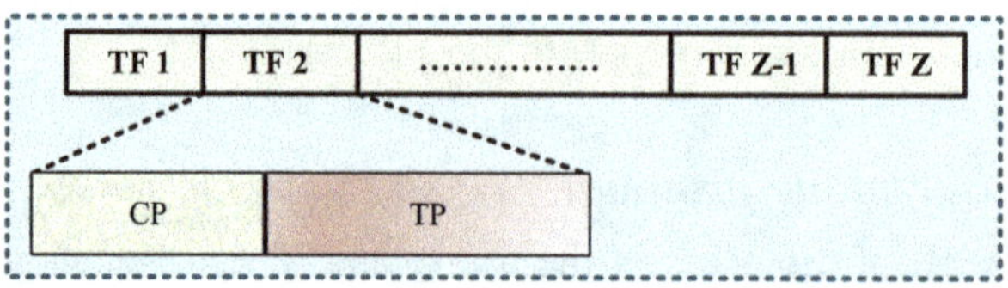

Figure 2.15 EV's radio access in time frame

2.5.4 Analytical model

Let an LTE system manage X number of EVs which can properly utilise radio resource under a base-station. Consider Y EVs have a possibility of transmitting data packet for the duration of one time frame. Assume that N number of EVs can succeed in making radio access with the LTE network while the remaining EVs fail to transmit packet, get blocked and finally transform into outage. Therefore, the total number of EVs in sleep mode is Y-N. Indeed, they make radio connection attempt in the next time frame and transmit application data.

An LTE time frame consists of a contention period (CP) and a transmission period (TP), as depicted in Figure 2.15. By following contention-based random-access procedures, the CP offers an equal opportunity to all Y EVs which are attempting to transmit. On the other hand, the time period TP works on a reservation basis. The transmission period can explicitly be divided into equal amount of time slots. In fact, M2M carrying each vehicle uses its own predefined slot for the packet transmission.

Let all Y vehicles that are trying to communicate during CP have an access probability a. This parameter determines the probability of using radio resource in a specific time frame.

The time interval between the *(p-1)th* and the p-th successful contention is t_p. If N_p^c collisions occur during a time period t_p, then by applying p-persistent algorithm to the MAC scheduling of EVs, this time can be expressed as

$$t_p = \sum_{k=1}^{N_p^c} col_k + \sum_{k=1}^{N_p^c} idle_k + msg_p + idle_{N_p^c} \tag{2.1}$$

For a channel busy period t_p, $idle_k$ and col_k represent the total time duration of the *kth* event and *kth* collision, respectively, and msg_p denotes the transmitted message length. A random variable T_{CP} can be introduced as

$$T_{CP} = \sum_{p=1}^{N} t_p \tag{2.2}$$

After N successful contention, the mean or expected value of T_{CP} can be written as

$$\begin{aligned} E[T_{CP}] &= \sum_{p=1}^{N} E[t_p] \\ &= \sum_{p=1}^{N} E\left[N_p^c\right]\ E[coll_p] + \sum_{p=1}^{N} E\left[N_p^c\right]\ E[idle_p] \\ &\quad + \sum_{p=1}^{N} E[idle_p] + \sum_{p=1}^{N} E\left[msg_p\right] \end{aligned} \tag{2.3}$$

If $E[coll_p] = \delta_{idle} = \delta_{coll}$ and $E[msg_p] = \delta_{succ}$, then p-persistent MAC scheduler suggests as

$$E\left[N_p^c\right] = \frac{1-(1-a)^{Y-p}\{1+(Y-p)a(1-a)^{-1}\}}{(Y-p)a(1-a)^{Y-p-1}} \tag{2.4}$$

$$E[idle_p] = \frac{(1-a)^{Y-p}}{1-(1-a)^{Y-p}}\delta_{idle} \tag{2.5}$$

Here, $E\ [N_p^c]$ and $E\ [idle_p]$ represent the expectation or mean of random variables N_p^c and $idle_p$ respectively. Combining (2.4) and (2.5) and substituting into (2.3), it can be simplified (after some numerical calculation) as,

$$T_{CP} = \sum_{p=1}^{N} \frac{(1-a)^{Y-p}}{(Y-p)a(1-a)^{Y-p-1}} \delta_{idle} - \sum_{p=1}^{N} \left[\left\{ 1 - \frac{1-(1-a)^{Y-p}}{(Y-p)a(1-a)^{Y-p-1}} \delta_{coll} \right\} + \delta_{succ} \right] \tag{2.6}$$

In LTE, the duration of one time frame (T_{frame}) can be resolved into a contention period and a transmission period. The former is a function of variables N and a, while the latter is constructed by some equally spaced time slots (T_{tran}). Indeed, these time slots determine the number of connected vehicles to the base-station. In short, it can be written as

$$T_{CP}(N,a) + N \;\; T_{tran} \leq \;\; T_{frame} \tag{2.7}$$

For a greater number of contending EVs, the parameter Y tends to become infinite (i.e. $Y \to \infty$). Therefore, an approximation of $(1-a)^{Y-p-1} \approx (1-a)^{Y-p}$can be considered. As a result, (2.6) can be simplified as

$$T_{CP}(N,a) + N \;\; T_{tran} \leq \;\; T_{frame} \tag{2.8}$$

$$T_{CP} = N \left[\frac{1}{Ya} \delta_{idle} - \delta_{coll} \left(1 + \frac{1}{Ya} - \frac{1}{Ya(1-a)^{Y-1}} \right) + +\delta_{succ} \right] \tag{2.9}$$

Substituting into (2.7), the number of EVs per time frame, which will succeed to integrate with the base-station, can be written as

$$N = \frac{T_{frame}}{\left[\frac{1}{Ya} \delta_{idle} - \delta_{coll} \left(1 + \frac{1}{Ya} - \frac{1}{Ya(1-a)^{Y-1}} \right) + \delta_{succ} \right] + T_{tran}} \tag{2.10}$$

where, $\delta_{succ} = Treq + Tack + SP + FD$ and $\delta_{succ} = Treq + FD$

Here, T_{req} denotes msg1 which denotes the length of the requested message sent by EV. T_{ack} represents the acknowledgement message duration (i.e. msg2). The switching period (SP) and frame delay (FD) indicate the delay of the radio channel for transmitting msg1 and the delay between two consecutive frames.

Equation (2.10) is very important to understand the scalability limit of EVs during one time frame of LTE; in other words, to compute the number of EVs which are simultaneously participating to M2M communication during an LTE time frame in a distributed EMSs. With the EVs number exceeding the limit under any base station, the outage or blocking of M2M communication will happen and fail to transmit the trip information. In that case, the participant EVs need to wait for the next time frame to perform the transmission process. For a probability of a = 1/500 and 0.5 ms transmitting time slots (defined by LTE standard), the total number of EVs capable of transmitting data logging information at a time is around 220–250.

2.5.5 Simulation and performance evaluation

This section describes the scalability limits of EVs in terms of delay and blocking rate. The limit is numerically simulated in the user-plane. For simulation, the inter-packet interval maintained by EV is 20 s. The simulation scenario was considered for the parameters enlisted in Table 2.2 and the environment depicted in Figure 2.16. A C++ and Python coded open source framework Network Simulator-3 (NS-3) was employed for the performance evaluation. Each node of the simulator network is assumed to convey the full protocol stack of LTE. All the EVs in the model show random speed where the maximum velocity is set at 120 km/h. The performance was evaluated for the Manhattan street scenario in the USA. To avoid collision, 10 m inter-spacing distance is deliberately provided from one vehicle to another.

To facilitate clear visualisation, the PDCP delay of EVs from the DLS to the base station is illustrated in Figure 2.17. In the beginning, it is noticeable that the

Table 2.2 Vehicle communication simulation parameters

System bandwidth	**20 MHz**
Cell radius	1.5 km
EVs maximum velocity	120 km/h
EVs inter-spacing distance	10 m (at least)
Inter-packet interval	20 s
Packet size	1.5 kB, UDP
Maximum transmission unit	1500 Bytes
Mobility	Random
Time frame	100 ms
Buffer size	Infinite
Frame delay	2.5 μs
Switching period	0.5 μs
Fast fading model	Enabled

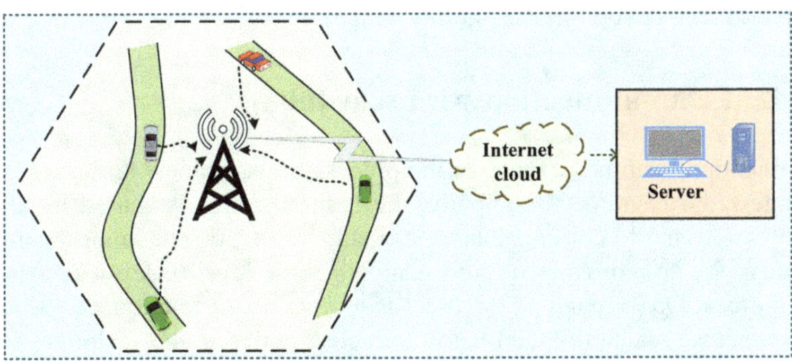

Figure 2.16 Simulation scenario in network simulator-3

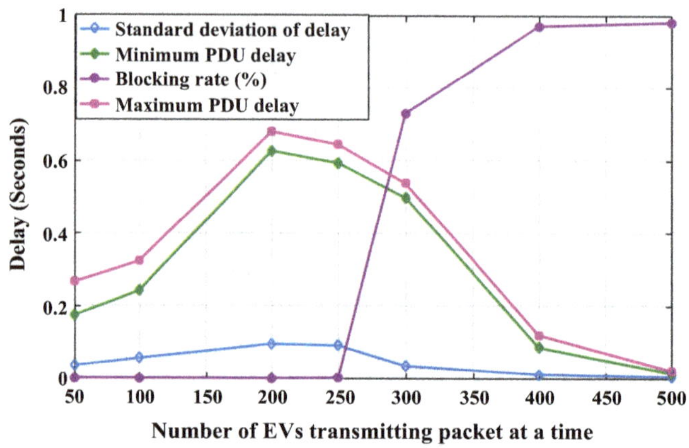

Figure 2.17 Delay vs number of EVs

radio link delay proportionally increases with the EV number. However, the delay result is significant for the EV number between 200 and 250. This is because after this range the collision probability increases with the EV number, resulting in a considerable blocking rate. As a result, the EVs, which are carrying the M2M device and simultaneously trying to connect with the base-station, go into the sleep mode. From Figure 2.17, the blocking rate is negligible up to 250 vehicles. Afterwards, the blocking rate rises considerably and reaches to around 95% for 500 vehicles. Interestingly, the simulation results support the findings of analytical model as explained in Section 2.5.4. The only difference of simulation with analytical modelling is that the numerical simulation carried out under the roaming condition of EVs, while the modelling is not.

For the varying packet size, the tine difference between the minimum PDU and maximum PDU maintains a consistent value in the PDCP layer. Moreover, the standard deviation delay in the scalability range also shows a constant trend.

Figure 2.18 shows the plot of signal-to-interference and noise ratio (SINR) with the increasing EVs. The results reveal that the performance can be significantly reduced beyond the scalability range.

2.6 M2M communication with scheduling

Packet scheduling can be defined as the process of assigning M2M packets to an appropriate radio resource. Principally, the radio resource is shared by all M2M users who are trying to communicate. Attributing an LTE scheduling mainly aims to maximise the system capacity and maintain some level of fairness. Also, confirming the best QoS among M2M users is of paramount importance. Section 2.5 investigated the scalability limit for vehicles under a 4G cellular network. However, if vehicle numbers exceed the scalability limit, it is recommended for M2M communication to apply different scheduling algorithms to achieve the best

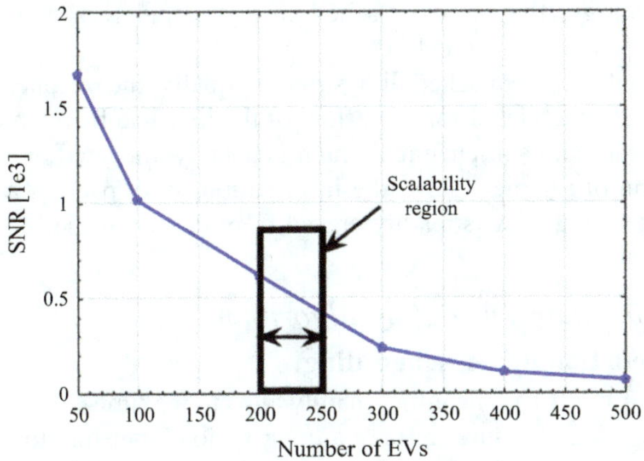

Figure 2.18 Delay vs number of EVs

service quality. To facilitate clear understanding of EVs, this section investigates the M2M performance under three popular scheduling algorithms.

2.6.1 LTE scheduling

As discussed earlier, the LTE radio frame has duration of 10 ms and mainly consists of ten sub-frames. Each sub-frame is 1 ms long and can be split into two transmission-time intervals (TTI) of 0.5 ms. When there are many M2M users with different channel qualities, the scheduler attempts to find the best user at a given TTI. Apart from that, the searching mechanism for offering radio resources dominantly depends on various channel conditions. For every transmission interval, the scheduler successively computes a given parameter for all flows that can be scheduled. The parameters especially represent any i_{th} flow for a j_{th} sub-channel. It is advisable that the schedule will prioritise an M2M user who has the highest w_{ij} to assign radio resource. Section 2.6.2 will describe the principle of some popular scheduling algorithms. For obvious reasons, the scheduling procedures for the uplink transmission can briefly be summarised as follows:

- Firstly, the scheduler residing in eNodeB creates a list of flows containing uplink packets which need to be transmitted in the current sub-frame.
- During the scheduling operation, the scheduler periodically measures the channel quality experienced by each M2M device. After every measurement, each M2M device generates some reference symbols associated with the channel quality, commonly known as a Channel Quality Indicator (CQI), and sends it to the base-station (eNodeB), which acknowledges the transmitted CQI symbols. Based on CQI feedbacks, the radio resources are mapped with the channel link and make correlation with the adaptation module to select the most suitable modulation and coding scheme (MCS). A right selection of MCS at the physical layer enhances the spectral efficiency. For M2M network in

LTE technology, this step for scheduling is popularly known as Adaptive Modulation and Coding (AMC).

- With the deployment of scheduling strategy, finally, the parametric value of w_{ij} is sequentially calculated for each flow of the list. The flow which has highest w_{ij} receives a radio sub-channel, which is usually allocated by the base. After transmission of a typical flow, the information of all packets associated with the corresponding flow is readily erased from the list at the MAC layer.

2.6.2 LTE popular scheduling algorithms

2.6.2.1 Proportional fair scheduling

The proportional fair (PF) scheduling mainly aims to maximise the total throughput of an LTE network and assure fairness among various transmission flows. For PF scheduling, the metric w_{ij} implies the ratio of the instantaneous data rate and the average data rate. Mathematically, the metric can be written as

$$w_{ij} = \frac{r_{ij}}{\overline{R_i}} \tag{2.11}$$

Here, $\overline{R_i}$ denotes the average data rate and r_{ij} represents the instantaneous data rate computed by the AMC modulating scheme. The computation is basically performed according to the feedback received by CQI. For any *j-th* sub-channel, the parameter $\overline{R_i}$ in every TTI can be defined by

$$\overline{R_i}(n) = 0.8\,\overline{R_i}(n-1) + 0.2\,\overline{R_i}(n) \tag{2.12}$$

where $\overline{R_i}(n)$ and $\overline{R_i}(n-1)$ represent the achieved data rate in the *n-th* TTI and the *(n-1)-th* TTI, respectively, for any *i-th* flow.

2.6.2.2 Modified largest weighted delay first scheduling

For modified largest weighted delay first (M-LWDF) scheduler, a packet delay of τ_i is deliberately provided to the largest weighted metric of *i-th* flow. In this case, the metric w_{ij} can be defined by

$$w_{ij} = \alpha_i \;\Delta_H \frac{r_{ij}}{\overline{R_i}} \tag{2.13}$$

$$\alpha_i = \frac{\log\ \delta_i}{\tau_i} \tag{2.14}$$

where Δ_H denotes the delay time of first packet in the scheduling queue and δ_i defines the maximum probability of first packet exceeding the delay threshold.

2.6.2.3 Exponential scheduling

In the exponential (EXP) scheduler, the head-of-line packet (i.e. the first packet) delay is approximated close to the attributed delay threshold. The resource metric can be computed as

$$w_{ij} = \exp\left(\frac{\alpha_i \Delta_H - \chi}{1 + \sqrt{\chi}}\right)\frac{r_{ij}}{\overline{R_i}} \tag{2.15}$$

$$\chi = \frac{1}{N}\sum_{i=1}^{N} \alpha_i \Delta_H \tag{2.16}$$

Principally, if the packets of M-LWDF and EXP schedulers fail to transmit before the deadline threshold, the information regarding certain flow will be removed from the scheduling queue.

2.6.3 Performance evaluation

The vehicular M2M performance with the discussed PF Scheduler, the M-LWDF Scheduler, and the EXP Scheduler has been investigated in this work. For numerical simulation, a software tool named as LTE-Sim version-5 software was used. The software is an open-source framework coded with the C++ language. In fact, LTE-sim allows research communities to measure the M2M performance in a 4G cellular network using various scheduling algorithms. The simulation scenario was considered for the Manhattan highway scenario in the USA. To inspect the performance, the scheduling algorithms examine the simulation environment for the same parameters listed in Table 2.2. The flow duration is counted for 300 s. Instead of considering the random speed, the only exception of two constant speeds of 60 and 120 km/h are predominantly reported in this numerical simulation.

Figure 2.19 illustrates the average end-to-end packet delays from M2M vehicles to the base station. The simulation is repeated by varying numbers of vehicles. Here, the delay of M2M communications is calculated up to 500 vehicles. It can be observed that, for transmitting a 1.5 kB packet, the average delays are 17 and 11 s for 60 and 120 km/h speeds, respectively.

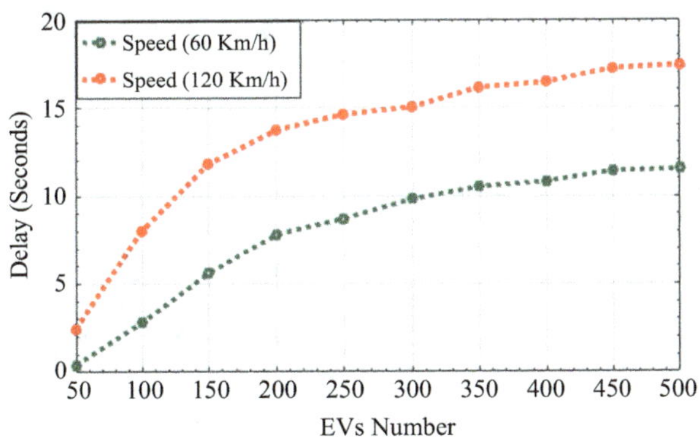

Figure 2.19 Delay vs number of EVs

This simulation is carried out for only the PF Scheduler since the other two schedulers always maintain a threshold delay of 0.5 s. The higher speed results in higher delay because of the fast fading effect. Also, with the increasing velocity, the procedures of radio link adaption take a considerable time for radio access.

Figure 2.20 shows the packet-loss ratio (PLR) of the three scheduling algorithms. The PLR is defined by the ratio of the total numbers of packets received by the base station and the number of packets transmitted by M2M vehicles. It is clearly seen from the figure that the loss-ratio is higher for the PF scheduler than for the other two schedulers. This is because, with the increasing number of simultaneous real-time flows, the probability of discarding packets due to deadline expiration can explicitly be increased. For ensuring a better M2M communication system to retrieve the transmitted information, a maximum packet loss of 10% has been considered as a threshold value. It is also seen that for every scheduler, the PLR is around 8–9% at both 60 and 120 km/h speeds.

Further investigation of PLR ratio up to 800 vehicles is also carried out to see how far PLR would go. It is found that the PLR approaches the threshold value and results in poor performance of M2M communications. The simulation results are illustrated in Figure 2.21. From analysis, it can be remarked that the exceeding scalability limit of EVs can introduce a reasonable amount of delay and packet loss, which consequently reflects the poor performance of the M2M communication system.

To sum up, it is worthwhile to describe that under an LTE base station, EVs in either case, i.e. without scheduling (explained in Sections 2.5.4 and 2.5.5) and with scheduling (PF, M-LWDF, EXP) show poor performance in M2M communication. With the increasing number of vehicles, the data packets start to be getting blocked, and communication also delays disrupt system performance. Since the radio resource of an LTE network is limited, it is suggested to deploy 5G network or other wireless communication standards to accommodate the increasing density of EVs.

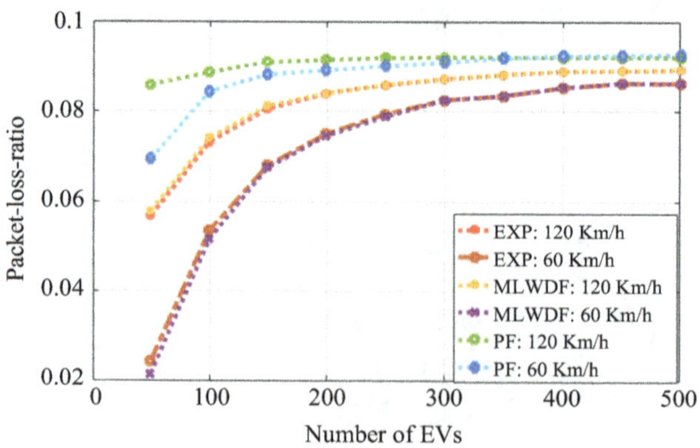

Figure 2.20 Packet-loss ratio plot for 60 and 120 km/h speeds

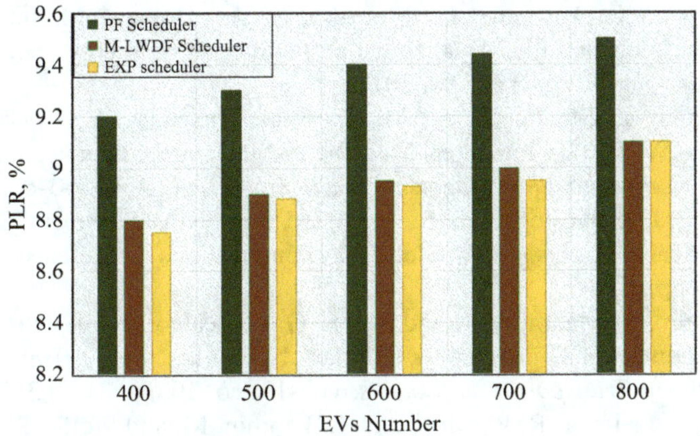

Figure 2.21 Packet-loss ratio for the higher number of vehicles (greater than 400)

2.7 Conclusion

This chapter describes the significant potential of wireless M2M communication for the monitoring and control necessary to manage the impact of EVs on energy systems. To demonstrate, an automated 4G M2M Raspberry Pi DLS, developed for monitoring vehicles wirelessly, was presented. The in-vehicle module sends trip information such as SOC and location of EVs to a remote server at regular intervals. Although the logging system was developed specifically for EVs, it could be readily adapted to any vehicle management system incorporating 'intelligent' or connected vehicles. The scalability of M2M in 4G networks for vehicles has also been investigated. The results from both numerical simulations and analytical calculations show that up to 250 vehicles can be logged per base station before delay and blocking disrupt system operation, assuming a specified information rate per vehicle and in the absence of other network communications. The case study provides guidance regarding the system design required to accommodate a given density of vehicles. It is expected that 5G networks and/or other modes of wireless communication will be needed to accommodate future intelligent transport systems. Finally, to facilitate better understanding, vehicle M2M performance (i.e. QoS) has been investigated using the Proportional Fair (PF) scheduler, Modified Largest Weighted Delay First (M-LWDF) scheduler and the Exponential (EXP) scheduler. In conclusion, this research benefits future development of 'Smart Infrastructure' and 'Intelligent Transport'.

References

[1] Amodu, O. A. and Othman, M. Machine-to-Machine Communication: An Overview of Opportunities. *Computer Networks*, vol. 145, pp. 255–276, 2018.

[2] Mahmud, K., Town, G. E., Morsalin, S. and Hossain, M. Integration of Electric Vehicles and Management in the Internet of Energy. *Renewable and Sustainable Energy Reviews*, 2017.

[3] Fadlullah, Z. M., Fouda, M. M., Kato, N., Takeuchi, A., Iwasaki, N. and Nozaki, Y. Toward Intelligent Machine-to-Machine Communications in Smart Grid. *Communications Magazine, IEEE*, vol. 49, no. 4, pp. 60–65, 2011.

[4] Rezgui J. and Cherkaoui, S. An M2M Access Management Scheme for Electrical Vehicles. IEEE Global Communications Conference, Singapore, 2017, pp. 1–6.

[5] Morsalin, S. Mahmud, K. and Town, G. E. Scalability of Vehicular M2M Communications in a 4G Cellular Network. *IEEE Transactions on Intelligent Transportation Systems*, vol. 19, no. 10, pp. 3113–3120, 2018.

[6] Jermyn, J., Jover, R. P., Murynets, I., Istomin, M. and Stolfo, S. Scalability of Machine to Machine Systems and the Internet of Things on LTE Mobile *Networks. IEEE16th Int. Symp.on World of Wireless, Mobile and Multimedia Networks (WoWMoM)*, pp. 1–9, 2015.

[7] Madueño, G. C., Nielsen, J. J., Kim, D. M., Pratas, N. K., Stefanovic, C. and Popovski, P. Assessment of LTE Wireless Access for Monitoring of Energy Distribution in the Smart Grid. *Information Theory; Networking and Internet Architecture*, 2015.

[8] Sivaraman S. and Trivedi, M. M. Looking at Vehicles on the Road: A Survey of Vision-Based Vehicle Detection, Tracking, and Behavior Analysis. *IEEE Transactions on Intelligent Transportation Systems*, vol. 14, no. 4, pp. 1773–1795, 2013.

[9] Cha, I., Shah, Y., Schmidt, A. U., Leicher, A. and Meyerstein, M. V. Trust in M2M Communication. *IEEEVehicular Technology Magazine*, vol. 4, no. 3, pp. 69–75, 2009.

[10] Moloisane, N. R., Malekian, R. and CapeskaBogatinoska, D. Wireless Machine-to-Machine Communication for Intelligent Transportation Systems: Internet of Vehicles and Vehicle to Grid. 40th IEEE International Convention on Information and Communication Technology, Electronics and Microelectronics (MIPRO), Opatija, 2017, pp. 411–415.

[11] Trigg, T., Telleen, P., Boyd, R., Cuenot, F., D'Ambrosio, D., Gaghen, R., Gagné, J., Hardcastle, A., Houssin, D. and Jones. Global EV Outlook: Understanding the Electric Vehicle Landscape to 2020. *International Journal of Energy Agency*, pp. 1–40, 2013.

[12] Palmer, K., Tate, J. E., Wadud, Z. and Nellthorp, J. Total Cost of Ownership and Market Share for Hybrid and Electric Vehicles in the UK, US and Japan. *Applied Energy*, vol. 209, pp. 108–119, 2018.

[13] Jenn, A., Springel, K. and Gopal, A. R. Effectiveness of Electric Vehicle Incentives in the United States. *Energy Policy*, vol. 119, pp. 349–356, 2018.

[14] AECOM. 'Electric Vehicle Uptake and Behaviour Modelling. EV0795 to AEMC, Tech. Rep., June 2012.

[15] AEMC. Energy Market Arrangements for Electric and Natural Gas Vehicles, Final Advice. AEMC, Tech. Rep., 11 December 2012.

[16] Nykvist B. and Nilsson, M. Rapidly Falling Costs of Battery Packs for Electric Vehicles. *Nature Climate Change*, vol. 5, no. 4, pp. 329–332, 2015.

[17] Mwasilu, F., Justo, J. J., Kim, E.-K., Do, T. D. and Jung, J.-W. Electric Vehicles and Smart Grid Interaction: A Review on Vehicle to Grid and Renewable Energy Sources Integration. *Renewable and Sustainable Energy Reviews*, vol. 34, pp. 501–516, 2014.

[18] Kempton W. and Letendre, S. E. Electric Vehicles as a New Power source for Electric Utilities. *Transportation Research Part D: Transport and Environment*, vol. 2, no. 3, pp. 157–175, 1997.

[19] Habib, S., Kamran, M. and Rashid, U. Impact Analysis of Vehicle-to-Grid Technology and Charging Strategies of Electric Vehicles on Distribution Networks–A Review. *Journal of Power Sources*, vol. 277, pp. 205–214, 2015.

[20] Sbordone, D., Bertini, I., Di Pietra, B., Falvo, M., Genovese, A. and Martirano, L. EV Fast Charging Stations and Energy Storage Technologies: A Real Implementation in the Smart Micro Grid Paradigm. *Electric Power Systems Research*, vol. 120, pp. 96–108, 2015.

[21] Fernandez, L. P., Román, T. G. S., Cossent, R., Domingo, C. M. and Frás, P. Assessment of the Impact of Plug-in Electric Vehicles on Distribution Networks. *IEEE Transactions onPower Systems*, vol. 26, no. 1, pp. 206–213, 2011.

[22] Qi, X., Wu, G., Boriboonsomsin, K. and Barth, M. J. Development and Evaluation of an Evolutionary Algorithm-Based Online Energy Management System for Plug-In Hybrid Electric Vehicles. *IEEE Transactions on Intelligent Transportation Systems*, vol. 18, no. 99, pp. 1–11, 2016.

[23] Vermesan, O., Blystad, L.-C., John, R., Hank, P., Bahr, R. and Moscatelli, A. Smart, Connected and Mobile: Architecting Future Electric Mobility Ecosystems. In Proceedings of the Conference on Design, Automation and Test in Europe. EDA Consortium, 2013, pp. 1740–1744.

[24] Wietfeld, C., Georg, H., Groening, S., Lewandowski, C., Mueller, C. and Schmutzler, J. Wireless M2M Communication Networks for Smart Grid Applications. In IEEE 17th European- Sustainable Wireless Technologies, April 2011, pp. 1–7.

[25] Lee C.-Y. and Yang, C.-S. Distributed Energy-Efficient Topology Control Algorithm in Home M2M Networks. *International Journal of Distributed Sensor Networks*, vol. 8, no. 9, p. 387192, 2012.

[26] Niyato, D., Xiao, L. and Wang, P. Machine-to-Machine Communications for Home Energy Management System in Smart Grid. *IEEE Communications Magazine*, vol. 49, no. 4, 2011.

Chapter 3

Electrical vehicles charging and discharging scheduling for the cloud-based energy management service

Yu-Wen Chen[1]

The electric vehicles (EVs) have been dramatically increased and popularized in recent years. The ability to export the power to the grid via the vehicle-to-grid (V2G) technology makes the EVs become the promising solutions for reducing the peak demand in the power grid but could also severely increase the fluctuated penetration if no scheduling mechanisms are deployed. Therefore, it is an essential task to provide the optimal EV charging and discharging scheduling. However, to practically reinforce the scheduling relies not only on the ability to offer it in the efficient, reliable and scalable approaches, but also on the willingness of the participation from EVs owners. The cloud-based energy management service (EMS) satisfies the needs to practically deploy the scheduling mechanisms for the heterogeneous EVs and provide the incentives for customers' participation. The cloud computing is introduced with its characteristics and is utilized for the design of an extensive cloud-based framework, which provides the energy management as a service (EMaaS) to suggest the optimal electricity usage and trading options for every participated customer. The framework and the procedure of the cloud-based EMS are illustrated, and the EVs charging and discharging scheduling for the cloud-based EMS is formulated and implemented. The scheduling results for both EVs with and without the V2G ability are discussed with various examples.

List of abbreviation

ADMS	Advanced Distribution Management System
API	Application Programming Interface
AWS	Amazon Web Service
DR	Demand Response
EMaaS	Energy Management as a Service

[1]New York City College of Technology, City University of New York, USA

EMS	Energy Management Service
EV	Electric Vehicle
GCE	Google Compute Engine
IaaS	Infrastructure as a Service
ISO	International Organizational and Standardization
NIST	National Institute of Standards and Technology
PaaS	Platform as a Service
SaaS	Software as a Service
V2G	Vehicle-to-Grid

3.1 Introduction

With the growing concerns in reducing the greenhouse emission and making the power grid environment-friendly, electric vehicles (EVs) have been dramatically increased and popularized in recent years. Impacts are aroused to the existing power grid when integrating the EVs due to the high fluctuated penetration charging request from EVs could overload the power system and increase the difficulty for the utilities achieving the optimal operation scheduling [1,2]. Although EVs are viewed as one of the promising solutions for reducing the peak demand [3] due to the ability of exporting the power back to the grid through the vehicle-to-grid (V2G) technology, it could also severely increase the fluctuated penetration by adding the discharging requests to the power grid and bring more significant impacts. To mitigate the problem and enhance the ability for integrating EVs as the promising solutions for reducing the peak demand, the optimal scheduling schemes for charging and discharging the EVs are required. However, two difficulties exist to reinforce the optimal EV charging and discharging scheduling practicality as it relies not only on the ability to offer the scheduling in the efficient, reliable and scalable approaches, but also on the willingness of the participation from EVs owners.

Literatures in EV charging scheduling and demand response (DR) have provided financial incentives for customers utilizing the controllable loads and storage to change the demands in response to the fluctuated electricity prices over time. Authors in [4] propose the DR schemes to minimize the global cost through a formulated game, and authors in [5] study the EV charging scheduling under a parking garage scenario. The EV aggregator absorbs the EVs penetration while minimizing costs for aggregator in [6], and authors in [7] introduced an incentive-based DR strategy with critical and controllable loads to accommodate EV charging while keeping the peak demand unchanged. The authors in [8] propose the robust optimization for the EV charging scheduling with the uncertain electricity prices. However, without the cloud-based framework, difficulty exists in above literatures for practically realizing the control mechanisms for the heterogeneous EVs due to the requirements of duplicated, dedicated control entity.

Cloud computing has been widely discussed in Smart Grid. The virtual-SCADA is proposed in [9], and the authors in [10] have highlighted the advantages of integrating the cloud computing to the connected vehicular network and the Internet of Thing technology. Authors in [11] implemented a novel Advanced Distribution Management System (ADMS) solution on a cloud infrastructure. The cloud-based framework is also successfully utilized for realizing the EV charging and discharging schemes in literatures effectively. A cloud-based charging-discharging model for the public supply stations is proposed in [12]. The cloud server is used in [13] to provide the mobile edge computing system for the EV charging reservations. Authors in [14] scheduled the EV charging-discharging service with a dynamic pricing model on the cloud architecture to reduce the peak load. However, customers in the above EV charging schemes are facing the insufficient trading options as the individual customer only interacts to the single aggregator or the public supply station. Providing more *"trading choices"* for customers, such as create the trading behaviors among other involved customers, more financial incentives can be brought to attract customers' participation in the EV charging and discharging scheduling.

To address the *"trading choices,"* the customer-orientated energy management is proposed in [11] to provide energy management as a service (EMaaS) from an extensive cloud-based framework for *"community."* The community is formed when customers involved in the same energy management service (EMS), and a new price indicator appears for customers performing the *"virtual trading"* among each other. Customers become the virtual retail electricity providers by themselves without the limitation of the existence of the physical power distribution lines. Customers' components, such as distributed energy resources (DER), storage system and the electricity demand, are efficiently facilitated within the community to minimize the global cost for the community and enhance the DER to operate grid-friendly. The cloud-based energy management is not only an extensive framework with the interoperability but also a business model for the DER integration. With the ability to satisfies the needs of efficiency, scalability, and flexibility and provides the incentives for customers' participation, realizing the EV charging and discharging scheduling on the cloud-based EMS is a potential solution and is the focus of this chapter.

This chapter introduces the cloud-based EMS and provides an overview of implementing the EV charging and discharging scheduling for the EMaaS from [15] and [16]. The remainder of this chapter is organized as follows. Section 3.2 discusses cloud computing and Section 3.3 introduces the cloud-based EMS. Section 3.4 illustrates the charging and discharging scheduling of EVs for the EMaaS. The scheduling results are discussed in Section 3.5 and Section 3.6 concludes this chapter.

3.2 Cloud computing

Cloud computing brings enormous advantages including avoiding capital investments, reducing maintenance expenses, providing secure management, and simplifying implementations [17, 18]. The definition of cloud computing from the

National Institute of Standards and Technology (NIST) is "the model for enabling ubiquitous, convenient, on-demand network access to a shared pool of configurable computing resources, which can be rapidly provisioned and released with minimal management effort or service provider interaction." [19] The International Organizational and Standardization (ISO) also defined cloud computing as *"a paradigm for enabling network access to a scalable and elastic pool of shareable physical or virtual resources with self-service provisioning and administration on-demand* [20]*,"* where examples of the resources include the networks, storage, servers, software, applications, operating systems, etc.

Cloud computing provides various service models, referred as selling "X as a Service." The most widely mentioned three service models are "*Software as a Service (SaaS)*," "*Platform as a Service (PaaS)*" and the "*Infrastructure as a Service (IaaS)*" [21]. The SaaS provides the ability for customers consuming the applications from the cloud infrastructure. Customers can access the applications through the thin client interface such as the web browser and application programming interface (API). The cloud service providers maintain the layers of the application, operating system, storage, networking and servers. Examples of the SaaS are Gmail, Google Docs, Dropbox, etc. The PaaS allows customers to build their application on the cloud infrastructure based on the provided programming language, services and tools that are supported by cloud service providers. Customers have the control to deploy the application and maintain data layer. The cloud service provider controls the operating system, servers, storage and networking layers that underly the cloud infrastructure. Most of the cloud computing services provided by major cloud service providers, such as IBM Smart Cloud, Amazon Web Services and Microsoft Azure, are the examples of PaaS. The IaaS allows customers to migrate their infrastructure to the cloud. Customers can provision the application, data, runtime, middleware and operating system layers. The cloud service provider maintains the servers, storage and networking layers. Examples for the IaaS are Amazon Web Service (AWS) and the Google Compute Engine (GCE).

Depending on who can access the cloud computing infrastructure, the cloud infrastructure is deployed with four models: private cloud, community cloud, public cloud and hybrid cloud. The private cloud could be used if the cloud infrastructure is operated solely for an organization. If several organizations shared the similar concerns and goals, the community cloud would be the option to deploy the cloud infrastructure. The public cloud allows the infrastructure available to the general public or used by the organization selling cloud services, and different deployment models can combine as the hybrid cloud.

NIST defines five essential characteristics for the cloud computing [21]:

- *On-demand self-service*: the computing ability can be provided automatically to customers without human interaction.
- *Broad network access*: the network accessibility is available by the heterogeneous thin or thick client platforms, such as the workstation, tablet, mobile phone, web browser or API.

- *Resource pooling*: the computing resources are pooled to serve multiple customers. Depending on the need from the customers, resources are dynamically assigned and reassigned.
- *Rapid elasticity*: the capability of cloud computing can be provisioned and released elasticity. It can be scaled rapidly outward and inward with customers' demands.
- *Measured service*: metering ability is provided for the resources. Customers pay when they use the resources and the provided resources are automatically controlled, analyzed, predicated and optimized by the cloud computing platform.

3.3 Cloud-based energy management service

Providing the EMS on the service-oriented cloud computing infrastructure can cope with the issues of efficiency, flexibility and scalability even when the complexity of coordinating the massive data is increased. Compared to providing the service without the cloud infrastructure for serving N groups of customers, the cloud-based framework reduces the capital investment cost $N-1$ times, which include the cost for establishing the local computational machine to execute the service for each of customers, the duplicated implementations and maintenance expenses for the individual service. Figure 3.1 presents the concept structure of the cloud-based EMS, which is provided from the service-oriented cloud computing infrastructure. The services could be various types that serve diverse purposes and are offered by different EMS providers for the particular group of customers.

The cloud-based EMS utilized the extensive cloud framework to provide the customer-oriented EMaaS for the DERs owners [11]. Similar to the existing mobile

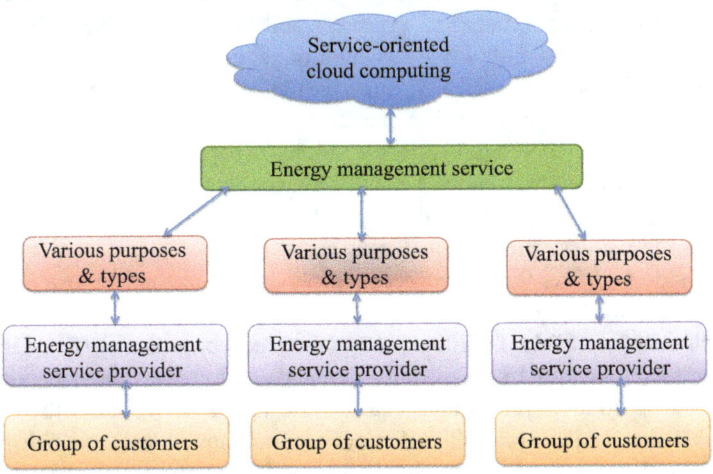

Figure 3.1 Concept structure of cloud-based energy management service

service plans offered by ATT or Verizon, when DERs owners agree to join the same service plan provided by the EMS providers, they become the customers of the EMaaS, and form the *"community"* with other customers who agree to involve in the same service plan. When the community is formed, a virtual platform is created with the new price indicator for customers performing the "virtual trading" for the produced renewable energy within the community. The trading is performed virtually due to the physical power distribution lines might not exist among customers or cannot support bidirection power distribution, and it can be realized efficiently by implementing a mapping mechanism within the cloud-based framework. With the trading ability within the community, customers become the virtual retail electric providers by themselves without the geographical limitation. The created trading choices allow customers benefit mutually and are determined in the EMS to deliver the incentives to customers by reducing the electricity and environmental costs. Customer's willingness is increased to involve the energy management and the investment of DER. Moreover, by grouping the DERs, the cloud-based EMS enhances the renewable energy integration by reducing the reserve capacity for the dispatchable conventional generators.

3.3.1 Framework and utilized data

The framework of the cloud-based EMS is shown in Figure 3.2. The EMaaS service providers offered the service based on the cloud infrastructure, which includes the service manager and the information pool. The manager and the information pool are accessible to the involved customers through the thin client interface (e.g., web browser or API) via the information exchanging lines. The EMaaS service provider offers different plans for customers; in other words, the EMaaS is provided to multiple communities.

Customers are the DERs owners, who equip the photovoltaic (PV) panels on the rooftop of various types of building (e.g., a single household, apartment, etc.,) and have the role as prosumer (i.e., producer and consumer). The DERs are connected to the power grid following the interconnection agreements with the local transmission and distribution utility, such as the electric substantive rule 25.211 addressed in [22]. The power grid is also connected by the different conventional generators (e.g., fossil-fueled generators), which are capable of providing sufficient energy to the customers' demands but not environment-friendly. The storage system is the essential component in the cloud-based EMS. As shown in Figure 3.2, the small home battery, such as the Tesla home battery [23], is equipped at each customer's side and was facilitated accordingly based on the suggested behaviors from the service manager.

The cloud-based EMS facilitates the connected components (i.e., DERs, storage systems, and customers' electricity demand) and creates more decision variables for the produced renewable energy, storage systems and the demands. Examples of the decision variables for each component are listed below, where variables could be extended for the EMS model when more components are included.

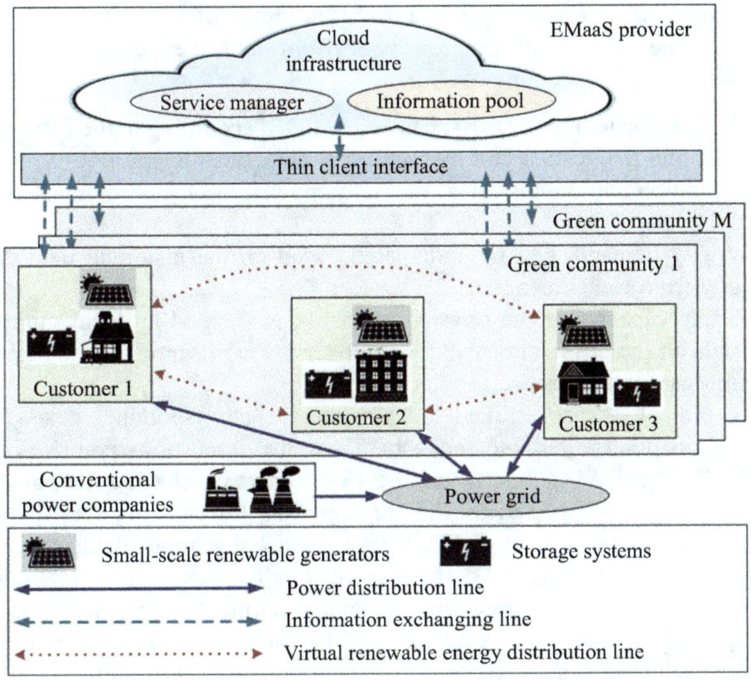

Figure 3.2 EMaaS framework, sources: [11]

- Decision variables for DERs:
  - E^{rm}: export the produced renewable energy to the power grid
  - E^{rc}: export the produced renewable energy to the community
  - E^{rs}: export the produced renewable energy to the storage system
  - E^{rd}: export the produced renewable energy for electricity demand
- Decision variables for the storage system:
  - I^{ms}: import the energy from power grid to storage
  - I^{cs}: import the energy from community to storage
  - I^{rs}: import the energy from DER to storage
  - E^{sm}: export the energy from storage to power grid
  - E^{sc}: export the energy from storage to community
  - E^{sd}: export the energy from storage for demands
  - S: state of charge for storage
- Decision variables for electricity demand:
  - I^{rd}: use the energy from DER
  - I^{md}: use the energy from the power grid
  - I^{cd}: use the energy from the community
  - I^{sd}: use the energy from storage

The sequential time series data are utilized as inputs of the scheduling and are gathered in the shared information pool from multiple sources via various approaches.

- Production capacities of DERs: can be predicted according to the local weather forecast and the system and measurement data, such as the PV panel system configuration and the historical measured power and meteorological information [24].
- Electricity demand: can be anticipated based on the historical data or input directly from each customer.
- Available capacity of the physical network: is decided by local utilities and depends on the physical distribution network, which supports loads from both non-EMaaS and EMaaS customers.
- Price indicators: are utilized to calculate the corresponding costs for fulfilling customers' demand requests. There are three primary prices indicators, P^m, P^r and $P^s \cdot P^m$ is the price for customers purchasing the power from the power grid supported by the conventional power generators. It is time-variant and is provided to EMaaS as a fixed know input value by the conventional power companies or local utilities based on their predictions or regulations. P^s is the price for customers selling the produced renewable energy back to the power grid. It excludes the environmental costs, such as CO_2 emission or other different environment penalties in each region, and is a smaller value than $P^m \cdot P^r$ is the new price appeared when the community is formed. It is used for customers performing the virtual trading within the community.

3.3.2 *Procedure*

The procedure of the EMaaS is presented in Figure 3.3. In the beginning, the time series utilized data are gathered for K time steps ahead from the information pool. Secondly, the EMS is run as an optimization model (i.e., linear programming model in [11]) for each community to minimize the global costs for the entire community with the determined decision variables. Based on the combination of the decision variables, each customer receives the operation suggestion and the corresponding costs from the EMaaS through APIs. The operation suggestions are offered as the results of the mapping mechanism. For each component, the decision variables for importing are summed and subtract the exporting decision variables. That is, the operation suggestion for the storage would be $I^{ms} + I^{cs} + I^{rs} - E^{sm} - E^{sc} - E^{sd}$, which indicates the charging operation with the positive value and the discharging operation with the negative value. After the suggestion, from the customer side, the actual amount of produced and acquired power from the real-time power distribution are checked. The difference amount is notified to the EMaaS for the self-adjusting process to cooperate the uncertainty issue.

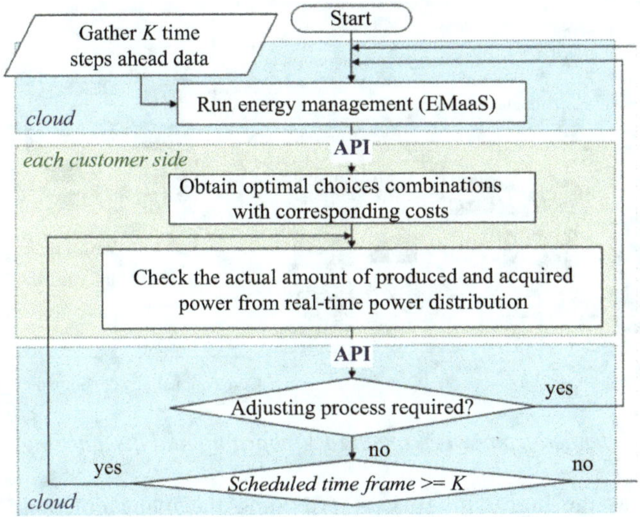

Figure 3.3 Procedure of the cloud-based energy management service, sources: [11]

3.4 Electrical vehicles for the cloud-based energy management service

Having the scalability, flexibility and efficiency characteristics for the management is essential to provide the efficient charging and discharging scheduling management for the emerging EVs in the heterogeneous environment. The implementation of the management needs to be efficient at different levels of the charging and discharging stations. For example, the public EV charging sites in [25], the commercial or hospital level [26], and the residential level [27] [28]. When the EV is equipped with the V2G ability and is connected at the charging/discharging site, EV can serve as a storage unit or the deferrable loads in the DR program, which rely on customer's willingness. Therefore, it is necessary to distribute the financial incentives along with the scheduling management to increase customer's participation.

To satisfy the needs of the charging and discharging scheduling management, providing the management service on the cloud-based energy management platform is a potential solution. The interoperability of EMaaS is successfully demonstrated in [16], where the EV charging and discharging scheduling are implemented for the cloud-based EMS along with the various load types for DR program. This section illustrates the extended EV charging and discharging scheduling model for the cloud-based EMS.

The cloud-based EMS platform supports the new components from the heterogeneous locations with the characteristic of scalability and efficiency. For the cloud-based EMS providers, including the EV charging and discharging scheme in

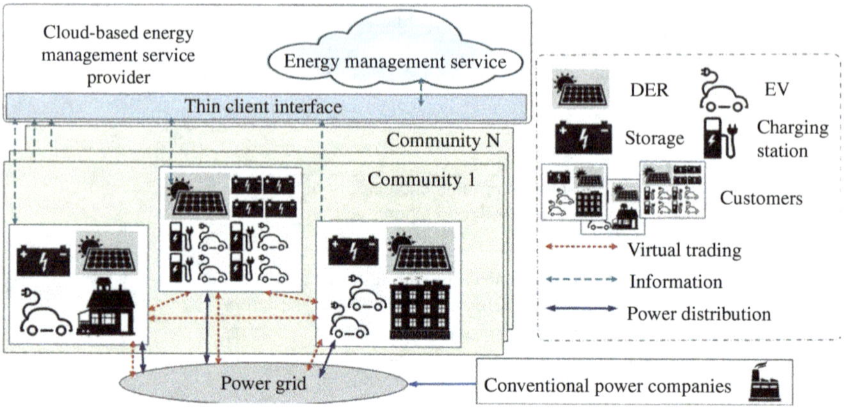

Figure 3.4 Extended framework for the EV charging and discharging, source: [16]

the service has the high similarity of providing the management service to the customers who are the EV owners. The extended framework is shown in Figure 3.4. EV involves in the EMS as a component like the DER or storage system discussed in the EMaaS. The extended framework is similar to Figure 3.2 with the involvement of the new types of customers. The EVs owners become the customers by involving in the EMS when connecting the EVs from different charging sites, such as the charging site at the single household, at the charging parking lot of a business building, or at a large-scale public charging aggregator. With the advantage of flexibility and scalability from the cloud computing framework, the EMaaS is also a beneficial attempt for supporting the business model for roaming EV charging [29] by allowing the owner to participate the virtual trading with the involved communities that have the collaboration among various cloud-based EMS providers without the geographical limitation as introduced in Section 3.3.

The discussed EVs are not limited to the charging scenario. The ability to provide energy back to the power grid which is known as V2G can also cooperate efficiently. Series of decision variables would be created when adopting the EVs components for the cloud-based EMS are listed below.

- Decision variables for the EV:
  - I^{me}: import the energy from the power grid to EV
  - I^{ce}: import the energy from the community to EV
  - Ire: import the energy from DER to EV
  - I^{se}: import the energy from storage to EV
  - E^{em}: export the energy from EV to the power grid
  - E^{ec}: export the energy from EV to the community
  - E^{ed}: export the energy from EV for demands
  - E^{es}: export the energy from EV to storage
  - S^{e}: state of charge for EV

- Extended decision variables for DER:
  - E^{re}: export the produced renewable energy to EV
- Extended decision variables for storage:
  - I^{es}: import the energy from EV to storage
  - E^{se}: export energy from storage to EV
- Extended decision variables for demands:
  - I^{ed}: use the energy from EV

Depending on the various types of EVs, driving behaviors and scheduling requirements, each EV feeds the unique parameters to the cloud-based EMS via the direct inputs from customers or the prediction from the historical data. The utilized parameters include:

- α: the arriving time when EV connects to the site
- D^{α}_{ev}: the initial capacity when EV connects to the site
- β: the leaving time when EV disconnects from the site
- D^{β}_{ev}: the required capacity when EV disconnects from the site
- η^{c}: the charging efficiencies
- η^{d}: the discharging efficiencies
- γ: the charging and discharging rate
- $S^{e,min}$: the minimum capacity allowed by the EV
- $S^{e,max}$: the maximum capacity allowed by the EV

The following constraints (3.1)–(3.9) are added to the EMaaS optimization model to satisfy the requirements and characteristics for each EV at each time step (e.g., an hour) [16]. The time step is denoted as the subscripts t. Constraints (3.1) and (3.2) notify the EMaaS management model the initial capacity and the required capacity of the EV at the arriving time and leaving time. The status of the EV appears 0 when it is not connected to the charging/discharging site in (3.3). Constraint (3.4) guarantees that EV is operated within the range between $S^{e,min}$ and $S^{e,max}$. Constraint (3.5) prevents the exported amount of energy exceeds the previous imported power from the DER and community as only the energy that is imported from the DERs and the community can be sold with the new appeared trading prices, P^{r}, that excludes the environmental cost. Constraint (3.6) prevents the status of EV smaller than $S^{e,min}$ after exporting the energy. Constraints (3.7) and (3.8) represent the relationship among the charging/discharging rate, efficiency and the variables. Constraint (3.9) indicates that during the connected time frame (α to β), the status in the next time step ($t+1$) depends on the status and the variables for importing and exporting with the consideration of efficiencies at the current time step (t). If the owned EV has no V2G ability, all the exporting related variables (i.e., E^{em}_{t}, E^{ec}_{t}, E^{es}_{t} and E^{ed}_{t}) equal to 0.

$$S^{e}_{t} = D^{\alpha}_{ev}, \quad t = \alpha \tag{3.1}$$

$$S^{e}_{t} \geq D^{\beta}_{ev}, \quad t = \beta \tag{3.2}$$

$$S_t^e = 0, t = 1, 2, \ldots, \alpha - 1, \quad and \quad \beta + 1, \beta + 2, \ldots, K \tag{3.3}$$

$$S^{e,max} \geq S_t^e \geq S^{e,min} \tag{3.4}$$

$$\sum_{t'=0}^{t-1} \left[\eta^c \left(I_{t'}^{re} + I_{t'}^{ce} \right) - E_{t'}^{ec} \right] - E_t^{ec} \geq 0, \quad \forall t \tag{3.5}$$

$$S_t^e - \left(E_t^{em} + E_t^{ec} + E_t^{es} + E_t^{ed} \right) \geq S^{e,min}, \quad \forall t \tag{3.6}$$

$$\eta^d \left(E_t^{em} + E_t^{ec} + E_t^{es} + E_t^{ed} \right) \leq \gamma \tag{3.7}$$

$$\eta^c \left(I_t^{me} + I_t^{ce} + I_t^{se} + I_t^{re} \right) \leq \gamma \tag{3.8}$$

$$S_{t+1}^e = S_t^e + \eta^c \left(I_t^{me} + I_t^{ce} + I_t^{se} + I_t^{re} \right) - \left(E_t^{em} + E_t^{ec} + E_t^{es} + E_t^{ed} \right), \quad t = \alpha, \alpha + 1, \ldots, \beta - 1, \quad \forall t \tag{3.9}$$

Constraints (3.10)–(3.13) indicate that the exported amounts among the customers' components are equal to the corresponding imported amount based on the charging/discharging efficiency rate of the EV (i.e., η^d and η^d) and storage (assumes it is ηs^d)

$$I_t^{re} - E_t^{re} = 0, \quad \forall t \tag{3.10}$$

$$I_t^{es} - \eta^d E_t^{es} = 0, \quad \forall t \tag{3.11}$$

$$I_t^{ed} - \eta^d E_t^{ed} = 0, \quad \forall t \tag{3.12}$$

$$I_t^{se} - \eta s^d E^{se} = 0, \quad \forall t \tag{3.13}$$

3.5 Scheduling results discussion

The EV charging and discharging scheduling is successfully adopted in the cloud-based EMS. To present the EV charging and discharging scheduling results, this section discusses the examples extended from the experiments in [16], where the fair DR with EV charging scheduling is efficiently implemented for the cloud-based EMS. A binary linear programming model is formulated for the management service, and the global cost for each community is minimized.

The experiments are conducted in the hourly day-ahead operation interval, where a time step is an hour and K equals to 24. The utilized price indicators of P^m, P^r and P^s are presented in Figure 3.5. Three types of customers are considered to form the community, as shown in Table 3.1, where customers may or may not own an EV, and the EV could has V2G ability or not. Every EV has various requirements (i.e., α, β, D_{ev}^{α} and D_{ev}^{β}) and is designed based on different uniform distribution functions listed in Table 3.2. The rest parameters of EV are assumed same for every EV that has the V2G ability in the experiments. The value of $S^{e,min}$ is designed as 15% of $S^{e,max}$ according to [30]. The values of γ, η^d, η^c, $S^{e,min}$ and

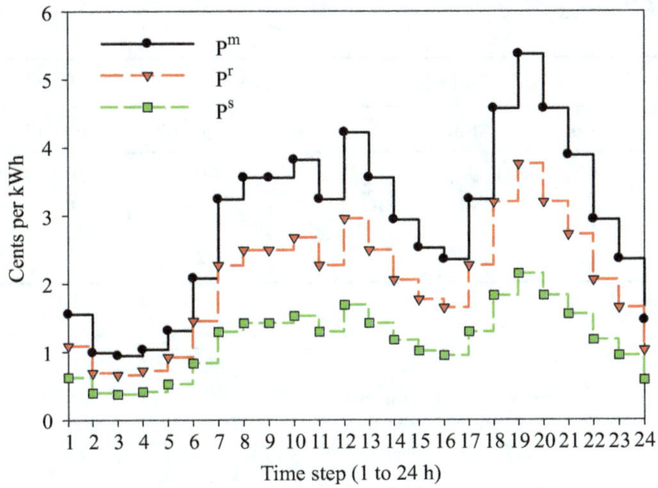

Figure 3.5 Utilized price indicators P^m, P^r and P^s

Table 3.1 Customer types

Customer types	EV	V2G ability
1	1	1
2	1	0
3	0	0

Table 3.2 Experiment parameters setting for each EV

α	$\mathscr{U}(8, 16)$
D_{ev}^{α}	$\mathscr{U}(1, 4)$
β	$\mathscr{U}(\alpha + \lceil (D_{ev}^{\alpha} - D_{ev}^{\beta}) \rceil, K)$
D_{ev}^{β}	$\mathscr{U}(D_{ev}^{\alpha} + 4, D_{ev}^{\alpha} + 20)$

$S^{e,max}$ are set as 4, 0.96, 0.96, 10 kW and 24 kW, respectively, according to [6]. For the EV without V2G ability, the η^d would be set as 0.

To address the status of the EV, three type 1 customers and three type 2 customers are extracted from a community with a size of 500. The details for these six customers and their EV requirements (i.e., α, β, D_{ev}^{α} and D_{ev}^{β}) are listed in Table 3.3. Customer nos. 1, 78 and 46 are type 2 customers, who own an EV that has no V2G ability. Customer nos. 69, 87 and 168 are type 1 customers, who own an EV that has the V2G ability.

Table 3.3 EV cases from different customers

V2G ability	**Not exist**			**Exist**		
Customer	**no. 1**	**no. 78**	**no. 46**	**no. 69**	**no. 87**	**no. 168**
α	15	9	8	10	10	11
β	20	21	9	15	22	14
D_{ev}^{α}	9	9	9	7	8	10
D_{ev}^{β}	23	24	10	22	23	14

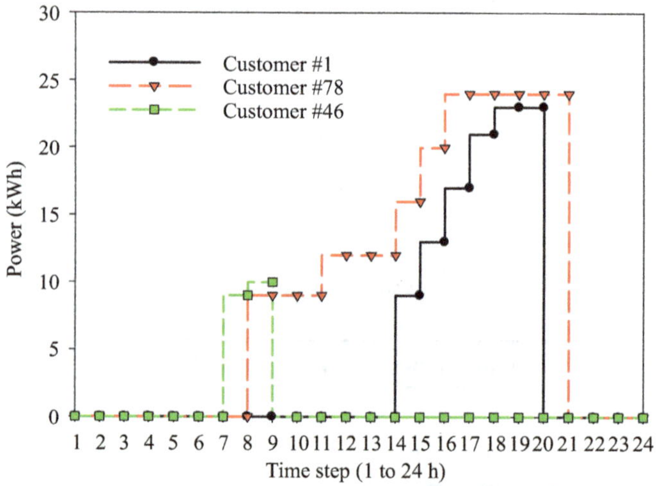

Figure 3.6 Examples for type 2 customers' EV charging schedules

The charging schedules for the type 2 customers are presented in Figure 3.6, and the charging and discharging schedules for type 1 customers are illustrated in Figure 3.7. All EVs are scheduled based on the requirements, where they connect to the system at the time step α with the initial capacity D_{ev}^{α}, and disconnect to the system at the time step β with the required capacity D_{ev}^{β}. While connecting to the system, the charging schedule for the EVs without the V2G ability is determined by the EMaaS as described in Section 3.3 to fulfill its requirement (i.e., the capacity D_{ev}^{β} at the leaving time β). As presented in Figure 3.6, customer no. 78 charges the EV at time steps 11, 14, 15 and 16 as the price indicators at those time steps are lower during the time frame that EV is connected to the system (i.e., time steps 9 to 21). Similarly, customer no. 46 charges the EV at time step 8 when it is connected to the system, and the customer no. 1 performs the EV charging at time steps 15, 16, 17 and 18.

For the EVs with V2G ability, EMaaS has the opportunity to operate the EVs as the storage when the EVs are connected to the system. However, if the EV only

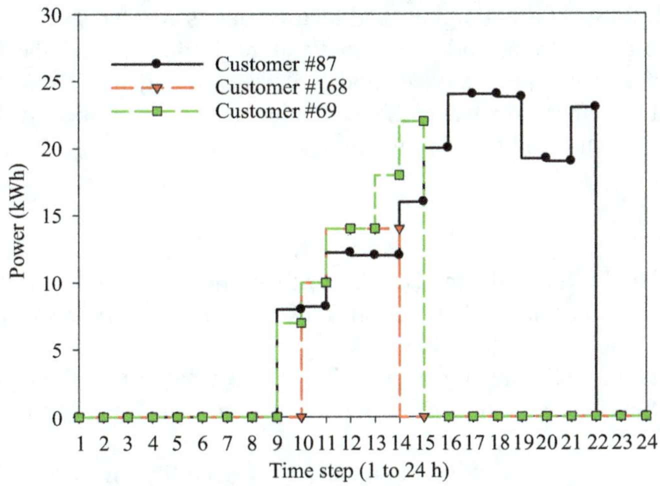

Figure 3.7 Examples for type 1 customers' EV charging and discharging schedules

connects to the system for a short time period, it will mainly obtain the charging schedule to fulfill its requirement first, like the customer nos. 69 and 168 in Figure 3.7. If the EV connects to the system for a longer time frame, it has the opportunity to achieve its requirement and export the power back to the grid to support the electricity demand or other trading options via the discharging schedule during the connected time frame. For example, in Figure 3.7, customer no. 87 charges the EV at time steps 12, 15, 16 and 17, and then discharges the EV to utilize the V2G ability at time steps 19 and 20. At time step 21, the customer charges the EV again to fulfill the required capacity for leaving at time step 22.

3.6 Conclusion

The EV charging and discharging scheduling is required for managing the high fluctuated penetration from EV. To practically reinforce the EV scheduling, the management tool has to be offered in the efficient, reliable and scalable approaches. Moreover, the management tool needs to provide financial incentives to customers for their participation in the management as well as operate their EV as a storage unit or cooperate with the DER and DR programs. Providing the EV charging and discharging scheduling from the cloud-based EMS satisfies the needs mentioned above. The emerged EVs in the heterogeneous environment can involve the EMS from various charging sites and cooperate with all the connected components (e.g., DER, storage, EV and electricity demands) from all the involved customers, who formed the community. The global multiperiod cost is minimized for each community. This chapter introduces cloud computing and the cloud-based EMS [11]. The extended model of EV charging and discharging scheduling for the cloud-based EMS[16] is highlighted and focused in this chapter. The characteristics of the cloud-based framework are beneficial to further extend the EMS with sophisticated and complex service models or the pricing models with the consideration of the emerged advance EV technologies (e.g., new specs of EV battery and the EV charging and discharging ability).

References

[1] Leou, R.-C., Su, C.-L. and Lu, C.-N. Stochastic Analyses of Electric Vehicle Charging Impacts on Distribution Network. *Transactions on Power Systems*, vol. 29, no. 3, pp. 1055–1063, 2014.

[2] Mukherjee J. C. and Gupta, A. A Review of Charge Scheduling of Electric Vehicles in Smart Grid. *IEEE Systems Journal*, vol. 9, no. 4, pp. 1541–1553, 2015.

[3] Ahmad M. S. and Sivasubramani, S. Feasibility of V2G Ideology in Developing Economy: Operation, Analysis and Impact. In *2016 National Power Systems Conference (NPSC)*, Bhubaneswar, 2016.

[4] Mohsenian-Rad, A.-H., Wong, V. W. S., Jatskevich, J., Schober, R. and Leon-Garcia, A. Autonomous Demand-Side Management Based on Game-Theoretic Energy Consumption Scheduling for the Future Smart Grid. *IEEE Transactions on Smart Grid*, vol. 1, no. 3, 2010.

[5] Wei, Z., Li, Y., Zhang, Y. and Cai, L. Intelligent Parking Garage EV Charging Scheduling Considering Battery Charging Characteristic. *IEEE Transactions on Industrial Electronics*, vol. 65, no. 3, pp. 2806–2816, 2018.

[6] Ortega-Vazquez, M., Bouffard, F. and Silva, V. Electric Vehicle Aggregator/System Operator Coordination for Charging Scheduling and Services Procurement. *IEEE Transactions on Power Systems*, vol. 22, no. 2, pp. 1806–1815, 2013.

[7] Shao, S., Pipattanasomporn, M. and Rahman, S. Grid Integration of Electric Vehicles and Demand Response With Customer Choice. *IEEE Transactions on Smart Grid*, vol. 3, no. 1, pp. 543–550, 2012.

[8] Korolko N. and Sahinoglu, Z. Robust Optimization of EV Charging Schedules in Unregulated Electricity Markets. *IEEE Transactions on Smart Grid*, vol. 8, no. 1, pp. 149–157, 2017.

[9] Alcaraz, C., Agudo, I., Nũnez, D. and Lopez, J. Managing Incidents in Smart Grids à la Cloud. In *IEEE Third International Conference on Cloud Computing Technology and Science*, Athens, 2011.

[10] Ibáñez, J. A. G., Zeadally, S. and Contreras-Castillo, J. Integration Challenges of Intelligent Transportation Systems with Connected Vehicle, Cloud Computing, and Internet of Things Technologies. *IEEE Wireless Communications*, vol. 22, no. 6, pp. 122–128, 2015.

[11] Popovic, N., Popovic, D. and Seskar, I. A Novel Cloud-Based Advanced Distribution Management System Solution. *IEEE Transactions on Industrial Informatics*., 6 Dec. 2017.

[12] Eldjalil C. D. A. and Lyes, K. Optimal Priority-Queuing for EV Charging-Discharging Service Based on Cloud Computing. Paris, 2017.

[13] Cao, Y., Song, H., Kaiwartya, O., Zhou, B., Zhuang, Y., Cao, Y. and Zhang, X. Mobile Edge Computing for Big-Data-Enabled Electric Vehicle Charging. *IEEE Communications Magazine*, vol. 56, no. 3, pp. 150–156, 2018.

[14] Chekired, D. A. E., Dhaou, S., Khoukhi, L. and Mouftah, H. T. Dynamic Pricing Model for EV Charging-Discharging Service Based on Cloud Computing Scheduling. In *13th International Wireless Communications and Mobile Computing Conference (IWCMC)*, Valencia, 2017.

[15] Chen Y.-W. and Chang, J. M. EMaaS: Cloud-Based Energy Management Service for Distributed Renewable Energy Integration. *Transactions on Smart Grid*, vol. 6, no. 6, pp. 2816–2824, 2015.

[16] Chen Y.-W. and Chang, J. M. Fair Demand Response with Electric Vehicles for the Cloud Based Energy Management Service. *Transactions on Smart Grid*, vol. 9, no. 1, pp. 458–468, 2018.

[17] Samaresh Bera, S. M. and Rodrigues, J. J. Cloud Computing Applications for Smart Grid: A Survey. *Transactions on Parallel and Distributed Systems*, vol. 26, no. 5, pp. 1477–1494, 2015.

[18] Baliga, J., Ayre, R. W. A., Hinton, K. and Tucker, R. S. Green Cloud Computing: Balancing Energy in Processing, Storage, and Transport. In *Proceedings of the IEEE*, 2011.

[19] Mell, P. and Grance, T. SP 800–145. The NIST Definition of Cloud Computing. National Institute of Standards & Technology, Gaithersburg, MD, 2011.

[20] ISO/IEC 17788: Information technology - Cloud computing - Overview and vocabulary. International Organization for Standardization, Geneva, Switzerland, 2014.

[21] Hogan, M., Liu, F., Sokol, A. and Tong, J. SP 500-291. NIST Cloud Computing Standards Roadmap-Version 1.0. National Institute of Standards & Technology, 2011.

[22] Public Utility Commission of Texas, Electric Substantive Rule 25.211. January 2017. [Online]. Available from https://www.puc.texas.gov/agency/rulesnlaws/subrules/electric/25.211/25.211.pdf. [Accessed May 2018].

[23] Tesla Powerwall. Tesla, [Online]. Available from https://www.tesla.com/powerwall. [Accessed May 2018].

[24] Wan, C., Zhao, J., Song, Y., Xu, Z., Lin, J. and Hu, Z. Photovoltaic and Solar Power Forecasting for Smart Grid Energy Management. *CSEE Journal of Power and Energy Systems*, vol. 1, no. 4, pp. 38–46, 2015.

[25] Green eMotion Project Results Executive Summary. Feb. 2015. [Online]. Available from http://www.greenemotion-project.eu/upload/pdf/deliverables/Project-Results-online.pdf. [Accessed May 2018].

[26] Liu, C., Chai, K. K., Zhang, X., Lau, E. T. and Chen, Y. Adaptive Blockchain-based Electric Vehicle Participation Scheme in Smart Grid Platform. *IEEE Access*, 2018.

[27] Torres-Sanz, V., Sanguesa, J. A., Martinez, F. J., Garrido, P. and Marquez-Barja, J. M. Enhancing the Charging Process of Electric Vehicles at Residential Homes. *IEEE Access*, vol. 6, pp. 22875–22888, 2018.

[28] Li, H., Alsolami, M., Yang, S., Alsmadi, Y. M. and Wang, J. Lifetime Test Design for Second-Use Electric Vehicle Batteries in Residential Applications. *IEEE Transactions on Sustainable Energy*, vol. 8, no. 4, pp. 1736–1746, 2017.

[29] Mustafa, M. A., Zhang, N., Kalogridis, G. and Fan, Z. Roaming Electric Vehicle Charging and Billing: An Anonymous Multi-User Protocol. In *2014 IEEE International Conference on Smart Grid Communications*, Venice, Italy, 2015.

[30] Doughty, D. H. Vehicle Battery Safety Roadmap Guidance. National Renewable Energy Laboratory, Oct. 2012. [Online]. Available from https://www.nrel.gov/docs/fy13osti/54404.pdf. [Accessed May 2018].

Chapter 4

Multi-criteria optimization of electric vehicle fleet charging and discharging schedule for secondary frequency control

Aleksandar Janjić[1] and Lazar Z. Velimirović[2]

Commercial fleet of electric vehicles (EV) is serving clients for different purposes, but each of those services is requiring some time to be fulfilled. When the vehicle is charging or discharging using the vehicle-to-grid (V2G) concept, the waiting time to service increases, and the vehicle owner is suffering losses due to the reduced service quality. In this chapter, we are exploring the scheduling optimization problem of an EV fleet controlled by the aggregator. The optimization of vehicle daily scheduling is needed in order to increase the revenues from offering ancillary services and reduce costs of service quality loss. New, practical multi-criteria decision-making methodologies for the daily scheduling of EV fleet are proposed.

Criteria that have to be fulfilled simultaneously are the minimization of the service waiting time (SWT), maximization of the revenues coming from the frequency regulation services and the minimization of the costs incurred by the vehicle charging, including the costs of battery degradation. The stochastic nature of vehicles driving patterns (time the car owner has to calculate for the service provision) is considered using the queuing theory. The proposed methodology has been successfully implemented on two cases of EV commercial fleet daily scheduling.

List of abbreviations

ACE	Area Control Error
AGC	Automatic Generator Control
AHP	Analytical Hierarchy Process
EMP	Energy Management Program
EV	Electric Vehicle

[1]University of Niš, Faculty of Electronic Engineering, Niš, Serbia
[2]Mathematical Institute of the Serbian Academy of Sciences and Arts, Belgrade, Serbia

EVA	Electric Vehicle Aggregator
MAUT	Multi-Attribute Utility Theory
MCDA	Multi-Criteria Decision Analysis
MINLP	Mixed Integer Nonlinear Programming
QT	Queuing Theory
SOC	State-of-Charge
TSO	Transmission System Operator
V2G	Vehicle-to-Grid

Nomenclature

Δs	- Deviation from the preferred number of vehicles
A	- Accuracy performance
C	- Battery degradation costs (€)
c_b	- Battery cost (€/kWh)
C_{cap}	- Regulation capacity (MW)
c_i	- Battery replacement labor and time (€)
$c_{i,max}$	- Maximal value of criteria i
$c_{i,min}$	- Minimal value of criteria i
DoD	- Depth of discharge (%)
E_s	- Total energy storage of the battery (kWh)
K	- Normalization constant
k_i	- Parameter from the trade-offs for component i,
M_{mi}	- Mileage
$M_{mil,i,j}$	- Forecasted mileage of i-th EV in hour j
N	- Total number of EVs
$p_{(rc)j}$	- Capacity price in hour j, (€/kW –h)
$p_{(rm)j}$	- Regulation mileage price in hour j, (€/kW –h)
P^{+}_{max}	- Maximal discharging power (kW)
p_0	- The probability that all vehicles are available
P_{avg}	- Average charging power (kW)
P_i	- Charging power of vehicle i (kW),
P^{-}_{max}	- Maximal charging power (kW)
p_{rc}	- Forecast price of regulation capacity at period t(€)
p_{rm}	- The forecast price of regulation mileage at period t (€)
R	- Decision-maker risk tolerance
r_{cap}/r_{mil}	- Regulation capacity/mileage revenue (€)
s_{pref}	- Preferred number of vehicles not connected to the grid
SWT	- Service waiting time (h)

$T_{max,\ on}$ - Latest allowable moment for plugging the vehicle i (h)
T - Number of time segments
$T_{max,\ plug}$ - Maximal duration of plugging time (h)
$T_{min,\ on}$ - Earliest moment for plugging the vehicle i (h)
$T_{min,\ plug}$ - Minimal duration of plugging time (h)
$t_{on,i}$ - The moment when the vehicle i is plugged on (h)
$t_{plug,i}$ - The duration of plugging time (h)
$u_i(x_i)$ - The single-attribute utility value for attribute i with value $x_i \in (0, 1)$,
w - Weighting factors
ΔP - Load variability
λ - Expected number of interventions per time unit
μ - Expected number of completed interventions per time unit
ρ - Utilization factor ($\rho = \lambda / s\mu$)

4.1 Introduction

4.1.1 Motivation

The participation of electric vehicles (EVs)* in the frequency regulation has been a subject of intensive research during latest years. Using EV fleet as a secondary frequency regulation resource, automatic generator control (AGC) should be able to issue raise or lower signals based on each EV's ability to provide the desired response in a reasonable amount of time, and on real-time economic dispatch results. These signals require a dedicated communications infrastructure, which also telemeters the state of all EVs in the aggregator's balancing area. EV, as a provider of frequency control that is ready to deliver regulating power with a short notice, can be paid a capacity price and energy price for the regulating energy that is actually delivered.

Vehicle-to-Grid (V2G) technology enables a controllable, bidirectional electrical energy flow between a vehicle and a power grid [1,2]. The major goal of V2G systems is to integrate large numbers of EVs into the power grid, reducing the impact of uncontrollable renewable energy sources. Primarily, this task is performed by an EV aggregator (EVA) – an intermediary inserted between the vehicles performing frequency regulation and the grid system operator. The communication architecture of EVA is presented in [3,4]. EVA receives ancillary service requests from the grid system operator and issues power commands to contracted vehicles that are both available and willing to perform the required services in the way that the aggregated EV fleet meets the EVA's day-ahead market

*EVs can be categorized into pure battery electric vehicles (BEVs), hybrid electric vehicles (HEVs) and plug-in hybrid electric vehicles (PHEV) differentiated by the alternative combination of an internal combustion engine in the vehicle and the availability of plug-in option.

requirements concerning energy and reserves. The aggregator role in the market is explained more in detail in [5] and in the demonstration project [6].

Scheduling objectives and constraints for the day-ahead market can be formulated from three independent levels: transmission system operator (TSO), EVA and EV users. From the operator's perspective, while balancing the power system he must consider least cost to minimize operational cost and the charging time of EV could be scheduled and limited. At an aggregator level, more flexible and practical charging strategies can be implemented as the aggregator manages the charging sequence by controlling the attached EV chargers directly. For instance, an aggregator revenue strategy can be determined via a fixed rate on energy delivered to the EV, the revenues from selling regulation and spinning reserve, and the revenues from selling energy to the grid [7]. Finally, scheduling can be optimized to maximize revenue for the EV owners [8]. Users utilize a smart controller at home to instruct the EV charger, allocating charging time slots to gain specific preferred benefits based on user behavior and choice.

Optimization of EV charge/discharge schedule has not been restricted to revenues and costs and car owner's utility and satisfaction. Emission reduction, power loss minimization and battery wear are also important factors in daily EV scheduling. The particular problem of EV fleet owner perspective requires new criteria, besides costs and revenues, like the service level of EV offering different services and the minimization of the SWT. Consequently, the multi-criteria approach in the case of uncertainty is required, because of the stochastic nature of vehicle usage. Furthermore, the globally optimal scheduling scheme is impractical due to the large number of EVs and the uncertainty about the EVs behavior.

4.1.2 Literature review

The V2G concept was introduced by [9] and it is still a subject of thorough research for grid support [10,11], and minimization of load variance [12]. This concept also introduces the function of an aggregator (EVA) – necessary for the proper balancing service operation. EVA, as an intermediary inserted between the vehicles offering different ancillary services and the grid system operator, receives service requests from the system operator and sends power commands to contracted vehicles using centralized or decentralized control. The profitability of EVs acting as regulating power has been treated in various studies. In [10], estimated profits are obtained for different types of ancillary services, including peak load power, spinning reserves and regulation. In [13,14], economic feasibility of V2G frequency regulation, which considers the battery wear, is analyzed. The maximum average profits in the range 30–80€ per month and vehicle are values obtained in [15–17]. Similar results are obtained for Sweden [18] and Portugal [19]. Results from the German market [20,21] show that the most lucrative service for EVs related to the economic potential for both the secondary and the tertiary regulation energy market represents the provision of negative regulation energy, implying the charging the EV at a lower price.

Flexible and practical charging strategies are proposed from EVA level by controlling the attached EV chargers [22]. In [23], the unidirectional V2G has been explored, and in [24] this approach has been extended to bidirectional V2G and in [25,26] to a combined provision of regulation and reserve. In these studies, the problem of optimization of V2G assets or vehicle scheduling has been treated as one-dimensional problem, with the single objective of aggregator revenues maximization. The analyzed sources of aggregator revenues were the revenues from selling regulation and spinning reserve capacity, regulation energy and the revenues from selling energy to EVs. In [27], optimal charging and discharging EV scheduling in a parking station coordinated with PV and energy storage system has been investigated.

However, an EV scheduling problem often involves trade-off on different objectives: operator or aggregator costs, customer satisfaction and CO_2 emissions, while also pursuing welfare maximization, battery cost minimization, grid support (power deviation minimization), etc. Therefore, EV optimization and modeling in reality should include multi-objective optimization in order to be realistic.

In [28], a multi-objective optimization was considered combining the capital cost of new plant construction and the operation cost in the unit commitment scheduling problem formulation. The savings controlling the EV charging were investigated and associated with the wind generation penetration situation. The results showed that the controlled EV charging reduces the cost by 5%–15% and even higher for capacity expansions. Objectives to improve system reliability by minimizing the system disruption cost and reducing the power loss were used in [29]. An objective function in [30] consists of electricity generation, imported/exported power costs and revenue components, as well as the environmental-credit components that are assigned to each EV added to Ontario's transport sector. Charging cost, power loss and departure penalty were considered in [31] and [32] divided the charging task into maximum revenue dynamic charge scheduling and minimum cost dynamic charge scheduling, in which an upper bound of cost is designed to trade-off aggregator revenue with customer benefit. The power purchase, generation and emission cost are minimized using the generalized Benders decomposition method from a distribution network side, where the Pareto optimal solution for economic cost and emission cost is achieved. The optimal allocation of available vehicles on a day-ahead basis using the queuing theory and a fuzzy multi-criteria methodology has been determined in [33]. In [34], the optimization is performed using the multi-attribute utility theory (MAUT) that allows different decision maker's attitude toward risk. A more detailed comprehensive review of other computational scheduling methods for integrating EVs in power systems is given in [35].

Some fleets have to perform tasks that may be known well in advance or that are sometimes repetitive, but others operate in a demand-responsive mode (the fleet has to be deployed and managed in real-time [36]). This stochastic nature of vehicle driving patterns and available regulation capacity can be modeled using Queuing theory (QT) designed to predict the queue lengths and waiting times to schedule individual service behaviors [37]. QT has been investigated in [38] using $M/M/\infty$

model for the estimation of each vehicle charging and discharging time. However, this approach calculates available energy implicitly, not being aware of required driving distance for each vehicle. [39] modeled the EV charging demand at each load bus using QT to simulate the stochastic charging characteristics, and real and reactive power demand followed the variation of charging time. In [40], the residential community and the charging stations were modeled by QT, respectively, taking into account that the EV charging slots in the residential community are owned by the residents and limited to be served. Fuzzy sets-based QT are used in [41] for modeling the EV charging queue. QT is definitely able to provide more reasonable charging profile for an EV aggregator. However, it might not be appropriate to model multiple aggregators for unit commitment and grid power dispatch.

Although multi-objective oriented, previous studies take into account the vehicle owner benefit from the operator, aggregator or individual car owner perspective. The special case of decentralized control of commercial vehicle fleet and fleet owner perspective, concerning its revenues and the availability of vehicles remains neglected. While the aggregator needs as much vehicles as possible connected to the network, the fleet owner has to balance between the service quality and the revenues from ancillary services offered by EVs.

4.1.3 Contribution

In this chapter, a decentralized scheduling scheme from the EV fleet owner perspective has been presented. Two multi-criteria decision-making methodologies for the daily scheduling of EV fleet have been analyzed. The main goal is optimal day-ahead schedule of EVs from the fleet owner perspective. Objectives that have to be fulfilled simultaneously are the minimization of the costs incurred from being parked, maximization of the revenues offering secondary regulation and the minimization of environmental impact through the maximization of the vehicle fleet charging station efficiency and minimization of battery wear.

In the proposed methodology, the queuing theory approach, modeling an uncertainty of EV usage for their daily activities has been used, enabling the more accurate prediction of fleet regulation capacity. The multi-criteria approach from the fleet owner perspective; decision maker attitude toward risk calculation and more accurate revenues calculation with hourly, instead average energy costs represent the major benefits of this methodology. The use of genetic algorithm has served as a basis for solving problems of multi-objective allocation of these resources, while the queuing theory served for the determination of the time the car owner has to calculate for the service provision.

4.1.4 Chapter structure

The chapter is organized in the following way. After the introduction and literature review, details about the model parameters together with two multi-criteria decision-making technique are explained (fuzzy multi-criteria min–max methodology and MAUT). Methodologies are illustrated on a day-ahead scheduling for small commercial vehicles fleet, giving the optimal time of grid connection of vehicles. Finally, conclusions are brought about the possibilities of model application and further research directions.

4.2 Optimization problem

Many EVs stay at the parking lots most of the time. An EVA combines these EVs to participate in the energy market to satisfy their charging demands. On the other hand, for profit maximization, the EVA uses the EVs' available battery capacity to take part in the regulation market. In a word, the EVA can simultaneously participate in the energy and regulation market.

4.2.1 Frequency regulation

There are several types of frequency control, according to new harmonized nomenclature of balancing services in Europe, namely [42]:

- FCR – Frequency Containment, that is, primary control
- aFRR – automated Frequency Restoration, that is, secondary control
- mFRR – manual Frequency Restoration, that is, fast tertiary control
- RR – Replacement Reserve, that is, slow tertiary control

First control type that responds to a frequency deviation is primary frequency control. Its response is automatic, within seconds of noted disturbance, and it stabilizes frequency. In primary control strategy, V2G participation is designed on adaptive frequency droop control, based on the initial state of charge (SOC) to maintain the residual battery energy [43]. After the primary control stabilizes the frequency, the secondary control is returning frequency to its nominal value, in order to recover the tie-line power between control areas, and to relieve the units participating in the primary frequency control. AGC calculations are usually based on a weighted sum of system frequency and unscheduled power flows, with the resulting signal called area control error (ACE). If EV fleet is used as a regulation resource, AGC should be able to issue raise or lower signals based on each EV ability to provide the desired response in a reasonable amount of time, and on real-time economic dispatch to minimize ACE. The relationship between the EV's charging and discharging power and the regulation capacity is presented on Figure 4.1. P^{+}_{max} and P^{-}_{max} stand for the maximal discharging and charging power, respectively.

In a traditional regulation market, EV as a provider of frequency control that is ready to deliver regulating power with short notice can be paid capacity price and energy price for the regulating energy that is actually delivered. The AGC signal can call for either a positive or a negative correction; and, in the case of regulation up, the energy price is paid to the actor that delivers energy. In the case of regulation down, the actor pays for the energy taken from the grid, but with a lower price than what would be the case if buying the energy on the normal power market. In a long run, the net value of the dispatched energy for regulation purposes will be

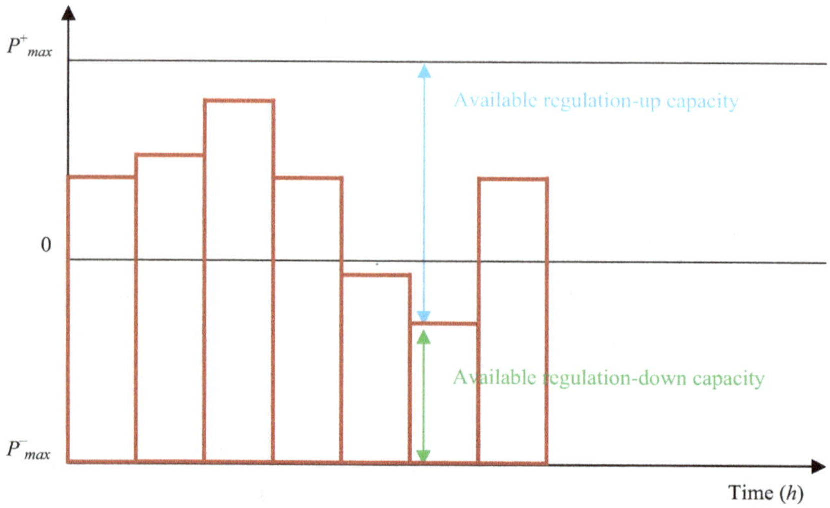

Figure 4.1 Charging/discharging and the regulation capacity relationship

zero, but the amount of dispatched power is not the same as the contracted power capacity. In [44], calculation yielded that the dispatched power is around 40% of the contracted power capacity. Therefore, the dispatch power in each direction would be around 20% of the contracted power capacity.

4.2.2 Business model

Payments for the secondary regulation energy include an energy payment per MWh for the actually provided regulation energy besides the capacity payment. For secondary and tertiary negative regulation, energy provision payments can also occur from provider to TSO. These bids are ranked before bids with payment from TSO to provider, reducing procurement costs and allocating surplus energy to flexible loads before other generators are paid to reduce their production.

In pay-for-performance regulation market, an EV participant is rewarded based on the regulation capacity it provides, as well as on the number of instructions to dispatch set-points by the TSO, and accuracy in following the regulation instructions. Pay-for-performance regulation markets have two-part offer and payment designs. A participant submits a regulation capacity offer and a regulation mileage offer, and then the system operator unifies the two offer prices into a single modified offer based on the participant's historical performance index and the expected dispatch regulation mileage. The settlement rules of the revenue of capacity (C_{cap}) and mileage (M_{mil}) can be summarized as [45]:

$$r_{cap} = C_{cap} \cdot p_{rc} \cdot A, \tag{4.1}$$

$$r_{mil} = M_{mil} \cdot p_{rm} \cdot A, \tag{4.2}$$

where A stands for accuracy performance, r_{cap}/r_{mil} is the regulation capacity/mileage revenue, p_{rc} the forecast price of regulation capacity at period t and p_{rm} is the forecast price of regulation mileage at period t.

In both market models, the EVA submits hourly bids (both energy and regulation) for the periods in the next day. It is therefore necessary to predict the available regulation capacity and reduce the uncertainty coming from the EV user or commercial EV fleet charging demands. In this chapter, we are exploring the problem of a daily scheduling of commercial fleet of EVs controlled by the aggregator. These vehicles are serving clients for different purposes, but each of this service is requiring some time to be fulfilled. If the service time increases, the vehicle owner is suffering losses due to the reduced service quality. At the same time, when the vehicles are parked, the vehicle is offering regulating services to the electricity network operator using V2G technology. Revenues from selling these services are directly related to the number of vehicles parked. The optimization of vehicle daily scheduling is needed in order to increase the revenues and reduce costs of quality loss.

4.2.3 ICT architecture

There are three main states of an EV: driving, charging and participation in frequency regulation (Figure 4.2). An EV enters the driving state after plugging out, with the sufficient SOC level to start the trip. Transitions between states of charging (state 2) and participating in frequency regulation (state 3) occur several times during the connection period, depending on the initial SOC. The determination of intraday strategy for these transitions is out of the scope of this study, and we will focus on the optimization of driving time (state 1) during the next day.

The presented methodology refers to companies that decide to install charging stations in their own parking areas. For instance, municipal fleets and the fleets of delivery services generally have parking areas fully equipped for charging their EVs. The main assumption is the existence of a parking central smart charging controller and EV energy management program (EMP). According to signals from aggregator side, EMP sends the chargers an order to start or stop charging, that is, to transit from state 2 to state 3 and vice versa. The total consumption of vehicle chargers is constantly transmitted to the charging station's controller, and the power demand from the charging stations is that actually available, not exceeding the aggregator contract or the charging station capacity. These signals (typically

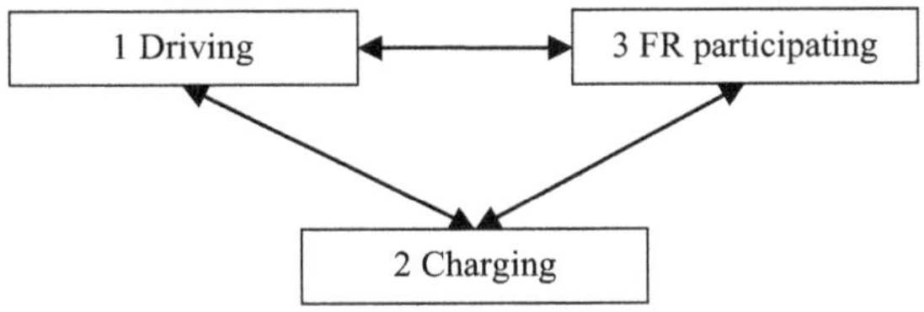

Figure 4.2 EV states

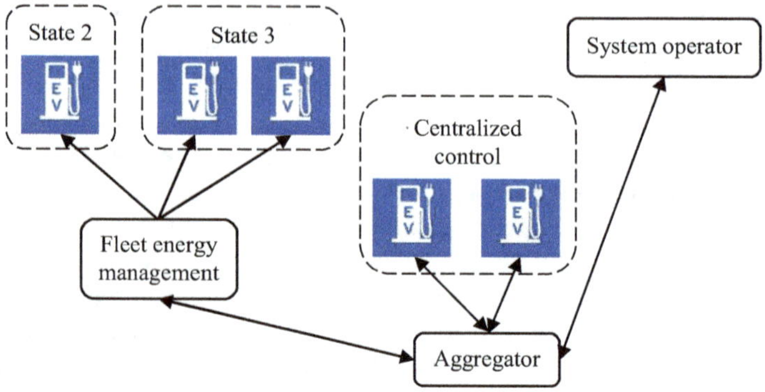

Figure 4.3 Charging station architecture

pulses of varying length, proportional to the requested output change) require a dedicated communications infrastructure, which also telemeters the state of all EVs in the aggregator's balancing area.

ICT architecture and the flow of information is presented on Figure 4.3. The grid control center allocates the frequency regulating power to the EVA according to the system frequency fluctuation and the predictive controllable capacity reported by the EVA. Then, the EVA instructs EVs in state 3 to output power charging or discharging to achieve the purpose of adjusting the frequency. EVs outside the EMP program are controlled centrally, from the EVA.

EVA collects the declared demand information of EVs which are accessing the grid in real time, including the residence time in the gird *Ti*, SOC and the predictive controllable capacity. According to the length of residence time in the grid and the SOC level, a signal from the EMP will be sent to the EVs to enter the state 2 or 3. When the regulation ancillary service is over, EVs will automatically switch into the energy demand group to meet the driving demand of vehicle owners. EVA reports the predictive controllable capacity of the EVs in regulation service group in the area in real time.

4.2.4 Optimization objectives

The optimization goal is to find the best combination of hours when the vehicle i is plugged on $t_{on,i}$ and the duration of plugging time (states 2 and 3 from Figure 4.2), denoted by $t_{plug,i}$. In order to minimize the *SWT*, load variability (ΔP) and costs due to the battery degradation (C) and at the same time maximize revenues from secondary regulation (C_{cap}). Formally, the problem can be represented in (4.3).

$$\begin{aligned} &min\, SWT(x, t_{on}, t_{plug}) \\ &max\, r_{cap}(x, t_{on}, t_{plug}) \\ &min\, \Delta P(x, t_{on}, t_{plug}) \\ &min\, C(x, t_{on}, t_{plug}) \end{aligned}, \tag{4.3}$$

subject to:

$$\begin{cases} x_{i,j} = 1, & j \in (t_{on,i}, t_{on,i} + t_{plug,i}), \quad \forall i \\ x_{i,j} = 0, & \text{otherwise} \end{cases}, \tag{4.4}$$

$$t_{on,i} + t_{plug,i} = T - 1, \quad \forall i, \tag{4.5}$$

$$T_{(\min,\, on),i} \leq t_{on,i} \leq T_{(\max,\, on)i}, \quad \forall i, \tag{4.6}$$

$$T_{(\min,\, plug),i} \leq t_{plug,i} \leq T_{(\max,\, plug)i}, \quad \forall i, \tag{4.7}$$

$$T_{(\min,\, plug)i} = \frac{W_i}{P_{\max,\, i} \cdot R_{d-c}}, \tag{4.8}$$

$$T_{(\max,\, plug)i} = T - T_{service,i}, \tag{4.9}$$

$$P_i \leq P_{i,\, \max}, \quad \forall i. \tag{4.10}$$

Constraint (4.4) represents output variables as binary values: $x_{i,j} = 1$, if the vehicle is connected to the grid, otherwise $x_{i,j} = 0$. Constraint (4.5) bounds the vehicle activity to 1 day (24 h). Constraint (4.6) gives the range of hours in which the decision maker allows the vehicle to be parked and not in service. The range of possible plugging times is set by constraint (4.7). The minimum (4.8) and maximum plugging time (4.9) of each vehicle depends on maximal active charger power $P_{max,i}$ and expected time of vehicle driving $T_{service,i}$. Finally, the charging/discharging power is limited with the active charger power $P_{i.max}$ in constraint (4.10). As stated before, criteria that will be simultaneously optimized are the service quality, revenues and environment issues.

4.2.4.1 Commercial service quality

The commercial quality is the crucial issue for any utility company maintaining the infrastructure for the public service. These services involve the extensive usage of vehicles to fulfill the required commercial quality aspect. The most frequent one is the timeliness of services requested by customers. If the company fails to provide the level of service required by the standard, it must compensate the customer affected. Problem of service time minimization is usually solved by the usage of QT and queuing models as an abstraction of Markov chain models. The M/M/s model assumes that all times between different service requests (inter-arrival times) are independently and identically distributed according to an exponential distribution, that all service times are independently and identically distributed according to another exponential distribution, and that the number of servers (crews or vehicles) is *s*.

QT has also been used to study the aggregate behavior of EVs. In [46], a simple M/M/queuing model for EV charging was presented and a similar idea was also used in [47] to determine V2G capacity. An M/M/∞queue with random interruptions was suggested in [48] to model the EV charging process and analyze the dynamics with time-scale decomposition. The service process adopted was assumed to be exponential and in [49], a smart charging mechanism to overcome this problem has been proposed. The transitions from states are modeled as an M/M/∞ queue with constant arrival rate and service rate.

When the service essentially consists of the same routine task, there is little variation in the service time required. The M/D/s model provides a representation of this situation, because it assumes that all service times actually equals, with the Poisson input process characterized with a fixed mean arrival time λ. The M/D/s model assumes zero variation in the service times, and the exponential service time distribution assumes a very large variation. Erlang distribution for the service time (M/Ek/s model) is lying between these two extremes.

The appropriate number of available vehicles assures service quality minimizing the *SWT*. In the system with s vehicles, the average waiting time to the service (*SWT*) is obtained from constraint (4.11):

$$SWT = \frac{(\lambda/\mu)^s \rho}{\lambda s!(1-\rho)^2} p_0 + \frac{1}{\mu}. \tag{4.11}$$

The preferred number of vehicles not connected to the grid (state 1 from Figure 4.2) and available for service can be calculated, based on customer maximum allowed waiting time to service $s_{pref} = f(WT_{max})$. If s_{pref} represents a vector of preferred number of vehicles for each hour, the goal is to minimize the deviations from this value, for all of N vehicles in T time segments (4.12):

$$\min(\Delta s) = \sum_{j=1}^{T} ((\sum_{i}^{N} x_{i,j}) - s_{pref,j})^2. \tag{4.12}$$

4.2.4.2 Revenues from the regulation services

The active power markets of V2G can be divided into four general groups [11]. These four groups are: base load power defined as the bulk power generation that is running most of the time; peak shaving – during predictable highest power demand hours; spinning reserves supplied by fast generators ready to respond in case of equipment or power supplier failures; and finally – active regulation used to keep the frequency and voltage steady. Typically, spinning reserve is called around 20 times a year. The duration of supply by a spin reserve is typically around 10 min but the source must be able to last up to 1 h. Regulation is called for only a few minutes at a time, but the number of times can be up to 400–500 times per day. The utility pays spinning reserves and regulation sources in part for just being available, per hour availability; however, base load and peak shaving are paid per kWh generated.

The formulas for calculating revenues depend on the market that the V2G power is sold into and the number of services the car owner contracts with the aggregator, paying for these services. This study assumes that a V2G vehicle performs frequency regulation service only, which previous studies have shown is the most lucrative and realizable ancillary service for V2G [1]. A second assumption is that a V2G vehicle contracting and performing both regulation-up and regulation-down services results in net zero energy transaction, avoiding capacity issues related to vehicle SOC.

For regulation services, the revenue derives from the payment for the maximum capacity contracted for the time duration, and the payment for the actual kWh produced. For regulation services, there can be 400 dispatches per day, varying in power (P_{disp}). For planning and scheduling purposes, to estimate revenue we approximate the sum of P_{disp} by using the average dispatch to contract ratio ($R_{d\text{–}c}$) defined in [10].

For both regulation up and down, the vehicle owners get paid for the contracted power and mileage, according to constraints (4.1) and (4.2). The total revenue for the period consisting of n segments is presented in constraint (4.13) should be maximized:

$$max(r_{cap} = \sum_{i=1}^{N}\sum_{j=1}^{T} x_{i,j}(p_{(\mathrm{rc})j}P_i + p_{(\mathrm{rm})j}M_{i,j})). \tag{4.13}$$

In this study, it is supposed that EV offers both regulation up and regulation down services, what is encompassed by the sole price p_{rcj} in (4.11). Another premise is that the measurement of energy is bidirectional, in order to register separately the energy for battery charging from the energy withdrawn from battery.

4.2.4.3 Environmental objectives

The main environmental objectives that have to be fulfilled are the minimization of battery wear and the maximization of the charger efficiency. The battery wear is in direct relationship to the absolute amount of transferred energy. The battery degradation is a function of the total energy storage of the battery, battery cost per kWh, battery replacement labor and time and the number of cycles during the battery life based on battery depth of discharge [33].

$$\min C = \sum_{i=1}^{N}\sum_{j=1}^{T} \frac{E_s c_b + c_L}{3x_{i,j} E_s DoD}. \tag{4.14}$$

Although it is possible to treat the total income as a net value, that is, the difference between the revenues and costs, in the general case these two values can be treated separately, as two independent criteria.

Another important factor that enables the efficient usage of charging station, reducing energy losses and, consequently, reducing CO_2 emission is the load factor defined as the ratio of the average demand P_{avg} to the maximum demand P_{max}. Load factor gives an indication of how well the charging station is being utilized. The optimal load factor would be 1, since the charging infrastructure has to be designed to handle the maximum demand. Alternatively, this criterion can be represented as the minimization of power deviations from the average power (4.15):

$$\min \Delta P = \sum_{i=1}^{N}\sum_{j=1}^{T}\left(x_{i,j}P_i - P_{\text{arg}}\right)^2. \tag{4.15}$$

After the single objective optimization of all of previously explained objectives, the multi-objective optimization is carried out.

4.3 Multi-objective optimization

The goal of multi-objective optimization is to minimize m functions $f_1, \ldots f_m$ of n-dimensional decision variables φ. A decision vector $\varphi_1 \in S$ is called Pareto-optimal if there is no other decision vector $\varphi_2 \in S$ that dominates it. Any vector φ_1 is said to dominate φ_2 if φ_1 is not worse than φ_2 in all of the objectives and it is strictly better than φ_2 in at least one objective. In case two solutions φ_1 and φ_2 do not dominate each other, they are indifferent to each other (or are non-dominated with respect to each other). To solve such problems, algorithms which can find a well-distributed set of trade-off and well-converged set of solutions with least computational expense are desired.

There are generally two approaches for the generation of solution sets for the multi objective optimization: scalarization and non-scalarization methods. Scalarization methods explicitly use a scalarization function to convert the problem into a single objective program. Both approaches are analyzed in this chapter. The general algorithm of the multi-criteria daily scheduling of EVs is presented on Figure 4.4.

The process starts with the definition of criteria of interest. Then, depending on the number of vehicles (N) and desired time segments (T) the set of possible alternatives (A) has been determined. For each criteria, minimal ($c_{i,min}$) and maximal ($c_{i,max}$) values are determined, whether by the single optimization and the recognition of best and worst alternative, whether by the direct elicitation of best and worst value. After the determination of two extreme values, it is possible to construct the utility function for each criterion.

One popular mathematical form that expresses the decision maker's attitude toward risk is the exponential utility function:

$$u(x) = R\left(1 - e^{-\frac{x}{R}}\right), \tag{4.16}$$

where R represents the decision-maker's risk tolerance. A great aversion to risk corresponds to a small value of R. In the next step, the relative importance and weight of criteria can be determined. In the proposed methodology, AHP analysis is used for the determination of criteria weighting factors. Alternatively, the trade-off

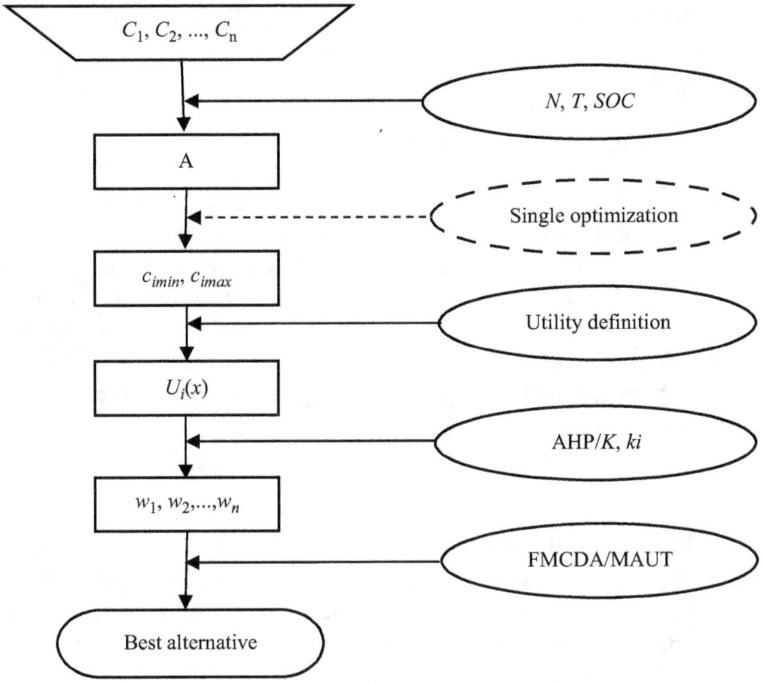

Figure 4.4 Multi-criteria EV scheduling process

between criteria can be established using MAUT approach, which be explained in Section 4.1.1. Finally, best alternative is determined by the Fuzzy MCDA in the case of the small number of alternatives, or using some heuristic technique (like Genetic algorithm in the proposed methodology) in the other case.

4.3.1 Fuzzy multi-criteria decision-making

In the first approach, the fuzzy multi-criteria decision-making (Bellman–Zadeh approach [50,51]) is as a basis for solving this optimization problem, where the maximum degree of implementing goals serves as a criterion of optimality. The following model includes several aspects of making decision under uncertainty. First, the decision-maker, from a set of available actions or alternatives, chooses an action x_i. Let $X = \{x_1, x_2,\ldots,x_m\}$ be a set of alternatives and $G = \{g_1, g_2,\ldots, g_n\}$ be a set of goals (criteria) to be achieved. A vector of objective functions $G(X) = \{g_1(X),\ldots,g_n(X)\}$ is examined, and the problem consists of simultaneous optimizing all objective functions. When the decision-making is employed in a fuzzy environment, each objective function $g_j(X)$ is replaced by a fuzzy objective function or a fuzzy set $\bar{G}_j$ for $j = 1, 2, \ldots, n$. The importance (weight) of goal j is expressed by w_j.

The achievement of goal j by alternative is illustrated by the membership function of x_{ki} in fuzzy set $\bar{G}_j$:

$$\mu_{G_i}(x) = \left[\frac{g_i(X) - \min(g_i(X))}{\max g_i(X) - \min g_i(X)}\right]^{w_i}, \tag{4.17}$$

By implementing this approach, the maximum degree of implementing goals serves as a criterion of optimality. This complies with the principle of guaranteed result and provides constructive lines in achieving harmonious solutions. This method provides a simple logic driven procedure to aggregate attribute attainments of alternatives and rank the alternatives on such an aggregation. In the proposed methodology, the vector of alternatives $X = \{x_1, x_2, \ldots, x_m\}$ is composed of multiple combinations of parked vehicles, used by the aggregator and vehicles used freely by the car owner.

4.3.2 MAUT

In the second approach, the multiplicative aggregation as one of the well-known scalarization technique is used. The optimization is performed using the MAUT that allows different decision-maker's attitude toward risk. Genetic algorithm is used for the minimization of multi-attribute utility function and daily scheduling of EVs. MAUT defines the utilities of multiple-attribute outcomes as a function of the utilities of each attribute taken individually. The theory specifies several possible functions and the conditions to be met under which each of these functions (multiplicative, additive and multi-linear) would be appropriate [52–54]. The multi-attribute utility function is of the following form if mutual utility independence exists:

$$U(x_1, x_2, \ldots, x_n) = \frac{\prod_i (1 + K k_i u_i(x_i)) - 1}{K}. \tag{4.18}$$

The three cases can be distinguished in terms of the multivariate risk posture of decision-maker. The first case represents multivariate risk aversion, second case risk neutrality and the third case risk seeking behavior. The method used in the proposed approach to determine the function is to measure each $u(x)$, determine the k_j values, and find the K value by iteratively solving (4.19):

$$1 + K = \prod_{i=1}^{n} (1 + K \cdot k_i). \tag{4.19}$$

The optimization problem of EV scheduling is reduced to single objective MINLP maximization of U, with $2N$ integer variables: $t_{on,i}$ and $t_{plug,i}$. Minimum and maximum plugging time of each vehicle depends on maximal charger power and expected time of vehicle driving.

4.4 Case studies

4.4.1 Belman–Zadeh approach

The case study elaborated in [33] is concerning the scheduling of EVs fleet consisting of nine EVs of one electricity distribution utility. Electric cars are used by faults repairing crews in the urban area, with supposedly three distinct periods during the day. The first one, between midnight and 8 AM is characterized by eight reported faults in average ($\lambda = 7/8$h), with the average of 2, 67 h per intervention (three repaired faults during this period $\mu = 3/8$h). The parameters of the second, more intensive period from 8 AM to 4 PM are $\lambda = 10/8$h, $\mu = 4/8$h, while in the third one, between 4 PM and midnight $\lambda = 8/8$h, $\mu = 3/8$h. Day-ahead spot energy and ancillary services prices are the actual prices used in the ERCOT system [55].

Calculated values for SWT, revenues and battery degradation costs are presented in Table 4.1. The first column represents the number of parked vehicles offering frequency regulation services (N_{reg}). The following columns are calculated regulation revenues (r_{cap}) battery degradation costs C, SWT and their fuzzy membership value of relative goal attainment ($\mu(r)^{w}{}_{1}$), $\mu(C)^{w}{}_{2}$ and $\mu(SWT)^{w}{}_{3}$, respectively, calculated using expression (4.17).

The weighting factors used in this example for revenues, costs and SWTs are $w_1 = 0.2$; $w_2 = 0.3$ and $w_3 = 0.5$, respectively. The optimal alternative, using max-min approach for each period is: six vehicles for ancillary services and three for the customer services in the first period, five vehicles for ancillary services in the second, and finally, four vehicles in the third period. Optimal values are bolded in Table 4.1.

4.4.2 MAUT methodology

The methodology for the optimization of day-ahead scheduling of EVs is illustrated on the example of EV fleet consisting of $N = 10$ vehicles. The service is modeled with M/M/s queuing model, with the average of $\mu = 2$ accomplished services per hour by one vehicle, and maximum allowable waiting time of $SWT_{max} = 30$ min. Input driving parameters are: the average consumption $d_d = 0.15$ kWh/km, $\eta_{charge/discharge} = 0.95$, average driving distance per service $d_s = 3$ km. In Table 4.2, individual charging characteristics for each vehicle in column (1) expected tasks (columns 2) and required energy and power (columns 3 and 4) are presented. Minimal required start and end of parking (columns 5 and 6), and minimal and maximal parking time (columns 7 and 8) are also presented. For the case of simplicity, the forecasted mileage has not been taken into account in this example.

Multi-objective optimization is performed using genetic algorithm using MATLAB® optimization package, for the compensatory case of decision-maker risk attitude. After the first step of the optimization, the following values are obtained (Table 4.3).

Table 4.1 Values for different criteria for different time periods and combinations

	Period I					
N_{reg}	r_{cap}(\$)	$\mu(r)^{w}{}_{1}$	C(\$)	$\mu(C)^{w}{}_{2}$	SWT (min)	$\mu(SWT)^{w}{}_{3}$
1	23.9	0.54	12.64	1.00	160	1.00
2	47.8	0.62	25.28	0.96	160	1.00
3	71.7	0.67	37.92	0.91	160	1.00
4	95.6	0.71	50.56	0.87	160	1.00
5	119.5	0.74	63.2	0.81	163	0.98
6	**143.4**	**0.77**	**75.84**	**0.75**	**177**	**0.90**
7	167.3	0.80	88.48	0.65	288	0
8	191.2	0.82	101.12	0.53	N.A.	0
9	215.1	0.84	113.76	0.00	N.A.	0
	Period II					
N_{reg}	r_{cap}(\$)	$\mu(r)^{w}{}_{1}$	C(\$)	$\mu(C)^{w}{}_{2}$	SWT (min)	$\mu(t)^{w}{}_{3}$
1	21.64	0.53	12	1.00	120	1.00
2	43.28	0.61	24	0.96	120	1.00
3	**64.92**	**0.66**	**36**	**0.92**	**121**	**0.99**
4	86.56	0.70	48	0.87	126	0.96
5	108.2	0.73	60	0.81	145	0.83
6	129.84	0.76	72	0.75	288	0
7	151.48	0.78	84	0.66	N.A.	0
8	173.12	0.80	96	0.54	N.A.	0
9	194.76	0.82	108	0.00	N.A.	0
	Period III					
N_{reg}	r_{cap}(\$)	$\mu(r)^{w}{}_{1}$	C(\$)	$\mu(C)^{w}{}_{2}$	SWT (min)	$\mu(t)^{w}{}_{3}$
1	57.7	0.64	19.04	1.00	160	1.00
2	115.4	0.74	38.08	0.96	160	1.00
3	173.1	0.80	57.12	0.92	162	1.00
4	230.8	0.85	76.16	0.87	171	0.99
5	**288.5**	**0.89**	**95.2**	**0.81**	**205**	**0.94**
6	346.2	0.92	114.24	0.74	542	0.00
7	403.9	0.95	133.28	0.66	N.A.	0.00
8	461.6	0.98	152.32	0.53	N.A.	0.00
9	519.3	1.00	171.36	0.00	N.A.	0.00

Individual utilities for r_{cap}, SWT and ΔP are represented in (4.20), (4.21) and (4.22), respectively.

$$U_R(x) = \frac{r_{cap}(x)}{\max r_{cap}(x)}, \tag{4.20}$$

$$U_{SC}(x) = \frac{\max SWT(x) - SWT(x)}{\max SWT(x) - \min SWT(x)}, \tag{4.21}$$

$$U_{LF}(x) = \frac{\Delta P(x)}{\min \Delta P(x)}, \tag{4.22}$$

Table 4.2 Input charging parameters

Vehicle	Nº of tasks	*Required daily energy W* (kWh)	P_{max} (kW)	$T_{min,on}$ (h)	$T_{max,on}$(h)	$T_{min,plug}$(h)	$T_{max,plug}$(h)	t_{on} (h)	t_{plug} (h)
(1)	(2)	(3)	(4)	(5)	(6)	(7)	(8)	(9)	(10)
1	10	4.5	4.7	1	16	10	16	12	13
2	14	6.3	5.1	1	16	13	16	1	13
3	14	6.3	5.1	1	16	13	16	1	13
4	16	7.2	6.3	1	16	12	16	1	12
5	16	7.2	6.3	1	16	12	16	10	15
6	16	7.2	6.3	1	16	12	16	7	16
7	18	8.1	7.1	1	16	12	16	13	12
8	18	8.1	7.1	1	16	16	16	1	16
9	20	9	5.92	1	16	16	16	1	16
10	22	9.9	6.51	1	16	16	16	1	16

Table 4.3 Input charging parameters

r_{cap}(€)	*SWT (min)*	ΔP
10.47	384	0.72

In the second step, multi-objective optimization is performed, using expression (4.18) and following utility function parameter: $k_1 = 0.5$, $k_2 = 0.5$, $k_3 = 0.5$, $K = -0.756$. The maximal generation number for the GA algorithm is 200, with population size set to the array length of 70. The optimal utility value is $U = 0.977$ ($U_1 = 0.7952$, $U_2 = 0.92$, $U_3 = 0.99$)

The time of plugging in each vehicle and the required time of being parking are represented in columns (9) and (10) in Table 4.2. Scheduled number of plugged EVs during the day and their available regulating capacities are represented on Figure 4.5.

Unlike the complementary case, when a good performance by one criterion is less important than balanced performance across the criteria, the simulation was carried out for the compensatory case, and the optimal overall utility value is obtained.

4.5 Conclusion

The optimal scheduling of EVs to their daily activities, including the offering of frequency regulation services using V2G technology, has been discussed.

The multi-objective decision-making methodology for the daily scheduling of commercial EV fleet has been proposed and successfully implemented. Objectives that have to be fulfilled simultaneously are the minimization of the costs incurred from being parked, maximization of the revenues offering secondary frequency

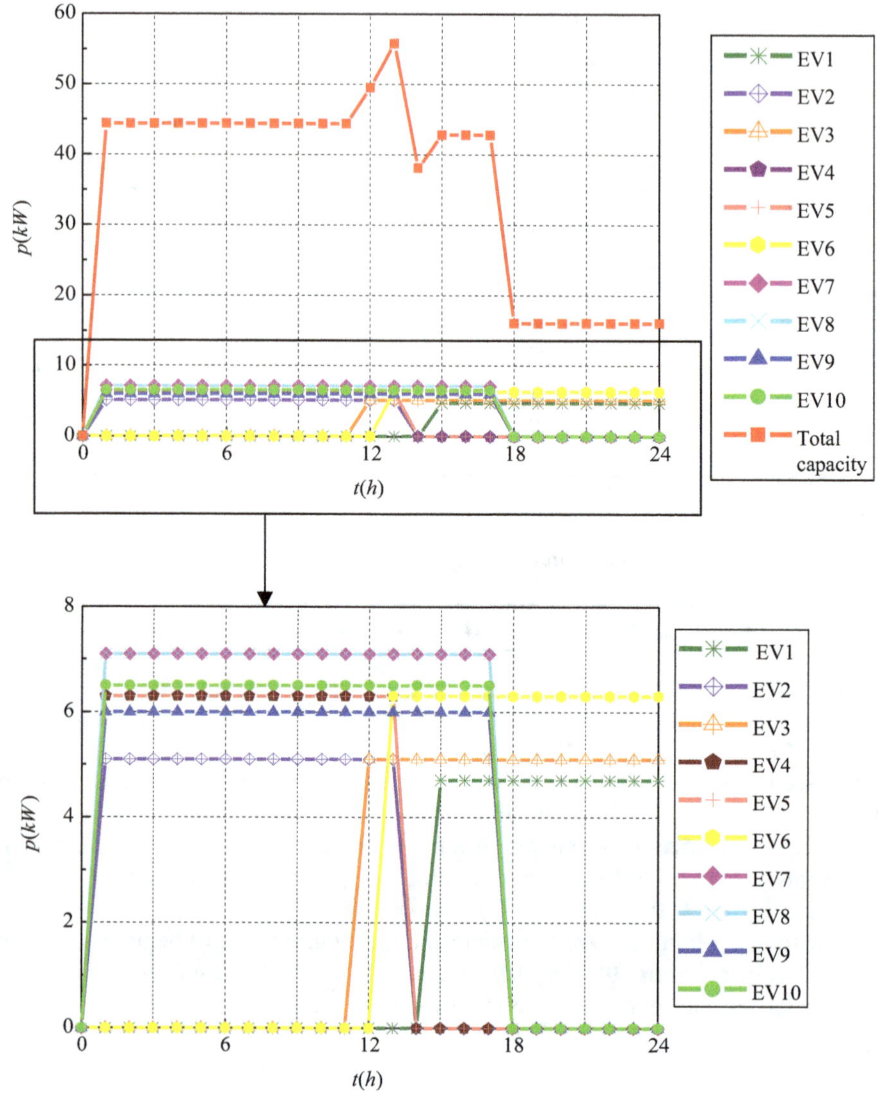

Figure 4.5 Scheduled regulation power during the day

regulation service and the maximization of the vehicle fleet charging station efficiency, including the battery degradation costs.

Bellman–Zadeh approach and MAUT were used to solve this optimization problem, where the maximum degree of implementing goals serves as a criterion of optimality. The first approach proves to be an effective (from the computational point of view) and rigorous (from the obtaining solutions perspective) method of analyzing multi-objective models. It allows us to keep a natural measure of uncertainty as well as to consider indices, criteria and constraints of qualitative character, while MAUT takes into account the decision-maker risk attitude and different trade-offs between criteria.

Optimization results confirmed the profitability of EVs as resources for the secondary regulation. Annual revenues per vehicle attain 360€, without substantial decrease of quality services offered to EV fleet customers. The optimization has been performed from the car owner perspective, not only from the aggregator's.

The stochastic nature of vehicles driving patterns is considered using the queuing theory for the determination of the time the car owner has to calculate for the service provision.

The multiplicative form of utility function modeling the decision maker risk attitude is taken into account.

The elaborated case encompassed only three criteria and offering of regulation services only, while the further research will be focused on the enlargement of both: criteria set (including environmental and social aspect) and ancillary services set.

Acknowledgments

This work was supported by the Ministry of Education, Science and Technological Development of the Republic of Serbia through Mathematical Institute SASA under Grant III 44006 and Grant III 42006.

References

[1] Guille, C., and Gross, G. 'A conceptual framework for the vehicle-to-grid (V2G) implementation'. *Energy Policy.* 2009;37(11):4379–90.

[2] Lopes J. A., Soares F. J., and Almeida P. M. 'Integration of electric vehicles in the electric power system'. *Proceedings of the IEEE.* 2011;99(1):168–83.

[3] Quinn C., Zimmerle D., and Bradley T. H. 'The effect of communication architecture on the availability, reliability, and economics of plug-in hybrid electric vehicle-to-grid ancillary services'. *Journal of Power Sources.* 2010;195(5):1500–9.

[4] Brooks A., Lu E., Reicher D., Spirakis C., and Weihl B. 'Demand dispatch'. *IEEE Power and Energy Magazine.* 2010;8(3):20–9.

[5] Brooks A.N. *Vehicle-to-grid Demonstration Project: Grid Regulation Ancillary Service with a Battery Electric Vehicle.* Sacramento, CA: California Environmental Protection Agency, Air Resources Board, Research Division; 2002 Dec 10.

[6] Bessa R.J., and Matos M.A. 'The role of an aggregator agent for EV in the electricity market'. 7th Mediterranean Conference and Exhibition on Power Generation, Transmission, Distribution and Energy Conversion (MedPower 2010); 2010

[7] Sortomme E., and El-Sharkawi M.A. 'Optimal scheduling of vehicle-to-grid energy and ancillary services'. *IEEE Transactions on Smart Grid*. 2012; 3(1):351–9.

[8] Han S., Han S., and Sezaki K. 'Optimal control of the plug-in electric vehicles for V2G frequency regulation using quadratic programming'. *InInnovative Smart Grid Technologies (ISGT)*, 2011 IEEE PES 2011 Jan 17 (pp. 1–6). IEEE.

[9] Kempton W., and Letendre S.E. 'Electric vehicles as a new power source for electric utilities'. *Transportation Research. Part D, Transport and Environment*. 1997;2(3):157–75.

[10] Tomić J., and Kempton W. 'Using fleets of electric-drive vehicles for grid support'. *Journal of Power Sources*. 2007;168(2):459–68.

[11] Lin J., Yu J.J., Leung K.C., and Li V.O. 'Optimal scheduling with vehicle-to-grid ancillary services'. *Energy Systems for Electric and Hybrid Vehicles*. 2016;2:395.

[12] Tan K.M., Ramachandaramurthy V.K., Yong J.Y., Padmanaban S., Mihet-Popa L., and Blaabjerg F. 'Minimization of load variance in power grids—investigation on optimal vehicle-to-grid scheduling'. *Energies*. 2017;10(11):1880.

[13] Han S., and Han S. 'Economic feasibility of V2G frequency regulation in consideration of battery wear'. *Energies*. 2013;6(2):748–65.

[14] Peterson S.B., Whitacre J.F., and Apt J. 'The economics of using plug-in hybrid electric vehicle battery packs for grid storage'. *Journal of Power Sources*. 2010;195(8):2377–84.

[15] Ciechanowicz D., Leucker M., and Sachenbacher M. 'Ökonomische bewertung von vehicle-to-grid in Deutschland'. Proceedings of the Multikonferenz Wirtschaftsinformatik (MKWI), Braunschweig. 2012 Feb.

[16] Dallinger D., Krampe D., and Wietschel M. 'Vehicle-to-grid regulation reserves based on a dynamic simulation of mobility behavior'. *IEEE Transactions on Smart Grid*. 2011;2(2):302–13.

[17] Mullan J., Harries D., Bräunl T., and Whitely S. 'The technical, economic and commercial viability of the vehicle-to-grid concept'. *Energy Policy*. 2012;48:394–406.

[18] Andersson S.L., Elofsson A.K., Galus M.D., *et al*. 'Plug-in hybrid electric vehicles as regulating power providers: Case studies of Sweden and Germany'. *Energy Policy*. 2010;38(6):2751–62.

[19] Camus C., Esteves J., and Farias T.L. 'Electric vehicles and the electricity sector regulatory framework: The Portuguese example'. Proceedings of EVS24, Stavanger. 2009 May 13.

[20] Hartmann N., and Özdemir E.D. 'Impact of different utilization scenarios of electric vehicles on the German grid in 2030'. *Journal of Power Sources*. 2011;196(4):2311–8.

[21] Schuller A., and Rieger F. 'Assessing the economic potential of electric vehicles to provide ancillary services: The case of Germany'. *Zeitschriftfür Energiewirtschaft*. 2013;37(3):177–94.

[22] Han S., Han S., and Sezaki K. 'Development of an optimal vehicle-to-grid aggregator for frequency regulation'. *IEEE Transactions on Smart Grid.* 2010;1(1):65–72.

[23] Sortomme E., and El-Sharkawi M.A. 'Optimal charging strategies for unidirectional vehicle-to-grid'. *IEEE Transactions on Smart Grid.* 2011; 2(1):131–8.

[24] Ortega-Vazquez M.A., Bouffard F., and Silva V. 'Electric vehicle aggregator/system operator coordination for charging scheduling and services procurement'. *IEEE Transactions on Power Systems.* 2013;28(2):1806–15.

[25] Zhang Q., Li Y., Li C., and Li C. 'Grid frequency regulation strategy considering individual driving demand of electric vehicle'. *Electric Power Systems Research.* 2018;163:38–48.

[26] Shaaban M.F., Ismail M., El-Saadany E.F., and Zhuang W. 'Real-time PEV charging/discharging coordination in smart distribution systems'. *IEEE Transactions on Smart Grid.* 2014;5(4):1797–807.

[27] Yao L., Damiran Z., and Lim W.H. 'Optimal charging and discharging scheduling for electric vehicles in a parking station with photovoltaic system and energy storage system'. *Energies.* 2017;10(4):550.

[28] Moradijoz M., Moghaddam M.P., Haghifam M.R., and Alishahi E. 'A multi-objective optimization problem for allocating parking lots in a distribution network'. *International Journal of Electrical Power & Energy Systems.* 2013;46:115–22.

[29] Hajimiragha A., Canizares C.A., Fowler M.W., and Elkamel A. 'Optimal transition to plug-in hybrid electric vehicles in Ontario, Canada, considering the electricity-grid limitations'. *IEEE Transactions on Industrial Electronics.* 2010;57(2):690–701.

[30] Xu J., and Wong V.W. 'An approximate dynamic programming approach for coordinated charging control at vehicle-to-grid aggregator'. In Smart Grid Communications (SmartGridComm), 2011 IEEE International Conference on 2011 Oct 17 (pp. 279–284). IEEE.

[31] Jin C., Tang J., and Ghosh P. 'Optimizing electric vehicle charging with energy storage in the electricity market'. *IEEE Transactions on Smart Grid.* 2013;4(1):311–20.

[32] Zakariazadeh A., Jadid S., and Siano P. 'Multi-objective scheduling of electric vehicles in smart distribution system'. *Energy Conversion and Management.* 2014;79:43–53.

[33] Janjic A., and Velimirovic L.Z. 'Optimal scheduling of utility electric vehicle fleet offering ancillary services'. *ETRI Journal.* 2015;37(2):273–82.

[34] Janjic A., Velimirovic L., Stankovic M., and Petrusic A. 'Commercial electric vehicle fleet scheduling for secondary frequency control'. *Electric Power Systems Research.* 2017;147:31–41.

[35] Yang Z., Li K., and Foley A. 'Computational scheduling methods for integrating plug-in electric vehicles with power systems: A review'. *Renewable and Sustainable Energy Reviews.* 2015;51:396–416.

[36] Bielli M., Bielli A., and Rossi R. 'Trends in models and algorithms for fleet management'. *Procedia-Social and Behavioral Sciences*. 2011;20:4–18.

[37] Lam A.Y., Leung K.C., and Li V.O. 'Capacity estimation for vehicle-to-grid frequency regulation services with smart charging mechanism'. *IEEE Transactions on Smart Grid*. 2016;7(1):156–66.

[38] Hillier F.S. *Introduction to Operations Research*. New York: McGraw-Hill Education; 2015.

[39] Garcia-Valle R., and Vlachogiannis J.G. 'Letter to the editor: Electric vehicle demand model for load flow studies'. *Electric Power Components and Systems*. 2009;37(5):577–82.

[40] Li G., and Zhang X.P. 'Modeling of plug-in hybrid electric vehicle charging demand in probabilistic power flow calculations'. *IEEE Transactions on Smart Grid*. 2012;3(1):492–9.

[41] Janjic A. 'Two-step algorithm for the optimization of vehicle fleet in electricity distribution company'. *International Journal of Electrical Power & Energy Systems*. 2015;65:307–15.

[42] ENTSO, Network Code on Load Frequency Control and Reserves (LFCR), 2013.

[43] Liu H., Hu Z., Song Y., and Lin J. 'Decentralized vehicle-to-grid control for primary frequency regulation considering charging demands'. *IEEE Transactions on Power Systems*. 2013;28(3):3480–9.

[44] Han S., and Han S. 'Economic feasibility of V2G frequency regulation in consideration of battery wear'. *Energies*. 2013;6(2):748–65.

[45] Howell D. 'Electric drive vehicle battery R&D and manufacturing initiatives at the US Department of Energy'. In Proceedings of 9th International Advanced Automotive Battery & EC Capacitor Conference, Long Beach, CA, USA 2009 Jun (Vol. 1012).

[46] Garcia-Valle R., and Vlachogiannis J.G. 'Letter to the editor: Electric vehicle demand model for load flow studies'.

[47] Chukwu U.C., and Mahajan S.M. 'V2G parking lot with PV rooftop for capacity enhancement of a distribution system'. *IEEE Transactions on Sustainable Energy*. 2014;5(1):119–27.

[48] Baek S.J., Kim D., Oh S.J., and Jun J.A. 'Modeling of electric vehicle charging systems in communications enabled smart grids'. *IEICE Transactions on Information and Systems*. 2011;94(8):1708–11.

[49] Lam A.Y., Leung K.C., and Li V.O. 'Capacity estimation for vehicle-to-grid frequency regulation services with smart charging mechanism'. *IEEE Transactions on Smart Grid*. 2016;7(1):156–66.

[50] Jones A., Kaufmann A., and Zimmermann H.J., editors. *Fuzzy Sets Theory and Applications*. Springer Science & Business Media; 2012 Dec 6.

[51] Bellmann R.E., and Zadeh L.A. 'Decision-making in a fuzzy environment'. *Management Science*, 1970.

[52] Keeney R.L., and Raiffa H. *Decisions with Multiple Objectives: Preferences and Value Trade-Offs*. Cambridge University Press; 1993.

[53] Sarin R.K. *Multi-attribute Utility Theory*. In Encyclopedia of Operations Research and Management Science 2013 (pp. 1004–1006). Boston, MA: Springer.
[54] Mateo J.R. *Multi-attribute Utility Theory*. In Multi-criteria Analysis in the Renewable Energy Industry 2012 (pp. 63–72). London: Springer.
[55] Electric Reliability Council of Texas, Market Information, 2013.

Chapter 5

Power-demand management in a smart grid using electric vehicles

Khizir Mahmud[1], Mohammad Sohrab Hasan Nizami[2], Jayashri Ravishankar[1] and M.J. Hossain[2]

The proliferation of intermittent renewable-energy-based power systems and the emergence of new types of loads are likely to introduce new power-quality and power-demand-management challenges in a smart grid [1–5]. An additional level of complication gets added when the system deals with a mass penetration of uncontrolled mobile energy sources and loads, that is, electric vehicles (EVs), to the grid. However, the use of an advanced EV management technique can overcome the challenges through an intelligent bidirectional energy transfer process [6–10]. This chapter highlights various control and optimization techniques to manage the power demand of both single and multiple customers in a smart grid using EVs. The techniques cover both the energy resource and load-management approaches. Energy-resource management technique for single customer coordinates between EV, photovoltaics (PV) and battery storages based on the peak and off-peak load conditions, to minimize the peak load and electricity cost with an increased efficiency. Likewise, the energy-resource management for multiple customers, controls the aggregated PVs, battery storage and aggregated EVs in a parking lot to flatten the energy demand curve, and reduce the peak load energy costs. In this process, a controller reads the real-time household power consumption data through smart meter, PV power generation under real environment, the state-of-charge (SOC) for both EVs and battery storage, and the EV availability, to intelligently control the power flow from/to the energy sources to reduce the grid load demand [11–23]. Additionally, an advanced charge management technique for both aggregated EVs and single EV is developed. On the contrary, the load-management technique models a load-scheduling technique for a demand response (DR)-based home energy management systems (HEMS) that minimizes the electricity cost for the consumer and incorporates operational constraints for individual loads and energy sources. A Mixed-Integer Linear Programming (MILP)-based optimization model is formulated to determine the optimal scheduling of operation

[1]School of Electrical Engineering and Telecommunications, University of New South Wales, Australia
[2]School of Engineering, Macquarie University, Australia

for residential loads and DERs according to a day-ahead time-of-use (TOU) electricity tariff. Peak load constraint is also incorporated into the optimization model to address grid reliability issues such as demand peaks, rebound peaks and congestion in the grid. The findings in this chapter suggest that an intelligent management technique can substantially reduce the power demand of the grid using EVs and reduce the impact of intermittent sources and thus improve the load factor.

List of abbreviations

DER	Distributed-energy resources
DG	Distributed generation
DR	Demand-response
DSM	Demand side management
DSO	Distribution system operators
EV	Electric vehicle
HEMS	Home energy management system
HVAC	Heating and ventilation air-conditioning
MILP	Mixed-integer linear programming
MPC	Model predictive control
PV	Photovoltaics
SOC	State-of-charge
ToU	Time-of-use
V2G	Vehicle-to-grid

5.1 Introduction

Increased electrification of loads and the rapid proliferation of renewable-based distributed-generation (DG) and distributed-energy resources (DERs) into residential buildings impose a number of capacity challenges to low-voltage (LV) distribution system operators (DSO) [18,19]. The challenges include overloading of grid assets (e.g., transformer, cables, etc.), voltage violations and phase unbalances [18–20]. Moreover, new forms of load consumptions and emerging technologies are being introduced at the grid-edge level behind the electricity meter such as heat pumps, heating and ventilation air-conditioning (HVAC) systems, etc. This makes the power demand of the end user significantly uncertain and unpredictable [21,22]. As a result, a higher generation reserve and capacity margin have to be maintained, as the power generators have to follow the demand of the end users. Additionally, conventional customers, who have various energy-storage devices, can sell energy back to the grid. They can take extra power during off-peak hours to charge their energy storages and sell-back to the grid during peak hours. In this way, customers can make a profit, and the utility companies can minimize stress on the power grid during off-peak hours. However, due to a complex power flow, the customers who are participating in this bidirectional energy transfer, alternatively known as prosumers, are making the power systems more complex and challenging to manage [23,24].

Recent trends in the automobile industry have led to a proliferation of vehicular electrification, for example, electric vehicle (EV), and so vehicular green technology has gained momentum. Moreover, an advanced integration of the prosumer-like concept to the EV, for example, vehicle-to-grid (V2G) bidirectional energy transfer, has enhanced the power-storage capability of the customers [25,26]. However, the mobility of the EV makes it significantly different from any other type of load or source. Therefore, unregulated integration of EVs to the grid adds complexity to the domestic-side demand management [25,26]. Additionally, the increasing battery capacity and decreasing charging duration add an extra level of challenge to the system stability and power management [27]. An intelligent management of EVs can meet these challenges and provide various benefits such as power regulation (i.e., power-flow balance and frequency stability), power quality management (voltage regulation, voltage flicker and voltage-rise minimization) and load management (peak load shaving through controlled charging and discharging) [28,29]. Therefore, this chapter describes various power-demand management techniques in a smart grid using EVs.

A number of demand-side-management (DSM) and demand-response (DR) strategies have been the focus of serious interest in recent years to adopt energy-management methodologies for building energy-management systems [28,30,31]. DR influences the consumer to alter their demand profile in response to time-varying electricity tariffs or economic incentives [32,33]. The most common case of DR is the price-based DR, where consumers are supplied with a time-varying price (TVP) for electricity and they can obtain economic benefits by properly managing their loads and DER operation. This, in return, assists the utility companies and grid operator, as the TVP is designed in such a way as to fit the needs of the power grid [32,33]. However, an appropriate energy-management system with a proper scheduling model is required to determine the ideal schedule of operation for different types of consumer loads and DER units [1].

There have been several research studies to optimize the energy use in residential buildings that minimize the electricity cost of the consumer by participating in price-based DR programs. For example, a residential load-scheduling model is developed in [34] that minimizes the energy cost by considering the uncertainty in load demand. The authors in [35] incorporate the user's dissatisfaction into the scheduling model to minimize electricity bills. The authors in [36] demonstrate a DR-based energy-management system for residential buildings that manages the energy consumption in the buildings to minimize the electricity bill. A model-predictive control (MPC)-based load-scheduling model is proposed in [37–39] to provide cost savings for the consumer under a time-varying electricity-tariff structure. However, most residential load-scheduling models and home energy-management models found in the literature only focus on minimizing the energy cost of the consumer, and do not consider the potential effect of the large-scale adoption of such energy-management schemes. For example, if not properly coordinated, such a HEMS can create rebound peaks during lower-price periods [40]. In addition, the operational constraints of residential appliances and associated user preferences for operation of appliances should also be incorporated into such a HEMS model. All these domestic-power-demand management techniques for smart grids can be classified into two broad categories: energy-resource management and load management.

The contributions of this chapter are as follows:

- Development of an advanced algorithm to reduce the grid peak load demand coordinating PV, EV and battery storage under real load, weather and seasonal conditions. It also develops a relationship between the percentage of peak load minimization and the state-of-charge of EVs.
- Investigate the peak load minimization process considering the weather-dependent load and energy resources, and their dynamics.
- Development of a charging-discharging algorithm for both single and aggregated EVs, and investigate their impact to the peak-load-reduction process.
- Model the peak-load-reduction process using an artificial neural network and compare with the proposed approach.
- Development of a load-scheduling technique for a DR-based HEMS that minimizes the electricity cost for the consumer. It also incorporates the operational constraints for individual loads and energy sources.
- Formulate a MILP-based optimization model to determine the optimal scheduling of operation for residential loads and DERs according to a day-ahead TOU electricity tariff.

Considering all of the above aspects for smart-grid Power Demand Management using EVs, this chapter has been divided into three major sections. These three sections discuss the Power Demand Management for both a single customer and multiple customers as well as both energy-resource management and load management.

Section 5.2 discusses the energy-resource-management technique of a single customer. It starts with an overview of the energy-resource-management technique followed by power-demand-management algorithms. It then analyses the EV charge-management technique and concludes with a case study and results.

Section 5.3 begins with an overview of load management and the HEMS model. It then describes the load-scheduling model and analyses the optimization technique. Finally, it provides case studies and results.

Section 5.4 starts with an overview of energy-resource management for multiple customers. Then, it provides an overview of power-demand management for multiple customers. The section ends with an analysis of various case studies and results.

5.2 Power-demand management for single customer: Energy-resource-management technique

In this section, the energy management of a single customer based on the energy-resource-management technique is discussed. The system comprises three different energy resources along with the grid supply, that is, EVs, battery storage and photovoltaics (PVs). As EVs are mobile and PVs are highly dependent on sunlight, a stationary battery storage is used to mitigate their intermittency impact [6,7]. A schematic of the system is in Figure 5.1.

A detailed schematic of the energy-resource management of a single customer is shown in Figure 5.2 [6,7]. All the energy sources are connected to the domestic AC bus through power-electronics converters. An intermediate DC bus connects the DC sources (PV and battery storage) to this AC bus. A controller collects domestic power-demand data from the smart meter. It also collects the battery and EV state-of-charge (SOC), EV availability and PV power-generation data. Based on this data, the controller with a predefined algorithm controls the power flow from/to the AC bus and storages.

In Figure 5.2, four different converters including both bidirectional and unidirectional converters (converter- 1, 2, 3, 4) are used to connect the energy resources. Converter 1 is a bidirectional DC-AC/AC-DC converter (a V2G-enabled EV charger) that connects the EV to the AC bus. Converter 2 is a unidirectional DC-DC converter that connects the PV to the intermediate DC bus. Converter 3 is a bidirectional DC-DC converter, used to charge the battery either from the grid or the excess energy of PV. Converter 4 is a bidirectional DC-AC/AC-DC converter, used to connect the intermediate DC bus to the main AC bus. The controller

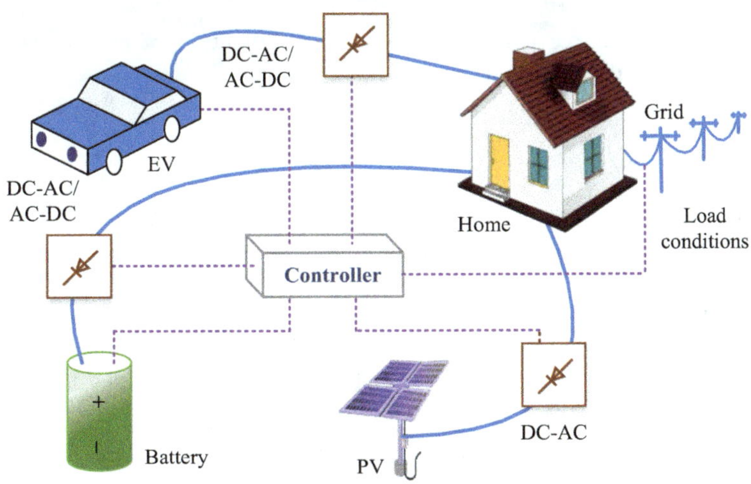

Figure 5.1 Schematic of the energy-resource management of a single customer [11,12]

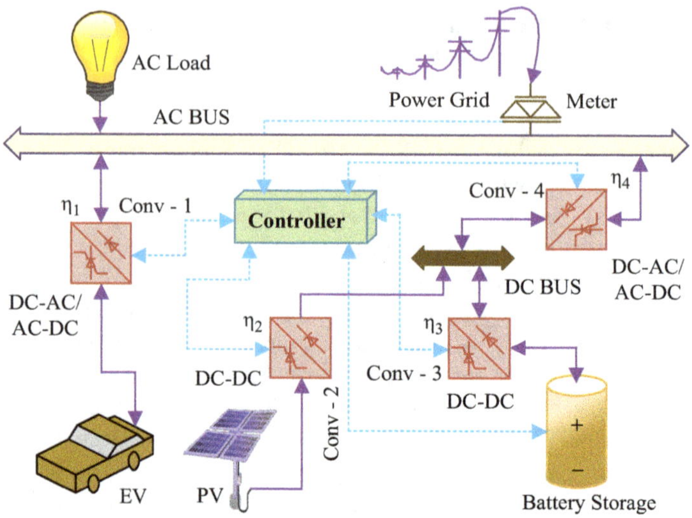

Figure 5.2 Detailed architecture of the energy-resource management of a single customer [11,12]

controls the switching of these converters based on the load conditions, SOC, and a few other constraints in the algorithms. Converter 2 turns on when the PV power generation is greater than zero, that is, $\psi^{i}_{pv} > 0$.

5.2.1 Load-management algorithm

The domestic load curve can be expressed as a function of customers' power demand at a particular time.

$$\mathcal{L}^{i}_{\hbar} = f\left(\mathcal{P}, t^{\ell}_{\hbar}\right) \tag{5.1}$$

The time $(t^{\ell}_{\hbar})$ in (5.1) is the duration of the load curve considered and the combination of both the peak $(t^{i}_{\mathcal{P}})$ and off-peak load (t^{i}_{ℓ}) periods as follows:

$$t^{\ell}_{\hbar} = \left(t^{i}_{p} \cup t^{i}_{b}\right) \tag{5.2}$$

$$t^{i}_{\mathcal{P}} = \mathcal{O}^{f}_{\mathcal{P}} * \left\{t^{i}_{\mathcal{P}} \quad for\, i = 0, 1, 2, \ldots, n\right\} \tag{5.3}$$

$$t^{i}_{\ell} = \mathcal{O}^{f}_{\ell} * \left\{t^{i}_{\ell} \quad for\, i = 0, 1, 2, \ldots, n\right\} \tag{5.4}$$

where $t^{0}_{\mathcal{P}}$ and t^{0}_{ℓ} are the peak load and off-peak load onset times, respectively. Likewise, $t^{n}_{\mathcal{P}}$ and t^{n}_{ℓ} are the peak load and off-peak load termination times, respectively. The frequency of the peak and off-peak hour occurrence is $\mathcal{O}^{f}_{\mathcal{P}}$ and

$\mathcal{O}^f_{\ell}$, respectively, for the duration of the load curve considered. The data acquisition rate of the controller is defined as follows:

$$\left(t^1_{\mathcal{P}} - t^0_{\mathcal{P}}\right), \left(t^1_{\ell} - t^0_{\ell}\right) \tag{5.5}$$

Let us assume that the desired load demand curve, a function of customers' power demand and time, is

$$\partial^t_d = f\left(\mathcal{P}_{\partial}, t^{\ell}_{p}\right) \tag{5.6}$$

The load curve is identified as a peak load when $(\mathcal{L}^i_p > \partial^t_d)$, and as off-peak load when $(\mathcal{L}^i_p < \partial^t_d)$. If the system comprises any renewables such as PVs, the desired value (∂^t_d) of the load curve can be set above the capacity of the renewables (i.e., PVs). By utilizing the available energy resources such as EVs, PVs and battery storage, the target of the system will be to achieve the desired load curve (∂^t_d). Let us denote the currents from the battery and EV to be, $i^{t_1}_b, i^{t_2}_b, i^{t_3}_b, \ldots\ldots, i^{t_n}_b$ and $i^{t_1}_{ev}, i^{t_2}_{ev}, i^{t_3}_{ev}, \ldots\ldots, i^{t_n}_{ev}$, respectively. The charging and discharging of the battery and EVs are identified from the sign of the current, for example, the battery and EV charges when $i^t_b, i^t_{ev} < 0$, and battery and EV discharges when $i^t_b, i^t_{ev} > 0$.

Let us denote the instantaneous load at a particular time to be, $\mathcal{L}^{i_1}_p$. So, the peak load, that is, the amount of power needed to provide from PV, battery and EV to shave the peak is

$$\mathcal{L}^s_p = \mathcal{L}^{i_1}_p - \partial^t_d, \quad \text{for } \mathcal{L}^i_p > \partial^t_d, \tag{5.7}$$

Likewise, in case of $\mathcal{L}^i_p < \partial^t_d$, the available power to charge the battery and EV is

$$\mathcal{L}^c_p = \partial^t_d - \mathcal{L}^{i_1}_p, \quad \text{for } \mathcal{L}^i_p < \partial^t_d \tag{5.8}$$

If the minimum discharging limit of the battery and EV is ψ^l_b and ψ^l_{ev}, respectively, and the charge of the battery and EV at a particular time is ψ^i_b and ψ^i_{ev} in percentages, respectively, the available power that the battery and EV can provide is given as

$$\psi^{av}_b = \kappa^c_b\left(\psi^i_b - \psi^l_b\right) \tag{5.9}$$

$$\psi^{av}_{ev} = \kappa^c_{ev}\left(\psi^i_{ev} - \psi^l_{ev}\right) \tag{5.10}$$

where κ^c_b represents the capacity of the battery and κ^c_{ev} represents the capacity of the EV. For the case of $\mathcal{L}^i_p > \partial^t_d$, the power required to be shaved $\left(\mathcal{L}^s_p\right)$ by the battery, EV and PV, as follows:

$$\mathcal{L}^s_p = \left\{\psi^{av}_{ev}\left(1 - e^{-\psi^{i=t_1,t_1+1,\ldots\ldots\ldots,t_2}_{ev}}\right)\right\} + \left\{\psi^{av}_b\left(1 - e^{-\psi^{i=t_1,t_1+1,\ldots\ldots\ldots,t_2}_b}\right)\right\} + \psi^i_{pv} \tag{5.11}$$

where, ψ^i_{pv} is the PV power generation at a particular time. $\psi^{i=t_1,t_1+1,\ldots\ldots\ldots,t_2}_{ev}$ and $\psi^{i=t_1,t_1+1,\ldots\ldots\ldots,t_2}_b$ are the SOC of the EV and battery at a particular time t. On the other hand, for $\mathcal{L}_{\mathscr{p}}^{i} < \partial_{d}^{t}$, the available power for charging the battery and EV is expressed as

$$\mathcal{L}_p^c = \left\{ \psi_{ev}^r \left(e^{\frac{1}{-\psi_{ev}^{i=t_1,t_1+1,\ldots\ldots,t_2}}} \right) \right\} + \left\{ \psi_b^r \left(e^{\frac{1}{-\psi_b^{i=t_1,t_1+1,\ldots\ldots,t_2}}} \right) \right\} \tag{5.12}$$

$$\psi_{ev}^r = \kappa_{ev}^c \left(\psi_{ev}^m - \psi_{ev}^i \right) \tag{5.13}$$

$$\psi_b^r = \kappa_b^c \left(\psi_b^m - \psi_b^i \right) \tag{5.14}$$

Here, ψ_b^m and ψ_{ev}^m are the maximum charging limits of the battery and EV, respectively, as shown in Figure 5.3.

The generated PV power is used to support the load and a portion is bypassed to the battery for charging. All the PV power is supplied to the AC bus and can be written as

$$\psi_{pv}^i = n.\wp_{pv}^g \cdot \eta_2 \quad \text{for } \mathcal{L}_{\mathscr{p}}^{i} = \psi_{pv}^i \tag{5.15}$$

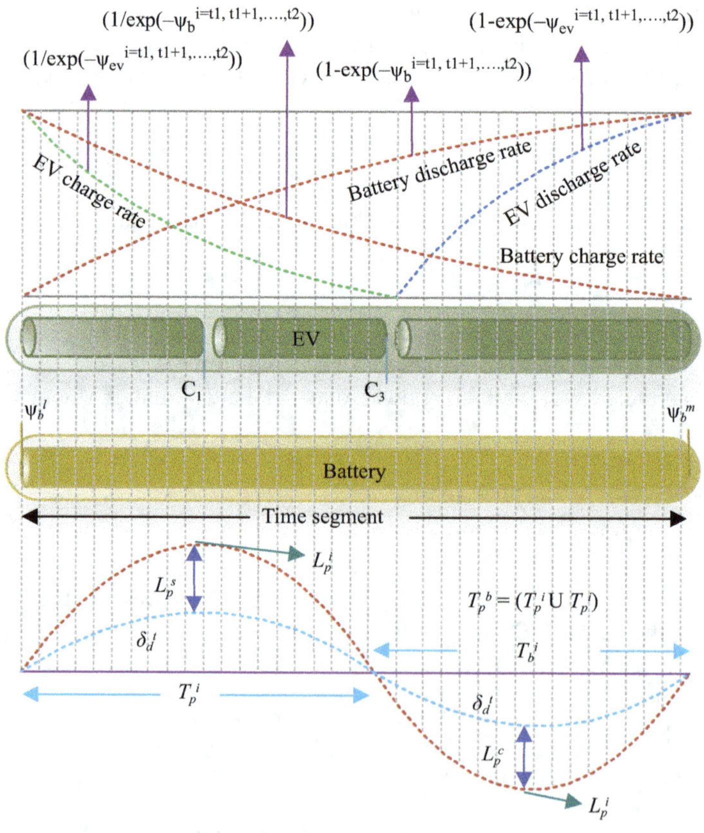

Figure 5.3 EV battery charging-discharging algorithm [11,12]

Here, $\wp^{g}_{pv}$ is the PV power generated from a single PV module, n is the number of modules and η_2 is the efficiency of converter 2. When $\mathcal{L}^{i}_{p} > \psi^{i}_{pv}$, the excess energy to reduce the peak is supplied by the battery and EV. If $\mathcal{L}^{i}_{p} < \psi^{i}_{pv}$, the excess energy of PV power generation is bypassed to the battery through converter 3 and is given as

$$\xi^{pv}_{ex} = \begin{cases} 0 & when\ \mathcal{L}^{i}_{p} \geq \psi^{i}_{pv}, i \in t^{b}_{p} \\ \psi^{i}_{pv} - \mathcal{L}^{i}_{p} & when\ \mathcal{L}^{i}_{p} < \psi^{i}_{pv}, i \in t^{b}_{p} \end{cases} \tag{5.16}$$

For ToU tariff system, the cost of electricity for both peak and off-peak load hours is

$$\left\{\Lambda^{p}_{e} * \left(f(\mathcal{P}, t^{i}_{p})\right)\right\} + \left\{\Lambda^{b}_{e} * f(\mathcal{P}, t^{i}_{b})\right\} \tag{5.17}$$

Here, Λ^{p}_{e} is the electricity cost during peak load hours and Λ^{b}_{e} is the electricity cost during base load hours. If the system does not have a provision to penetrate power beyond the smart meter, that is, the house does not provide power to the grid, the summation of the PV, battery and EV power must not exceed the energy demand. Therefore, the system will have the following constraints:

$$\mathcal{L}^{i}_{p} - \partial^{t}_{d} \geq 0 \quad for\ t^{b}_{p} \tag{5.18}$$

For the battery and EV charge limit, the system constrains are as follows:

$$\psi^{l}_{b} < \psi^{i}_{b} < \psi^{m}_{b} \tag{5.19}$$

$$\psi^{l}_{ev} < \psi^{i}_{ev} < \psi^{m}_{ev} \tag{5.20}$$

Both the stationary and the EV battery have a charge loss, which can be found from the Peukert's law [6,7,41].

5.2.2 EV charge management

The considered EV has a V2G capability, with a battery capacity of 24 kWh, equivalent to about 140 km (approximately, 4.88-6.18 km/kWh) [6,8,42]. The owner of the EV may follow either the regular or random trip plans. Some EVs charge their battery only at home, and some EVs have a charging facility at work. In this EV charge-management algorithm, we have shown various charging constraints and levels in case an EV does not have a charging facility at work and sometimes needs to charge rapidly in case of low SOC. The EV battery charge-management decision under various constraints is shown in Figure 5.4.

The EV battery capacity is divided into five different segments: C_0, C_1, C_2, C_3, C_4. The segment C_0 is the emergency battery-capacity reserve and the SOC must not drop beyond this limit. If the battery SOC comes to this stage and the owner has an emergency trip plan, the battery will charge rapidly regardless of the load conditions. If the SOC remains at this stage, the battery will charge based on the load conditions. The segment C_3 is used for load-support through V2G

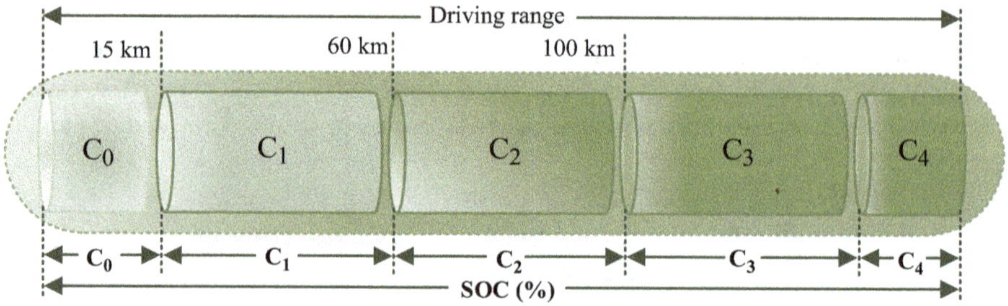

Figure 5.4 EV battery SOC management [11,12]

operation. As over-charging of the lithium-ion battery is harmful for its lifetime, the segment C_4 represents the maximum charging limit. The maximum charging-discharging limit must follow:

$$C_0 \geq emergency\ reserve \tag{5.21}$$

$$C_4 \leq 100\%\,SOC \tag{5.22}$$

These segments can be defined based on the SOC and the driving range.

SOC based: The EV owner can define their EV battery emergency reserve and the amount of energy transactions based on the percentage of SOC. For example, the owner can define the system to always preserve 10% of the SOC for any emergency and use any SOC range more than 80% in any V2G operation.

Driving-range based: The driving-range equivalent to the percentage of SOC can also be used to define the segments of EV capacity. For example, the owner can define the system to always preserve 15 km of driving range for any emergency and use any driving range more than 100 km in any V2G operation.

In various stages of the EV SOC based on the load conditions, EV charging-discharging will be controlled based on the following:

$$\textbf{Scenario A}\left(\psi^i_{ev} < C_1\, for\ load\ conditions\ \mathcal{L}^i_p > \partial^t_d \quad and \quad \mathcal{L}^i_p < \partial^t_d\right)$$

In this scenario, EV SOC is at the C_0 stage and less than the acceptable limit. Therefore, rapid charging is preferable in this condition. The EV charging rate $(\phi(t))$ is

$$\phi(t) = \phi(t)_{max} for\ \psi^i_{ev} < C_1 \tag{5.23}$$

For the load condition $\mathcal{L}^i_p < \partial^t_d$, the grid will supply $\phi(t)_{max}$ from $\mathcal{L}^c_p$. However, for the load conditions $\mathcal{L}^i_p > \partial^t_d$, the required power $(\phi(t)_{max})$ will be supplied by the battery and grid.

$$\textbf{Scenario B}\left(C_0 < \psi^i_{ev} < C_2\, for\ load\ conditions\ \mathcal{L}^i_p > \partial^t_d\ \ and\ \ \mathcal{L}^i_p < \partial^t_d\right)$$

Although EV SOC is greater than C_0, rapid charging is preferable if the owner has a next-day long-trip plan. The charging rate is given as

$$\phi(t) = \phi(t)_{max} \quad for\ C_0 < \psi^i_{ev} < C_2, \mathcal{L}^i_p > \partial^t_d \quad and \quad \mathcal{L}^i_p < \partial^t_d \tag{5.24}$$

If the owner does not have a next-day long-trip plan, the EV can charge flexibly based on the load conditions:

$$\mathcal{L}^c_p = \underbrace{\left\{\psi^r_{ev}\left(\frac{1}{e^{-\psi_{ev}^{i=t_1,t_1+1......,t_2}}}\right)\right\}}_{\phi(t)} + \underbrace{\left\{\psi^r_b\left(\frac{1}{e^{-\psi_b^{i=t_1,t_1+1......,t_2}}}\right)\right\}}_{batterychargerate} \tag{5.25}$$
$$for\ C_0 < \psi^i_{ev} < C_2, \mathcal{L}^i_p < \partial^t_d$$

Scenario C $\left(C_1 < \psi^i_{ev} < C_3\, for\ load\ conditions\ \mathcal{L}^i_p > \partial^t_d \ and \ \mathcal{L}^i_p < \partial^t_d\right)$

In this scenario, flexible charging is preferable and the charging rate during the $\mathcal{L}^i_p < \partial^t_d$ condition is given as

$$\phi(t) = \underbrace{\left\{\psi^r_{ev}\left(\frac{1}{e^{-\psi_{ev}^{i=t_1,t_1+1,.........,t_2}}}\right)\right\}}_{for\ C_1 < \psi^i_{ev} < C_3,\ \mathcal{L}^i_p < \partial^t_d} = \left\{\mathcal{L}^c_p - \psi^r_b\left(\frac{1}{e^{-\psi_b^{i=t_1,t_{1+1},.........t_2}}}\right)\right\} \tag{5.26}$$

Scenario D $\left(C_2 < \psi^i_{ev} < C_4\, for\ load\ conditions\ \mathcal{L}^i_p > \partial^t_d \ and \ \mathcal{L}^i_p < \partial^t_d\right)$

In this scenario, the battery will discharge if the peak load conditions occur. The EV discharge rate ($\Theta(t)$) is

$$\begin{aligned}\Theta(t) &= \left\{\psi^{av}_{ev}\left(1 - e^{-\psi_{ev}^{i=t_1,t_1+1,.........,t_2}}\right)\right\} \\ &= \left\{\mathcal{L}^s_p - \left\{\psi^{av}_b\left(1 - e^{-\psi_b^{i=t_1,t_{1+1},.........t_2}}\right)\right\} - \psi^i_{pv}\right\}\ for\ \mathcal{L}^i_p > \partial^t_d\end{aligned} \tag{5.27}$$

If $\mathcal{L}^i_p < \partial^t_d$, the EV will charge up to C_4, *where* $\psi^i_{ev} = C_4$ *and* $C_4 \leq 100\%$ *of SOC*. In this analysis, the charging rate $\phi(t)_{max}$= 3 A current.

5.2.3 *Case studies*

Two case studies are carried out to manage the domestic power demand of a single customer in a smart grid. A detailed management algorithm is explained in Sections 5.2.3.1 and 5.3.2.2. The considered EV in both case studies has a battery capacity of 24 kWh.

5.2.3.1 Case study 1

In this case study, the EV and PV are used to provide load support. For the EV, it is assumed that the EV leaves home at 7 AM and arrives home at 5 PM, and no charging facility is available at work. It is also assumed that the daily average travel distance of the EV is at most 35 km. The parameters considered in this case study are listed in Table 5.1.

When an EV is connected to the system and participates in a V2G operation, a substantial peak-load reduction occurs. Figure 5.5 shows the power-demand management when an EV is connected. Since the EV does not stay home all the time, it can only participate when it is at home and connected to the system. Additionally, if the trip duration of the EV increases, SOC may drop below the V2G allowance limit and thus it may not discharge the energy.

5.2.3.2 Case study 2

Since the EV can participate in the power demand depending on its availability, the addition of a small capacity of fixed battery storage can improve the performance.

Table 5.1 EV and PV specification used in case study 1

Parameter	Value
Minimum SOC level of EV (C_0)	20%
EV SOC level (C_2)	40%
EV SOC level to start V2G (C_3)	85%
Maximum SOC level of EV (C_4)	99%
Efficiency of converter 1, 2, 3, 4 ($\eta_1 = \eta_2 = \eta_3 = \eta_4$)	92%
Maximum charging rate of EV	$\phi(t)_{max}$= 3.5 kW
PV module power supply (ψ^i_{pv})	2.5 kW
Area of PV module	20.6 m^2
Tilt angle of PV module	45^0
Efficiency of PV	13.5%
Annual degradation of PV module	1%

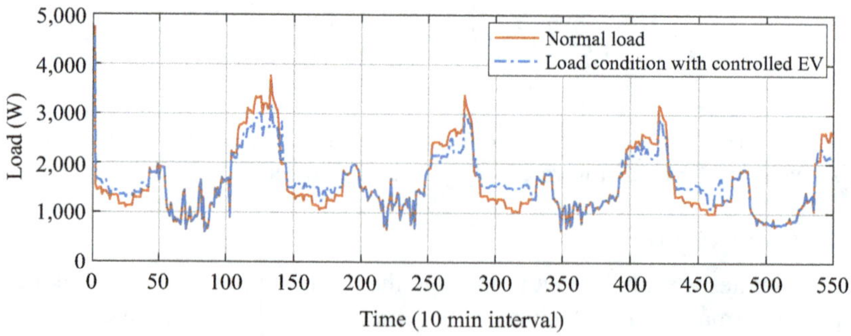

Figure 5.5 Power-demand management of a single customer using an EV and PV

In this next case study, a battery storage is attached along with the EV and PV to investigate the performance. The parameters considered in this case study is listed in Table 5.2.

The power-demand management with and without the controlled PV, EV and battery is shown in Figure 5.6. A small fixed battery improves the overall performance as it can charge using excess PV power generation during off-peak hours and discharge during peak hours.

From the case studies 1 and 2, it is clear that a strategic use of battery storage and EVs can reduce the peak power demand. In a research [8,43], the extra energy from the peak load management is shared with the neighbors to get extra advantages.

5.2.4 *Comparison with an artificial neural network technique*

The performance of the proposed energy-resource-management approach is compared with that of an artificial neural network (ANN). ANN is known to perform better for nonlinear control systems [11,12,44]. In this ANN, a feed forward neural network is used. The ANN is trained using a data set of 5×2100 matrix, which contains the PV power generation, household load demand, EV availability, EV

Table 5.2 Battery specification used in case study 2

Parameter	**Value**
Lithium-ion battery life cycles	>10k
Lithium-ion battery power density	~ 1.8 kW/kg
Lithium-ion battery self-discharge (per day)	0.1%–0.3%
Lithium-ion battery charging-discharging efficiency	98%
Initial SOC of the battery	40%

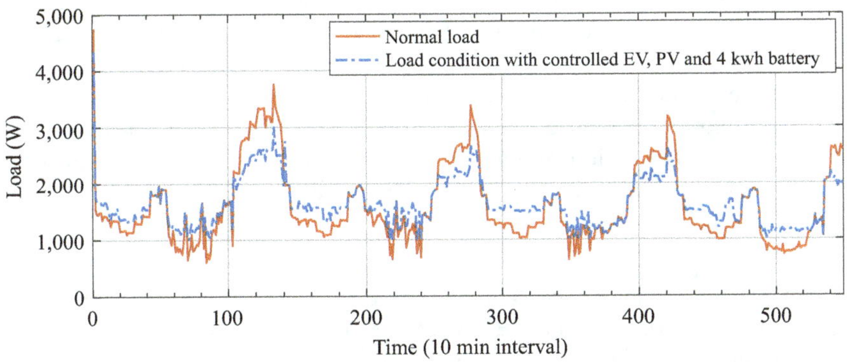

Figure 5.6 Power-demand management of a single customer using an EV, PV and battery

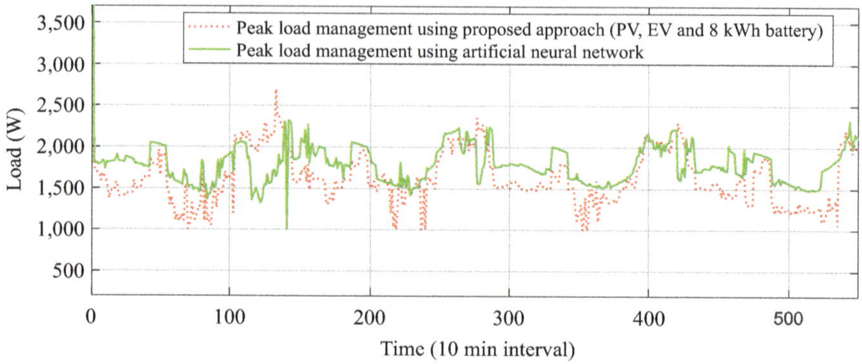

Figure 5.7 Comparison of the proposed method with an artificial neural network–based approach

trip plan and its SOC, and battery SOC. Another dataset of a matrix 1×2100 is used to train the network as a target. To validate the method, various post-processing tests (mean-squared error, confusion matrix, cross-entropy error and error histogram) are done. The cross-entropy error in training, validation and testing state were 1.70%, 2.45% and 2.85%, respectively. The result of the ANN-based technique is shown in Figure 5.7. From the figure, it is clear that the proposed method and the ANN-based approach shows similar performance.

5.3 Power-demand management for single customer: Load-management technique

The main focus of this section is to discuss a load-scheduling model for a DR-based HEMS that minimizes the electricity cost for the consumer and incorporates operational constraints for individual loads and DERs, user preference and dissatisfaction into the scheduling model. A MILP-based optimization model is formulated to determine the optimal scheduling of operation for residential loads and DERs according to a day-ahead TOU electricity tariff. Peak load constraint is also incorporated into the optimization model to address grid reliability issues such as demand peaks, rebound peaks and congestion in the grid [2–5,9,10,13–17,48–55].

5.3.1 HEMS model

Figure 5.8 depicts the conceptual system overview for the proposed HEMS model. It is considered that the HEMS receives DR signals, in terms of day-ahead TOU tariff and demand flexibility request, from the regional utility company or via

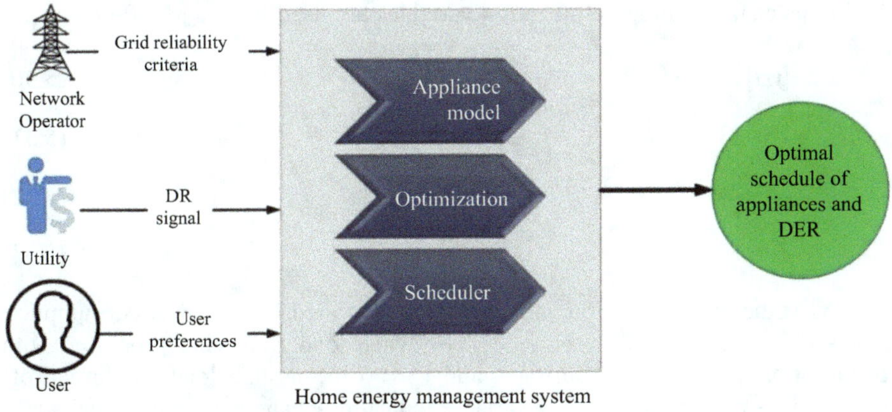

Figure 5.8 System overview of the HEMS [1]

third-party entities, such as an Aggregator. Grid reliability criteria are also fed to the HEMS to address grid reliability situations such as demand peaks, rebound peaks and overloading of grid assets (e.g., transformer, cable, etc.). HEMS also takes user preferences into consideration, such as optimal comfort settings and operation of residential appliances or DERs. Based on this information, a mathematical model of residential appliances and DERs along with a MILP-based optimization model are developed to minimize the electricity bill for the consumer by determining the optimal schedule of operation for appliances and DERs in the building.

5.3.2 Scheduling model

The scheduling model for the DR-based HEMS is formulated as a MILP problem. The primary objective of the scheduling model is to minimize the electricity bill for the consumer while maintaining an optimal comfort level and satisfying operational constraints and user preferences in terms of appliance operation. Sections 5.3.2.1–5.3.2.3 discuss the optimization model in detail.

5.3.2.1 Optimization variables

A MILP optimization problem can be formulated as

$$\min_x C^T F_0(x) \tag{5.28}$$

$$st{:}F_1(x) = 0, \quad F_2(x) \leq 0 \tag{5.29}$$

where $F_0(x)$, $F_1(x)$ and $F_2(x)$ represent the objective function, equality and inequality constraints function, respectively, whereas x is the optimization variable and C is a constant.

In this chapter, three optimization variables are used for the MILP model:

$$x = [yzt]^T \tag{5.30}$$

$$y_{ij}^k \in \{0, 1\} \tag{5.31}$$

$$z_{ij} \in \{0, 1\} \tag{5.32}$$

$$t_i \in \mathbb{Z} \tag{5.33}$$

where y_{ij}^k is the binary decision variable assigned to load j with power consumption level k during period i. Therefore, $y_{ij}^k = 1$ when load j is being processed with power-consumption level k during i, and $y_{ij}^k = 0$ means that load j is turned off during i. z_{ij} is an auxiliary binary variable indicating changes in operational states for load j during period i. If it turned on from off states then $z_{ij} = 1$ and $z_{ij} = 0$ otherwise. On the other hand, the auxiliary variable t_i indicates the indoor temperature during period i and it can have positive integer values only.

5.3.2.2 Objective function

The total execution length of 24 h is taken into account for the day-ahead scheduling model and divided into m time periods of ΔT hours. Sets M, N and L are used to indicate the set of time periods, residential devices (including both loads and DERs) and power consumption level of residential devices, respectively. The scheduling model minimizes the electricity bill of the consumer according to the day-ahead TOU tariff (λ_i) while satisfying user preferences and maintaining an optimal comfort level. The objective function for the scheduling model is defined as

$$\min_{y,z,t} \sum_{i \in M} \sum_{j \in N} \sum_{k \in L} \left[P_{ij}^k\right]^T y_{ij}^k \Delta T \lambda_i + \rho \sum_{i \in M} \sum_{j \in N} z_{ij} + \omega \sum_{i \in M} \left[t_i - T_i^d\right] \tag{5.34}$$

where P_{ij}^k represents the power-demand matrix of residential loads and DERs, whereas ρ and ω indicate the penalty cost parameter and dissatisfaction cost parameter, respectively. The first part of the objective function represents the electricity-bill component while the second and third parts indicate the penalty cost and dissatisfaction cost components. The penalty cost component ensures uninterrupted operation of devices where interruption in operation can result in economic loss. For example, devices such as a washing machine, clothes dryer and dishwasher should not be interrupted once they are turned on, as these appliances have to start over from the beginning if their operation is interrupted. From the operating cycles indicated in Figure 5.9, it is evident that these appliances consume high energy in their first few cycles of operation. Therefore, if their operation is interrupted after a few cycles, financial loss may be incurred to restart these appliances from the beginning. The discomfort of a user in terms of

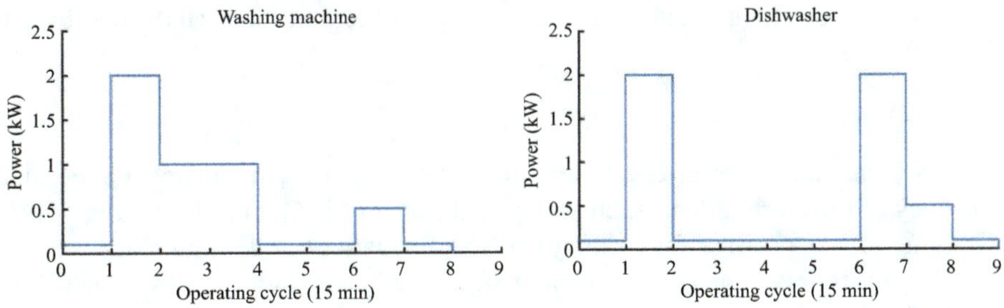

Figure 5.9 Operating cycles of washing machine and dishwasher [1]

indoor temperature is represented in the dissatisfaction cost component. It ensures that the indoor temperature stays close to the desired indoor temperature according to user preference (T_i^d) to minimize user dissatisfaction in terms of thermal comfort level.

5.3.2.3 Constraints

This section discusses the constraints of the optimization model. The constraints related to optimization variables are represented as follows:

$$\sum_{k \in L} y_{ij}^k \leq 1 \tag{5.35}$$

$$\sum_{k \in L} y_{ij}^k \leq 1 \tag{5.36}$$

$$y_{i+1,j}^k - y_{ij}^k - z_{ij} \leq 0 \, \forall \, i,j \tag{5.37}$$

Equation (5.34) ensures that any residential device (loads or DERs) can only operate at one power-consumption level. On the other hand, (5.35) indicates that the value of optimization variable z_{ij} should only be 1 if residential device j changes its operating state from 'off' to 'on' during period i. The maximum and minimum thresholds for the power-demand variable (P_{ij}) and indoor temperature are represented as follows:

$$P_j^{min} \leq P_{ij} \leq P_j^{max} \tag{5.38}$$

$$P_{min}^d \leq t_i \leq T_{max}^d \tag{5.39}$$

Each residential device needs a certain energy to complete its operation and this is represented as

$$\sum_{i \in M} P_{ij}^k \Delta T = E_j^k \, \forall j, k \tag{5.40}$$

Changes in indoor temperature due to operation of thermal loads (indicated by set N_{th}) are written as

$$\Delta t_i = \alpha\left(T_i^{out} - t_{i-1}\right) + \beta P_{ij} \, \forall j \in N_{th} \tag{5.41}$$

where α is the heat-transfer coefficient of the building and β indicates the thermal efficiency of the thermal load such as air conditioner (Air-Con) or heat pump (HP). The value of β is positive for heating loads and negative for cooling loads.

The constraints related to residential storage devices (indicated by set N_{st}) are formulated as follows:

$$SoC_j^{min} \leq SoC_{ij} \leq SoC_j^{max} \, \forall j \in N_{st} \tag{5.42}$$

$$\Delta SoC_{ij} = \frac{P_{ij}^k \Delta T}{E_j^{cap}} \tag{5.43}$$

where SoC_{ij} indicates the SOC of storage device j during i, whereas E_j^{cap} represents the maximum storage capacity of storage j.

Power consumption from the grid or injection into the grid should follow the maximum and minimum threshold according to grid reliability requests to satisfy grid constraints, that is

$$P_{grid,i}^{min} \leq \sum_{j \in N} \sum_{k \in L} P_{ij}^k \leq P_{grid,i}^{max} \tag{5.44}$$

5.3.3 Case study

A case study has been conducted with typical energy data for a typical Australian house to evaluate the effectiveness of the proposed scheduling model. Sections 5.3.3.1–5.3.3.3 discuss the description of the case study and the numerical results of the case study.

5.3.3.1 Description of the case study

Noncontrollable loads of the building are considered as base loads and the base-load profile used for the case study is indicated in Figure 5.10. Table 5.3 lists the input parameters used in the case study in terms of ratings, user preferences and operational constraints for controllable residential loads, DG and DERs.

5.3.3.2 Simulation setup

Based on the input parameters discussed above, the proposed optimization model is simulated for a typical summer day. As simulation time interval of 15 min is used. The optimization model is developed and solved in MATLAB®. An arbitrary value is assumed to indicate the grid-reliability request, as quantifying grid constraints to address grid reliability is out of the scope of this paper. The objective function defined in (5.34) is optimized in MATLAB subject to the constraints defined in (5.35)–(5.43).

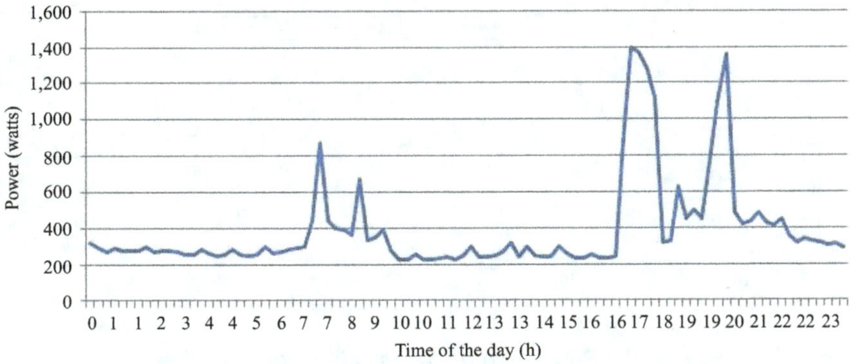

Figure 5.10 Base-load profile [1]

Table 5.3 Input parameters for case study

Device name	**Type**	**Nominal ratings**	**User preferences and operational constraints**
Air conditioner (Air-Con)	Thermal load	1.5 kW	• Optimal temperature range: 18–25°C • Not required during 10 AM to 12 PM and 2 PM to 3 PM
Washing machine (WM)	Time-shiftable load	2 kW (max)	• Must run once a day • Should follow the operating cycle
Electric vehicle (EV)	Storage device	20 kWh	• At least 70% charged at 10 AM • Away during 10 AM to 5 PM
Solar PV	Distributed generation	2.5 kW_p	• Maximum utilization

5.3.3.3 Results and discussion

According to the simulation results, a total of 21 kWh of energy is consumed by the building for the duration of 24 h, and the power profile of the building for the simulation window is depicted in Figure 5.11. Here, a negative value of power indicates that the building is injecting power back into the grid. As the base load is considered as noncontrollable, the HEMS does not schedule the base-load profile. However, the controllable devices are scheduled according to the input parameters listed in Table 5.3.

The operational schedule for controllable devices, according to the proposed scheduling model, is represented in Figures 5.12–5.14. From the WM schedule in Figure 5.12, it can be seen that the WM is scheduled to operate when the TOU rate of electricity is lower, to minimize the electricity bill. It also satisfies the user preferences and operational constraints of the WM while determining the optimal schedule for WM.

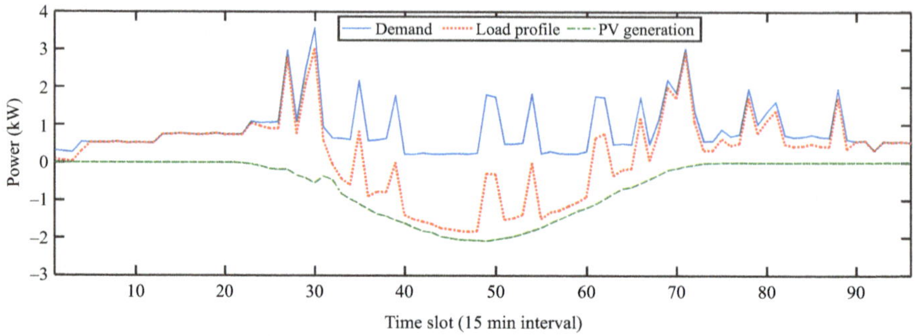

Figure 5.11 Power profile for the simulation window [1]

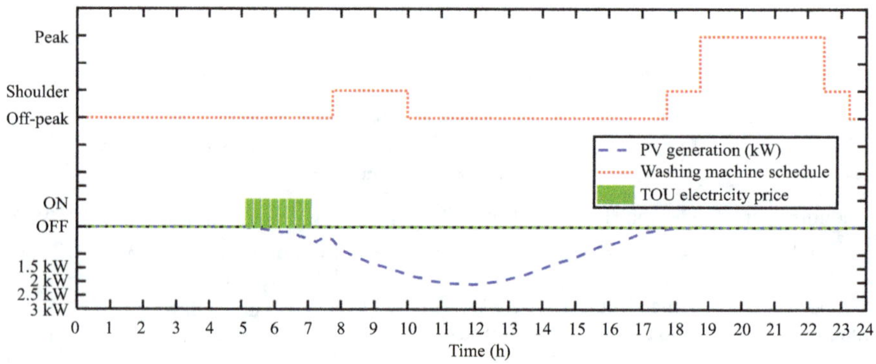

Figure 5.12 WM schedule [1]

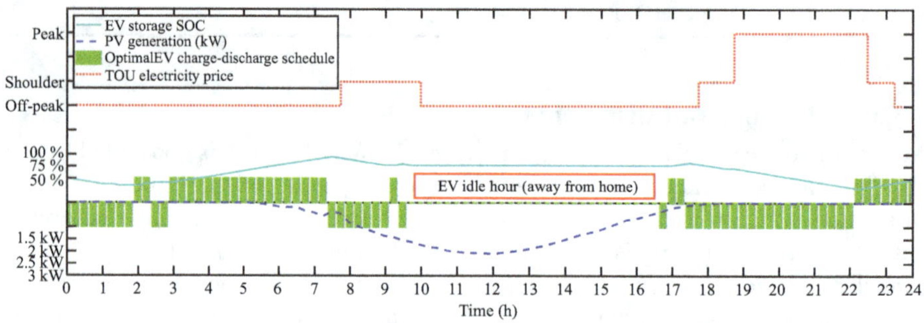

Figure 5.13 EV charging-discharging schedule with SOC of EV battery [1]

Figure 5.13 depicts the optimal schedule of EV battery charging and discharging and the associated SOC levels. EV idle hours are also indicated in the figure. It can be seen that the EV charging-discharging is scheduled in such a way that the SOC level is at least 70% at 10 AM to satisfy the user preference. It can also be seen

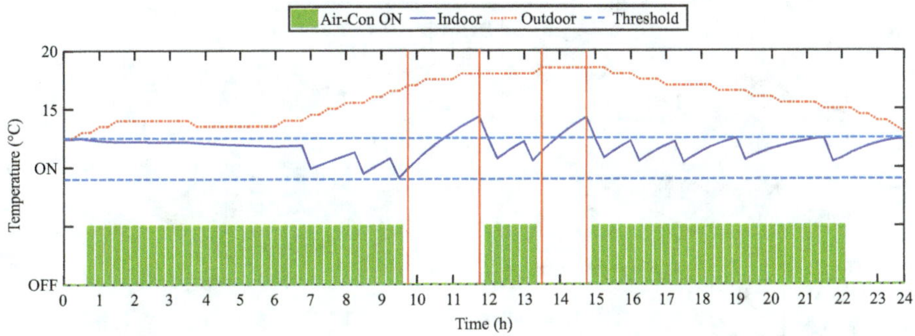

Figure 5.14 Air-Con schedule with indoor temperature profile [1]

that the EV does not charge during higher price periods; rather, it supplies a portion of the building load to reduce the electricity bill for the consumer.

From the optimal schedule of the Air-Con in Figure 5.14, it can be seen that the Air-Con is scheduled in such a way that the indoor temperature always stays within the desired levels according to user preference. From the temperature profile, it can also be identified that there are a few instances when Air-Con turns off the compressor and only runs the internal fan to reduce the electricity bill. Even though the room temperature increases a little in such periods, it stays within the desired levels as the Air-Con turns its compressor back on whenever the room temperature tends to violate the desired threshold level.

The power profile of the building changes from the optimal case when the grid reliability criteria are also incorporated as an optimization constraint, as seen in Figure 5.14. From the case study, it is found that the electricity bill can be reduced by up to 18% if the grid reliability constraint is not considered, and electricity-bill reduction reduces slightly from the optimal case if the grid reliability constraint is satisfied. Therefore, proper incentives should be provided to encourage consumers to maintain such grid reliability criteria. Nevertheless, the proposed load-scheduling model is found to be quite effective in terms of electricity-bill reduction even with grid-reliability constraints.

5.4 Power-demand management for multiple customers

5.4.1 General overview

In this section, energy-resource management of multiple customers, for example, building and commercial complex, is discussed. The power-demand management for multiple customers on a large scale is significantly different from that of a single customer, on a small scale. A schematic overview of the power-demand management for multiple customers is shown in Figure 5.15.

For a single customer, a single EV is considered. Likewise, the battery and PV capacity are also small. However, the power demand for a large-scale group, for

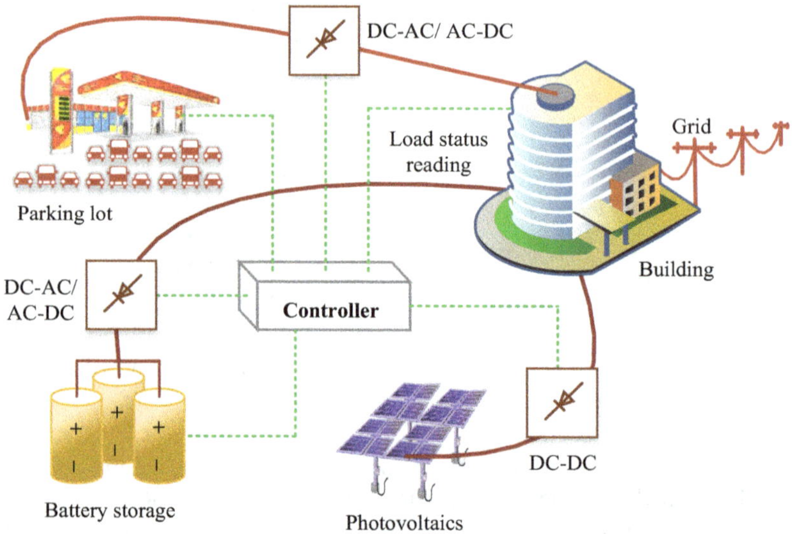

Figure 5.15 Schematic of the energy-resource management of multiple customers [11,12]

example, multiple customers in a building or a commercial customer, is significantly higher in capacity and different in shape from that of a single customer [45,46]. Additionally, EVs are considered on a large scale, that is, aggregated in a parking lot. Therefore, an advanced EV charge-management algorithm is essential to manage the EVs in an aggregated location. Since, the power demand is higher, several PV cells are also integrated in an aggregated way [47]. A more detailed architecture of the power-demand management for multiple customers in a building is shown in Figure 5.16.

5.4.2 Aggregated EV and power-demand management algorithm

In this section, large-scale power-demand management and aggregated EV management in a parking lot are discussed. It is considered that the controller reads the decision-making parameters (EVs' availability at parking lot, SOC of battery and EV, load conditions, etc.) on a sub-second time scale. The charging and discharging of both EV and battery storage are dependent on the peak and off-peak load conditions. Let us assume that the load curve as a function of power demand and time, at the common bus of a building where aggregated EVs, battery storage and PVs are connected as $\mathcal{L}^{i}_{\hbar} = f(\mathcal{P}, t^{\ell}_{\hbar})$, as described in (5.1). The time $(t^{\ell}_{\hbar})$ represents both the peak and off-peak load periods in a particular duration. Let us assume the expected load curve as a function of customer power demand and time to be

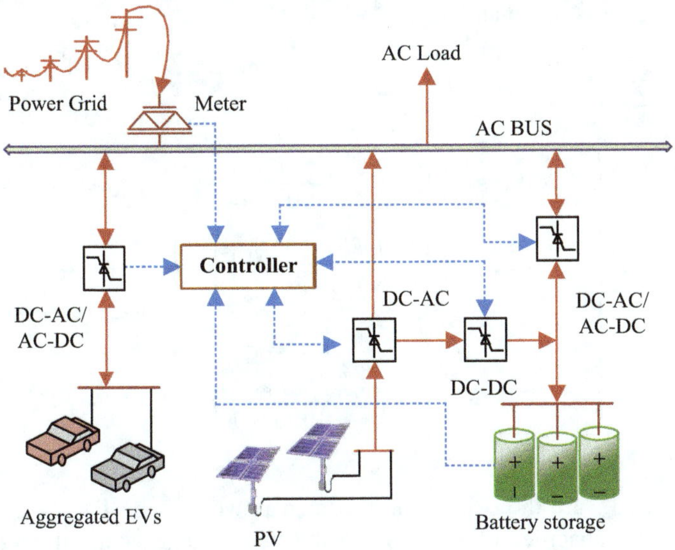

Figure 5.16 Detailed architecture of the energy-resource management of multiple customers [11,12]

$\partial^t_d = f(\mathcal{P}_{\partial}, t^b_p)$. The peak and off-peak loads occur during the $\mathcal{L}^i_p > \partial^t_d$ and $\mathcal{L}^i_p < \partial^t_d$ conditions, respectively. The amount of peak load is identified as $\mathcal{L}^s_p = \mathcal{L}^{i_1}_p - \partial^t_d$, for $\mathcal{L}^i_p > \partial^t_d$, as shown in (5.7). The amount of available power during off-peak hours to charge the batteries and EVs is $\mathcal{L}^c_p = \partial^t_d - \mathcal{L}^{i_1}_p$, for $\mathcal{L}^i_p < \partial^t_d$, as shown in (5.8). The amount of peak load $(\mathcal{L}^s_p)$ at a particular time *(t)* is requested from the PV, battery storage and EVs. In this case, PVs, battery storage and EVs provide power to shave the peak $(\mathcal{L}^s_p)$ based on their own constraints. Likewise, the available power $(\mathcal{L}^c_p)$ is shared by the battery and EV to charge based on their power demand and is written as

$$\mathcal{L}^c_p = \mathcal{L}^c_b + \mathcal{L}^c_{ev} \tag{5.45}$$

where $\mathcal{L}^c_b$ and $\mathcal{L}^c_{ev}$ are the available power to charge the battery and EV. The maximum requested power from the available power $(\mathcal{L}^c_p)$ is $\psi^r_b = \kappa^c_b(\psi^m_b - \psi^i_b)$ as shown in (5.14). So, the available power to charge the EVs is given as

$$\mathcal{L}^c_{ev} = \left(\mathcal{L}^c_p - \mathcal{L}^c_b\right) \tag{5.46}$$

Let us assume the instantaneous charge of the EVs to be $\psi^i_{ev_1}, \psi^i_{ev_2}, \psi^i_{ev_3}, \ldots\ldots\ldots\ldots, \psi^i_{ev_n}$, and the maximum charging limit of the EVs to be $\psi^m_{ev_1}, \psi^m_{ev_2}, \psi^m_{ev_3}, \ldots\ldots\ldots\ldots, \psi^m_{ev_n}$. The power allocated among the available EVs in the parking lot is given as

$$\mathcal{L}^{c}_{ev} = \sum_{ev=1}^{n} \mathfrak{I}^{p}_{ev} \tag{5.47}$$

$$\mathcal{L}^{c}_{ev} = \sum \begin{cases} \mathfrak{I}^{p}_{ev_1} = \left\{1 - \left(\dfrac{\psi^{m}_{ev_1} - \psi^{i}_{ev_1}}{100}\right)\right\} * \kappa^{c}_{ev_1} & for\, EV_1 \\ \mathfrak{I}^{p}_{ev_2} = \left\{1 - \left(\dfrac{\psi^{m}_{ev_2} - \psi^{i}_{ev_2}}{100}\right)\right\} * \kappa^{c}_{ev_2} & for\, EV_2 \\ \cdots\cdots\cdots\cdots\cdots\cdots\cdots\cdots\cdots\cdots \\ \cdots\cdots\cdots\cdots\cdots\cdots\cdots\cdots\cdots\cdots \\ \mathfrak{I}^{p}_{ev_n} = \left\{1 - \left(\dfrac{\psi^{m}_{ev_n} - \psi^{i}_{ev_n}}{100}\right)\right\} * \kappa^{c}_{ev_n} & for\, EV_n \end{cases} \tag{5.48}$$

$\mathfrak{I}^{p}_{ev_1}, \mathfrak{I}^{p}_{ev_2}, \ldots.., \mathfrak{I}^{p}_{ev_n}$ are the maximum allocated power among EVs. $\kappa^{c}_{ev_1}, \kappa^{c}_{ev_2}, \kappa^{c}_{ev_n}$ are the battery capacities of EVs. Since the bus voltage at the domestic bus is constant, the current will vary and be allocated based on individual EV demand. The current allocation among EVs is expressed as

$$\sum_{ev=1}^{n} \mathcal{A}^{i}_{ev_1} = \sum \begin{cases} \mathcal{A}^{i}_{ev_1} = \dfrac{\left(\mathcal{L}^{c}_{ev} * \mathfrak{I}^{p}_{ev_1}\right)}{\sum_{ev=1}^{n} \mathfrak{I}^{p}_{ev}} & for\, EV_1 \\ \mathcal{A}^{i}_{ev_2} = \dfrac{\left(\mathcal{L}^{c}_{ev} * \mathfrak{I}^{p}_{ev_2}\right)}{\sum_{ev=1}^{n} \mathfrak{I}^{p}_{ev}} & for\, EV_2 \\ \cdots\cdots\cdots\cdots\cdots\cdots\cdots\cdots \\ \cdots\cdots\cdots\cdots\cdots\cdots\cdots\cdots \\ \mathcal{A}^{i}_{ev_n} = \dfrac{\left(\mathcal{L}^{c}_{ev} * c^{p}_{cn}\right)}{\sum_{ev=1}^{n} \mathfrak{I}^{p}_{ev}} & for\, EV_n \end{cases} \tag{5.49}$$

Here, $\mathcal{A}^{i}_{ev_1}$ is the total allocated current proportional to the allocated power $\mathcal{L}^{c}_{ev}$, as the bus voltage is constant. A schematic of the charge management is shown in Figure 5.17.

The amount of peak power needed during peak-load hours is $\mathcal{L}^{s}_{p}$. This power will be shared by the PV, battery storage and EVs. The system will first utilize power from the PV, then from the battery storage and then from the EVs.

$$\mathcal{L}^{s}_{p} = \mathcal{L}^{s}_{pv} + \mathcal{L}^{s}_{b} + \mathcal{L}^{s}_{ev} \quad for\, \mathcal{L}^{i}_{p} > \eth^{t}_{d} \tag{5.50}$$

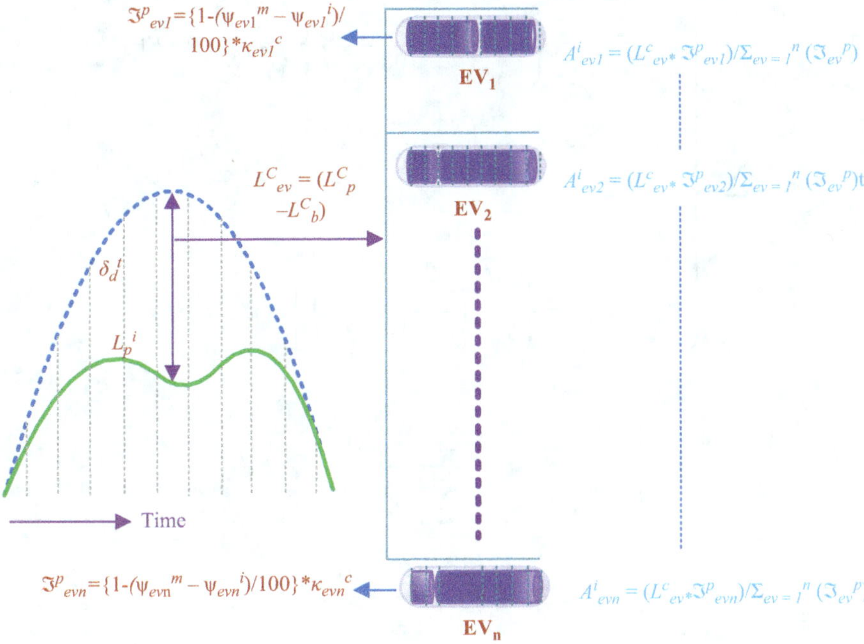

Figure 5.17 Schematic of the charge management of EVs [11,12]

$\mathcal{L}^s_{pv}$, $\mathcal{L}^s_b$, $\mathcal{L}^s_{ev}$ are the load-support power provided by the PV, battery storage and EVs, respectively. $\mathcal{L}^s_{pv}$ is equal to the PV power generation as follows:

$$\mathcal{L}^s_{pv} = \psi^i_{pv} \tag{5.51}$$

The power provided by the battery is given as

$$\psi^r_b = \kappa^c_b\left(\psi^i_b - \psi^l_b\right) \quad \mathit{for}\ \psi^i_b > \psi^l_b \quad \mathit{and} \quad \mathcal{L}^i_p > \partial^t_d \tag{5.52}$$

ψ^l_b is the lower SOC constraints. If the instantaneous charge is lower than the lower SOC limit, that is, $\psi^i_b < \psi^l_b$, the power supplied by the battery storage is zero, that is, $\psi^r_b = 0$. Then, the required power will be supplied by the EVs and is given as

$$\mathcal{L}^s_{ev} = \mathcal{L}^s_p - \mathcal{L}^s_{pv} - \mathcal{L}^s_b \quad \mathit{for}\ \mathcal{L}^i_p > \partial^t_d \tag{5.53}$$

$$\mathcal{L}^s_{ev} = \sum_{ev=1}^{n} \mathfrak{D}^p_{ev} \tag{5.54}$$

$$\mathcal{L}_{ev}^{s} = \sum \begin{cases} \mathfrak{D}_{ev_1}^{p} = \left\{1 - \left(\dfrac{\psi_{ev_1}^{i} - \psi_{ev_1}^{l}}{100}\right)\right\} * \kappa_{ev_1}^{c} & for\ EV_1 \\ \mathfrak{D}_{ev_2}^{p} = \left\{1 - \left(\dfrac{\psi_{ev_1}^{i} - \psi_{ev_2}^{l}}{100}\right)\right\} * \kappa_{ev_2}^{c} & for\ EV_2 \\ \cdots\cdots\cdots\cdots\cdots\cdots\cdots\cdots \\ \cdots\cdots\cdots\cdots\cdots\cdots\cdots\cdots \\ \mathfrak{D}_{ev_n}^{p} = \left\{1 - \left(\dfrac{\psi_{ev_1}^{i} - \psi_{ev_n}^{l}}{100}\right)\right\} * \kappa_{ev_n}^{c} & for\ EV_n \end{cases} \tag{5.55}$$

$$\sum_{ev=1}^{n} \mathcal{A}_{ev_1}^{i} = \sum \begin{cases} \mathcal{A}_{ev_1}^{i} = \dfrac{\left(\mathcal{L}_{ev}^{s} * \mathfrak{D}_{ev_1}^{p}\right)}{\sum_{ev=1}^{n} \mathfrak{D}_{ev}^{p}} & for\ EV_1 \\ \mathcal{A}_{ev_2}^{i} = \dfrac{\left(\mathcal{L}_{ev}^{s} * \mathfrak{D}_{ev_2}^{p}\right)}{\sum_{ev=1}^{n} \mathfrak{D}_{ev}^{p}} & for\ EV_2 \\ \cdots\cdots\cdots\cdots\cdots\cdots\cdots\cdots \\ \cdots\cdots\cdots\cdots\cdots\cdots\cdots\cdots \\ \mathcal{A}_{ev_n}^{i} = \dfrac{\left(\mathcal{L}_{ev}^{s} * \mathfrak{D}_{ev_n}^{p}\right)}{\sum_{ev=1}^{n} \mathfrak{D}_{ev}^{p}} & for\ EV_n \end{cases} \quad for \mathcal{L}_{\mathcal{p}}^{i} > \partial_{d}^{t} \tag{5.56}$$

$\mathfrak{D}_{ev_1}^{p}, \mathfrak{D}_{ev_2}^{p}, \ldots\ldots, \mathfrak{D}_{ev_n}^{p}$ are the power supplied by the individual EVs in a parking lot. $\mathcal{A}_{ev_1}^{i}, \mathcal{A}_{ev_2}^{i}, \ldots\ldots, \mathcal{A}_{ev_n}^{i}$ are the current supplied by the individual EVs in a parking lot based on their SOC condition and battery capacity, since the bus voltage is constant. The charging and discharging of the EVs and the battery storage is determined by the polarity of the current from/to battery and EV.

$$\forall i \in f\left(\mathcal{P}, t_{\mathcal{p}}^{\ell}\right) = \begin{cases} \mathcal{A}_{ev}^{i} > 0 & for\ \mathcal{L}_{\mathcal{p}}^{i} < \partial_{d}^{t} \\ \mathcal{A}_{ev}^{i} < 0 & for\ \mathcal{L}_{\mathcal{p}}^{i} > \partial_{d}^{t} \\ \mathcal{A}_{b}^{i} > 0 & for\ \mathcal{L}_{\mathcal{p}}^{i} < \partial_{d}^{t} \\ \mathcal{A}_{b}^{i} < 0 & for\ \mathcal{L}_{\mathcal{p}}^{i} > \partial_{d}^{t} \end{cases} \tag{5.57}$$

5.4.3 Case studies

Two case studies are carried out to manage the power demand of multiple customers in a smart grid. A detailed management algorithm is explained in Section 4.2.

The EV considered has a V2G capability. Six EVs are considered for the system in a parking lot, where three EVs are actively connected. Among these EVs 50% are Nissan Leaf type (24 kWh) and 50% are Tesla EV type (90 kWh). The parameters considered for this case study are listed in Table 5.4.

When the EVs in a parking lot participate in a V2G operation, they reduce the customers' power demand significantly. Figure 5.18 shows the power-demand management when aggregated EVs and PVs are connected. In this case study, the battery discharge limit of the EVs are kept within 85%–95% of the SOC. Reducing the lower limit of SOC will improve the power-demand management performance. The power demand further improves if a fixed battery storage is used in the system. It can intelligently utilize excess PV power generation during off-peak hours and discharge during peak hours, which improves the overall power-demand management performance. Figure 5.19 shows the power-demand management performance when an extra 20 kWh battery is added with the aggregated EVs and PVs.

Table 5.4 Parameters considered for power demand management of multiple customers

Parameter	**Value**
Number of EVs	6 (three EVs are actively connected)
Discharging (V2G) SOC of EVs	Between 85% and 95%
EV charger efficiency	95%
Battery storage capacity	20 kWh
Battery storage lifecycle	~10k
Charging-discharging efficiency of battery storage	95%
Specific energy of battery storage	~ 60 MJ/kg
Self-discharge rate of battery storage	~ 0.2% (per day)

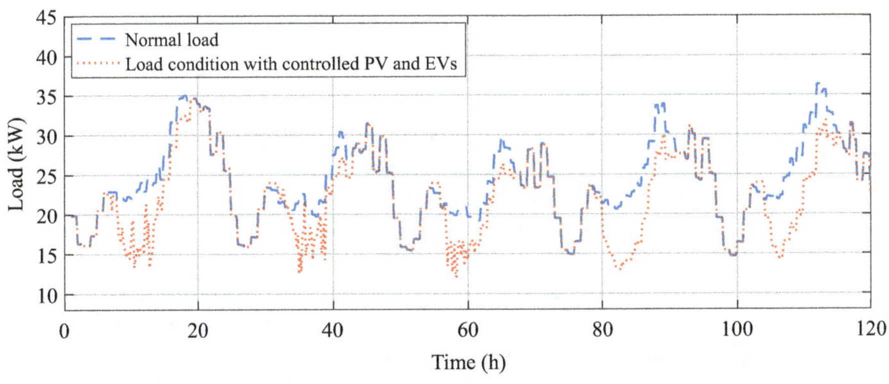

Figure 5.18 Power-demand management of multiple customers using aggregated EVs and PVs

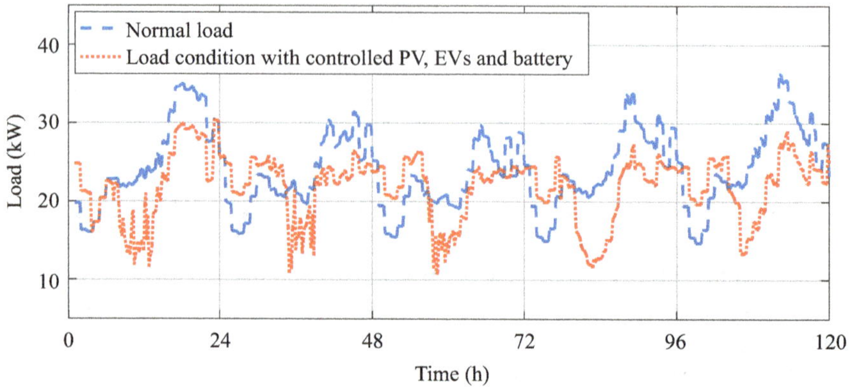

Figure 5.19 Power-demand management of multiple customers using aggregated EVs, PVs and battery storage (20 kWh)

5.5 Conclusion

This chapter presents a method to deal with the power-demand management of a smart grid that is heavily dependent on intermittent mobile energy sources. It provides a comprehensive idea of energy-resource and load-management approaches at various scales for both single and multiple customers. The findings of this study suggest that an intelligent management technique can substantially reduce the power demand of the grid using EVs and reduce the impact of intermittent sources. Additionally, the insights gained from this study may be of assistance to deal with the controlled bidirectional charging-discharging of EVs using renewable energy sources considering peak and off-peak hours. This finding will prove useful in expanding our understanding of how to manage the power demand in a smart grid and reduce the impact of intermittent sources by utilizing EVs.

References

[1] M. S. H. Nizami and J. Hossain, 'Optimal scheduling of electrical appliances and DER units for home energy management system', in *Universities Power Engineering Conference (AUPEC), 2017 Australasian*, 2017, pp. 1–6.

[2] Z. Lu, J. Qi, J. Zhang, L. He, and H. Zhao, 'Modelling dynamic demand response for plug-in hybrid electric vehicles based on real-time charging pricing', *IET Gener. Transm. Distrib.*, vol. 11, no. 1, pp. 228–235, 2017.

[3] M. H. K. Tushar, A. W. Zeineddine, and C. Assi, 'Demand-side management by regulating charging and discharging of the EV, ESS, and utilizing renewable energy', *IEEE Trans. Ind. Informatics*, vol. 14, no. 1, pp. 117–126, 2018.

[4] A. Khalid, N. Javaid, M. Guizani, and M. Alhussein, K. Aurangzeb, and M. Ilahi, 'Towards dynamic coordination among home appliances using

multi-objective energy optimization for demand side management in smart buildings', *IEEE Access*, vol. 6, pp. 19509–19529, 2018.

[5] M. H. Yaghmaee, M. S. Kouhi, A. Saeedi, and M. Zabihi, 'Demand side management controlling with personalised pricing method', *CIRED - Open Access Proc. J.*, vol. 2017, no. 1, pp. 2666–2669, 2017.

[6] K. Mahmud, S. Morsalin, Y. R. Kafle, and G. E. Town, 'Improved peak shaving in grid-connected domestic power systems combining photovoltaic generation, battery storage, and V2G-capable electric vehicle', in *2016 IEEE International Conference on Power System Technology (POWERCON)*, 2016, pp. 1–4.

[7] K. Mahmud, S. Morsalin, M. J. Hossain, and G. E. Town, 'Domestic peak-load management including vehicle-to-grid and battery storage unit using an artificial neural network', in *2017 IEEE International Conference on Industrial Technology (ICIT)*, 2017, pp. 586–591.

[8] K. Mahmud, M. S. H. Nizami, M. J. Hossain, and J. Ravishankar, 'A home-to-home energy sharing process for domestic peak load management', in *2018 IEEE International Conference on Environment and Electrical Engineering and 2018 IEEE Industrial and Commercial Power Systems Europe (EEEIC / I&CPS Europe)*, 2018, pp. 1–5.

[9] M. S. H. Nizami, A. N. M. M. Haque, P. H. Nguyen, and F. W. Bliek, 'HEMS as network support tool: Facilitating network operator in congestion management and overvoltage mitigation', in *2016 IEEE 16th International Conference on Environment and Electrical Engineering (EEEIC)*, 2016, pp. 1–6.

[10] K. Mahmud, G. E. Town, and M. J. Hossain, 'Mitigating the impact of rapid changes in photovoltaic power generation on network voltage', in *2017 IEEE Power and Energy Conference at Illinois (PECI)*, 2017, pp. 1–6.

[11] K. Mahmud, M. J. Hossain, and J. Ravishankar, 'Peak-load management in commercial systems with electric vehicles', *IEEE Syst. J.*, vol. 13, no. 2, pp. 1872–1882, 2018.

[12] K. Mahmud, M. J. Hossain, and G. E. Town, 'Peak-load reduction by coordinated response of photovoltaics, battery storage, and electric vehicles', *IEEE Access*, p. 1, 2018.

[13] J. C. Muñoz, J. F. Ordoñez, V. Gutierrez, C. A. Cañizares, K. Bhattacharya, and S. Paudyal, 'Applications of optimal industrial load management modeling in smart grids', in *2017 IEEE Manchester PowerTech*, 2017, pp. 1–6.

[14] W. A. Johnson, A. M. Maher, and T. M. Devaney, 'Experimental results of a load management system for large commercial customers', *IEEE Power Eng. Rev.*, vol. PER-5, no. 9, pp. 27–28, 1985.

[15] C. O. Bjork and B. G. Karlsson, 'Load management applications for industrial loads', *IEEE Power Eng. Rev.*, vol. PER-5, no. 8, p. 35, 1985.

[16] M. A. Rahman, R. Islam, K. F. Sharif, and T. Aziz, 'Developing demand side management program for commercial customers: A case study', in *2016 3rd International Conference on Electrical Engineering and Information Communication Technology (ICEEICT)*, 2016, pp. 1–6.

[17] S. Ashok and R. Banerjee, 'An optimization mode for industrial load management', *IEEE Trans. Power Syst.*, vol. 16, no. 4, pp. 879–884, 2001.
[18] K. Mahmud, G. E. Town, S. Morsalin, and M. J. Hossain, 'Integration of electric vehicles and management in the internet of energy', *Renew. Sustain. Energy Rev.*, vol. 82, pp. 4179–4203, 2018.
[19] Y. R. Kafle, K. Mahmud, S. Morsalin, and G. E. Town, 'Towards an internet of energy', in *2016 IEEE International Conference on Power System Technology (POWERCON)*, 2016, pp. 1–6.
[20] Y. Wang, W. Yang, and T. Liu, 'Appliances considered demand response optimisation for smart grid', *IET Gener. Transm. Distrib.*, vol. 11, no. 4, pp. 856–864, 2017.
[21] Y. Ozturk, D. Senthilkumar, S. Kumar, and G. Lee, 'An intelligent home energy management system to improve demand response', *IEEE Trans. Smart Grid*, vol. 4, no. 2, pp. 694–701, 2013.
[22] M. Yalcintas, W. T. Hagen, and A. Kaya, 'An analysis of load reduction and load shifting techniques in commercial and industrial buildings under dynamic electricity pricing schedules', *Energy Build.*, vol. 88, pp. 15–24, 2015.
[23] R. Zafar, A. Mahmood, S. Razzaq, W. Ali, U. Naeem, and K. Shehzad, 'Prosumer based energy management and sharing in smart grid', *Renew. Sustain. Energy Rev.*, vol. 82, pp. 1675–1684, 2018.
[24] S. Choi and S. W. Min, 'Optimal scheduling and operation of the ESS for prosumer market environment in grid-connected industrial complex', *IEEE Trans. Ind. Appl.*, vol. 54, no. 3, pp. 1949–1957, 2018.
[25] L. Igualada, C. Corchero, M. Cruz-Zambrano, and F. J. Heredia, 'Optimal energy management for a residential microgrid including a vehicle-to-grid system', *IEEE Trans. Smart Grid*, vol. 5, no. 4, pp. 2163–2172, 2014.
[26] A. S. Masoum, S. Deilami, P. S. Moses, M. A. S. Masoum, and A. Abu-Siada, 'Smart load management of plug-in electric vehicles in distribution and residential networks with charging stations for peak shaving and loss minimisation considering voltage regulation', *IET Gener. Transm. Distrib.*, vol. 5, no. 8, pp. 877–888, 2011.
[27] S. Manzetti and F. Mariasiu, 'Electric vehicle battery technologies: From present state to future systems', *Renew. Sustain. Energy Rev.*, vol. 51, pp. 1004–1012, 2015.
[28] K. L. López, C. Gagné, and M. A. Gardner, 'Demand-side management using deep learning for smart charging of electric vehicles', *IEEE Trans. Smart Grid*, vol. 10, no. 3, pp. 2683–2691, 2019.
[29] E. Kamal and L. Adouane, 'Hierarchical energy optimization strategy and its integrated reliable battery fault management for hybrid hydraulic-electric vehicle', *IEEE Trans. Veh. Technol.*, vol. 67, no. 5, pp. 3740–3754, 2018.
[30] N. Ahmed, M. Levorato, and G. P. Li, 'Residential consumer-centric demand side management', *IEEE Trans. Smart Grid*, vol. 9, no. 5, pp. 4513–4524, 2018.
[31] L. Labib, M. Billah, G. M. S. M. Rana, M. N. Sadat, M. G. Kibria, and M. R. Islam, 'Design and implementation of low-cost universal smart energy meter

with demand side load management', *IET Gener. Transm. Distrib.*, vol. 11, no. 16, pp. 3938–3945, 2017.

[32] A. N. M. M. Haque, P. H. Nguyen, F. W. Bliek, and J. G. Slootweg, 'Demand response for real-time congestion management incorporating dynamic thermal overloading cost', *Sustain. Energy, Grids Networks*, vol. 10, pp. 65–74, 2017.

[33] Q. Qdr, 'Benefits of demand response in electricity markets and recommendations for achieving them', *US Dept. Energy*, Washington, DC, USA, *Tech. Rep*, 2006.

[34] P. Samadi, H. Mohsenian-Rad, V. W. S. Wong, and R. Schober, 'Adaptive energy consumption scheduling with load uncertainty for the smart grid', in *Communications (ICC), 2013 IEEE International Conference on*, 2013, pp. 4244–4249.

[35] C. H. Antunes, A. Soares, and Á. Gomes, 'An energy management system for residential demand response based on multiobjective optimization', in *Smart Energy Grid Engineering (SEGE), 2016 IEEE*, 2016, pp. 90–94.

[36] O. Abrishambaf, M. A. F. Ghazvini, L. Gomes, P. Faria, Z. Vale, and J. M. Corchado, 'Application of a home energy management system for incentive-based demand response program implementation', in *Database and Expert Systems Applications (DEXA), 2016 27th International Workshop on*, 2016, pp. 153–157.

[37] C. Chen, J. Wang, Y. Heo, and S. Kishore, 'MPC-based appliance scheduling for residential building energy management controller', *IEEE Trans. Smart Grid*, vol. 4, no. 3, pp. 1401–1410, 2013.

[38] R. Bálint, K. M. Hangos, and A. Magyar, 'Optimal scheduling of a household refrigerator using adaptive model predictive technique', in *2017 International Symposium on Power Electronics (Ee)*, 2017, pp. 1–6.

[39] A. Parisio, C. Wiezorek, T. Kyntäjä, J. Elo, and K. H. Johansson, 'An MPC-based energy management system for multiple residential microgrids', in *2015 IEEE International Conference on Automation Science and Engineering (CASE)*, 2015, pp. 7–14.

[40] M. Muratori and G. Rizzoni, 'Residential demand response: Dynamic energy management and time-varying electricity pricing', *IEEE Trans. Power Syst.*, vol. 31, no. 2, pp. 1108–1117, 2016.

[41] J. Leithon, T. J. Lim, and S. Sun, 'Battery-aided demand response strategy under continuous-time block pricing', *IEEE Trans. Signal Process.*, vol. 64, no. 2, pp. 395–405, 2016.

[42] M. Safari, 'Battery electric vehicles: Looking behind to move forward', *Energy Policy*, vol. 115, pp. 54–65, 2018.

[43] M. S. H. Nizami, M. J. Hossain, K. Mahmud, and J. Ravishankar, 'Energy cost optimization and der scheduling for unified energy management system of residential neighborhood', in *2018 IEEE International Conference on Environment and Electrical Engineering and 2018 IEEE Industrial and Commercial Power Systems Europe (EEEIC / I&CPS Europe)*, 2018, pp. 1–6.

[44] S. Morsalin, K. Mahmud, and G. Town, 'Electric vehicle charge scheduling using an artificial neural network', in *2016 IEEE Innovative Smart Grid Technologies - Asia (ISGT-Asia)*, 2016, pp. 276–280.

[45] C. O. Bjork and B. G. Karlsson, 'Load management applications for industrial loads', *IEEE Trans. Power Appar. Syst.*, vol. PAS-104, no. 8, pp. 2058–2063, 1985.

[46] J. A. Pinzon, P. P. Vergara, L. C. P. da Silva, and M. J. Rider, 'Optimal management of energy consumption and comfort for smart buildings operating in a microgrid', *IEEE Trans. Smart Grid,* vol. 10, no. 3, pp. 3236–3247, 2019.

[47] L. Martirano, E. Habib, G. Parise *et al.*, 'Demand side management in mixed residential/commercial buildings with PV on site generation', in *2017 IEEE/IAS 53rd Industrial and Commercial Power Systems Technical Conference (I&CPS)*, 2017, pp. 1–7.

[48] N. Javaid, I. Ullah, M. Akbar *et al.*, 'An intelligent load management system with renewable energy integration for smart homes', *IEEE Access*, vol. 5, pp. 13587–13600, 2017.

[49] Q. Yang and X. Fang, 'Demand response under real-time pricing for domestic households with renewable DGs and storage', *IET Gener. Transm. Distrib.*, vol. 11, no. 8, pp. 1910–1918, 2017.

[50] M. Jarnut, S. Wermiński, and B. Waśkowicz, 'Comparative analysis of selected energy storage technologies for prosumer-owned microgrids', *Renew. Sustain. Energy Rev.*, vol. 74, pp. 925–937, 2017.

[51] S. Deilami, A. S. Masoum, P. S. Moses, and M. A. S. Masoum, 'Real-time coordination of plug-in electric vehicle charging in smart grids to minimize power losses and improve voltage profile', *IEEE Trans. Smart Grid*, vol. 2, no. 3, pp. 456–467, 2011.

[52] E. Kamal and L. Adouane, 'Intelligent energy management strategy based on artificial neural fuzzy for hybrid vehicle', *IEEE Trans. Intell. Veh.*, vol. 3, no. 1, pp. 112–125, 2018.

[53] G. Wu, K. Boriboonsomsin, and M. J. Barth, 'Development and evaluation of an intelligent energy-management strategy for plug-in hybrid electric vehicles', *IEEE Trans. Intell. Transp. Syst.*, vol. 15, no. 3, pp. 1091–1100, 2014.

[54] Z. Wei, Y. Li, Y. Zhang, and L. Cai, 'Intelligent parking garage EV charging scheduling considering battery charging characteristic', *IEEE Trans. Ind. Electron.*, vol. 65, no. 3, pp. 2806–2816, 2018.

[55] E. Akhavan-Rezai, M. F. Shaaban, E. F. El-Saadany, and F. Karray, 'Online intelligent demand management of plug-in electric vehicles in future smart parking lots', *IEEE Syst. J.*, vol. 10, no. 2, pp. 483–494, 2016.

Chapter 6

Energy management of a small-size electric energy system with electric vehicles, flexible demands, and renewable generating units

Luis Baringo[1]

This chapter describes the energy-management problem of a small-size electric energy system (SSEES), which comprises a set of electric vehicles (EVs); a set of flexible demands that can shift their energy consumption according to the system needs; and renewable generating units, such as solar- or wind-powered units.

We consider an energy-management system (EMS) in charge of all the components in the SSEES. This EMS determines the power scheduling of each component and guarantees that all the technical and economic constraints of these components are satisfied. Moreover, the EMS guarantees that the technical constraints of the SSEES are met.

In order to compensate the energy deviations, the SSEES is connected to the main grid from which the EMS can buy/sell the deficit/surplus energy to the electricity market. Particularly, we model the day-ahead (DA) market.

The considered problem is characterized by a number of uncertainties including the energy market prices and the production of renewable generating units. These uncertainties are modeled using a set of scenarios. Therefore, the resulting problem is formulated as a two-stage stochastic programming problem. In the first stage, the EMS decides the power bought from/sold to the DA market. In the second stage, and once the actual realization of the different uncertainty sources is known, the EMS decides the actual power scheduling of each component in the SSEES.

6.1 Introduction

Electric energy systems have experimented a deep transformation in the last decades. Traditionally, generating units were mainly based on fossil fuels and nuclear energy. These generating units were used to supply the demands, which used to be

[1]Departamento de Ingeniería Eléctrica, Electrónica, Automática y Comunicaciones, Universidad de Castilla-La Mancha, Ciudad Real, Spain

inelastic. It was not until the last decade of the last century that renewable generating units based on wind, solar, and photovoltaic energy started to be developed with an aim of reducing the greenhouse gas emissions. However, despite their benefits, the production of most renewable generating units is uncertain and variable since the renewable energy depends on the meteorological phenomena. This supposes a handicap on the operation of electric energy systems [1].

Although the increased penetration of renewable energy has probably been the main modification in electric energy systems, there have been other important changes in recent years, such as, the development of demand-response programs to incentive consumers to change their consumption patterns [2] and the increasing penetration of EVs. Although the market share of EVs is still negligible, EVs are expected to replace part of the conventional vehicles in the future [3].

The above changes suppose new challenges for electric energy systems. For example, the flexibility of consumers that can shift their energy consumption may be used to reduce the negative impact of the uncertain and variable renewable production. This requires the development of new methods for the optimal operation of current and future electric energy systems.

Within this context, this chapter describes a method for the optimal operation of a small-size electric energy system (SSEES) that includes EVs, flexible demands, and renewable generating units. All these elements are managed by an energy-management system (EMS) that acts as a central entity that aims at operating the SSEES at minimum cost while complying with all the technical and economic constraints of the different components. The SSEES is connected to the main grid from/to which it can buy/sell energy to compensate the energy deficit/surplus in the system. To do this, the EMS participates in the DA market [4].

The energy-management problem considered in this chapter is solved 1 day in advance for a planning horizon of 1 day divided in hourly time steps. Thus, in order to obtain informed operating decisions, it is essential to model the different sources of uncertainty in the problem, such as the market prices and the production of renewable generating units. In particular, we model these uncertainties using a number of scenarios which represent different realizations of the uncertain parameters. This allows us to formulate the energy-management problem as a two-stage stochastic programming model [5]. In the first stage, the EMS decides the participation of the SSEES in the DA market. In the second stage and once the actual realization of the uncertain parameters is known, the EMS decides the optimal operation of the SSEES.

The remainder of this chapter is organized as follows. Section 6.2 defines the main notation used in the chapter. Section 6.3 describes the modeling of EVs. Section 6.4 provides the formulation of the energy-management problem considering a deterministic approach. This problem does not take into account the different sources of uncertainty. These uncertainties are characterized in Section 6.5, which allows formulating the energy-management problem as a two-stage stochastic programming problem in Section 6.3. Sections 6.3–6.6 include a number of illustrative examples to show the working of the different models.

Section 6.7 summarizes the chapter and provides some relevant remarks. Finally, Section 6.8 provides the GAMS codes used to obtain the results of some of the illustrative examples in Sections 6.3–6.6.

6.2 Notation

The main notation used in this chapter is defined below for a quick reference. Some of the variables and parameters below may include subindexes t and ω, which denote the values of these variables and parameters at time period t and scenario ω, respectively.

6.2.1 Indices

d Demands

e EV aggregations

g Renewable generating units

ℓ Lines

t Time periods

t_f Last time period of the planning horizon

v Individual EV

ω Scenarios

6.2.2 Sets

$r(\ell)$ Receiving-end node of line ℓ

$s(\ell)$ Sending-end node of line ℓ

Ω^{MG} Node connected to the main grid

ψ_n^{D} Demands connected at node n

ψ_n^{E} EV aggregations connected at node n

ψ_n^{G} Renewable generating units connected at node n

6.2.3 Parameters

E_{e0}^{A} Energy stored (state of charge) in the virtual battery representing the EV aggregation e at the beginning of the planning horizon [MWh]

E_e^{AA} Energy contribution of EVs arriving at the virtual battery representing the EV aggregation e [MWh]

E_e^{AD} Energy drop of EVs departing from the virtual battery representing the EV aggregation e [MWh]

E_d^{D} Minimum daily energy consumption of demand d [MWh]

E_v^{IC} Energy consumption of the individual EV battery v [kWh]

E_{v0}^{I} Energy stored (state of charge) in the individual EV battery v at the beginning of the planning horizon [kWh]

$\overline{P}_e^{\mathrm{A}}$ Maximum charging power of the virtual battery representing the EV aggregation e [MW]
$\underline{P}_d^{\mathrm{D}}$ Minimum load consumption of demand d [MW]
$\overline{P}_d^{\mathrm{D}}$ Maximum load consumption of demand d [MW]
P_g^{GA} Available capacity of renewable generating unit g [MW]
$\overline{P}^{\mathrm{MG}}$ Maximum power that can be purchased from/sold to the main grid [MW]
X_ℓ Reactance of line ℓ [Ω]
Δt Duration of time steps [h]
η_e^{A} Charging efficiency of the virtual battery representing the EV aggregation e [p.u.]
η_v^{I} Charging efficiency of the individual EV battery v [p.u.]
λ^{DA} DA market price [€/MWh]
π_ω Probability of scenario ω [p.u.]

6.2.4 Optimization variables

e_e^{A} Energy stored (state of charge) in the virtual battery representing the EV aggregation e [MWh]
e_v^{I} Energy stored (state of charge) in the individual EV battery v [kWh]
P_e^{A} Charging power of the virtual battery representing the EV aggregation e [MW]
P_d^{D} Load consumption of demand d [MW]
P^{DA} Power purchased from (if positive) or sold to (if negative) the DA market [MW]
P_g^{G} Power production of renewable generating unit g [MW]
P_v^{I} Charging power of the individual EV battery v [kW]
P_ℓ^{L} Power flow through line ℓ [MW]
θ_n Voltage angle at node n [rad]

6.3 Modeling of electric vehicles

This section describes the mathematical modeling of EVs; first, by using an individual modeling that takes into account the driving requirements of each individual EV user; second, by using the figure of the so-called *EV aggregator* that manages a set of EVs.

6.3.1 Individual modeling

Each individual EV v can be represented the following set of equations [6]:

$$e_{vt}^{\mathrm{I}} = e_{v(t-1)}^{\mathrm{I}} + p_{vt}^{\mathrm{I}}\eta^{\mathrm{I}}\Delta t - E_{vt}^{\mathrm{IC}}, \quad \forall v, \forall t, \tag{6.1a}$$

$$e^{\mathrm{I}}_{vt_f} \geq E^{\mathrm{I}}_{v0}, \quad \forall v, \tag{6.1b}$$

$$\underline{E}^{\mathrm{I}}_{v} \leq e^{\mathrm{I}}_{vt} \leq \overline{E}^{\mathrm{I}}_{v}, \quad \forall v, \forall t, \tag{6.1c}$$

$$0 \leq p^{\mathrm{I}}_{vt} \leq \overline{P}^{\mathrm{I}}_{vt}, \quad \forall v, \forall t. \tag{6.1d}$$

Equation (6.1a) represents the energy evolution in the battery of each individual EV v. The energy stored at time period t is equal to the energy stored at the previous time period plus the charging power minus the energy consumption at that time period. Note that the charging power is multiplied by the charging efficiency to account for charging losses, while the energy consumption represents the driving needs of the EV user. Equation (6.1b) imposes that the energy stored at the end of the planning horizon (time period t_f) must be at least equal to the energy stored at the beginning of the planning horizon. Otherwise, the energy stored in the battery would tend to be depleted over the planning horizon. Equation (6.1c) represents the energy limits in the battery of each individual EV v. Finally, (6.1d) imposes limits on the charging power of each individual EV v. If EV v is parked in a charging station at time period t, that is, the EV is available for charging, the maximum charging power is equal to the charging rate of the EV. Otherwise, this maximum charging power is 0.

Illustrative Example 6.1 Individual EV modeling

We consider an EV whose battery has the characteristics provided in Table 6.1. The EV aims at identifying its charging profile for the following day, which is divided into hourly time periods (i.e., $\Delta t = 1\,\mathrm{h}$).

Figure 6.1 depicts the maximum charging power at each time period. Note that if this charging power at a given time period is equal to 0, it means that the EV is not available for charging at that time period.

Figure 6.2 depicts the consumption energy of the EV at each time period, which represents the driving requirements of the EV user.

We consider that the price paid for charging the EV is different throughout the day. Particularly, we consider the price profile depicted in Figure 6.3.

Table 6.1 Illustrative Example 6.1: EV battery data

Minimum energy content	0
Maximum energy content	30 kWh
Initial energy stored	10 kWh
Charging efficiency	0.9
Maximum charging rate	6.6

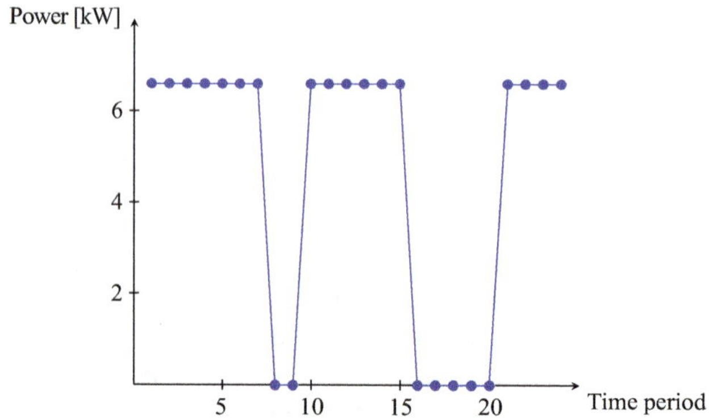

Figure 6.1 Illustrative Example 6.1: maximum charging power

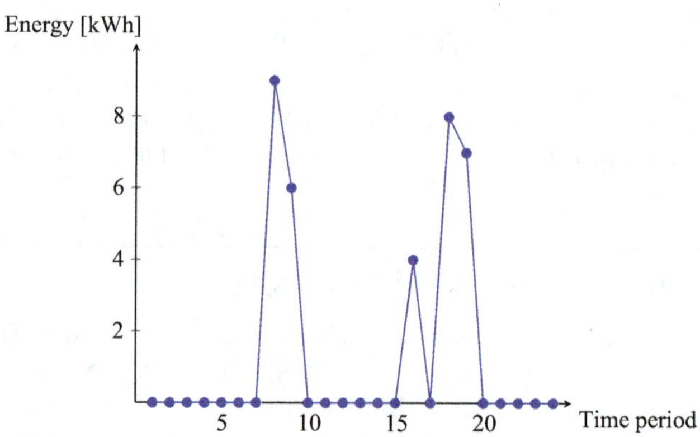

Figure 6.2 Illustrative Example 6.1: energy consumption

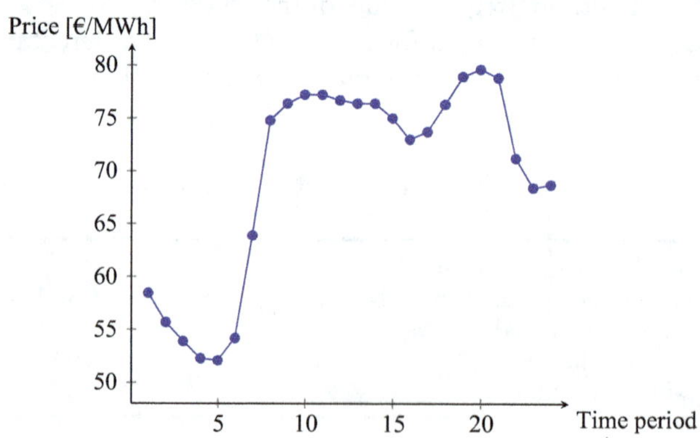

Figure 6.3 Illustrative Example 6.1: market prices

Using these data, the EV aims at determining its charging profile during the day that minimizes the total charging costs. To do so, the following problem is solved:

$\min_{\Theta^{\text{IE1}}}$

$$\sum_t \lambda_t^{\text{DA}} p_{vt}^{\text{DA}} \Delta t$$

subject to

$$\text{Constraints (6.1)},$$

where set $\Theta^{\text{IE1}} = \{p_{vt}^{I}, e_{vt}^{I}, v = 1, \forall t\}$ includes the optimization variables of the problem. In the above optimization problem, the total charging costs are minimized while complying with the technical constraints of the EV (6.1). Note that in this example, we consider just one EV and, thus, subscript v may be suppressed.

Results are provided in Figures 6.4 and 6.5, which depict the charging power and energy profiles of the EV, respectively. Note that the EV is parked and

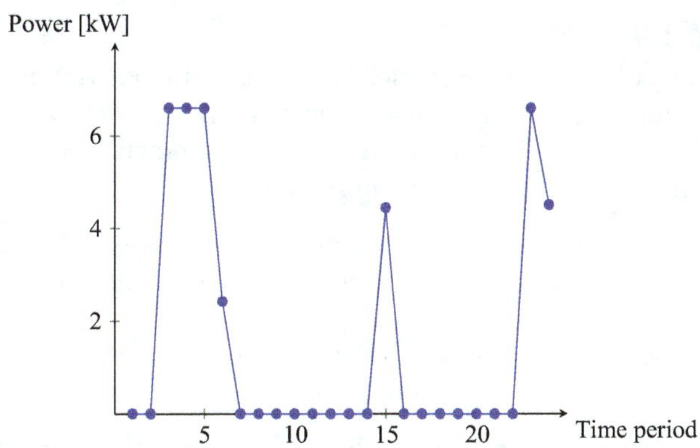

Figure 6.4 Illustrative Example 6.1: charging power

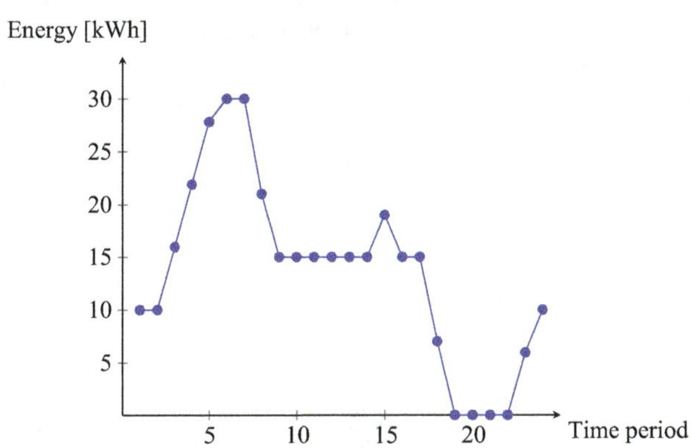

Figure 6.5 Illustrative Example 6.1: energy evolution in the EV battery

available for charging from time period 1 to time period 7; however, it does not start charging until period 3 since the price at this time period is lower than the prices at time periods 1 and 2. Then, at time periods 8 and 9, the EV departs from the charging point and is back at time period 10, when the EV is parked until time period 15 (see Figures 6.1 and 6.2). However, the EV is not charged until time period 15 since the price at this time period is lower than at time periods 10–14. Something similar occurs at the last time periods of the day.

The total charging cost is 2.27€. Note that the EV is parked most of the time. By postponing the charging to those time periods with the lowest charging costs, the EV is able to reduce the total charging costs.

The above problem is a Linear Programming (LP) problem [7] that can be easily solved using commercial software. For example, we can use CPLEX [8] under GAMS [9] to solve the problem. A GAMS code to solve this example is provided in Section 6.8.

6.3.2 EV aggregator

Section 6.3.1 describes a simple model to represent the energy evolution of the battery of each individual EV. This model can be used to model the energy needs of EVs in an electric energy system. However, it is not practical to model each EV individually, mainly due to the following reasons:

1. If there is a large number of EVs in the electric energy system, the problem may become intractable if we model the energy evolution in the battery of each individual EV.
2. The driving needs of individual EVs are subject to uncertainty, since EV users may decide to change their driving plans.

To deal with the above two issues, in this section we define the concept of EV aggregator, which acts as a central entity that manages a set of EVs. We consider that EVs are charged in a parking location, for example, the garage of a work or shopping center. Then, instead of modeling each individual battery, we model the different charging locations as virtual batteries that take into account all the EVs connected at that place in a given time.

The equations that define the EV aggregation are provided below [6]:

$$e_{et}^{\mathrm{A}} = e_{e(t-1)}^{\mathrm{A}} + p_{et}^{\mathrm{A}}\eta_e^{\mathrm{A}}\Delta t + E_{et}^{\mathrm{AA}} - E_{et}^{\mathrm{AD}}, \quad \forall e, \forall t, \tag{6.2a}$$

$$e_{et_f}^{\mathrm{A}} \geq E_{e0}^{\mathrm{A}}, \quad \forall e, \tag{6.2b}$$

$$\underline{E}_{et}^{\mathrm{A}} \leq e_{et}^{\mathrm{A}} \leq \overline{E}_{et}^{\mathrm{A}}, \forall e, \quad \forall t, \tag{6.2c}$$

$$0 \leq p_{et}^{\mathrm{A}} \leq \underline{P}_{et}^{\mathrm{A}}, \quad \forall e, \forall t. \tag{6.2d}$$

Equation (6.2a) models the energy content in the virtual battery modeling each EV aggregation e. At each time period t and for each EV aggregation e, the energy in the virtual battery is equal to the energy stored in the previous period, plus the charging energy, plus the energy contribution of the batteries of EVs arriving to this EV aggregation, minus the energy drops of EVs departing from this EV aggregation. Equation (6.2b) avoids energy drop at the end of the planning horizon. Finally, (6.2c) and (6.2d) impose limits on the energy content and charging power of EV aggregations, respectively. Note that (6.2c) is related to the driving requirements of EVs, while (6.2d) is related to the number of EVs connected in each parking location (and, thus, available for charging).

The use of an EV aggregation reduces the two problems of the individual EV modeling. On the one hand, the computational burden of the problem is reduced since we use just one virtual battery per parking location. On the other hand, the uncertainty in the problem reduces if the number of EVs is large enough since changes in the driving plans of different EV user are compensated.

Illustrative Example 6.2 Aggregator EV modeling

We consider a set of EVs managed by an aggregator. The main data defining this EV aggregation are provided in Table 6.2.

Figure 6.6 provides the energy contributions and drops of EVs arriving to and departing from the EV aggregation, respectively. These energy profiles are related to the number of EVs that arrive to and depart from the parking location.

Figure 6.7 provides the maximum charging power. Note that these limits are related to the number of EVs connected, and thus available for charging, at the EV aggregation.

Figure 6.8 provides the energy limits. In this case, these limits are related to the driving requirements of EVs at each time period. For example, we observe a peak in the lower energy limit at hour 7 that represents that at the end of this time period most of EVs parked in this parking location must be charged for their next trips.

The price paid for charging is different throughout the day. We consider the price profile provided in Illustrative Example 6.1 (Figure 6.3).

Considering these data, the EV aggregator aims at determining the charging profile during the day that minimizes the total charging costs. To do so, the following problem is solved:

$\min_{\Theta^{\mathrm{IE2}}}$

$$\sum_{t} \lambda_t^{\mathrm{DA}} p_{et}^{\mathrm{A}} \Delta t$$

subject to

$$\text{Constraints (6.2)},$$

Table 6.2 Illustrative Example 6.2: EV aggregation data

Initial energy stored	330 MWh
Charging efficiency	0.9

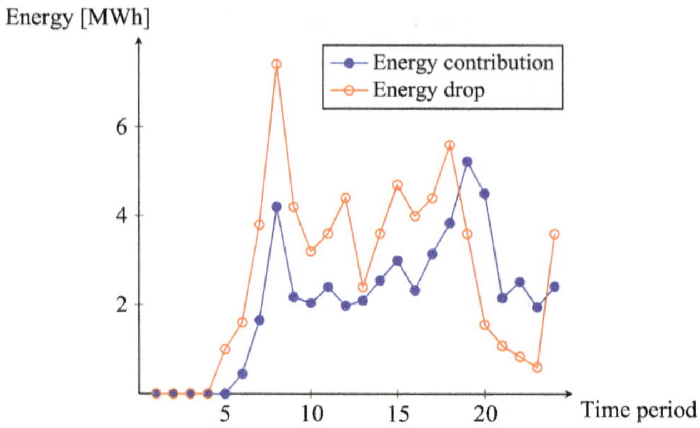

Figure 6.6 Illustrative Example 6.2: energy contribution and drop of EVs arriving to and departing from the charging station

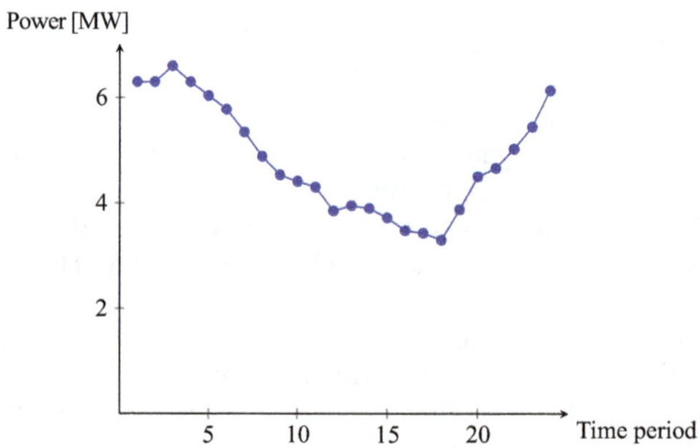

Figure 6.7 Illustrative Example 6.2: maximum charging power in the charging station

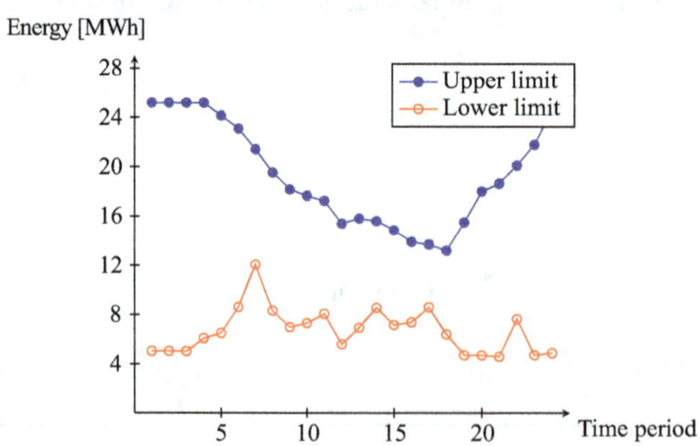

Figure 6.8 Illustrative Example 6.2: energy limits in the charging station

where set $\Theta^{\text{IE2}} = \{p^A_{et}, e^A_{et}, e = 1, \forall t\}$ includes the optimization variables of the problem. In the above optimization problem, the total charging costs are minimized while complying with the technical constraints of the EV aggregation (6.2). Note that in this example, we consider just one EV aggregation and thus subscript *e* may be suppressed.

Results are provided in Figures 6.9 and 6.10 that, respectively, represent the charging power and energy profiles. The charging power profile of Figure 6.9 exhibits a similar pattern to that obtained in Illustrative Example 6.1: the virtual battery representing the EV aggregation is preferably charged at those time periods with the lowest market prices. On the other hand, the energy profile of Figure 6.10 shows that the energy content of the virtual battery is always within the considered lower and upper bounds. The total charging cost in this case is 988.41€.

A GAMS [9] code to solve this example is provided in Section 6.8.

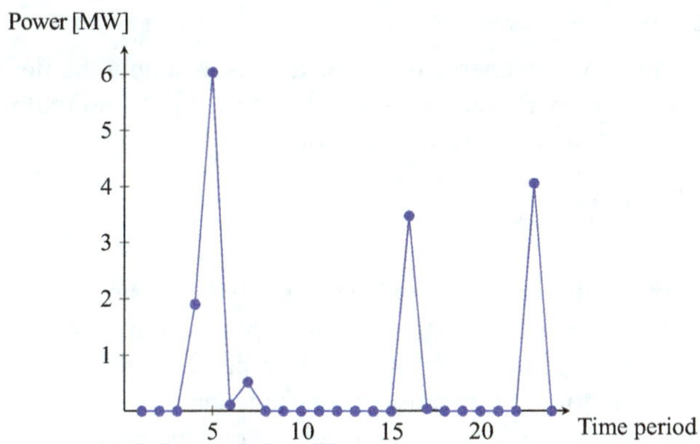

Figure 6.9 Illustrative Example 6.2: charging power in the charging station

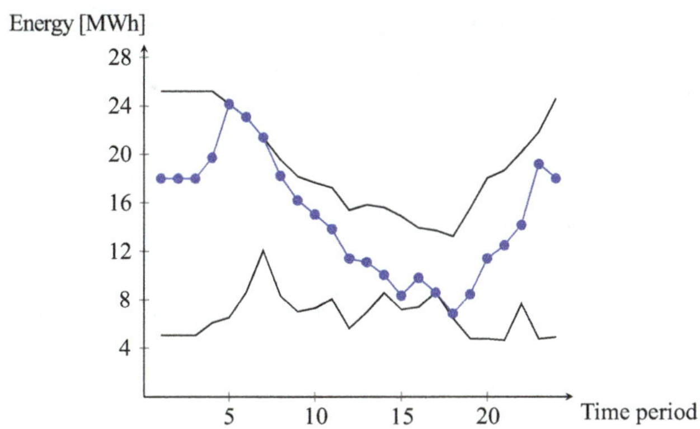

Figure 6.10 Illustrative Example 6.2: energy evolution in the virtual battery representing the charging station

6.4 Deterministic energy-management problem

This section describes the energy management of a SSEES that includes flexible demands, EV aggregations, and renewable generating units. A central entity acts as an EMS and decides the power scheduling of each system component taking into account their economic and technical constraints. Moreover, the SSEES is connected to the main grid from which it can buy/sell the energy deficit/surplus. The planning horizon of the problem is 1 day divided in hourly time periods.

The problem is formulated in this section using a deterministic approach that considers perfect information of all the parameters of the problem. Next, we describe the objective function and the technical and economic constraints of all the components and the SSEES.

6.4.1 Objective function

The objective of the EMS in charge of the SSEES is to supply the flexible demands and the power demanded by the EV charging stations at minimum cost. This is represented by the following objective function:

$$\sum_{t} \lambda_t^{\mathrm{DA}} p_t^{\mathrm{DA}} \Delta t. \tag{6.3}$$

The EMS can supply the demands and EV aggregation using the renewable generating units, which are assumed to have null production costs, and purchasing energy from the main grid. Therefore, (6.3) represents the cost of purchasing energy from the main grid. In this sense, we assume that the EMS participates in the DA market. Moreover, the EMS can also sell the energy surplus to this market. In such a case, variables p_t^{DA} would be negative, and (6.3) would represent the minus revenues achieved by the EMS for selling energy.

6.4.2 Power balance constraints

At each time period t, the power balance at each node of the system must be satisfied:

$$p_t^{\mathrm{DA}} + \sum_{g \in \Psi_n^{\mathrm{G}}} p_{gt}^{\mathrm{G}} + \sum_{\ell | r(\ell)=n} p_{\ell t}^{\mathrm{L}} = \sum_{d \in \Psi_n^{\mathrm{D}}} p_{dt}^{\mathrm{D}} + \sum_{e \in \Psi_n^{\mathrm{E}}} p_{et}^{\mathrm{A}} + \sum_{\ell | s(\ell)=n} p_{\ell t}^{\mathrm{L}}, \quad n \in \Omega^{\mathrm{MG}}, \forall t, \tag{6.4a}$$

$$\sum_{g \in \Psi_n^{\mathrm{G}}} p_{gt}^{\mathrm{G}} + \sum_{\ell | r(\ell)=n} p_{\ell t}^{\mathrm{L}} = \sum_{d \in \Psi_n^{\mathrm{D}}} p_{dt}^{\mathrm{D}} + \sum_{e \in \Psi_n^{\mathrm{E}}} p_{et}^{\mathrm{A}} + \sum_{\ell | s(\ell)=n} p_{\ell t}^{\mathrm{L}}, \quad \forall n \backslash n \in \Omega^{\mathrm{MG}}, \forall t. \tag{6.4b}$$

Constraint (6.4a) models the power balance at the node connected to the main grid, while constraint (6.4b) imposes the power balance at the remaining nodes of the system. At each node, the power generated plus the power injected from the lines arriving at that node must be equal to the power consumption of demands and charging stations, plus the power flows of lines departing from that node. In case of the node connected to the main grid, we also consider the power exchanged with the main grid.

6.4.3 Network constraints

At each time period t, the following network constraints must hold:

$$-\overline{P}^{\mathrm{MG}} \leq p_t^{\mathrm{DA}} \leq \overline{P}^{\mathrm{MG}}, \quad \forall t, \tag{6.5a}$$

$$p_{\ell t}^{\mathrm{L}} = \frac{1}{X_\ell}\left(\theta_{s(\ell)t} - \theta_{r(\ell)t}\right), \quad \forall \ell, \forall t, \tag{6.5b}$$

$$-\overline{P}_\ell^{\mathrm{L}} \leq p_{\ell t}^{\mathrm{L}} \leq \overline{P}_\ell^{\mathrm{L}}, \quad \forall \ell, \forall t, \tag{6.5c}$$

$$\underline{\Theta} \leq \theta_{nt} \leq \overline{\Theta}, \quad \forall n, \forall t. \tag{6.5d}$$

$$\theta_{nt} = 0 \quad \mathrm{n: ref.}, \quad \forall t. \tag{6.5e}$$

Constraint (6.5a) imposes bounds on the power bought from/sold to the main grid. Constraint (6.5b) defines the power flow through network lines. We use a DC power flow [10] without losses for the sake of simplicity. These power flows are bounded by the capacity of the lines in (6.5c). Constraint (6.5d) imposes bounds on voltage angles. Finally, constraint (6.5e) defines the voltage angle at the reference node.

6.4.4 Electric vehicle constraints

At each time period t, the driving needs of EV users, as well as the technical constraints of the EV charging stations must be satisfied. This is ensured by imposing the EV aggregator constraints (6.2) described in Section 6.3.

6.4.5 Demand constraints

We consider that flexible demands are connected at the different nodes of the SSEES. The technical constraints of these demands are represented using the following equations:

$$\underline{P}_{dt}^{\mathrm{D}} \leq p_{dt}^{\mathrm{D}} \leq \overline{P}_{dt}^{\mathrm{D}}, \quad \forall d, \forall t, \tag{6.6a}$$

$$\sum_t p_{dt}^{\mathrm{D}} \Delta t \geq E_d^{\mathrm{D}}, \quad \forall d. \tag{6.6b}$$

Constraint (6.6a) imposes bounds on the hourly power consumption of each demand. Constraint (6.6a) imposes a minimum daily energy consumption on each demand.

6.4.6 Renewable production constraints

At each time period t, the renewable power production must be below the maximum available production:

$$0 \leq p^{\mathrm{G}}_{gt} \leq P^{\mathrm{GA}}_{gt}, \quad \forall g, \forall t. \tag{6.7}$$

6.4.7 Formulation

Considering all the technical and economic constraints of the SSEES described in the previous sections, the energy-management problem can be finally formulated using the following deterministic model:

$$\min_{\Theta^{\mathrm{D}}}$$

$$\text{Objective function (6.3)} \tag{6.8a}$$

subject to

$$\text{Power balance constraints (6.4),} \tag{6.8b}$$

$$\text{Network constraints (6.5),} \tag{6.8c}$$

$$\text{Electric vehicles constraints (6.2),} \tag{6.8d}$$

$$\text{Demand constraints (6.6),} \tag{6.8e}$$

$$\text{Renewable production constraints (6.7),} \tag{6.8f}$$

where set $\Theta^{\mathrm{D}} = \{p^{\mathrm{DA}}_t, \forall t;\ p^{\mathrm{G}}_{gt}, \forall g, \forall t;\ p^{\mathrm{L}}_{\ell t}, \forall \ell, \forall t;\ p^{\mathrm{D}}_{dt}, \forall d, \forall t;\ p^{\mathrm{A}}_{et}, e^{\mathrm{A}}_{et}, \forall e, \forall t;\ \theta_{nt}, \forall n, \forall t\}$ includes the optimization variables of problem (6.8). Note that problem (6.4) is an LP problem [7].

Illustrative Example 6.3 Deterministic energy management

We consider the four-node electric energy system depicted in Figure 6.11.

This SSEES includes four lines whose data are provided in Table 6.3. The system comprises three demands located at nodes 2, 3, and 4. The hourly lower and upper bounds for these demands are provided in Figure 6.12 and are the same for the three demands. The minimum daily energy consumption of each demand is 70 MWh. There is a solar-power unit located at node 2, whose maximum hourly

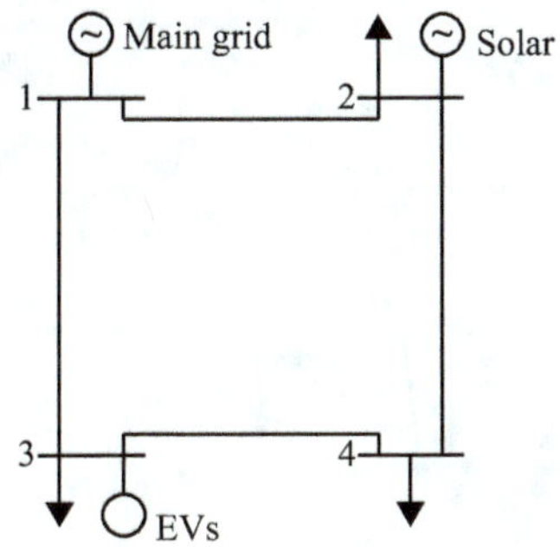

Figure 6.11 Illustrative Example 6.3: four-node system

Table 6.3 Illustrative Example 6.3: line data

Line	From node	To node	Reactance [Ω]	Capacity [MW]
1	1	2	0.20	6
2	1	3	0.30	6
3	2	3	0.25	6
4	2	4	0.30	6

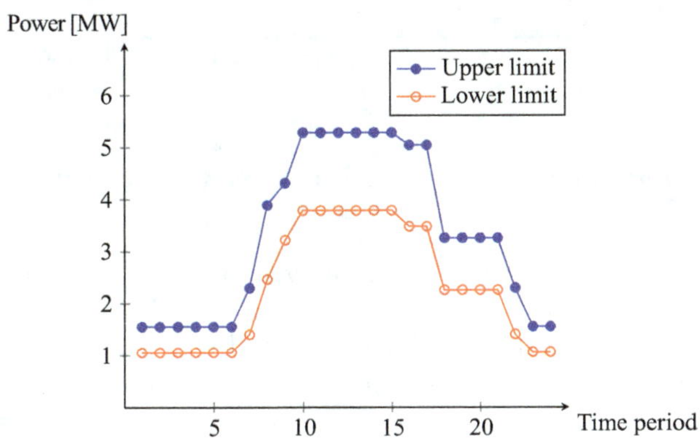

Figure 6.12 Illustrative Example 6.3: demand bounds

production is provided in Figure 6.13. An EV aggregation is located at node 3. The data defining this EV aggregation are those provided in Illustrative Example 6.2. The SSEES is connected to the main grid through node 1. The maximum power that can be interchanged with the main grid is 10 MW. The DA

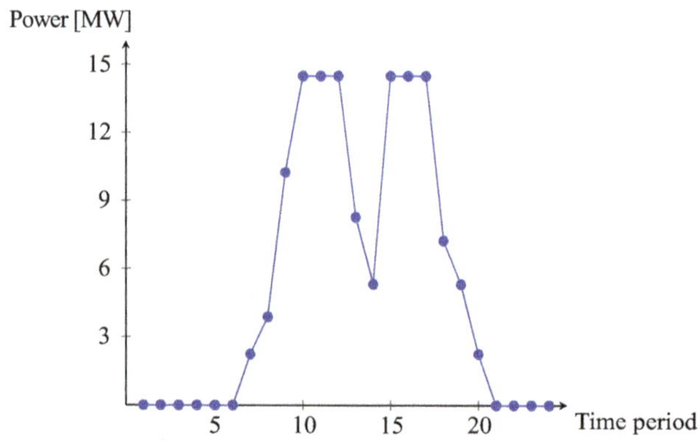

Figure 6.13 Illustrative Example 6.3: solar power availability

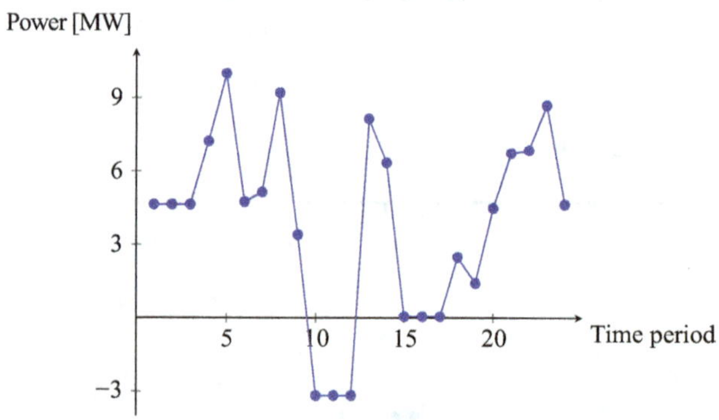

Figure 6.14 Illustrative Example 6.3: power interchanged with the main grid

market prices are those considered in Illustrative Example 6.1 (Figure 6.3). Finally, the reference node is node 1 and the voltage angle bounds are $-\pi$ and π rad.

Using these data, we solve problem (6.4) and obtain the results provided in Figures 6.14, 6.15, and 6.16, which depict the power interchanged with the main grid, the hourly demand consumptions, and the power used to charge the EV aggregation, respectively.

Figure 6.14 shows that the EMS in charge of the SSEES buys energy from the main grid at most of the time periods. However, at time periods 10, 11, and 12, it sells the surplus energy to the market. Note that these time periods correspond with time periods with comparatively high market prices (see Figure 6.3) and the highest solar power availability (see Figure 6.13).

Figures 6.15 and 6.16 show the power consumptions in the SSEES, including both demands and EV charging stations. Note that the EMS uses the flexibility of

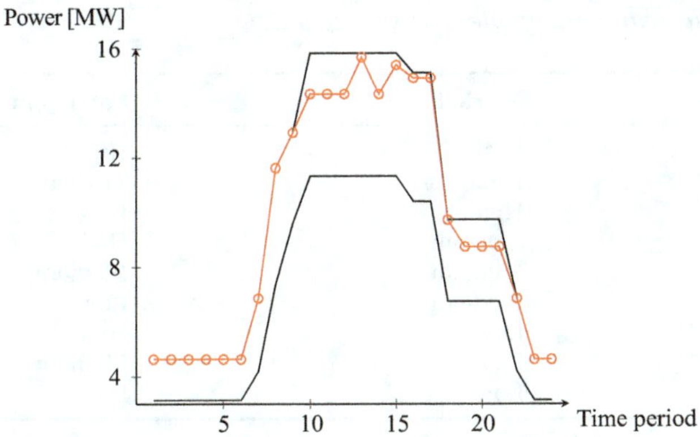

Figure 6.15 Illustrative Example 6.3: demand bounds

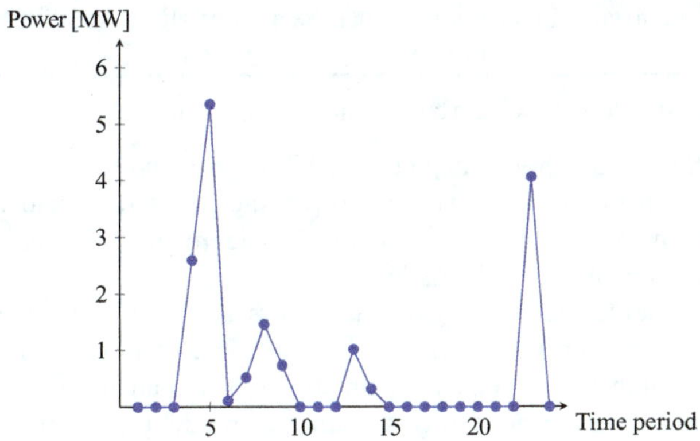

Figure 6.16 Illustrative Example 6.3: charging power

the demands and EVs to supply them at those time periods with the lowest market prices and/or highest solar power production.

The total cost is €6210.73.

A GAMS [9] code to solve this example is provided in Section 6.8.

6.5 Uncertainty characterization

The deterministic model described in Section 6.4 assumes that all parameters are known at the time the energy-management problem is solved. However, this is not the case since the market prices and the available production of renewable generating units are uncertain variables.

Table 6.4 Illustrative Example 6.4: scenario data

Scenario	Market price	Solar power availability
1	High	High
2	High	Medium
3	High	Low
4	Medium	High
5	Medium	Medium
6	Medium	Low
7	Low	High
8	Low	Medium
9	Low	Low

One possible way of representing this uncertainty is using a set of scenarios [11]. Each scenario represents a realization of the uncertain parameters and has associated a probability of occurrence in such a way that the sum of all probabilities is equal to 1.

Illustrative Example 6.4 Uncertainty characterization

We consider the market prices depicted in Figure 6.3 and the solar power availability provided in Figure 6.13. These two parameters were considered known in Illustrative Example 6.3, where the energy-management problem, considering a deterministic approach, was solved.

Next, we assume that these two parameters are subject to uncertainty. We consider that the market prices can be high, average, or low, meaning that these prices are 20% higher than, equal to, or 20% lower than the prices depicted in Figure 6.3. That is, we consider three scenarios to model the uncertainty in market prices. Similarly, we use three scenarios to model the uncertainty in the solar power availability. We assume that it can be high, average, or low, meaning that the solar productions are 20% higher than, equal to, or 20% lower than the productions depicted in Figure 6.13. Then, we assume that both parameters are independent for the sake of simplicity so that we consider all the possible scenario combinations, which results in a total of nine scenarios whose information is summarized in Table 6.4. For the sake of simplicity, we consider that all scenarios have the same probability of occurrence (1/9).

6.6 Stochastic energy-management problem

The deterministic problem (6.8) described in Section 6.4 is extended here to account for the uncertainties in market prices and production of renewable generating units. These uncertainties are modeled using a set of scenarios as described in Section 6.5. This allows formulating the energy-management problem as the following two-stage stochastic programming problem:

$$\min_{\Theta^{\mathrm{S}}}$$

$$\sum_{\omega} \pi_{\omega} \left[\sum_{t} \lambda_{t\omega}^{\mathrm{DA}} p_t^{\mathrm{DA}} \Delta t \right] \tag{6.9a}$$

subject to:

$$p_t^{\mathrm{DA}} + \sum_{g \in \Psi_n^{\mathrm{G}}} p_{gt\omega}^{\mathrm{G}} + \sum_{\ell | r(\ell)=n} p_{\ell t\omega}^{\mathrm{L}} = \sum_{d \in \Psi_n^{\mathrm{D}}} p_{dt\omega}^{\mathrm{D}} + \sum_{e \in \Psi_n^{\mathrm{E}}} p_{et\omega}^{\mathrm{A}}$$

$$n \in \Omega^{\mathrm{MG}}, \forall t, \forall \omega, \tag{6.9b}$$

$$\sum_{g \in \Psi_n^{\mathrm{G}}} p_{gt\omega}^{\mathrm{G}} + \sum_{\ell | r(\ell)=n} p_{\ell t\omega}^{\mathrm{L}} = \sum_{d \in \Psi_n^{\mathrm{D}}} p_{dt\omega}^{\mathrm{D}} + \sum_{e \in \Psi_n^{\mathrm{E}}} p_{et\omega}^{\mathrm{A}} + \sum_{\ell | s(\ell)=n} p_{\ell t\omega}^{\mathrm{L}},$$

$$\forall n \backslash n \in \Omega^{\mathrm{MG}}, \forall t, \forall \omega, \tag{6.9c}$$

$$-\overline{P}^{\mathrm{MG}} \leq p_t^{\mathrm{DA}} \leq \overline{P}^{\mathrm{MG}}, \quad \forall t, \tag{6.9d}$$

$$p_{\ell t\omega}^{\mathrm{L}} = \frac{1}{X_\ell} (\theta_{s(\ell)t\omega} - \theta_{r(\ell)t\omega}), \quad \forall \ell, \forall t, \forall \omega, \tag{6.9e}$$

$$-\overline{P}_\ell^{\mathrm{L}} \leq p_{\ell t\omega}^{\mathrm{L}} \leq \overline{P}_\ell^{\mathrm{L}}, \quad \forall \ell, \forall t, \forall \omega, \tag{6.9f}$$

$$\underline{\Theta} \leq \theta_{nt\omega} \leq \overline{\Theta}, \quad \forall n, \forall t, \forall \omega, \tag{6.9g}$$

$$\theta_{nt\omega} = 0, \quad n : \text{ref.}, \forall t, \forall \omega, \tag{6.9h}$$

$$e_{et\omega}^{\mathrm{A}} = e_{e(t-1)\omega}^{\mathrm{A}} + p_{et\omega}^{\mathrm{A}} \eta^{\mathrm{A}} \Delta t + E_{et}^{\mathrm{AA}} - E_{et}^{\mathrm{AD}}, \quad \forall e, \forall t, \forall \omega, \tag{6.9i}$$

$$e_{et_f\omega}^{\mathrm{A}} \geq E_{e0}^{\mathrm{A}}, \quad \forall e, \forall \omega, \tag{6.9j}$$

$$\underline{E}_{et}^{\mathrm{A}} \leq e_{et\omega}^{\mathrm{A}} \leq \overline{E}_{et}^{\mathrm{A}}, \quad \forall e, \forall t, \forall \omega, \tag{6.9k}$$

$$0 \leq p_{et\omega}^{\mathrm{A}} \leq \overline{P}_{et}^{\mathrm{A}}, \quad \forall e, \forall t, \forall \omega, \tag{6.9l}$$

$$\underline{P}_{dt}^{\mathrm{D}} \leq p_{dt\omega}^{\mathrm{D}} \leq \overline{P}_{dt}^{\mathrm{D}}, \quad \forall d, \forall t, \forall \omega, \tag{6.9m}$$

$$\sum_t p_{dt\omega}^{\mathrm{D}} \Delta t \geq E_d^{\mathrm{D}}, \quad \forall d, \forall \omega, \tag{6.9n}$$

$$0 \leq p_{gt\omega}^{\mathrm{G}} \leq P_{gt\omega}^{\mathrm{GA}}, \quad \forall g, \forall t, \forall \omega, \tag{6.9o}$$

where set $\Theta^S = \{p_t^{\mathrm{DA}}, \forall t;\ p_{gt\omega}^{\mathrm{G}}, \forall g, \forall t, \forall \omega;\ p_{\ell t\omega}^{\mathrm{L}}, \forall \ell, \forall t, \forall \omega;\ p_{dt\omega}^{\mathrm{D}}, \forall d, \forall t, \forall \omega;\ p_{et\omega}^{\mathrm{A}}, e_{et\omega}^{\mathrm{A}}, \forall e, \forall t, \forall \omega;\ \theta_{nt\omega}, \forall n, \forall t, \forall \omega\}$ includes the optimization variables of problem (6.9). Note that problem (6.9) is an LP problem [7].

Problem (6.9) is analogous to problem (6.8) with the following differences:

1. The aim of this problem is minimizing the expected cost (6.9a). Thus, the cost corresponding to each scenario ω is multiplied by the weight of the corresponding scenario.
2. Constraints (6.9b)–(6.9c) and (6.9e)–(6.9o) must be satisfied for all scenarios.

Problem (6.9) is a two-stage stochastic programming problem. In the first stage, we determine the power bought from/sold to the main grid, that is, variables p_t^{DA}, $\forall t$. These decisions are *here-and-now* since they do not depend on the future scenario realization. Therefore, variables p_t^{DA}, $\forall t$, do not include subscript ω. Then, in the second stage and once known the actual scenario realization, we determine the optimal operation of the problem, that is, variables $\Theta^{\mathrm{S}} \backslash p_t^{\mathrm{DA}}$. These decisions are *wait-and-see* since they do depend on the future scenario realization. Therefore, variables $\Theta^{\mathrm{S}} \backslash p_t^{\mathrm{DA}}$ do include subscript ω.

Illustrative Example 6.5 Stochastic energy management

We consider the four-node system described in Illustrative Example 6.3. We consider the same data but, in this case, we assume that market prices and the production of the solar-power unit are subject to uncertainty. In particular, we model this uncertainty using the nine scenarios described in Illustrative Example 6.4. Then, we solve the two-stage stochastic programming problem (6.9).

Results are provided in Figure 6.17, which depicts the power traded in the DA market for both the stochastic and deterministic problems. Note that both problems use the same data and that the uncertain parameters in the stochastic version of the problem have the same expected value that in the deterministic problem. However, the results in Figure 6.17 show that the decisions for both problems are different. The power traded in the DA market follows a similar pattern for both problems; however, decisions are different. For example, the amount of energy sold to the main grid at time periods 10, 11, and 12 is reduced in the stochastic problem. This highlights the importance of modeling the uncertainty in the problem. Using a deterministic problem where all parameters are modeled using their expected values simplifies the problem; however, decisions may be suboptimal.

In this case, the expected cost is 8192.60€, which is 31.91% higher than the cost obtained in the deterministic problem solved in Illustrative Example 6.3.

A GAMS [9] code to solve this example is provided in Section 6.8.

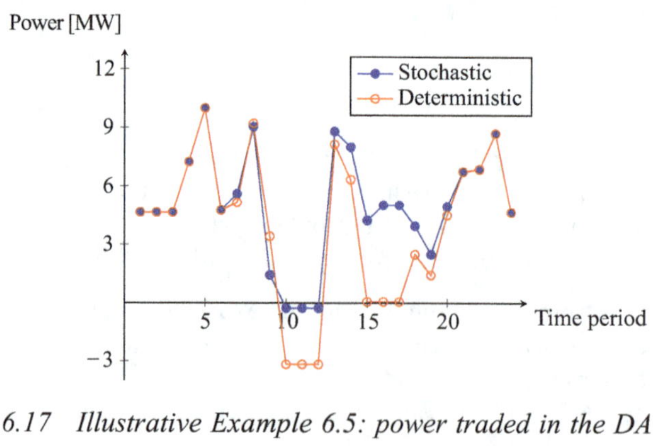

Figure 6.17 Illustrative Example 6.5: power traded in the DA market

---

6.7 Summary and conclusions

This chapter analyzes the optimal operation of a SSEES with EVs, flexible demands, and renewable generating units. First, we model EVs using the so-called EV aggregator. Second, we formulate a deterministic instance of the operation problem in which all parameters are known. Third, we model the uncertainty in the renewable production and market prices using a set of scenarios. Finally, the operation problem is formulated using a two-stage stochastic programming model.

Considering the theoretical framework and the numerical experiments carried out, the conclusions below are in order:

1. EV fleets can be effectively modeled using the so-called EV aggregator. This reduces both the computational complexity of the problem and the uncertainty associated with the EV driving needs.
2. The flexibility of consumers and EVs can be used to compensate the negative effects of the variability and uncertainty in the production of most renewable generating units.
3. An accurate modeling of the uncertainties is essential to obtain informed operation decisions.

Some suggestions to extend the models developed in this chapter are as follows:

1. To model EVs with Vehicle-To-Grid (V2G) technology, that is, EVs that can also inject energy into the network.
2. To consider also storage facilities in the SSEES.
3. To model the participation of the SSEES in other trading markets.
4. To model uncertain parameters using alternative methods such as robust optimization.

6.8 GAMS codes

This section provides the GAMS [9] codes used to solve some of the Illustrative Examples analyzed in this chapter.

6.8.1 *Illustrative example 6.1*

The GAMS [9] code used to solve Illustrative Example 6.1 is as follows:

```
*************************************************
***Illustrative Example: Individual EV modeling***
*************************************************
*Definition of indexes:
SETS
t          time periods /t1*t24/;
*Definition of parameters:
SCALARS
EI_min    minimum energy content /0/
EI_max    maximum energy content /30/
```

```
EI0        energy content at the beginning of planning horizon /10/
EFFI       charging efficiency /0.9/
DeltaT     duration of time steps /1/;
PARAMETER EI_con(t)        energy consumption
/t1   0.00,  t2   0.00,  t3   0.00,  t4   0.00,
 t5   0.00,  t6   0.00,  t7   0.00,  t8   9.00,
 t9   6.00,  t10  0.00,  t11  0.00,  t12  0.00,
 t13  0.00,  t14  0.00,  t15  0.00,  t16  4.00,
 t17  0.00,  t18  8.00,  t19  7.00,  t20  0.00,
 t21  0.00,  t22  0.00,  t23  0.00,  t24  0.00/;
PARAMETER PI_max(t)        maximum charging power
/t1   6.60,  t2   6.60,  t3   6.60,  t4   6.60,
 t5   6.60,  t6   6.60,  t7   6.60,  t8   0.00,
 t9   0.00,  t10  6.60,  t11  6.60,  t12  6.60,
 t13  6.60,  t14  6.60,  t15  6.60,  t16  0.00,
 t17  0.00,  t18  0.00,  t19  0.00,  t20  0.00,
 t21  6.60,  t22  6.60,  t23  6.60,  t24  6.60/;
PARAMETER price(t)         prices
/t1   58.44, t2   55.69, t3   53.90, t4   52.29,
 t5   52.10, t6   54.20, t7   63.89, t8   74.78,
 t9   76.39, t10  77.23, t11  77.23, t12  76.69,
 t13  76.39, t14  76.39, t15  75.00, t16  73.00,
 t17  73.73, t18  76.31, t19  78.94, t20  79.62,
 t21  78.81, t22  71.22, t23  68.45, t24  68.73/;
*Prices in     /kWh:
price(t)=0.001*price(t);
*Definition of optimization variables:
VARIABLES
eI(t)      energy content in the battery
pI(t)      charging power
z          value of the objective function;
*Definition of equations:
EQUATIONS
OF         objective function
E_EV1      energy evolution in the first time period
E_EV       energy evolution
E_F        energy at the end of the planning horizon
E_LIM1     minimum energy limit
E_LIM2     maximum energy limit
P_LIM1     minimum charging limit
P_LIM2     maximum charging limit;
OF..                z=e=sum(t,price(t)*pI(t)*DeltaT);
E_EV1..             eI('t1')=e=EI0+pI('t1')*EFFI*DeltaT-EI_con('t1');
E_EV(t)$(ORD(t) GT 1)..   eI(t)=e=eI(t-1)+pI(t)*EFFI*DeltaT-EI_con(t);
E_F..               eI('t24')=g=EI0;
E_LIM1(t)..         eI(t)=g=0;
E_LIM2(t)..          eI(t)=l=EI_max;
P_LIM1(t)..          pI(t)=g=0;
P_LIM2(t)..          pI(t)=l=PI_max(t);
*Definition of the model:
MODEL     IndividualEV     /all/;
*Solution of the model:
SOLVE IndividualEV USING LP MINIMIZING Z;
*Display of results:
DISPLAY z.l, PI.l, EI.l;
```

6.8.2 Illustrative example 6.2

The GAMS [9] code used to solve Illustrative Example 6.2 is as follows:

```
**************************************************
***Illustrative Example: Aggregator EV modeling***
**************************************************
*Definition of indexes:
SETS
t         time periods /t1*t24/;
*Definition of parameters:
SCALARS
EA0       energy content at the beginning of planning horizon /18/
EFFA      charging efficiency /0.9/
DeltaT    duration of time steps /1/;
PARAMETER EA_A(t)        energy contribution of arrival EVs
/t1  0.00,  t2  0.00,  t3  0.00,  t4  0.00,
 t5  0.00,  t6  0.45,  t7  1.65,  t8  4.20,
 t9  2.18,  t10 2.04,  t11 2.40,  t12 1.98,
 t13 2.10,  t14 2.55,  t15 3.00,  t16 2.33,
 t17 3.15,  t18 3.84,  t19 5.22,  t20 4.50,
 t21 2.16,  t22 2.52,  t23 1.95,  t24 2.42/;
PARAMETER EA_D(t)        energy drop of departing EVs
/t1  0.00,  t2  0.00,  t3  0.00,  t4  0.00,
 t5  1.00,  t6  1.60,  t7  3.80,  t8  7.40,
 t9  4.20,  t10 3.20,  t11 3.60,  t12 4.40,
 t13 2.40,  t14 3.60,  t15 4.70,  t16 4.00,
 t17 4.40,  t18 5.60,  t19 3.60,  t20 1.56,
 t21 1.08,  t22 0.84,  t23 0.60,  t24 3.60/;
PARAMETER PA_max(t)      maximum charging power
/t1  6.30,  t2  6.30,  t3  6.30,  t4  6.30,
 t5  6.04,  t6  5.78,  t7  5.35,  t8  4.89,
 t9  4.54,  t10 4.42,  t11 4.31,  t12 3.85,
 t13 3.95,  t14 3.90,  t15 3.72,  t16 3.48,
 t17 3.43,  t18 3.30,  t19 3.88,  t20 4.51,
 t21 4.67,  t22 5.03,  t23 5.45,  t24 6.14/;
PARAMETER EA_max(t)      maximum energy content
/t1  25.20, t2  25.20, t3  25.20, t4  25.20,
 t5  24.15, t6  23.10, t7  21.42, t8  19.53,
 t9  18.17, t10 17.66, t11 17.24, t12 15.39,
 t13 15.81, t14 15.60, t15 14.87, t16 13.92,
 t17 13.71, t18 13.21, t19 15.52, t20 18.04,
 t21 18.67, t22 20.14, t23 21.82, t24 24.57/;
PARAMETER EA_min(t)      minimum energy content
/t1  5.04,  t2  5.04,  t3  5.04,  t4  6.09,
 t5  6.51,  t6  8.61,  t7  12.05, t8  8.32,
 t9  6.99,  t10 7.31,  t11 8.07,  t12 5.60,
 t13 6.94,  t14 8.56,  t15 7.18,  t16 7.40,
 t17 8.62,  t18 6.42,  t19 4.74,  t20 4.74,
 t21 4.62,  t22 7.66,  t23 4.74,  t24 4.91/;
PARAMETER price(t)       prices
/t1  58.44, t2  55.69, t3  53.90, t4  52.29,
 t5  52.10, t6  54.20, t7  63.89, t8  74.78,
 t9  76.39, t10 77.23, t11 77.23, t12 76.69,
 t13 76.39, t14 76.39, t15 75.00, t16 73.00,
 t17 73.73, t18 76.31, t19 78.94, t20 79.62,
 t21 78.81, t22 71.22, t23 68.45, t24 68.73/;
```

```
*Definition of optimization variables:
VARIABLES
eA(t)      energy content in the battery
pA(t)      charging power
z          value of the objective function;
*Definition of equations:
EQUATIONS
OF         objective function
E_EV1      energy evolution in the first time period
E_EV       energy evolution
E_F        energy at the end of the planning horizon
E_LIM1     minimum energy limit
E_LIM2     maximum energy limit
P_LIM1     minimum charging limit
P_LIM2     maximum charging limit;
OF..                   z=e=sum(t,price(t)*pA(t)*DeltaT);
E_EV1..                eA('t1')=e=EA0+pA('t1')*EFFA*DeltaT+EA_A('t1')-EA_D('t1');
E_EV(t)$(ORD(t) GT 1)..  eA(t)=e=eA(t-1)+pA(t)*EFFA*DeltaT+EA_A(t)-EA_D(t);
E_F..                  eA('t24')=g=EA0;
E_LIM1(t)..            eA(t)=g=EA_min(t);
E_LIM2(t)..            eA(t)=l=EA_max(t);
P_LIM1(t)..            pA(t)=g=0;
P_LIM2(t)..            pA(t)=l=PA_max(t);
*Definition of the model:
MODEL     AggregationEV     /all/;
*Solution of the model:
SOLVE AggregationEV USING LP MINIMIZING Z;
*Display of results:
DISPLAY z.l, PA.l, EA.l;
```

6.8.3 Illustrative example 6.3

The GAMS [9] code used to solve Illustrative Example 6.3 is as follows:

```
****************************************************
***Illustrative Example: Deterministic problem****
****************************************************
*Definition of indexes:
SETS
a          EV aggregations         /a1*a1/
d          demands                 /d1*d3/
g          renewable units         /g1*g1/
l          lines                   /l1*l4/
n          nodes                   /n1*n4/
t          time periods            /t1*t24/
se(l,n)    sending node lines      /l1.n1,l2.n1,l3.n2,l4.n3/
re(l,n)    receiving node lines    /l1.n2,l2.n3,l3.n4,l4.n4/
OmeA(a,n)  EV connections          /a1.n3/
OmeD(d,n)  demand connections      /d1.n2,d1.n3,d1.n4/
OmeG(g,n)  renewable connections   /g1.n2/
OmeMG(n)   main grid connection    /n1/;
```

```
*Definition of parameters:
SCALARS
PMGmax    maximum power bought from or sold to the main grid /10/
ED        minimum daily energy consumption /70/
EA0       energy content at the beginning of planning horizon /18/
EFFA      charging efficiency /0.9/
DeltaT    duration of time steps /1/;
PARAMETER X(l)              reactances of lines
/l1  0.2,    l2  0.3,    l3  0.25,   l4  0.3/;
PARAMETER PLmax(l)          capacities of lines
/l1  6,    l2  6,    l3  6,    l4  6/;
PARAMETER EA_A(t)           energy contribution of arrival EVs
/t1  0.00,  t2  0.00,  t3  0.00,  t4  0.00,
 t5  0.00,  t6  0.45,  t7  1.65,  t8  4.20,
 t9  2.18,  t10 2.04,  t11 2.40,  t12 1.98,
 t13 2.10,  t14 2.55,  t15 3.00,  t16 2.33,
 t17 3.15,  t18 3.84,  t19 5.22,  t20 4.50,
 t21 2.16,  t22 2.52,  t23 1.95,  t24 2.42/;
PARAMETER EA_D(t)           energy drop of departing EVs
/t1  0.00,  t2  0.00,  t3  0.00,  t4  0.00,
 t5  1.00,  t6  1.60,  t7  3.80,  t8  7.40,
 t9  4.20,  t10 3.20,  t11 3.60,  t12 4.40,
 t13 2.40,  t14 3.60,  t15 4.70,  t16 4.00,
 t17 4.40,  t18 5.60,  t19 3.60,  t20 1.56,
 t21 1.08,  t22 0.84,  t23 0.60,  t24 3.60/;
PARAMETER PA_max(t)         maximum charging power
/t1  6.30,  t2  6.30,  t3  6.30,  t4  6.30,
 t5  6.04,  t6  5.78,  t7  5.35,  t8  4.89,
 t9  4.54,  t10 4.42,  t11 4.31,  t12 3.85,
 t13 3.95,  t14 3.90,  t15 3.72,  t16 3.48,
 t17 3.43,  t18 3.30,  t19 3.88,  t20 4.51,
 t21 4.67,  t22 5.03,  t23 5.45,  t24 6.14/;
PARAMETER EA_max(t)         maximum energy content
/t1  25.20, t2  25.20, t3  25.20, t4  25.20,
 t5  24.15, t6  23.10, t7  21.42, t8  19.53,
 t9  18.17, t10 17.66, t11 17.24, t12 15.39,
 t13 15.81, t14 15.60, t15 14.87, t16 13.92,
 t17 13.71, t18 13.21, t19 15.52, t20 18.04,
 t21 18.67, t22 20.14, t23 21.82, t24 24.57/;
PARAMETER EA_min(t)         minimum energy content
/t1  5.04,  t2  5.04,  t3  5.04,  t4  6.09,
 t5  6.51,  t6  8.61,  t7  12.05, t8  8.32,
 t9  6.99,  t10 7.31,  t11 8.07,  t12 5.60,
 t13 6.94,  t14 8.56,  t15 7.18,  t16 7.40,
 t17 8.62,  t18 6.42,  t19 4.74,  t20 4.74,
 t21 4.62,  t22 7.66,  t23 4.74,  t24 4.91/;
PARAMETER price_DA(t)          prices
/t1  58.44, t2  55.69, t3  53.90, t4  52.29,
 t5  52.10, t6  54.20, t7  63.89, t8  74.78,
 t9  76.39, t10 77.23, t11 77.23, t12 76.69,
 t13 76.39, t14 76.39, t15 75.00, t16 73.00,
 t17 73.73, t18 76.31, t19 78.94, t20 79.62,
 t21 78.81, t22 71.22, t23 68.45, t24 68.73/;
```

```
PARAMETER PDmin(t)          demand lower limit
/t1  1.05,  t2  1.05,  t3  1.05,  t4  1.05,
 t5  1.05,  t6  1.05,  t7  1.40,  t8  2.46,
 t9  3.21,  t10 3.78,  t11 3.78,  t12 3.78,
 t13 3.78,  t14 3.78,  t15 3.78,  t16 3.47,
 t17 3.47,  t18 2.25,  t19 2.25,  t20 2.25,
 t21 2.25,  t22 1.40,  t23 1.05,  t24 1.05/;
PARAMETER PDmax(t)          demand upper limit
/t1  1.55,  t2  1.55,  t3  1.55,  t4  1.55,
 t5  1.55,  t6  1.55,  t7  2.29,  t8  3.88,
 t9  4.31,  t10 5.28,  t11 5.28,  t12 5.28,
 t13 5.28,  t14 5.28,  t15 5.28,  t16 5.04,
 t17 5.04,  t18 3.25,  t19 3.25,  t20 3.25,
 t21 3.25,  t22 2.29,  t23 1.55,  t24 1.55/;
PARAMETER PGmax(t)          renewable power limit
/t1  0,     t2  0,     t3  0,     t4  0,
 t5  0,     t6  0,     t7  2.24,  t8  3.88,
 t9  10.25, t10 14.50, t11 14.50, t12 14.50,
 t13 8.28,  t14 5.31,  t15 14.50, t16 14.50,
 t17 14.50, t18 7.25,  t19 5.31,  t20 2.25,
 t21 0,     t22 0,     t23 0,     t24 0/;
*Definition of optimization variables:
VARIABLES
p_DA(t)  power in the DA market
pD(d,t)  power demand
pG(g,t)  power from renewable units
pL(l,t)  power flow through lines
ANG(n,t) voltage angles
eA(a,t)  EV energy content in the battery
pA(a,t)  EV charging power
z        value of the objective function;
ANG.fx('n1',t)=0;
*Definition of equations:
EQUATIONS
OF       objective function
PB1      power balance (main grid connection)
PB2      power balance (remaining nodes)
NetL1    network limits I
NetL2    network limits II
NetL3    network limits III
NetL4    network limits IV
PF       power flow through lines
PFL1     power flow limits I
PFL2     power flow limits II
E_EV1    EV energy evolution in the first time period
E_EV     EV energy evolution
E_F      EV energy at the end of the planning horizon
E_LIM1   EV minimum energy limit
E_LIM2   EV maximum energy limit
P_LIM1   EV minimum charging limit
P_LIM2   EV maximum charging limit
PDL1     demand limits I
PDL2     demand limits II
PDL3     demand limits II
PGL1     renewable power limits I
PGL2     renewable power limits II;
```

```
OF..                          z=e=sum(t,price_DA(t)*p_DA(t)*DeltaT);
PB1(n,t)$OmeMG(n)..           p_DA(t)+sum(g$OmeG(g,n),pG(g,t))+sum(l$re(l
     ,n),pL(l,t))=e=sum(d$OmeD(d,n),pD(d,t))+sum(a$OmeA(a,n),pA(a,t))+
     sum(l$se(l,n),pL(l,t));
PB2(n,t)$(not OmeMG(n)).. sum(g$OmeG(g,n),pG(g,t))+sum(l$re(l,n),pL(l
     ,t))=e=sum(d$OmeD(d,n),pD(d,t))+sum(a$OmeA(a,n),pA(a,t))+sum(l$se
     (l,n),pL(l,t));
NetL1(t)..                    -PMGmax=l=p_DA(t);
NetL2(t)..                    p_DA(t)=l=PMGmax;
NetL3(n,t)..                  -3.1416=l=ANG(n,t);
NetL4(n,t)..                  ANG(n,t)=l=3.1416;
PF(l,t)..                     pL(l,t)=e=(1/X(l))*(sum(n$se(l,n),ANG(n,t))
     -sum(n$re(l,n),ANG(n,t)));
PFL1(l,t)..                    -PLmax(l)=l=pL(l,t);
PFL2(l,t)..                    pL(l,t)=L=PLmax(l);
E_EV1(a)..                     eA(a,'t1')=e=EA0+pA(a,'t1')*EFFA*DeltaT+
     EA_A('t1')-EA_D('t1');
E_EV(a,t)$(ORD(t) GT 1).. eA(a,t)=e=eA(a,t-1)+pA(a,t)*EFFA*DeltaT+
     EA_A(t)-EA_D(t);
E_F(a)..                       eA(a,'t24')=g=EA0;
E_LIM1(a,t)..                  eA(a,t)=g=EA_min(t);
E_LIM2(a,t)..                  eA(a,t)=l=EA_max(t);
P_LIM1(a,t)..                  0=l=pA(a,t);
P_LIM2(a,t)..                  pA(a,t)=l=PA_max(t);
PDL1(d,t)..                    PDmin(t)=l=pD(d,t);
PDL2(d,t)..                    pD(d,t)=L=PDmax(t);
PDL3(d)..                      sum(t,pD(d,t))=G=ED;
PGL1(g,t)..                    0=l=pG(g,t);
PGL2(g,t)..                    pG(g,t)=l=PGmax(t);
*Definition of the model:
MODEL     DeterministicEMS   /all/;
*Solution of the model:
SOLVE DeterministicEMS USING LP MINIMIZING Z;
*Display of results:
Parameter Dtotal(t);
Dtotal(t)=sum(d,pd.l(d,t));
DISPLAY z.l, p_DA.l, Dtotal, pA.l;
```

6.8.4 Illustrative example 6.5

The GAMS [9] code used to solve Illustrative Example 6.5 is as follows:

```
*************************************************
***Illustrative Example: Stochastic problem ******
*************************************************
*Definition of indexes:
SETS
a          EV aggregations          /a1*a1/
d          demands                  /d1*d3/
g          renewable units          /g1*g1/
l          lines                    /l1*l4/
n          nodes                    /n1*n4/
w          scenarios                /w1*w9/
t          time periods             /t1*t24/
```

```
se(l,n)    sending node lines       /l1.n1,l2.n1,l3.n2,l4.n3/
re(l,n)    receiving node lines     /l1.n2,l2.n3,l3.n4,l4.n4/
OmeA(a,n)  EV connections           /a1.n3/
OmeD(d,n)  demand connections       /d1.n2,d1.n3,d1.n4/
OmeG(g,n)  renewable connections    /g1.n2/
OmeMG(n)   main grid connection     /n1/;
*Definition of parameters:
SCALARS
PMGmax   maximum power bought from or sold to the main grid /10/
ED       minimum daily energy consumption /70/
EA0      energy content at the beginning of planning horizon /18/
EFFA     charging efficiency /0.9/
DeltaT   duration of time steps /1/;
PARAMETER X(l)              reactances of lines
/l1  0.2,   l2  0.3,   l3  0.25,  l4  0.3/;
PARAMETER PLmax(l)          capacities of lines
/l1  6,   l2  6,   l3  6,   l4  6/;
PARAMETER EA_A(t)           energy contribution of arrival EVs
/t1  0.00,  t2  0.00,  t3  0.00,  t4  0.00,
 t5  0.00,  t6  0.45,  t7  1.65,  t8  4.20,
 t9  2.18,  t10 2.04,  t11 2.40,  t12 1.98,
 t13 2.10,  t14 2.55,  t15 3.00,  t16 2.33,
 t17 3.15,  t18 3.84,  t19 5.22,  t20 4.50,
 t21 2.16,  t22 2.52,  t23 1.95,  t24 2.42/;
PARAMETER EA_D(t)           energy drop of departing EVs
/t1  0.00,  t2  0.00,  t3  0.00,  t4  0.00,
 t5  1.00,  t6  1.60,  t7  3.80,  t8  7.40,
 t9  4.20,  t10 3.20,  t11 3.60,  t12 4.40,
 t13 2.40,  t14 3.60,  t15 4.70,  t16 4.00,
 t17 4.40,  t18 5.60,  t19 3.60,  t20 1.56,
 t21 1.08,  t22 0.84,  t23 0.60,  t24 3.60/;
PARAMETER PA_max(t)         maximum charging power
/t1  6.30,  t2  6.30,  t3  6.30,  t4  6.30,
 t5  6.04,  t6  5.78,  t7  5.35,  t8  4.89,
 t9  4.54,  t10 4.42,  t11 4.31,  t12 3.85,
 t13 3.95,  t14 3.90,  t15 3.72,  t16 3.48,
 t17 3.43,  t18 3.30,  t19 3.88,  t20 4.51,
 t21 4.67,  t22 5.03,  t23 5.45,  t24 6.14/;
PARAMETER EA_max(t)         maximum energy content
/t1  25.20, t2  25.20, t3  25.20, t4  25.20,
 t5  24.15, t6  23.10, t7  21.42, t8  19.53,
 t9  18.17, t10 17.66, t11 17.24, t12 15.39,
 t13 15.81, t14 15.60, t15 14.87, t16 13.92,
 t17 13.71, t18 13.21, t19 15.52, t20 18.04,
 t21 18.67, t22 20.14, t23 21.82, t24 24.57/;
PARAMETER EA_min(t)         minimum energy content
/t1  5.04,  t2  5.04,  t3  5.04,  t4  6.09,
 t5  6.51,  t6  8.61,  t7  12.05, t8  8.32,
 t9  6.99,  t10 7.31,  t11 8.07,  t12 5.60,
 t13 6.94,  t14 8.56,  t15 7.18,  t16 7.40,
 t17 8.62,  t18 6.42,  t19 4.74,  t20 4.74,
 t21 4.62,  t22 7.66,  t23 4.74,  t24 4.91/;
PARAMETER price_DA(t)          prices
/t1  58.44, t2  55.69, t3  53.90, t4  52.29,
 t5  52.10, t6  54.20, t7  63.89, t8  74.78,
 t9  76.39, t10 77.23, t11 77.23, t12 76.69,
```

```
 t13 76.39, t14 76.39, t15 75.00, t16 73.00,
 t17 73.73, t18 76.31, t19 78.94, t20 79.62,
 t21 78.81, t22 71.22, t23 68.45, t24 68.73/;
Parameter price_DASc(t,w);
price_DASc(t,'w1')=1.2*price_DA(t);
price_DASc(t,'w2')=1.2*price_DA(t);
price_DASc(t,'w3')=1.2*price_DA(t);
price_DASc(t,'w4')=1*price_DA(t);
price_DASc(t,'w5')=1*price_DA(t);
price_DASc(t,'w6')=1*price_DA(t);
price_DASc(t,'w7')=0.8*price_DA(t);
price_DASc(t,'w8')=0.8*price_DA(t);
price_DASc(t,'w9')=0.8*price_DA(t);
PARAMETER PDmin(t)            demand lower limit
/t1  1.05,  t2  1.05,  t3  1.05,  t4  1.05,
 t5  1.05,  t6  1.05,  t7  1.40,  t8  2.46,
 t9  3.21,  t10 3.78,  t11 3.78,  t12 3.78,
 t13 3.78,  t14 3.78,  t15 3.78,  t16 3.47,
 t17 3.47,  t18 2.25,  t19 2.25,  t20 2.25,
 t21 2.25,  t22 1.40,  t23 1.05,  t24 1.05/;
PARAMETER PDmax(t)            demand upper limit
/t1  1.55,  t2  1.55,  t3  1.55,  t4  1.55,
 t5  1.55,  t6  1.55,  t7  2.29,  t8  3.88,
 t9  4.31,  t10 5.28,  t11 5.28,  t12 5.28,
 t13 5.28,  t14 5.28,  t15 5.28,  t16 5.04,
 t17 5.04,  t18 3.25,  t19 3.25,  t20 3.25,
 t21 3.25,  t22 2.29,  t23 1.55,  t24 1.55/;
PARAMETER PGmax(t)            renewable power limit
/t1  0,     t2  0,     t3  0,     t4  0,
 t5  0,     t6  0,     t7  2.24,  t8  3.88,
 t9  10.25, t10 14.50, t11 14.50, t12 14.50,
 t13 8.28,  t14 5.31,  t15 14.50, t16 14.50,
 t17 14.50, t18 7.25,  t19 5.31,  t20 2.25,
 t21 0,     t22 0,     t23 0,     t24 0/;
PARAMETER PGmaxSc(t,w);
PGmaxSc(t,'w1')=1.2*PGmax(t);
PGmaxSc(t,'w2')=1*PGmax(t);
PGmaxSc(t,'w3')=0.8*PGmax(t);
PGmaxSc(t,'w4')=1.2*PGmax(t);
PGmaxSc(t,'w5')=1.0*PGmax(t);
PGmaxSc(t,'w6')=0.8*PGmax(t);
PGmaxSc(t,'w7')=1.2*PGmax(t);
PGmaxSc(t,'w8')=1.0*PGmax(t);
PGmaxSc(t,'w9')=0.8*PGmax(t);
PARAMETER weight(w);
weight(w)=0.111111111111;
*Definition of optimization variables:
VARIABLES
p_DA(t)      power in the DA market
pD(d,t,w)    power demand
pG(g,t,w)    power from renewable units
pL(l,t,w)    power flow through lines
ANG(n,t,w)   voltage angles
eA(a,t,w)    EV energy content in the battery
pA(a,t,w)    EV charging power
z            value of the objective function;
ANG.fx('n1',t,w)=0;
```

```
*Definition of equations:
EQUATIONS
OF          objective function
PB1         power balance (main grid connection)
PB2         power balance (remaining nodes)
NetL1       network limits I
NetL2       network limits II
NetL3       network limits III
NetL4       network limits IV
PF          power flow through lines
PFL1        power flow limits I
PFL2        power flow limits II
E_EV1       EV energy evolution in the first time period
E_EV        EV energy evolution
E_F         EV energy at the end of the planning horizon
E_LIM1      EV minimum energy limit
E_LIM2      EV maximum energy limit
P_LIM1      EV minimum charging limit
P_LIM2      EV maximum charging limit
PDL1        demand limits I
PDL2        demand limits II
PDL3        demand limits II
PGL1        renewable power limits I
PGL2        renewable power limits II;
OF..                          z=e=sum(w,weight(w)*sum(t,price_DASc(t,w)
     *p_DA(t)*DeltaT));
PB1(n,t,w)$OmeMG(n)..         p_DA(t)+sum(g$OmeG(g,n),pG(g,t,w))+sum(
     l$re(l,n),pL(l,t,w))=e=sum(d$OmeD(d,n),pD(d,t,w))+sum(a$OmeA(a,n)
     ,pA(a,t,w))+sum(l$se(l,n),pL(l,t,w));
PB2(n,t,w)$(not OmeMG(n)).. sum(g$OmeG(g,n),pG(g,t,w))+sum(l$re(l,n),
     pL(l,t,w))=e=sum(d$OmeD(d,n),pD(d,t,w))+sum(a$OmeA(a,n),pA(a,t,w)
     )+sum(l$se(l,n),pL(l,t,w));
NetL1(t)..                    -PMGmax=l=p_DA(t);
NetL2(t)..                    p_DA(t)=l=PMGmax;
NetL3(n,t,w)..                -3.1416=l=ANG(n,t,w);
NetL4(n,t,w)..                ANG(n,t,w)=l=3.1416;
PF(l,t,w)..                   pL(l,t,w)=e=(1/X(l))*(sum(n$se(l,n),ANG(n
     ,t,w))-sum(n$re(l,n),ANG(n,t,w)));
PFL1(l,t,w)..                 -PLmax(l)=l=pL(l,t,w);
PFL2(l,t,w)..                 pL(l,t,w)=L=PLmax(l);
E_EV1(a,w)..                  eA(a,'t1',w)=e=EA0+pA(a,'t1',w)*EFFA*
     DeltaT+EA_A('t1')-EA_D('t1');
E_EV(a,t,w)$(ORD(t) GT 1).. eA(a,t,w)=e=eA(a,t-1,w)+pA(a,t,w)*EFFA*
     DeltaT+EA_A(t)-EA_D(t);
E_F(a,w)..                    eA(a,'t24',w)=g=EA0;
E_LIM1(a,t,w)..               eA(a,t,w)=g=EA_min(t);
E_LIM2(a,t,w)..               eA(a,t,w)=l=EA_max(t);
P_LIM1(a,t,w)..               0=l=pA(a,t,w);
P_LIM2(a,t,w)..               pA(a,t,w)=l=PA_max(t);
PDL1(d,t,w)..                 PDmin(t)=l=pD(d,t,w);
PDL2(d,t,w)..                 pD(d,t,w)=L=PDmax(t);
PDL3(d,w)..                   sum(t,pD(d,t,w))=g=ED;
PGL1(g,t,w)..                 0=l=pG(g,t,w);
PGL2(g,t,w)..                 pG(g,t,w)=l=PGmaxSc(t,w);
*Definition of the model:
MODEL      DeterministicEMS     /all/;
*Solution of the model:
SOLVE DeterministicEMS USING LP MINIMIZING z;
*Display of results:
DISPLAY z.l, p_DA.l;
```

References

[1] Morales, J.M., Conejo, A.J., Madsen, H., Pinson, P., and Zugno, M.: Integrating Renewables in Electricity Markets: Operational Problems. Springer, New York (2014)

[2] Conejo, A.J., Morales, J.M., and Baringo, L.: Real-Time Demand Response Model. IEEE Transactions on Smart Grid. 1(3), 236–242 (2010)

[3] Zakariazadeh, A., Jadid, S., and Siano, P.: Integrated Operation of Electric Vehicles and Renewable Generation in a Smart Distribution System. Energy Conversion and Management. 89, 99–110 (2015)

[4] Shahidehpour, M., Yamin, H., and Li, Z.: Market Operations in Electric Power Systems.Wiley, New York (2002)

[5] Birge, J.R., and Louveaux, F.: Introduction to Stochastic Programming, 2nd ed. Springer, New York (2011)

[6] González Vayá, M., Baringo, L., Krause, T., Andersson, G., Almeida, P., Rapoport, S., and Geth, F.: EV Aggregation Models for Different Charging Scenarios. In 23rd International Conference and Exhibition on Electricity Distribution (CIRED), Lyon, France (2015)

[7] Sioshansi, R., and Conejo, A.J.: Optimization in Engineering. Models and Algorithms. Springer, New York (2017)

[8] IBM ILOG CPLEX. Available at: www.ibm.com/analytucs/cplex-optimizer (2018)

[9] Rosenthal, R.E.: GAMS, A User's Guide. GAMS Development Corporation, Washington (2012)

[10] Conejo, A.J., and Baringo, L: Power System Operations. Springer, Switzerland (2018)

[11] Conejo, A.J., Carrión, M., and Morales, J.M.: Decision Making Under Uncertainty in Electricity Markets. Springer, New York (2010)

Chapter 7

Peer-to-peer energy market between electric vehicles

Roberto Álvaro-Hermana[1,2], Julia Merino[3], Jesús Fraile-Ardanuy[4], Sandra Castaño-Solis[5] and David Jiménez[4]

Electric vehicles (EVs) are a major component of future electric grids, both for the increase in electricity demand and the flexibility they can add to the grid. Vehicle-to-grid and vehicle-to-building pilots have been tested and some have been approved by grid operators, but the EVs' possibilities shall be further enhanced. In previous works, the authors proposed a peer-to-peer energy market between EVs that largely reduced the expenses of their costly day-charging. This chapter further expands the model by taking into account the long-term effects of the market, which reduce the impact of the electric grid prices forecast on the market. The ratio between EVs that offer energy and those that demand energy is shown to be a good indicator for the market price forecast. Almost all energy demand occurs in pairs zone-time in which the number of offering EVs is more than five times the number of demanding EVs, for which the market price is very close to the electricity price at night.

List of abbreviations

EU	European Union
EV	Electric vehicle
FEATHERS	*Forecasting Evolutionary Activity Travel of Households and their Environmental RepercussionS*
GHG	Greenhouse gas
ICE	Internal combustion engine

[1]Orkestra – Basque Institute of Competitiveness, Bilbao, Vizcaya, Spain
[2]University of Deusto, Bilbao, Vizcaya, Spain
[3]TECNALIA, Parque Científico y Tecnológico de Bizkaia, Derio, Spain
[4]Information Processing and Telecommunications Center (IPTC), Universidad Politécnica de Madrid, Madrid, Spain
[5]Escuela Técnica Superior de Ingeniería y Diseño Industrial (ETSIDI), Universidad Politécnica de Madrid, Madrid, Spain

IMOB	*Instituut Moor Mobiliteit*
OVG	Research Mobility Behaviour (*OnderzoekVerplaatsingsgedrag*)
P2P	Peer-to-peer
P2PEM	P2P energy market
SOC	State of charge
t_0	Start time
t_1	End time
TAZ	Traffic zone
V2G	Vehicle-to-grid
WHO	World Health Organization

List of parameters and variables

$\boldsymbol{c}$	vector that defines $f(\boldsymbol{x})$
c_i	total day travel consumption of the i-th EV
D	total amount of energy demanded by all vehicles from set B in a specific TAZ at a particular time slot t in P2PEM (kWh)
$e_{ch}(t)$	energy extracted from the electric grid to charge the vehicle battery (kWh)
E^{max}	charging station power in a time slot t (kWh)
$f(\boldsymbol{x})$	average cost function of the P2PEM problem
$g_j(\boldsymbol{x})$	linear restriction inequality of the quadratic P2PEM
$h(\boldsymbol{x})$	linear restriction equality of the quadratic P2PEM
$i(t)$	energy injected into the battery [or extracted from it] (kWh)
J	a subset of indexes for which $x_j^* > x_i^{max} \forall j \in J$
n	number of EVs from set A parked at time slot t in a particular TAZ in P2PEM
$O(t)$	energy consumption due to driving activity (kWh)
p_i	final price in a specific TAZ at a particular time slot t in P2PEM (kWh)
P_{max}	charging station power (kW)
p^{max}	maximum electricity price for the time slot t in P2PEM [equal to the grid electricity price] (€/kWh)
p^{min}	minimum electricity price in P2PEM (€/kWh)
$Pr(t)$	grid electricity hourly price (€/kWh)
$\boldsymbol{Q}$	diagonal matrix that defines $f(\boldsymbol{x})$
$R_{ev}(t, TAZ)$	the ratio between the energy available for sharing and the energy demanded by the EVs from set B at time slot t and TAZ*TAZ* (-)
$soc(t)$	effective battery SOC (kWh)
SOC_{max}	SOC upper limit (kWh)
SOC_{min}	SOC lower limit (kWh)

T_{arr}	final arrival time (time slot number)
T_{dep}	initial departure time (time slot number)
$\boldsymbol{x}^*$	solution of the quadratic P2PEM problem
x_i	energy extracted from the *i*-th EV in P2PEM (kWh)
x_i^0	maximum amount of energy to share trading by the *i*-th EV in time slot *t* in P2PEM
x_i^{max}	maximum energy deliverable by the *i*-th EV at a time slot *t* in P2PEM
α_i	slope of the *i*-th EV's price curve
γ_{eff}	battery discharge rate (-)
Δt	length of the time slot (h)
η_c	charging efficiency rate (-)
μ_j	Karush–Kuhn–Tucker coefficients of the P2PEM quadratic problem.
Ω	feasible compact region of the P2PEM problem
Ω_1	set of points of Ω that satisfies the P2PEM problem

7.1 Introduction

The burning of fossil fuels releases a large diversity of unhealthy substances into the air. These substances, gas or solids, highly affect population health, causing asthma and other cardiorespiratory diseases. This is accentuated in urban environments. According to the World Health Organization (WHO), approximately 7 million premature deaths per year are directly related to air pollution worldwide [1]. These pollutants also contribute to global warming, the retreating of glaciers, the increase of the sea level and affecting the rainfall, among other consequences [2]. Globally, the most important causes of greenhouse gas (GHG) emissions are electricity and heat (31%) and transportation (15%) [3].

Governments are promoting different initiatives to reduce GHG emissions as well as improve air quality in urban areas. In particular, the European Union (EU) has developed ambitious plans to reduce GHG emissions up to 85%–90% compared to 1990 levels by 2050, and has also set different roadmaps to reduce air pollutants by 2020 (20%) and 2030 (40%). The EU is committed to cut these GHG emissions through domestic reductions alone, rather than relying on international credits [4].

The electrification of the road transportation represents an essential part of these plans, since the transport sector is one of the largest GHG and air pollutant producers in Europe, with almost a quarter of Europe's GHG; it is also the main responsible for air pollution in cities, according to [5]. By 2050, GHG from transport should be at least 60% lower than in 1990. In order to accelerate the deployment of electric vehicles (EVs), EU Governments are promoting them through different initiatives such as taxation policies, consumer subsidies and other local incentives (free parking, access to bus lanes, etc.) [6].

Despite the diverse advantages of the electrification of the road transportation (zero tail-pipe emissions and noise reduction in urban areas, lower global emission of different air pollutants, etc.), it has to be considered how a large-scale deployment of EVs will affect the current power system. In particular, how this large-scale deployment will affect the grid, leading to an increase in electric power losses, voltage drops, voltage unbalances, overloading of transformers and power lines, generating harmonics, and, in general, decreasing the power quality in the distribution grid, where they will be connected. In the generation side of the electric system, this deployment will increase the peak demand and the balancing needs of the power system [7–11].

The traditional way to reduce these negative effects is developing smart charging algorithms for EVs aiming at flattening of the global electric load profile. From the electric generation perspective, EV charging schedule should draw on valley-filling, avoiding peak demands, which makes better use of base load units. A step ahead is to use the EVs, through a vehicle-to-grid (V2G) configuration, to provide ancillary services, such as frequency regulation, voltage control, load following, etc. [12], and remove barriers allowing their participation in the ancillary services markets. This participation could generate additional revenues for their owners, promoting the adoption of this type of vehicles.

Peer-to-peer (P2P) applications began to grow in the 1990s [13], allowing millions of people around the world to share music, video and other digital contents over the internet [14,15]. After this first phase, these systems have evolved to lay the foundations for a sharing economy, interconnecting supply and demand in different markets, such as online auctions [16], car-sharing [17,18], taxi cabs [19,20], parking lots [21–23], spare rooms [24], etc. Recently, some new companies have broken the centralized infrastructure of the electric grid, allowing to connect power producers and consumers directly, establishing a new landscape of the electric power sector [25–28]. First P2P-based energy applications are connecting small-distributed renewable energy generators with consumers interested in buying cheaper and cleaner (with a higher share of renewables) energy.

In [29], a P2P energy trading between EVs to reduce the demand on the electric system during peak tariff hours was proposed. This system was further developed in [30], which combines the interaction between the electrical grids and the EVs studied in [12], with the aspects of P2P energy trading, which are discussed in [31]. In [32], a blockchain system was developed to facilitate this P2P exchange, in a similar way to that proposed in [28]. In [33], different P2P electricity trading business cases are discussed; the model proposed required additional privacy considerations, as studied in [34].

This chapter provides a revision of the results presented in [30] with the addition of the long-term results coming from the P2P energy market. To reach them, the main outcomes aforementioned have been taken into account, resulting in a demand-offer equilibrium between the analysed drivers. It also makes clear the decoupling between the P2P market and the electrical grid.

Section 7.1 reviews the context and the groundwork developed in previous research publications. In Section 7.2, the activity-based mobility model for the

Flanders region (Belgium) employed in the analysis is described. In Section 7.3, a consumption model for the EVs is presented; it assumes from Section 7.2 that all EVs have been completely recharged during the off-peak night period. This model classifies the driver agents into three different sets:

- Set A: drivers capable of completing all their daily trips without extra battery charging (with energy surplus)
- Set B: drivers needing extra battery charging (and are able to do it) to reach their final destination (with energy scarcity)
- Set C: drivers unable to use EVs for their routes without modifying their mobility habits (insufficient energy)

Only set A and set B will make the transition from an internal combustion engine (ICE) vehicle to an EV at the first stage, as set C would need to make deep changes in their mobility schedule to have enough energy with a daily recharge.

The P2P trading system is designed in two independent steps. First, an optimization algorithm, detailed in Section 7.4, for the intermediate recharging of each individual driver from set B is developed. It minimizes the electricity cost to be paid by a single driver, determining the time and location for its EV charging during the business hours at a minimum cost, in accordance with the mobility restrictions set by the driver's agenda. Second, since the EVs from sets A and B are parked in the same zone at the same time slots during the day, a P2P energy market (P2PEM) among the two sets of EVs is proposed in Section 7.5. The results of this P2PEM are described in Section 7.6, further extended to the long-term application of the market in Section 7.7. Section 7.8 presents the summary and main conclusions of the chapter.

7.2 Activity-based model

The first task of this work is to determine the total energy available in the batteries of the EVs with a surplus of energy (set A) at the end of their daily trips, and the total energy demanded by other EVs with lack of energy at the end of their daily trips which will require intermediate recharging during the day (set B).

To do this, it is necessary to determine the mobility patterns in a specific region. The transportation problem is multidimensional, and different types of transportation models have been proposed in the literature to support decision-making and forecasting under different scenarios considering changes in the infrastructure, socioeconomic circumstances, people's behaviour, etc. [35]. The aim of this type of models is to portray reality as accurate as possible.

In activity-based models, travel demand is derived from activities that individuals of a population need (or wish) to perform. The analysis is based on sequences or patterns of behaviour instead of individual trips. In this model, households and other community infrastructures (such as schools and shops) influence travel and activity behaviour. The model predicts which activities are conducted when, where, for how long, who do them and the transportation model used.

The activity-based model used in this work is FEATHERS (*Forecasting Evolutionary Activity Travel of Households and their Environmental RepercussionS*), developed by The Transportation Research Institute (*Instituut Moor Mobiliteit*-IMOB) from Hasselt University (Belgium). This model is fully operational in Flanders (Belgium), but it has been also applied in other regions from other countries, such as the UK, South Korea, and Slovenia [36]. This model was originally employed in transportation research to foresee the effect that could have the implementation of different types of policies before its application. Additionally, it has also been used for other purposes such as quantification of emissions and EVs deployment. In [37], this model was used to predict the time-dependent electric power demand for Flanders under specific assumptions of the EV market share and considering several scenarios for vehicle charging behaviour.

FEATHERS generates a daily agenda for each individual of the synthetic population as it is shown in Figure 7.1. Every single agenda consists of a sequence of episodes, each of them formed by exactly one trip and the subsequent activity. The trips within their agendas provide a different travel demand for each of the agents.

The trips are performed from a certain location to another (shown in different colour boxes in the agenda). Each trip belongs to a certain traffic zone (TAZ) in the Flanders region, with an average area of 5 km^2. For each activity, its type (sleep, in-home, work, daily shopping, leisure, etc.) as well as its start (t_0) and end (t_1) times, are specified. The transportation mode (car, walk, bike, bus, train, etc.) is also provided in this agenda, where the start and end trip times are derived as the difference between the consecutive activity times.

The daily agenda is forecasted as a function of several parameters. The FEATHERS model predicts the outcome of the decisions taken by each individual while building a daily agenda with a behavioural model. The model covers both the activity planning and the scheduling stages for those activities. Planning refers to deciding the composition of the set of activities to be completed during a given day (which activities are daily performed). Scheduling determines the sequence order and the timing for the selected activities (when these activities are done) and the transportation mode required. Some of the daily activities are mandatory such as work, school, pick/drop children, etc., and other activities are discretionary such as leisure, shop, etc.

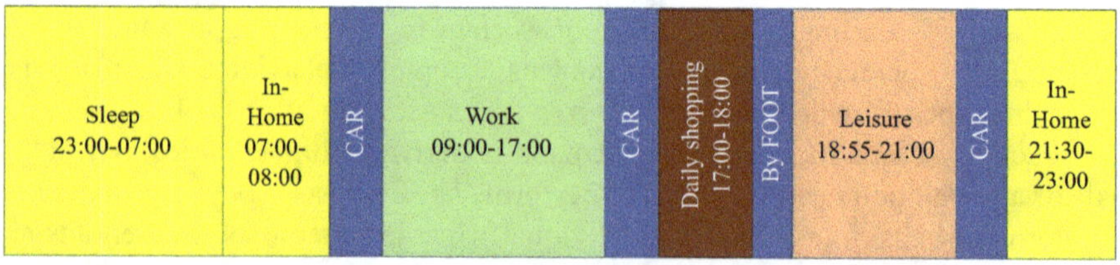

Figure 7.1 Daily agenda for an individual of the FEATHERS activity-based model

The behavioural model is composed of several different sub-models: the most important (and complex) sub-models are the location choice, the transport mode selection and the activity sequence decision models. Simultaneously, these models make use of more basic components, such as gravity models for distance selection.

The agenda for each individual of this synthetic population is generated by a series of interdependent decisions, which are settled in a specific order. Due to the computational complexity and in order to limit it, the FEATHERS model considers individuals to be independent of each other except for members within the same household. These coordination constraints are reflected in the input data and hence captured by the data mining techniques used in FEATHERS and are reflected in the agendas of the agents both in their transportation mode (e.g., travel as car passenger) and their trip timing.

The decisions taken by the agents are data-driven: a stochastic process that makes use of decision trees trained by means of survey data determines their outcome. Figure 7.2 shows the main data flows involved in the schedule prediction. The most important data sets, represented by rectangles, are the following:

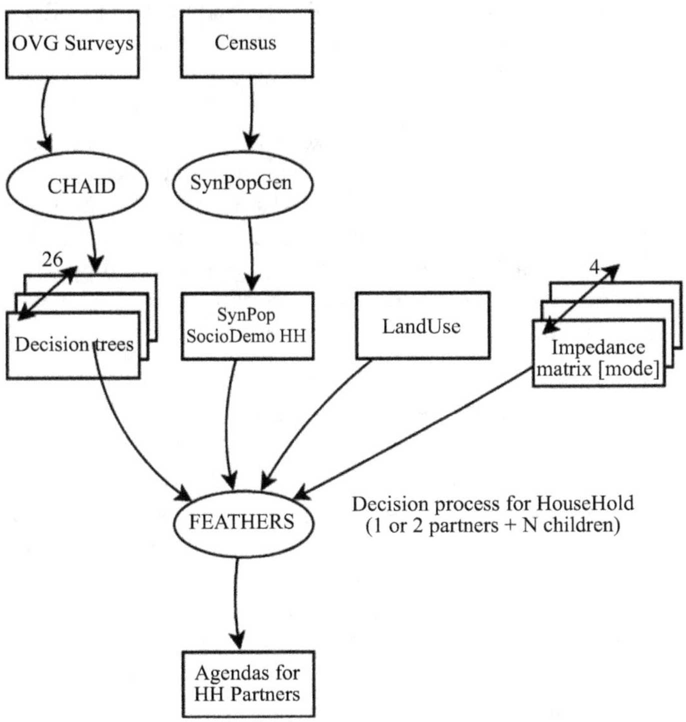

Figure 7.2 Data flow for the FEATHERS activity-based model. Ovals represent processes; rectangles represent data sets

1. A set of decision trees is trained using the Research Mobility Behaviour (*OnderzoekVerplaatsingsgedrag*-OVG) survey data collected by periodic travel surveys in Flanders [38]. The fourth OVG survey campaign covers the period 2008–2013 and involves approximately 8,000 participants. The outcome of a specific decision is given by a decision tree according to socio-economic data and to the partial agenda already built (hence, the interdependency of decisions).
2. Census data are used to generate the synthetic population of the model, where the marginal distributions for specific quantities in each TAZ are similar for real and synthetic populations.
3. Land-use data, listing TAZ properties such as number of inhabitants in each TAZ, number of shops, number and size of schools, job opportunities, etc.
4. Transportation impedance matrix; the performance characteristics of the transportation networks for each travel mode (car, train, bus, bike . . .) are expressed as expected travel times (impedance) between TAZ centroids for both peak and off-peak cases.

Activity-based models are stochastic micro-simulators (based on Monte Carlo simulations) whose predictions are validated by comparing data aggregated at a given level to real observations. FEATHERS model was validated by aggregating traffic flows between TAZ for each hour of the day and loading the derived demand into the transportation networks. The resulting traffic flows were compared with time-dependent hourly flows derived from traffic count. The foreseen arrivals for commuting trips were validated by comparing to census data that apply to the complete population and specify the work zone for each individual [39].

Once all daily agendas, for each individual of the synthetic population, have been generated, the next step is to evaluate the number of commuting trips that can be driven by battery-only EV for people living in a particular TAZ and being able to charge at the work location. Then, the total energy demanded by the aggregation of these consumers for each TAZ during these working hours is also evaluated, as it will be presented in Section 7.3.

7.3 Consumption model and drivers classification

7.3.1 Consumption model

There are different methodologies in the literature to estimate the average consumption (kWh/km) of EVs [40]. The first type of EV consumption model employs a physical description of the longitudinal forces actuating on the vehicle to estimate the energy consumption, evaluating their power and energy requirements over a real trip defined by its GPS traces [41]. The second method uses recorded real consumption data to estimate the energy consumption through different regression techniques, such as artificial neural networks, to establish a correlation between the input variables (distance, current speed, acceleration, etc.) and the EV's consumption [42].

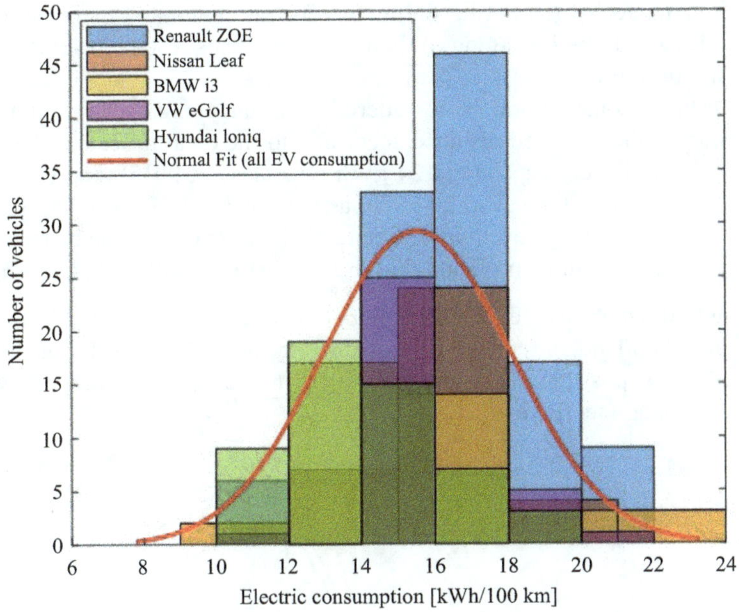

Figure 7.3 Electric consumption for different models of EVs

The mobility model used in this work only provides the travelled distance as mobility output data. Therefore, it is not possible to use any of the previous methods proposed in the literature to estimate vehicle consumption. In previous work, the authors assumed a constant energy consumption for all involved EVs, using a conservative value of 0.18 kWh/km [29]. Later, a more realistic consumption model was derived and used in [30], based on real data extracted from [43].

In this work, the authors have followed a similar procedure, generating a consumption model based on the real consumption of five different models of most commercial EVs sold in Europe and analysing the real consumption from more than 270 EVs. The consumption histogram for each vehicle is presented in Figure 7.3 and the calculated average consumption value was 15.5 kWh/100 km. The total consumption distribution is approximately fitted by a normal distribution and, to generate different consumption profiles for different agents in the mobility model, a randomly selected consumption sample has been extracted from this distribution, obtaining an electric consumption value to assign to any single agent in the mobility model.

7.3.2 Drivers classification

Four initial assumptions have been taken into account:

1. Only drivers that can fulfil their daily activities without any modification of their scheduled trips will make the transition from an ICE vehicle to an EV (set A and set B).

2. All EVs are recharged during the night off-peak hours and their batteries are fully charged at the beginning of their first daily trips (used to evaluate the EV penetration rate).
3. The vehicle consumption is considered constant along the day and it only depends on the travelled distance according to the data in Section 7.3.1.
4. The nominal battery capacity is 24 kWh (Nissan Leaf 2015), and the effective battery capacity is limited to 20 kWh, as it is shown in Figure 7.4. Therefore, when in the following equations, the state of charge (SOC) of the battery reaches 0%, it means it is in its minimum but in fact, there are still 4 kWh remaining.

According to the mobility FEATHERS model, there are 1,141,735 vehicles driving daily around the Flanders region. After evaluating the daily range for each vehicle in the model, it is possible to classify the drivers in one of these three different sets as aforementioned (see Figure 7.5):

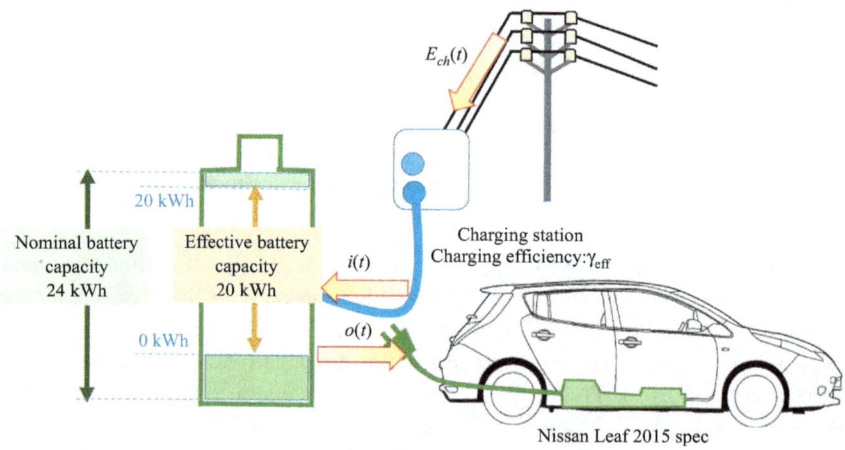

Figure 7.4 Nominal battery capacity versus effective battery capacity

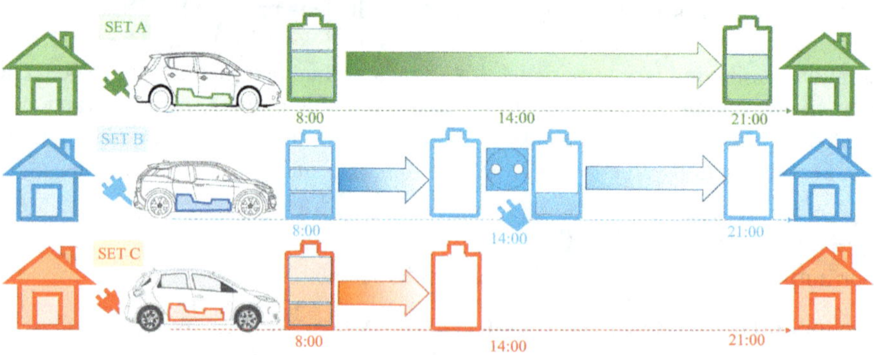

Figure 7.5 Drivers' classification depending on their daily mobility behaviour

1. Set A is the largest set and it is composed by all drivers who can complete all their daily activities using their EVs with the energy recharged during the night period, avoiding any intermediate charging along the day and with remaining energy in their batteries at the end of the day.
2. Set B is composed of those drivers who require, at least, one intermediate charge along the day. Those intermediate charging periods are done during the time when these vehicles are parked during their daily activities. Therefore, these drivers do not need to modify their mobility behaviour. It is important to notice that drivers from this set must charge their vehicles during high electricity price (peak period) in order to complete all daily scheduled trips. For that reason, these vehicles will be charged with the minimal energy enough to reach their final destination while avoiding recharging them with more expensive energy than necessary, expecting the off-peak night period to be fully recharged.
3. Set C includes those drivers who cannot complete their daily trips with an EV without modifying their current mobility behaviour. There are two possible reasons for that:
   (a) The total distance of, at least, one daily trip is higher than the maximum range of the EV.
   (b) The time available during the day to charge the vehicle is sort and the vehicle will require additional time charging.

In both cases, the drivers will require modifying their daily activities and they will not initially adopt an EV. For that reason, these drivers are not considered in this work.

Figure 7.6 shows the histogram of the surplus energy stored in the EV batteries from set A at the end of the daily trips. It is deduced from this figure that

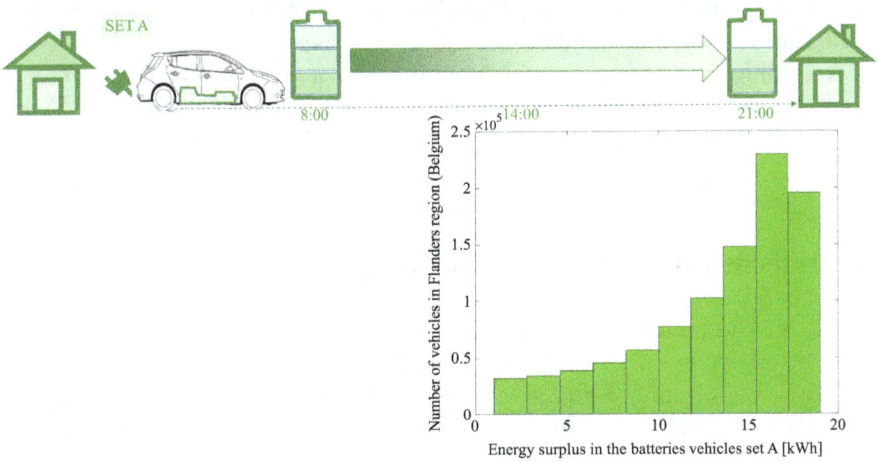

Figure 7.6 Total energy stored in batteries of set A vehicles at the end of the day

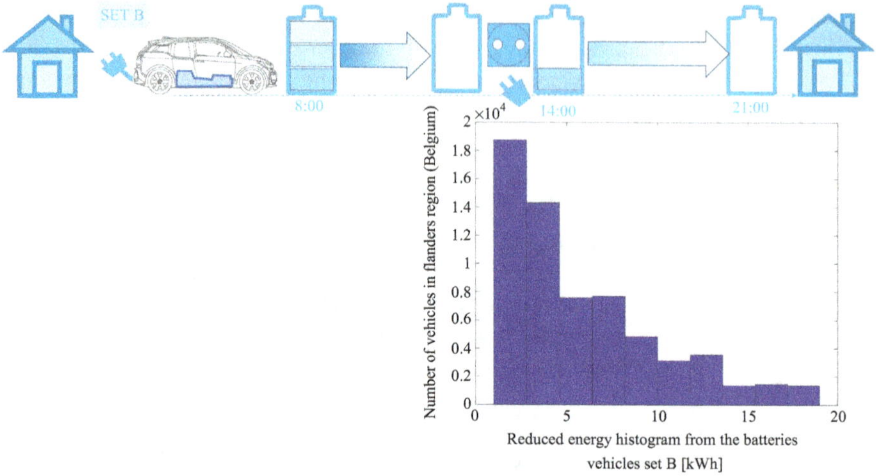

Figure 7.7 Total energy demanded by vehicles from set B to fulfil all scheduled daily trips

most of these vehicles travel very short distances during their daily trips, consuming a small amount of energy and having a high amount of remaining energy stored at their final destination. More than 200,000 vehicles store more than 18 kWh each at the end of the day, resulting in a total extra stored energy of 13.246 MWh.

Similarly, the analysis from vehicles of set B shows the additional energy required by vehicles to complete their daily trips reaching their final destinations (Figure 7.7). It is observed that most of these vehicles only require a small amount of extra energy to fulfil their daily trips. The total energy demanded by this set of EVs during the day is 363.67 MWh.

According to the energy surplus stored in the batteries of vehicles from set A, there is 36 times more energy stored in these batteries (of vehicles from set A) than the total energy demanded by vehicles from set B to finish their daily trips around the day.

7.4 Intermediate charging process optimization

To promote the transition to electromobility, the battery recharging needs from drivers belonging to set B should be attended.

Due to the electricity prices vary each hour, it is important for the users to know when (at what time slot) and where (in which TAZ) is most convenient to recharge their vehicles in advance, fulfilling their daily agenda without needing to modify their mobility behaviour.

Assuming that the variable hourly grid electricity price is known in a day-ahead time horizon [44] and taking into account the daily agenda of each driver (provided by the FEATHERS mobility model), car owners can optimize their daily energy costs, scheduling where and when they have to charge their EV at minimum cost. Each driver of set B will look for the minimization the following objective function:

$$\text{Min}\left[\sum_{t=T_{arr}}^{T_{dep}} e_{ch}(t)\cdot Pr(t)\right] \tag{7.1}$$

where $Pr(t)$ is the grid electricity hourly price (in €/kWh), $e_{ch}(t)$ is the energy extracted from the electric grid to charge the vehicle battery (in kWh) and T_{dep} and T_{arr} are the initial departure time and the final arrival time, respectively, (the start of the first scheduled trip and the end of the last scheduled trip). This objective function is subject to the following restrictions:

$$SOC_{min} \leq soc(t) \leq SOC_{max} \tag{7.2}$$

$$0 \leq i(t) \leq \eta_c \cdot P_{max} \cdot \Delta t \cdot VA(t) \tag{7.3}$$

$$soc(t) = soc(t-1) + i(t) - O(t) \tag{7.4}$$

$$e_{ch}(t) = i(t)/\eta_c \tag{7.5}$$

$$soc\left(T_{dep}\right) = 20\ kWh \tag{7.6}$$

where (7.2) ensures that the effective battery SOC represented by the variable over time $soc(t)$ in this set of equations, is within the lower and upper limits (SOC_{min} and SOC_{max}).

Equation (7.3) imposes the charging rate limit, with P_{max} representing the charging station power (in kW), Δt the length of the time slot employed for the algorithm to charge the vehicle and η_c the charging efficiency rate [45]. The variable $i(t)$, represents the energy extracted from the grid and injected to the battery during the charging process. This variable depends on the vehicle charging availability provided by FEATHERS mobility model through the binary variable $VA(t)$, (where $VA(t) = 1$ if the car is parked at this time slot and can be charged, or $VA(t) = 0$ if the car is not connected).

Equation (7.4) represents the dynamic evolution of the battery SOC, where the current SOC is equal to the previous SOC plus $i(t)$ (energy injected in the battery during the charging process) minus the energy consumption due to driving activity, $O(t)$ (in kWh). This value is provided by a consumption model developed in Section 7.3.1.

Table 7.1 Charging optimization problem constants

Symbol	Description	Value
T_{dep}	Departure time slot	1
T_{arr}	Arrival time slot	96
SOC_{max}	Maximum effective State of Charge	20 kWh
SOC_{min}	Minimum effective State of Charge	0 kWh
P_{max}	Charging station power limit	3.3 kW
η_c	Charging efficiency rate	0.95
Δt	Time slot length	15 min

The charging process is not 100% efficient, and for that reason, (7.5) determines the relationship between the energy effectively injected to the battery and the amount of energy extracted from the electric grid, $e_{ch}(t)$, which is the total energy that the consumer must pay.

Finally, (7.6) is the initial condition of the battery SOC, which is fully charged just before departure.

Equations (7.1)–(7.6) constitute a linear optimization problem that is solved using GAMS® and CPLEX Solver optimizer for each of the EVs from set B. The main decision variable is $i(t)$ which depends on the availability of the EV to charge, $VA(t)$, as it is shown Figure 7.4. If the same optimization problem is applied to vehicles from set A, the solution will be, $i(t) = 0 \forall t$ since no energy is required to fulfil the daily agenda of the driver.

Table 7.1 shows the constants used for this optimization problem. The day has been divided into 96 time periods of 15 min each, so the first period is from 00:00 to 00:15 and the last one (96) is from 23:45 to 00:00, although it can be extended for a certain amount of EVs that arrive at later times. The limits for the effective SOC are set between 0 and 20 kWh. The charging station power P_{max} is set to 3.3 kW, corresponding to a 16 A socket in a 230 V single-phase installation. Assuming this power limit, the maximum energy extracted from the grid in each period of 15 min will be 0.825 kWh. Finally, the charging efficiency rate η_c is 0.95, as derived from [45].

Figure 7.8 shows the solution of this linear optimization problem for a single EV belonging to set B. The background colours represent the different zones where the car is parked when the user is performing his/her activities (at home, at work, shopping, etc.). In this example, this driver performs only three daily trips.

The first trip takes place from 06:45 to 07:30, leaving the blue area (home) and arriving at the red one (work); a total travel distance of 12.73 km. It is observed that during the trip the battery SOC (blue line) decreases. A second trip occurs from 12:30 to 13:45, leaving the red area (work) and arriving at the grey one (shopping), travelling 59.48 km, which is the longest trip of the day. During the way back home, from 18:30 to 19:15, the user has to drive 44.85 km.

The total distance covered during the day is 117.06 km, with a total consumption of 22.12 kWh. This value is above the effective battery capacity

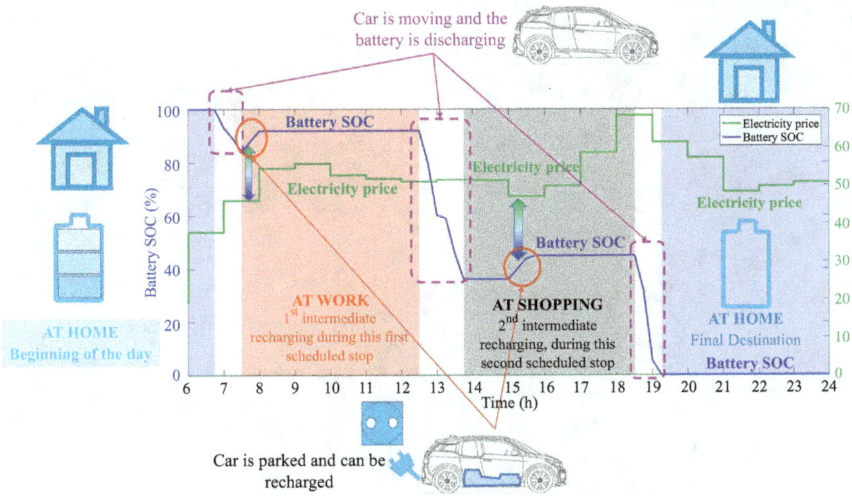

Figure 7.8 Effective battery SOC evolution for a vehicle belonging to set B (blue line), grid electricity price (green line) and TAZ zones (background colours)

(20 kWh) and therefore an intermediate recharge will be required. Taking into account the variable electricity price (shown by a green line in this figure), there are time slots more appropriate than others to perform this recharge.

Solving the linear optimization problem defined by (7.1)–(7.6) for this particular user, the solution presents two different periods to recharge: the first one, from 07:30–08:00, is just after arriving at the first destination (work). The second charging period is from 15:00 to 16:00, in the middle of its second scheduled stop at shopping. The hourly electricity price there reaches the lowest values within both periods, minimizing the cost of the required recharging.

This way, the owners of these vehicles can schedule their daily charging periods in advance, determining when and where they will recharge in order to minimize their charging cost, given by (7.1).

It is possible to evaluate the total energy demanded at each TAZ and in each time slot aggregating the solution of each linear optimization problem from all EVs belonging to the set B.

Figure 7.9 shows all TAZ in Flanders region where it has been highlighted a specific TAZ, number #904 in yellow. In the lower part of the figure, it is presented the temporal variation of the total energy demanded by all EVs from set B in this TAZ (blue line) and the hourly electricity price (in green).

It is observed that the maximum demand (1.391 MWh) occurs at 15:00 h when the hourly electricity price is the lowest along central day hours. There also are other additional peaks of high-energy demand such as one at 8:00 h in the morning, as soon as the vehicles arrive at their first scheduled destination and the electricity price is still low, and at noon, coincident with a small reduction in the hourly

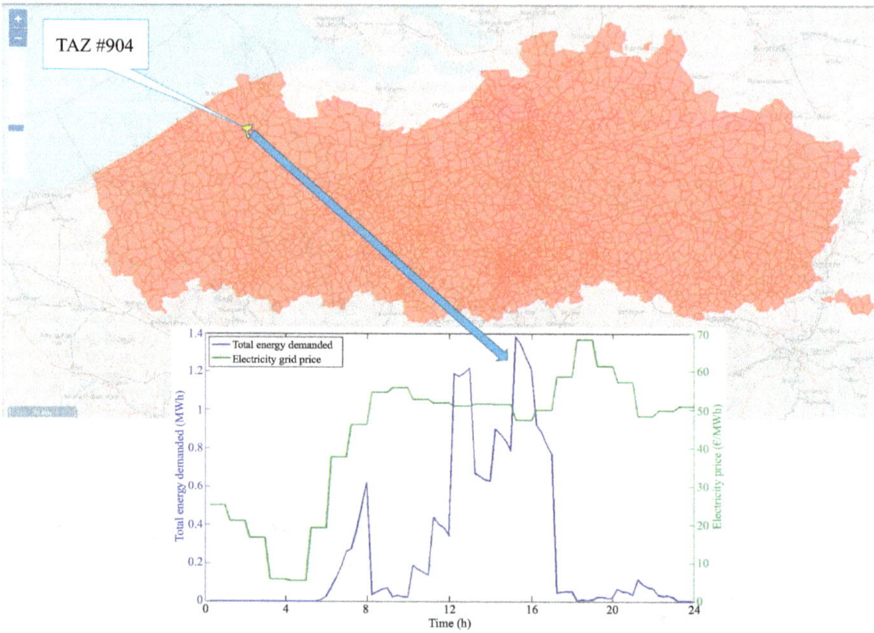

Figure 7.9 TAZ locations and temporal energy demand for TAZ #904 (highest energy demand) with grid electricity price (green line)

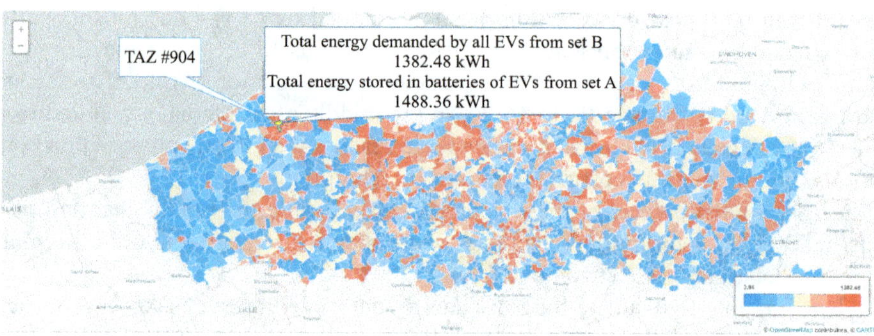

Figure 7.10 Spatial energy demand per each TAZ in Flanders region. Scale in kWh (more info available at: http://cdb.io/1BjL9pX)

electricity price compared to precedent and later hours (relative minimum price during that activity).

Figure 7.10 shows the spatial energy demand distribution for each TAZ in the Flanders region for a specific time of the day. The selected hour for this representation is at 15:00 h, coinciding with the lowest hourly electricity price value during business hours, which generates the highest total energy peak demand.

The higher demand, marked by a red area, is produced in the main cities such as Brussels, Gent, Brugge, Antwerpen, Leuven and Hasselt, whereas the total energy demanded is lower at the periphery and countryside areas in Flanders. According to this figure, the total energy demanded at 15:00 h in TAZ number #904 was 1382.48 kWh.

7.5 Peer-to-peer trading system

In this section, a P2P trading system to reduce the impact of recharging EVs on the electric grid during day hours is proposed. This trading system works as a trading platform similar to other P2P sharing economy businesses. The platform links both market actors: the 'producers' (vehicles with an excess of energy in their batteries, EVs from set A) and the 'consumers' (vehicles demanding energy to finish their daily trips, EVs from set B).

The proposed market system operates as follows: at the beginning of the day, drivers from set B optimize their daily charging cost individually by programming when and where they should charge their vehicles, considering the day-ahead electricity prices provided by the market operator and their mobility constraints. All these individual electricity demands are added, and a total aggregated spatio-temporal energy demand is generated. This total demanded energy can be provided to these vehicles by two different sources: the conventional power grid at the current grid tariff (which is different for each day-hour) or it can be obtained from those EVs belonging to set A which are parked at the same time slot at the same TAZ, as it is shown in Figure 7.11. In this case, both actors can agree on a common

Figure 7.11 Peer-to-peer energy interchange among electric vehicles parked at the same time in the same parking lot

electricity price for this energy exchange at each particular time slot to complete the bilateral transaction

For drivers belonging to set A, there are two limits for selling the energy stored in their vehicles' batteries. The lower limit will be the maximum value of the off-peak tariff, which represents the price that they paid for night charging, and an additional cost, which consider all efficiency losses produced during the charging-recharging process and the extra cost of the battery degradation due to overcycling. This lower limit is denoted by p^{min} and it is considered to be the same for all drivers and constant for that day. Drivers belonging to set A want to maximize their profit; therefore, they are willing to sell the stored energy at the highest possible price. This value is limited by the electricity grid price at each time slot, referred previously as $Pr(t)$. If they try to sell the energy at a higher price than $Pr(t)$ buyers from set B will buy the demanded energy directly from the electricity grid. Since this process is repeated for each time slot t, $Pr(t)$ will be further referred to as p^{max}, whose values varies for each of the time slots analysed.

Additionally, drivers from set A must fix a maximum amount of energy to share in this trading (named x_i^0) to ensure that they have enough energy stored in their batteries to finish their daily trips (plus the battery security margin previously established). All these defined values are presented in Figure 7.12.

On the other hand, drivers belonging to set B are willing to pay as little as possible, assuming a preference for charging from the P2P trading system in case that both prices (from the P2P market and the electricity grid price) are equal while charging directly from the grid if the P2P trading price were higher than the electricity grid price.

The total amount of energy demanded by all vehicles from set B in a specific TAZ at a particular time slot t is denoted by D, as it is shown in Figure 7.13. It is important to notice that this demand is inelastic because all vehicles belonging to this set B will require an intermediate charging to complete their daily trips, as scheduled at the beginning of the day. Therefore, this demand will not depend on the final P2P delivery price.

A second optimization algorithm is used to evaluate the final price (denoted by p_i) to be paid in this P2P trading system for each TAZ and for each period. In this new optimization problem, the objective function is to minimize the total cost of the energy exchanged by all EVs at every TAZ and every time slot.

7.5.1 Determining the final price for the P2P trading system each TAZ and each time period

For each time slot, the price at which each vehicle from set A offers its shared energy is provided by (7.8) (see Figure 7.12):

$$p_i(x_i) = a_i p_i + p^{min}, x_i \in \left[0, x_i^0\right], \quad \forall i = 1, \ldots, n \tag{7.8}$$

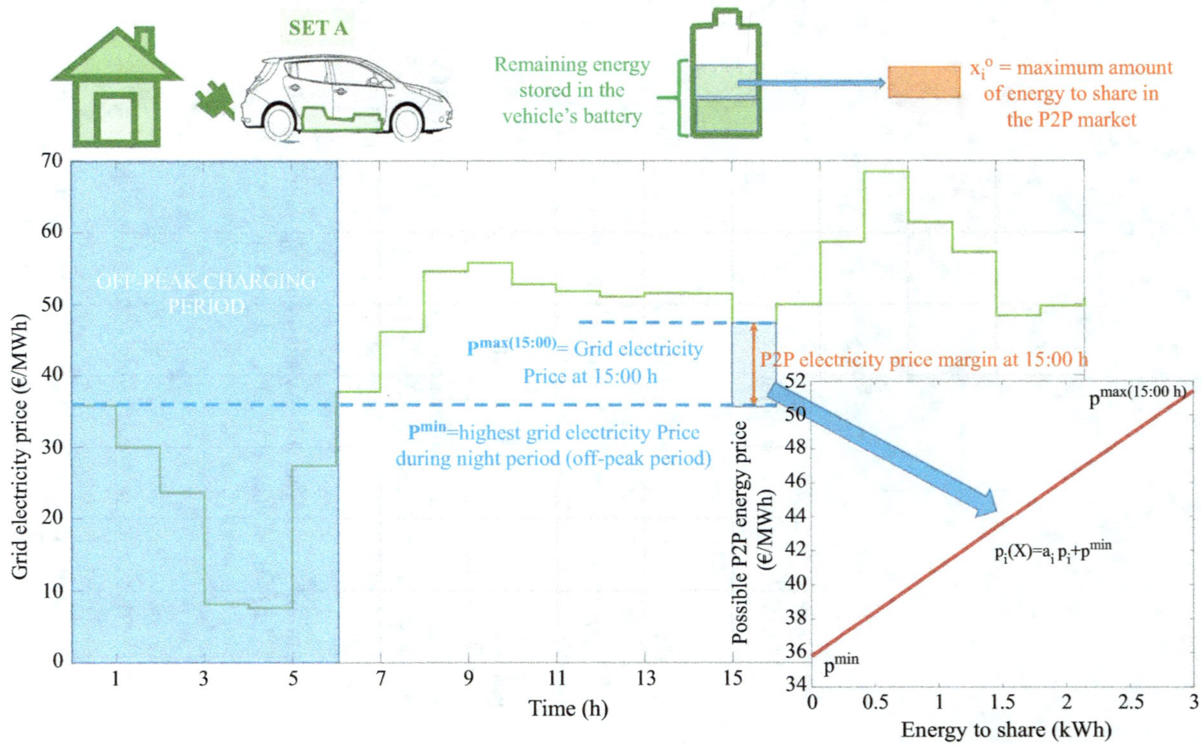

Figure 7.12 P2P price margins and maximum amount of energy to share in this P2P market

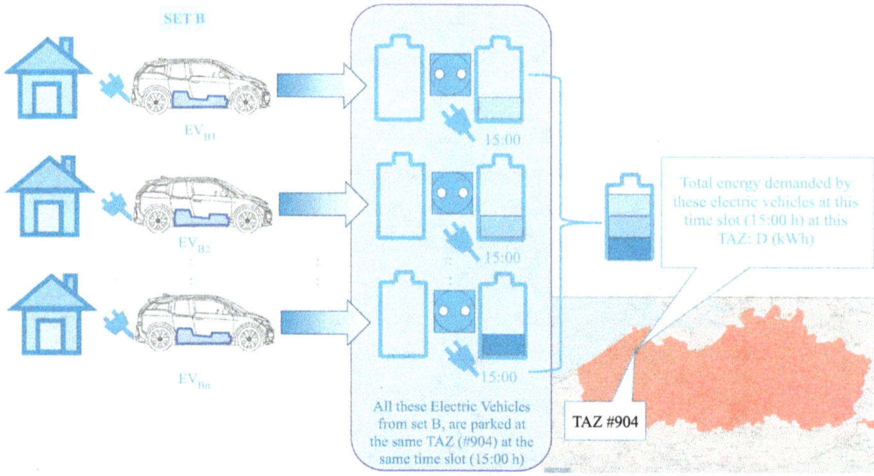

Figure 7.13 P2P energy demanded by vehicles from set B at a particular time slot t = 15:00 h and at a particular TAZ (#904)

Where:

- p_i represents the price paid to the i-th EV from set A (in €/kWh)
- x_i denotes the energy extracted from the i-th EV (in kWh)
- $\alpha_i = (p^{max} - p^{min})/x_i^o$ is the slope of the i-th EV's price curve
- p^{max} represents the maximum price to be paid to the i-th EV in € (the grid price)
- p^{min} denotes the minimum price to be paid to the i-th EV in € (the maximum night price, $p^{max} > p^{min}$)
- x_i^o indicates the energy available to exchange by the i-th EV (in kWh) in the trading system, that is, its offer ($x_i^o > 0$, $\forall i = 1, \ldots, n$)
- n represents the number of EVs from set A parked at this time slot in a particular TAZ

Given the maximum energy deliverable by the charging point for the considered time step $E^{max} = P^{max} \cdot \Delta t > 0$, in each time slot, the maximum energy deliverable by an EV i at a time slot t is defined as

$$x_i^{max} = \min\{x_i^0, E^{max}\}, \quad \forall i = 1, \ldots, n \tag{7.9}$$

The total combined offer relies on the feasible compact region: $\Omega = [0, x_1^{max}] \times [0, x_2^{max}] \times \ldots \times [0, x_n^{max}] \subset \Re^n$.

Given the total energy demanded by all EVs parked in this particular TAZ during the considered time slot belonging to set B, as $D \geq 0$; the average cost function, $f(\boldsymbol{x})$, can be defined as

$$f(\boldsymbol{x}) = f(x_1, \ldots, x_n) = \frac{1}{D} \sum_i x_i p_i(x_i), \quad x \in \Omega \tag{7.10}$$

The equilibrium between energy supply and demand for each time slot gives a solution to the following optimization problem:

$$\min_{x \in \Omega} f(x) \text{ subject to : } \sum_{i=1}^{n} x_i = D \tag{7.11}$$

Where Ω_I is the set of points of Ω that satisfies equation (7.11) and $\Omega_I \neq \emptyset$ if and only if:

$$\sum_{i=1}^{n} x_i^{max} \geq D \tag{7.12}$$

7.5.2 Quadratic programming formulation

To solve the optimization problem defined by (7.11), it is necessary to reformulate it as a quadratic programming problem, where the cost function $f(\boldsymbol{x})$ is defined as

$$\begin{aligned} f(\boldsymbol{x}) &= \boldsymbol{x}^T \boldsymbol{Q}\boldsymbol{x} + c^T \boldsymbol{x} \\ \boldsymbol{Q} &= \frac{1}{D} diag(a_1, \ldots, a_n), \quad c = \frac{p^{min}}{D}(1, \ldots, 1)^T \end{aligned} \tag{7.13}$$

With the following linear restrictions $g_j(\boldsymbol{x}) \leq 0$, $j=1,\ldots,n$ and $h(\boldsymbol{x})=0$:

$$g_j(\boldsymbol{x}) = x_j - x_j^{max}, \quad j = 1, \ldots, n; \quad h(\boldsymbol{x}) = \sum_{i=1}^{n} x_i - D \tag{7.14}$$

A solution $\boldsymbol{x}^*$ of the system defined by (7.11)–(7.13)–(7.14) will implicitly satisfy $\boldsymbol{x}^*>0$ (i.e., $x_i^*>0 \forall i$). The Karush–Kuhn–Tucker conditions to be fulfilled by $\boldsymbol{x}^*$ are:

$$\left.\frac{\partial f}{\partial x_i}\right|_{\boldsymbol{x}^*} = \sum_{j=1}^{n} \mu_j \left.\frac{\partial g}{\partial x_i}\right|_{\boldsymbol{x}^*} + \left.\frac{\partial h}{\partial x_i}\right|_{\boldsymbol{x}^*}, \quad i = 1, \ldots, n \quad \mu_j \leq 0, \quad j = 1, \ldots, n$$

$$\mu_j g_j(\boldsymbol{x}^*) \leq 0, \quad j = 1, \ldots, n \quad h(\boldsymbol{x}^*) = 0 \tag{7.15}$$

Which, for (7.13)–(7.14), take the form:

$$\frac{2\alpha_i x_i + p^{min}}{D} = \mu_i + \lambda, \quad i = 1, \ldots, n \tag{7.16}$$

$$\mu_i \leq 0, \quad i = 1, \ldots, n \tag{7.17}$$

$$\mu_i (x_i - x_i^{max}) = 0, \quad i = 1, \ldots, n \tag{7.18}$$

$$\sum_{j=1}^{n} x_i = D \tag{7.19}$$

7.5.3 Proposed algorithm for solving the P2P market trading

The proposed problem is computationally feasible due to $\boldsymbol{Q}$ is a diagonal positive definite matrix. In this section, a simple algorithm for solving it is proposed.

1. Initially, it is assumed that the minimum is reached in the interior of Ω. Thus, $\mu_j=0\ \forall j=1,\ldots,n$. and:

$$x_i^* = D\frac{x_i^0}{\sum_{k=1}^{n} x_k^0}\ , \quad i = 1,\ldots,n \tag{7.20}$$

2. If $x_i^* \leq x_i^{max} \forall i=1,\ \ldots,n$ then x^* is in Ω and solves the problem. If $x_i^* > x_i^{max}$ for some i, continue to step 3.
3. Let $J \subset \{1,\ldots,n\}$ be the subset of indexes for which $x_j^* > x_i^{max} \forall j \in J$. Those values are redefined as $x_j^* = x_i^{max}$ and remain fixed for the rest of the algorithm. The system (7.16) and (7.19) is solved again for the remaining components x_i^*, $i \in \{1,\ldots,n\} \backslash J$. Its solution satisfies that $\mu_j < 0\ \forall j \in J$.
4. If $x_i^* \leq x_i^{max} \forall i \in \{1,\ \ldots,n\} \backslash J$, then $\boldsymbol{x}^*$ solves the problem. Otherwise, those indexes for which $x_j^* > x_i^{max}$ are added to the previous set of indexes J and step 3 is repeated.

   The algorithm requires a maximum number of n steps.

7.6 Results of the P2P energy market

In this section, the different effects of the proposed P2PEM are analysed. The analysis is initially focused on the individual behaviours of a restricted sample of vehicles from set A and set B. Afterwards, the mobility TAZ areas are also analysed and categorized according to the market behaviour.

7.6.1 Individual analysis of vehicles from sets A and B

Figure 7.14 shows the application of the P2P trading system from the perspective of a specific EV from set A. The vehicle will move during its daily trips among three different TAZ zones, denoted by the background colours blue (departure zone-HOME), red and grey. There are shown three different additional lines: the green one represents the hourly grid electricity price, the red one represents the P2P electricity price at which the shared energy is paid and the blue one is the energy remaining in the EV vehicle. In this example, this price will be variable throughout the day, but it will never exceed the grid electricity price. It is also important to notice that vehicles from set A are not able to share energy with other vehicles when they are moving. For that reason, the P2P electricity price during the trips goes to zero during these periods. Finally, the blue line represents the total energy remaining in the EV battery.

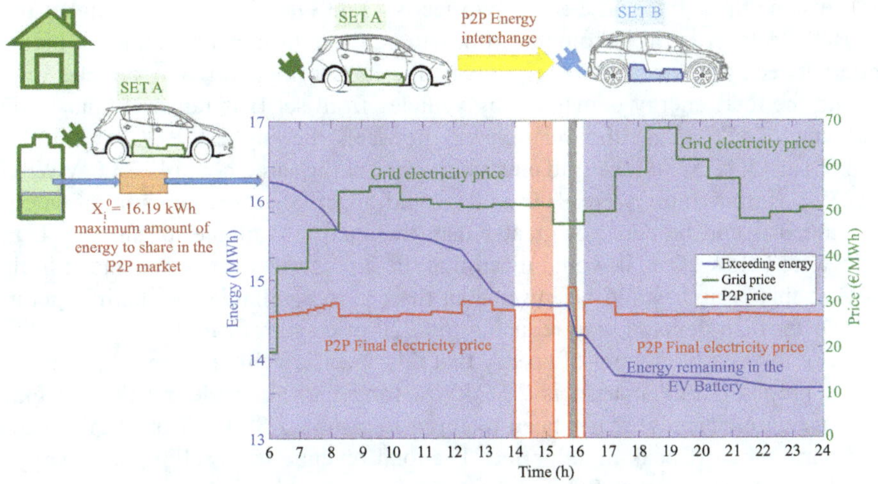

Figure 7.14 P2P trading behaviour of a vehicle from to set A, representing the effective battery SOC surplus (blue line), hourly grid electricity price (green line), P2P delivery price (red line) and visited TAZ during the daily trips (background colours)

The vehicle travels during three very short trips during the day, consuming c_i=2.95 kWh. Considering the losses in battery discharge through the battery discharge rate, γ_{eff}, its energy available for sharing in the trading process is $x_i^o= (SOC^{max}-c_i)\cdot\gamma_{eff} =(20–2.95)\cdot 0.95 = 16.19$ kWh.

The P2P market for this particular EV starts at 06:00 h. The EV from set A is initially parked at its departure zone (represented by blue background colour) until 14:00 h, arriving at zone 2 (represented by red background colour) at 14:30 h. During this initial stop, there are two periods of great energy exchange activity: from 06:00 to 08:00 h and from 12:00 to 14:00 h, where the vehicle sells 0.9 kWh of stored energy to other vehicles from set B, which are currently parked at the same zone during the same period. It is observed that during these two moments, the P2P price, shown as a red line in Figure 7.14, is higher than the p^{min} value, but it is significantly lower than the grid electricity price, indicating that drivers from set B are paying much less than if they recharged directly from the electricity grid.

The vehicle is parked in zone 2 (red background colour) for 45 min, leaving at 15:15 h and reaching zone 3 (represented in grey in Figure 7.14) at 15:45 h, where it remains for 15 min. It is observed that from 15:00 to 16:00 h the grid electricity price is the lowest during the period from 9:00 to 21:00 h. Therefore, during this period, the greatest demand for buying energy by vehicles of set B will be produced, as it was previously shown in Figure 7.9. For this reason, despite being available during a short time during this period (15 min), the vehicle from set A sells 0.37 kWh at the highest price (33.24 €/MWh) during the stop in the grey zone.

At the same time, the grid electricity price is 47.33 €/kWh, remaining higher than the P2P electricity price. From this information, it can be concluded that the highest ratio between the amount of energy available for sharing, stored in vehicles of set A, and the total energy demanded by vehicles from set B is rather similar in this particular TAZ. Thus, the relationship between the highest P2P price (33.24 €/MWh at 15:45 h) and the grid electricity price at the same hour (47.33 €/MWh) is 70.23%. If this ratio were 100%, it would mean that the amount of energy demanded would be equal or greater than the energy available for sharing. If the ratio were below 50%, it would mean that the amount of energy demanded in this TAZ at this period would be lower than the energy available for sharing, getting more offer than demand, and reducing the P2P price significantly.

The total daily amount of energy that this specific vehicle from set A has sold in the proposed P2P market is 2.56 kWh, for which, considering the charging-discharging efficiency, it has been required to charge 2.89 kWh from the electric grid during the off-peak night period. The total revenue from selling this energy is 0.077 € and the net benefit, if the energy has been charged at an average 16.5 €/MWh, is 0.029 €. The total daily vehicle consumption is 6.37 kWh, which implies a cost of 0.105 € and therefore the total cost of charging this vehicle has been reduced by 27.6%.

Figure 7.15 shows the day-charging schedule for the same EV (from set B) analysed previously in Figure 7.8. The vehicle is recharged in the same time slots than those provided by the solution of the optimization algorithm presented in Section 7.4, but the cost of these two intermediate recharges has been decreased significantly, from an initial value of 0.156 € (when the vehicle was recharged from

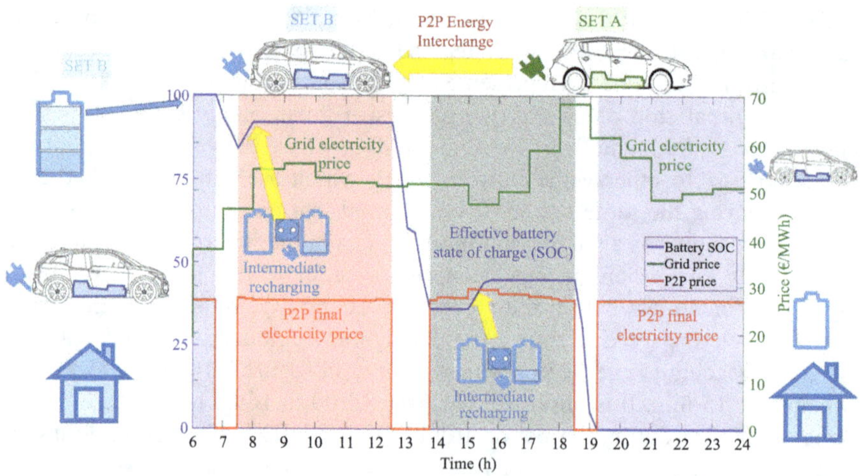

Figure 7.15 Effective battery SOC evolution for a vehicle belonging to set B (blue line), grid electricity price (green line), TAZ zones (background colours) and P2P delivery price (red line)

the electric grid) to a final cost of 0.0934 € from the proposed P2P market, producing a reduction of 40.1% in the day-charging cost. Taking into account that fully charging the battery at night implies a total cost of 0.3056 €, the expected savings reach up to 13.6% of the total daily cost.

7.6.2 Electricity price analysis at TAZ level

Figure 7.16(a) shows the energy ratio between the total energy demanded (from vehicles belonging to set B) and the total energy available to share (from vehicles from set A, which are parked in the same TAZ and at the same time period) in the three zones, whereas Figure 7.16(b) shows the resulting P2P price in each of those zones.

In areas with low energy ratio such as TAZ #2350 (represented by a blue line), there are many more EVs offering energy than EVs demanding energy for recharging, with an average value of 1,052 vehicles from set A and 90 vehicles from set B from 6:00 to 23:00, resulting in an average ratio of 8.6%. The energy available for sharing, provided by vehicles from set A, varies between 739 and 985 kWh, while the maximum demanded energy (by vehicles from set B), during this period is 29 kWh. The P2P price in this area is almost constant and it barely varies, taking values between 27.00 and 27.65€/MWh. The maximum price is

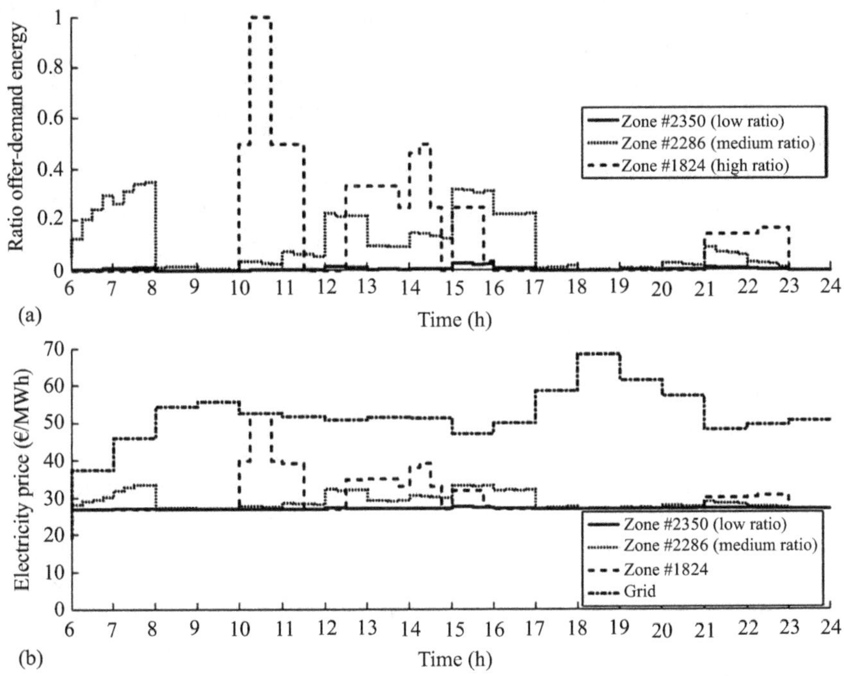

Figure 7.16 (a) P2P energy ratio at three different TAZ zones, (b) P2P electricity price at three trading zones

reached at 15:00, when the grid electricity price is the lowest, and the demand for intermediate recharging increases. Nevertheless, this P2P price is still 71% lower than the grid electricity price (27.62€/MWh for P2P price at 15:00 versus 47.22 €/MWh for grid electricity price at the same hour).

In areas with medium energy ratio such as TAZ #2286 (represented by a green line in Figure 7.16) this ratio is slightly increased, particularly during certain periods, obtaining an average ratio of 7.49%, but with a peak value of 34.2%. The available energy varies between 485 and 4,441 kWh; with a maximum demanded energy of 1,230 kWh. The maximum P2P price (33.66€/MWh) is reached early in the morning (at 07:45), while the grid electricity price at the same time is 46.04 €/MWh (44% higher). Similar P2P prices are reached during other periods (from 12:00 to 13:00 and from 15:00 to 17:00), whereas the electric demand is kept low during other periods of the day (from 08:00 to 10:00 and from 17:00 to 24:00), when the P2P price is kept close to the minimum value, p^{min}.

Finally, in areas with a high-energy ratio such as TAZ #1824 (denoted by a red line in Figure 7.16), the average ratio is 9.41%, but reaching 100% (the highest value) at 10:15. In this case, the demanded energy and the available energy for sharing are both very low, because the number of vehicles parked in this area at those specific time periods is very less as well. The maximum energy available for sharing varies from 0.825 (provided by a single vehicle from set A) to 5.775 kWh (provided by seven vehicles from the same set). At the same time, the demanded energy varies between 0 (no vehicles are demanding) and 1.65 kWh (demanded by two vehicles from set B). During the period 10:15–10:45, the energy available for sharing is equal to the total energy demand, which implies that the P2P delivery price should be equal to the grid electricity price, as it is shown in Figure 7.16(b). At other time periods in which the energy ratio is still high, from 12:30 to 14:45, when this ratio varies between 25% and 50%, the obtained P2P price is lower than the grid electricity price but higher than the previous P2P prices obtained in the other zones previously analysed.

7.7 Long-term peer-to-peer energy market

It is important to notice that the proposed P2PEM is designed for short-term planning. The P2P prices at each time slot and each TAZ are obtained assuming that the initially available information is the day-ahead electricity grid price and the daily mobility, as it was shown in Sections 7.4 and 7.5. Unfortunately, if this P2P market is implemented, it could happen that, after a while, the drivers from set B would try to optimize equations (7.1)–(7.6) taking into account the P2P prices obtained in previous days instead of the hourly grid electricity price, which is significantly higher than the P2P prices. In the medium term, higher demand will gather on those TAZ with lower prices, raising them and decreasing the prices at those TAZ whose costs was previously higher.

In the long term, these estimations are not realistic. EVs will adapt their schedules to the forecast facilitated by the P2P market, and although there may be small differences between TAZ prices, the exceeding energy scattered distribution guarantees that most EVs will benefit from the proposed P2P market without needing to modify their agendas.

In order to evaluate the results of this long term P2P market, it is supposed that, in each TAZ, the final equilibrium price will depend on the ratio between the EVs energy available for sharing (provided by vehicles from set A) and the energy demanded by the EVs from set B, $R_{ev}(t,TAZ)$. Following the results seen in Figure 7.16, three prices are chosen for the TAZs at each time slot: p^{min}, $p^{max,}$ and their average value.

Table 7.2 shows its application for the scenario previously studied at the end of Section 7.6. It is observed that, analysing all TAZ areas and all time slots defined in this problem, 96.3% of these (time slot, TAZ) periods have a high ratio (R_{ev}>5), which means that most of the time (and in most of the TAZ areas), there is more energy available for sharing than the required energy demanded by vehicles from set B. There are 3.3% of the (time slot, TAZ) periods that have a medium ratio (R_{ev} between 2 and 5) and 0.4% of the (time slot, TAZ) periods have a low ratio (R_{ev} lower than 2).

The algorithm to schedule the EV charging is the one employed previously in Section 7.5. Since the prices at most times and locations are the same, most EVs will choose its charging time only constrained by their mobility behaviour, reducing the impact of variable price. This causes that there are not great peaks of aggregated EV charging along the day, although the previous main peaks are kept in this new procedure. For example, in Figure 7.17 it is shown the grid electricity price (in green), the initial (short-term) energy demanded by set B in this TAZ, represented by a blue line (which is equal to the blue line in Figure 7.6), and the new proposal P2P price (represented by a red line). It can be observed that both energy demanded profiles (blue and red lines) have two main peaks at the same time periods: from 12:00 to 13:00 and from 15:00 to 16:00, but the new (long-term) proposal reduces the first peak to the half and the second peak by to two-thirds. Consequently, the difference between consecutive periods, such as between 16:00–17:00 and 17:00–18:00, has also decreased severely.

Finally, considering this new model of energy demand, the P2P market is implemented again and the price assumptions are checked against the corresponding P2P long-term market price. Only 5,919 pairs (t,TAZ) present a higher difference between the assumed price and the forecasted price, none of them verifying $R_{ev}(t,TAZ) \geq 5$. During this time at that TAZ, no energy is exchanged.

Table 7.2 P2P price and ratio between EVs

Ratio between EVs	Price	Number of (time-slot,TAZ)
$R_{ev}(t,TAZ) \geq 5$	p^{min}	165443
$5>R_{ev}(t,TAZ) \geq 2$	$(p^{min} + p^{max})/2$	5703
$R_{ev}(t,TAZ)<2$	$p^{maxx^{o}}$	646

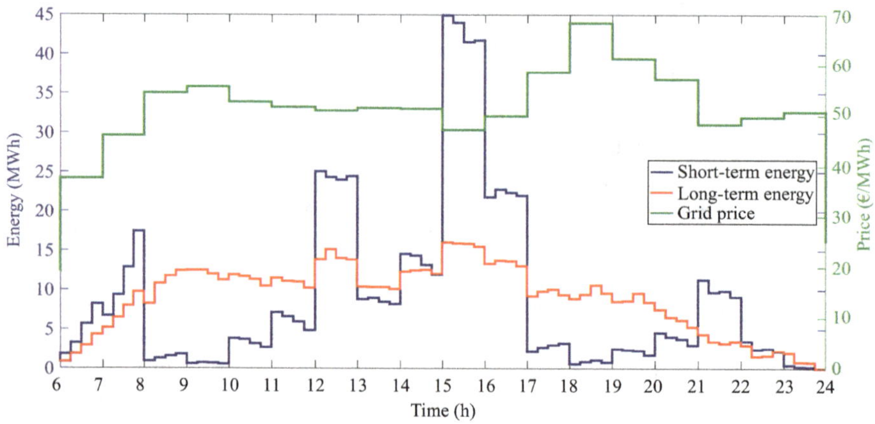

Figure 7.17 Aggregated energy demand for short-term market and long-term market

Furthermore, at zones in which the ratio R_{ev} is lower than 1 (i.e., more demanding EVs than offering EVs) only 7.5 kWh are demanded, which implies an almost complete decoupling between the P2P market and the electric grid.

This market equilibrium is a consequence of the higher number of EVs from set A than from set B: EVs from set A do not require discharging deeply their batteries and, furthermore, they are easily replaceable by other EVs in case of departure, ensuring the stability of the solution.

7.8 Conclusions

This chapter has presented a P2PEM between EVs that can be used to reduce the impact of EV charging on the power grid during the peak periods. The main findings of this market can be summarized as follows:

- The electrification of a fleet of private vehicles in a region-wide scale has been analysed, classifying the drivers into different categories depending on the possibility of using effectively an EV and their daily intermediate charge requirements. It has been observed that most drivers can use an EV for their daily activities and that the number of those EVs that would require intermediate charge is much smaller than those which would not require it.
- Two separate optimization problems have been solved. First, an individual optimal charging algorithm has been developed to obtain the best daily charging schedule (in time and space) according to public day-ahead grid electricity prices for those EVs requiring an intermediate charge. With this result, the total expected demand per each time slot and per TAZ is determined.
- Second, a P2P energy trading mechanism between EVs which are parked in the same TAZ at the same period has been developed. This mechanism determines the optimal P2P electricity price to be paid at every location and during each

time slot. The proposed P2PEM severely modifies the price currently paid by those vehicles, which require an intermediate charge, reducing the total daily energy cost. It also provides some savings to the offering EVs derived from the benefits of the additional energy charged at night.

- Decoupling both optimization procedures allows for a simple practical implementation where the decision of where and when to charge is independent of the consequent P2P market-saving benefits.
- The ratio between offering and demanding EVs has been proposed to determine the P2P market equilibrium in the long term. Most demanding EVs will charge their energy at TAZ where the ratio between demanding and offering EVs is low during the charging time, since there is a higher number of offering EVs than demanding EVs and thus the market price will be lower. This long-term equilibrium provides larger electricity trade than the short-term market initially proposed since most trading takes place at a lower price. Day charging along the day is also more distributed, reducing charging congestion.

Future works include the practical implementation of the system for a given fleet of EVs, studying the impact of the night electricity price in the general problem and the introduction of additional electric-power vehicles (such as bikes and buses) in the general problem.

References

[1] European Environment Agency, 'Air Quality in Europe – 2014 Report', Publications Office of the European Union, EEA Report No 5/2014, 2014.

[2] V. Ramanathan, and Y. Feng, 'Air Pollution, Greenhouse Gases and Climate Change: Global and Regional Perspectives', *Atm. Environ.*, vol. 43, no. 1, pp. 37–50, 2009, ISSN 1352–2310, https://doi.org/10.1016/j.atmosenv.2008.09.063

[3] World Resources Institute, 2017, Climate Analysis Indicators Tool, 2018, http://cait.wri.org/

[4] European Commission, 'Climate Strategies & Targets: 2050 Low-carbon Economy', https://ec.europa.eu/clima/policies/strategies/2050_en

[5] European Commission, 'A European Strategy for Low-emission Mobility', https://ec.europa.eu/clima/policies/transport_en

[6] Airuse EU project. 'Strategies to Encourage Use of Electric, Hybrid Electric and Gas Vehicles in Northern and Central Europe', Report 18, Dec. 2016. http://airuse.eu/wp-content/uploads/2013/11/R18_AIRUSE-Encourage-clean-vehicles-CNE.pdf

[7] K. Clement-Nyns, E. Haesen, and J. Driesen, 'The Impact of Charging Plug-In Hybrid Electric Vehicles on a Residential Distribution Grid', *IEEE Trans. Power Syst.*, vol. 25, no.1, pp. 371–380, 2010

[8] H. Turker, S. Bacha, D. Chatroux, and A. Hably, 'Low-Voltage Transformer Loss-of-Life Assessments for a High Penetration of Plug-In Hybrid Electric Vehicles (PHEVs)', *IEEE Trans. Power Del.*, vol. 27, no. 3, pp. 1323–1331, 2012

[9] P. T. Staats, W. M. Grady, A. Arapostathis, and R. S. Thallam, 'A Statistical Analysis of the Effect of Electric Vehicle Battery Charging on Distribution System Harmonic Voltages', *IEEE Trans. Power Deliv.*, vol. 13, no. 2, pp. 640–646, 1998
[10] P. Richardson, D. Flynn, and A. Keane, 'Impact Assessment of Varying Penetrations of Electric Vehicles on Low Voltage Distribution Systems', in Proc. IEEE Power and Energy Society General Meeting, 2010, pp. 1–6, 25–29 July 2010
[11] F. Geth, K. Willekens, K. Clement, J. Driesen, and S. De Breucker, 'Impact-Analysis of the Charging of Plug-in Hybrid Vehicles on the Production Park in Belgium', in Proc. MELECON 2010, pp. 425–430, 26–28 April 2010
[12] W. Kempton, and S. Letendre, 'Electric Vehicles as a New Source of Power for Electric Utilities', *Transport. Res. Part D: Transport. Environ*. vol. 2, no. 3, pp. 157–175, 1997.
[13] 'The rise of sharing economy', The Economist, 9 Mar. 2013
[14] Napster. [Online]. Available: http://napster.com
[15] Bittorrent. [Online]. Available: http://www.bittorrent.com
[16] Ebay. [Online]. Available: http://www.ebay.com/
[17] Getaround. [Online]. Available: https://www.getaround.com/
[18] Blablacar. [Online]. Available: https://www.blablacar.es/
[19] Uber. [Online]. Available: https://www.uber.com
[20] Lyft. [Online]. Available: https://www.lyft.com/
[21] Justpark. [Online]. Available: https://www.justpark.com/
[22] Appapark. [Online]. Available:http://www.apparcar.com/
[23] Aparca & go. [Online]. Available:http://www.aparcandgo.com
[24] Airbnb. [Online]. Available: https://www.airbnb.com
[25] Peer Energy Cloud. [Online]. Available: http://www.peerenergycloud.de
[26] Vanderbron. [Online]. Available: https://vandebron.nl/
[27] ePower. [Online]. Available: http://www.epowerauctions.co.uk/
[28] E. Mengelkamp, J. Gärttner, K. Rock, S. Kessler, L. Orsini, Christ of Weinhardt, 'Designing Microgrid Energy Markets: A Case Study: The Brooklyn Microgrid', *App. Ener.*, vol. 210, no. 2018, pp. 870–880, ISSN 0306–2619, https://doi.org/10.1016/j.apenergy.2017.06.054.
[29] R. Alvaro, J. Gonzalez, C. Gamallo, J. Fraile-Ardanuy, and D. L. Knapen, 'Vehicle to Vehicle Energy Exchange in Smart Grid Applications', in 2014 International Conference on Connected Vehicles and Expo (ICCVE), Vienna, pp. 178–184, 3–7 Nov. 2014
[30] R. Alvaro-Hermana, J. Fraile-Ardanuy, P. J. Zufiria, L. Knapen, and D. Janssens, 'Peer to Peer Energy Trading with Electric Vehicles', in *IEEE Intell. Transport. Sys. Mag.*, vol. 8, no. 3, pp. 33–44, 2016.
[31] O. Jogunola, A. Ikpehai, K. Anoh, *et al.*, 'State-Of-The-Art and Prospects for Peer-To-Peer Transaction-Based Energy System', in *Energies 2017*, vol. 10, no. 12, 2016.

[32] J. Kang, R. Yu, X. Huang, S. Maharjan, Y. Zhang, and E. Hossain, 'Enabling Localized Peer-to-Peer Electricity Trading among Plug-in Hybrid Electric Vehicles Using Consortium Blockchains', *IEEE Trans. Indus. Inform.*, vol. 13, no. 6, pp. 3154–3164, 2017. https://doi.org/10.1109/TII.2017.2709784

[33] C. Park, and T. Yong, 'Comparative Review and Discussion on P2P Electricity Trading', in *Energy Procedia*, vol. 128, pp. 3–9, 2017. [Online] Available: https://doi.org/10.1016/j.egypro.2017.09.003

[34] S. Karnouskos, and F. Kerschbaum, 'Privacy and Integrity Considerations in Hyperconnected Autonomous Vehicles', in *Proc. of the IEEE*, vol. 106, no. 1, pp. 160–170, 2018. https://doi.org/10.1109/JPROC.2017.2725339

[35] J. Castiglione, M. Bradley, and J. Gliebe, 'Activity-Based Travel Demand Models: A Primer. S2-C46-RR-1', the National Academies of Sciences, Engineering and Medicine. Washington D.C., 2015, ISBN 978–0–309–27399–2, DOI 10.17226/22357

[36] T. Bellemans, B. Kochan, D. Janssens, G. Wets, T. Arentze, and H. Timmermans, 'Implementation Framework and Development Trajectory of FEATHERS Activity-Based Simulation Platform', in Transport. Res. Rec.: J. Transport. Res. Board, vol. 2165, pp. 111–119, 2010 [Online] Available: http://trb.metapress.com/content/e5018jp286371r46/

[37] L. Knapen, B. Kochan, T. Bellemans, D. Janssens, and G. Wets, 'Using Activity-Based Modeling to Predict Spatial and Temporal Electric Vehicle Power demand in Flanders', in *Transport. Res. Rec.*, 2012.

[38] OVG (OnderzoekVerplaatsingsgedragVlaanderen) survey results available from http://mobielvlaanderen/ovg/ovg04.php

[39] B. Kochan, T. Bellemans, D. Janssens, and G. Wets 'Validation of an Activity-Based Traffic Demand Model for Flanders Implemented in the FEATHERS Simulation Platform', in *Comput. Intel. Traffic Mobility*, vol. 8, pp. 80–105, 2012

[40] J. Fraile-Ardanuy, S. Castano-Solis, R. Álvaro-Hermana, J. Merino, and Á. Castillo, 'Using Mobility Information to Perform a Feasibility Study and the Evaluation of Spatio-temporal Energy Demanded by an Electric Taxi Fleet', *Energy Conv. Manag.*, vol. 157, pp. 59–70, 2018, ISSN 0196–8904, https://doi.org/10.1016/j.enconman.2017.11.070

[41] T. Jurik, A. Cela, R. Hamouche, *et al.*, 'Energy Optimal Real-Time Navigation System', in *IEEE Intell. Transport. Sys. Mag.*, vol. 6, no. 3, pp. 66–79, 2014

[42] C. Bingham, C. Walsh, and S. Carroll, 'Impact of Driving Characteristics on Electric Vehicle Energy Consumption and Range', in *IET Intell. Transport. Syst.*, vol. 6, no. 1, pp. 29–35, 2012

[43] MPG and Cost Calculator and Tracker. [Online] Available: http://www.spritmonitor.de/en/

[44] Belgian Power Exchange, [Online]. Available: http://www.belpex.be

[45] J. Sears, D. Roberts, and K. Glitman, 'A Comparison of Electric Vehicle Level 1 and Level 2 Charging Efficiency', in Proc. IEEE Conf. on Technol. Sustain, SusTech2014, pp. 255, 258, 24–26 July 2014.

Chapter 8

Dispatch of vehicle-to-grid battery storage using an analytic hierarchy process

Lu Wang[1,2], *Suleiman Sharkh*[1] *and Andy Chipperfield*[1]

The number of electric vehicles (EVs) is expected to increase significantly in the future to combat air pollution and reduce reliance on fossil fuels. This will impact the power system. However, with appropriate charging and discharging through vehicle-to-grid (V2G) operation, EV batteries could replace some stationary energy storage to provide support for the power system and benefit EV owners. This raises the questions of when and how EV battery storage should be dispatched, taking into account both vehicle users' and power system requirements and priorities, as well as the constraints of the battery system. This chapter proposes a decentralized dispatch strategy based on the analytic hierarchy process (AHP) taking into account the relative importance of the different criteria such as cost, battery state of charge (SOC), power system contingency and load levelling. The proposed AHP-based dispatch strategy is tested on an IEEE Reliability Test System (RTS) with different EV numbers and capacities to investigate the efficacy of such an approach. The simulation results demonstrate the feasibility and benefits of the strategy.

8.1 Introduction

Growing concerns over energy savings, emissions and the desire to reduce reliance on fossil fuels have resulted in ambitious plans for expanding the use of electric vehicles (EVs) [1,2]. The large-scale penetration of EVs will challenge the power system as an additional load due to the substantial need for battery charging. However, due to the significant storage/generation capacity that EV batteries can offer to the grid if vehicle-to-grid (V2G) is enabled, EVs, with an appropriate dispatch strategy, will be capable of providing grid operational support and improvements of power system performance such as load levelling, spinning reserves and voltage regulation [3,4].

In this chapter, a novel decentralized dispatch strategy for EV batteries is developed based on analytic hierarchy process (AHP) [5] taking into account the concerns and requirements of both EV users and the grid, including cost, sufficient

[1]Faculty of Engineering and Physical Sciences, University of Southampton, Southampton, United Kingdom
[2]Global Energy Interconnection Development and Cooperation Organization, Beijing, China

SOC for the next journey, load levelling and alleviation of the thermal overload caused by N-1 contingency (i.e. an outage of only one component of the transmission system). Using the proposed strategy, each EV determines the dispatch action by itself depending on its own situation and the information gathered from the power grid. The proposed dispatch strategy is then tested on an IEEE Reliability Test System (RTS) for its effectiveness in satisfying the requirements of both EV users and the grid. Comparisons with the rule-based dispatch strategy [6] are made to demonstrate the potential benefits of the proposed AHP-based dispatch strategy over previous approaches.

The rest of this chapter is organized as follows. A description of typical EV battery characteristics is presented in Section 8.2. In Section 8.3, the proposed dispatch strategy of EV battery storage is presented in detail. The results of simulations using MATLAB, to verify its feasibility and efficacy, are presented and discussed in Section 8.4. This is followed by a presentation and discussion of the results of parameter sensitivity analysis in Section 8.5. Finally, the conclusions of the work are presented in Section 8.6.

This work has been published in the *IEEE Transactions on Vehicular Technology*, vol. 66, no. 4, pp. 2952–2965, April 2017 [7].

8.2 Battery characteristics

The capacity of a battery varies with the discharging current. This can be modelled based on the Peukert equation [8,9], as follows:

$$C_p = I^k T \tag{8.1}$$

where C_p is the Peukert Capacity, k is the Peukert Coefficient (e.g. $k = 1.2$ for a lead acid battery), I is the constant discharging current in Amperes and T is the corresponding time the battery will last in hours. Furthermore, C_p can be calculated by substituting I and T with the nominal discharging current I_n and rated discharge time T_r, as follows:

$$C_p = I_n^k T_r \tag{8.2}$$

Therefore, given a constant discharging current i_d in Amperes, the corresponding available capacity C_a of a battery in Ampere-hours (Ah) can be obtained as

$$C_a = \frac{i_d C_p}{i_d^k} = \frac{i_d I_n^k T_r}{i_d^k} \tag{8.3}$$

The state of charge (SOC) of a battery after being discharged for t hours at a constant discharging current i_d is calculated by (8.4).

$$SOC(t) = 1 - \frac{i_d t}{C_a} \tag{8.4}$$

Table 8.1 Battery voltage versus released capacity at different charging/discharging current (240 V, 100 Ah)

Released capacity (Ah)	Voltage(V)/Current(A)				
0	259.4/2	...	258.9/10	...	257.7/30
...	...	...	...	...	...
20	244.6/2	...	244.2/10	...	243/30
...	...	...	...	...	...
60	237.5/2	...	237/10	...	235.8/30
...	...	...	...	...	...
80	225.4/2	...	224.9/10	...	223.7/30
...	...	...	...	...	...
100	123.2/2	...	122.8/10	...	121.6/30

Table 8.2 Characteristics of the typical EV battery (240 V, 100 Ah)

Rated capacity (Ah)	Nominal current (A)	Nominal voltage (V)
100	20	240
Peukert Capacity (Ah)	Charge/Discharge Current (A)	Effective Available Capacity (Ah)
182.06	2	158.5
	10	114.87
	30	92.21

As for the voltage at each time step when calculating charge/discharge power of a battery, the method proposed in [6], based on the simplified generic rechargeable battery model described in [10], is used due to the limited information on the detailed charge/discharge characteristics of actual EV batteries. A lookup table of the voltages of a typical battery (240 V, 100 Ah) corresponding to the different pairs of released capacities and charge/discharge currents is built and presented in Table 8.1. Therefore, given current SOC and charge/discharge current, the corresponding voltage can be obtained by linear interpolation of the data in Table 8.1. The main parameters of a typical battery that is used as an EV battery here are calculated using (8.2) and (8.3) and shown in Table 8.2.

8.3 Dispatch strategy of electric vehicle battery storage

As discussed above, by running the proposed V2G dispatch strategy for each EV, its dispatch action (i.e. whether charge or discharge and at what dispatch current) is determined by the power system operator (PSO) at the beginning of each time interval. In preparation for that, the PSO gathers real-time data, such as real-time pricing (RTP) data, network fixed load demand (i.e. load without EVs) and overload information, a couple of minutes prior to the beginning of each time interval. In this work, an AHP-based dispatch strategy is designed for V2G batteries to support power grid operations including load levelling and the alleviation of thermal overload caused by N-1 contingencies, and to reduce cost to users while ensuring that the battery SOC is sufficient for the next journey. In order to jointly consider the different requirements of EV users and the grid, they should be weighed depending on their relative importance in determining how and when EV batteries should be dispatched and thus a hierarchy model is developed first.

8.3.1 AHP hierarchy process

AHP is a multi-criteria decision-making method developed by Saaty [5], which has been widely used in various areas including power systems as mentioned earlier. The main steps of the AHP methodology can be summarized as follows:

- Step 1: A hierarchy model is established based on an analysis of the problem.
- Step 2: Pairwise comparison matrices are formed at each level of the hierarchy model. Each element $A(i,j)$ in the pairwise comparison matrix A depicts the expert's judgment of relative importance between the pair of *ith* and *jth* factors using the ratio scale method [5,11]. $A(i,j)$ is the reciprocal of $A(j,i)$.
- Step 3: For each pairwise comparison matrix, the maximum eigenvalue and the corresponding eigenvector are calculated. The normalization of the calculated eigenvector gives the normalized principle eigenvector with elements equal to the priority scales (weightings) of the factors.
- Step 4: The consistency of judgments is checked. The Consistency Index (CI) and Consistency Ratio (CR) are measured for each pairwise comparison matrix. The CI is defined as

  $$CI = \frac{\lambda_{max} - n}{n - 1} \tag{8.5}$$

  where λ_{max} and n are the maximum eigenvalue and the dimension of the corresponding pairwise comparison matrix, respectively. The CR is defined as

  $$CR = \frac{CI}{RI} \tag{8.6}$$

  where RI is a set of given average random consistency indices which can be obtained using statistical calculations [12]. If CR is no larger than 10%, the inconsistency of judgments is acceptable. Otherwise, the judgments need to be revised.

Based on the principles of AHP as presented above, a five-level hierarchy model for the dispatch of an EV battery is made, as shown in Figure 8.1.

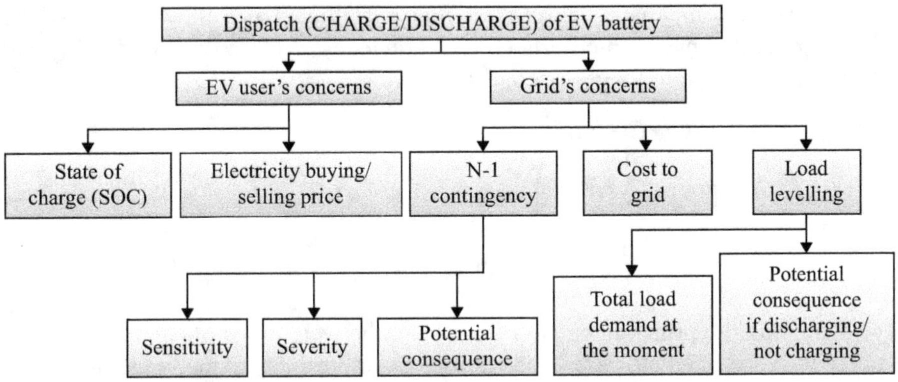

Figure 8.1 AHP hierarchy model for dispatch of EV battery

The top level is the objective: dispatch of the EV battery, which should take into account the requirements and concerns of both EV users and the grid (second-level criteria) as discussed earlier. EV users' main concerns are the SOC of the EV batteries and the cost to dispatch them (third-level sub-criteria). The grid might utilize the EVs to support system load levelling (i.e. valley filling and peak shaving) or relieve thermal overload after an N-1 contingency while allowing for the costs of employing the storage/generation capacity of EV batteries (third-level sub-criteria). Several factors need to be considered regarding the support that an EV can provide in an N-1 contingency such as the sensitivity of the bus to which the EV is connected in alleviating thermal stress of the overloaded lines, the severity of the overload and the potential consequence if the contingency is not dealt within 15 min (fourth-level sub-criteria). Furthermore, both current system demand and potential consequence of discharging/not charging the EV battery (fourth-level sub-criteria) should be taken into account when determining the preferred support that an EV can provide for load levelling. Finally, the battery can be either charged or discharged (alternatives of dispatch actions) as a result of this decision making process.

It is important to note, as mentioned earlier, that the dispatch system is owned by the PSO, not the EV owner. The PSO may adjust the priorities to maximize their profit, which may change the weighting towards the grid or vice versa. Here, the concerns of EV users and the grid are considered as equally important and thus both weighed at 50%. SOC and electricity buying/selling price are also weighed equally with 50% each. Among the three sub-criteria under grid's concerns, the cost to the grid is considered to be more important than the other two (i.e. twice as important), because the grid can choose other kinds of energy storage for grid operational support if it is more expensive to use EV batteries. The overload alleviation under N-1 contingency and load levelling are two different kinds of grid support and thus are deemed to be equally important.

A pairwise comparison matrix A_g is thus constructed as follows, to derive the weightings of the three sub-criteria (N-1 contingency (NC), cost to grid (CG) and load levelling (LL)) under grid's concerns:

$$A_g = \begin{matrix} \\ NC \\ CG \\ LL \end{matrix} \begin{matrix} NC & CG & LL \\ \end{matrix} \begin{pmatrix} 1 & \frac{1}{2} & 1 \\ 2 & 1 & 2 \\ 1 & \frac{1}{2} & 1 \end{pmatrix} \tag{8.7}$$

By calculating the principle eigenvector of A_g, the weightings of NC, CG and LL, namely w_{NC}, w_{CG} and w_{LL}, are 25, 50 and 25%, respectively. Among the three sub-criteria under N-1 contingency, sensitivity is considered to be the most important, because there is no need for the EV battery to deal with a contingency that occurs far from the bus to which it is connected, no matter how severe the overload and the potential consequence are. Potential consequence of the N-1 contingency is considered to be a bit more important than severity, because a contingency that does not cause severe overloading but has severe potential consequences is surely more important and needs more attention than one that causes severe overload but no severe potential consequence. Therefore, sensitivity is determined to be four times as important as severity and three times as important as potential consequence, with potential consequence twice as important as severity. The pairwise comparison matrix A_{NC} for weighing the three sub-criteria (Sensitivity (*Sen*), Severity (*Sev*) and Potential consequence (*PoC*)) under N-1 contingency is thus:

$$A_{NC} = \begin{matrix} Sen \\ Sev \\ PoC \end{matrix} \overset{\begin{matrix} Sen & Sev & PoC \end{matrix}}{\begin{pmatrix} 1 & 4 & 3 \\ \frac{1}{4} & 1 & \frac{1}{2} \\ \frac{1}{3} & 2 & 1 \end{pmatrix}} \tag{8.8}$$

Calculating the principal eigenvector of A_{NC} gives the weightings of *Sen*, *Sev* and *PoC*, namely w_{Sen}, w_{Sev} and w_{PoC}, which are 62.5, 13.7 and 23.9%, respectively. Moreover, the two sub-criteria under load levelling, viz. total load demand and potential consequence, are weighed 60 and 40%, respectively, with total load demand considered more important than potential consequence.

The factors that can be included in the AHP hierarchy model are not limited to those in Figure 8.1. Other factors such as battery degradation, voltage constraints and integration of renewables can be readily included. For example, voltage limits, related to the distribution system, can be included by adding it as another grid concern in Figure 8.1. Indeed, additional sub-criteria specific to the distribution system can be added. But increasing the number of factors would increase complexity and implementation cost. Additionally, some factors may encompass others indirectly. For example, what an EV owner cares about most are the vehicle's basic function (i.e. driving) and the direct cost of utilizing this function (i.e. charging cost). But she or he implicitly understands (or should be told by the PSO as part of their duty of care) that participating in a V2G will accelerate battery degradation. If he or she decides to participate in the V2G scheme, he or she would need to be convinced that the savings of charging cost, or even better receiving some payment, an adequate compensation for battery accelerated degradation. A PSO may also offer some incentives such as paying a battery degradation charge to convince the driver that participating in V2G is worth doing. In other words, the cost model includes some elements related to battery degradation. Similarly, the criteria which set a minimum and a maximum state of charge indirectly reduces the rate of battery degradation; battery life is known to improve if the depth of discharge is reduced [13,14], for example, it is well known that the cycle life of some lithium-ion batteries improves from 2,000 deep cycles to more than 15,000 shallow cycles (from 30–70% state of charge). The driver can have the option to limit the depth of discharge (by setting the minimum and maximum SOC), to reduce battery degradation.

8.3.2 Determination of the dispatch action

The priorities of alternatives are determined by evaluating how suitable it is to charge/discharge the EV battery with respect to every sub-criterion as follows:

State of Charge (SOC): The upper limit of SOC S_{max} is selected to avoid overcharge, while the lower limit S_{min} is set to ensure that sufficient electric energy remains within the battery for the next journey and for the protection of the health of the battery. In V2G mode, S_{max} is set to be 0.8 and $S_{min} = 0.4 + S_n$ where S_n represents the per-unit capacity that is going to be consumed on the road to the next destination. It is clear that the lower the SOC the more favourable it is to charge the battery. The higher the SOC the more favourable it is to discharge it. Therefore, the priorities of charging and discharging with respect to SOC are defined as follows:

Charge:

$$P^c_{SOC} = \begin{cases} 1 & SOC \leqslant S_{min} \\ 1 - \dfrac{SOC - S_{min}}{S_{max} - S_{min}} & SOC > S_{min} \end{cases} \tag{8.9}$$

Discharge:

$$P^d_{SOC} = \begin{cases} 1 & SOC \geqslant S_{max} \\ 1 - \dfrac{S_{max} - SOC}{S_{max} - S_{min}} & SOC < S_{max} \end{cases} \tag{8.10}$$

Electricity Buying/Selling Price: In order to reduce cost, it is better to charge the battery when the buying price is low and discharge it when the selling price is high. Therefore, the priorities of charging and discharging in terms of electricity price are determined as follows:

Charge:

$$P^c_{EP} = \begin{cases} 1 & bp \leqslant hbp \\ 1 - \dfrac{bp - hbp}{0.1 \times mbp} & bp > hbp \end{cases} \tag{8.11}$$

Discharge:

$$P^{d}_{EP} = \begin{cases} 1 & sp \geqslant hsp \\ 1 - \dfrac{hsp - sp}{0.6 \times msp} & sp < hsp \end{cases} \tag{8.12}$$

where *bp* and *sp* are the electricity buying and selling prices determined by the RTP information. *hbp* and *hsp* are the high buying and selling prices, respectively. *hbp* is defined as 90% of the maximum buying price *mbp*, while *hsp* is defined as 60% of the maximum selling price *msp* [6]. These four price values (*hbp*, *hsp*, *mbp* and *msp*) can be determined by using the day-ahead pricing information. In this work, real pricing data recorded in [15] are used to define these four values.

N-1 Contingency: The priority of charging is zero with respect to an N-1 contingency that requires discharging corrective action at the bus to which the EV is connected. Similarly, the priority of discharging is zero if the contingency requires a charging corrective action. The priorities of these actions are determined as follows:

Sensitivity: The sensitivity of load flow through a branch b overloaded by an N-1 contingency C to changes of power injection at a particular bus j can be evaluated using the sensitivity factor defined in [16]:

$$S^{j}_{C} = \frac{\Delta F_b}{\Delta P_j} \tag{8.13}$$

where ΔF_b is the change of power flow through the branch b resulting from the change in power injection at bus $j(\Delta P_j)$. As certain buses have relatively higher sensitivity factors with respect to the overloaded branch than others, the same power injections at these buses could make a more obvious difference to the alleviation of the overload. Thus, an EV connected to a bus with a high-sensitivity factor will have a high priority of dispatch during an N-1 contingency, while the dispatch of an EV connected to a bus with a very low-sensitivity factor is not recommended due to its small contribution. By comparing the sensitivity of a bus to the bus with the highest sensitivity factor, the priority of charging/discharging an EV at a specific bus with respect to sensitivity, P_{sen}, can be defined as:

$$P_{sen} = \frac{|S^{j}_{C}|}{max\{|S^{1}_{C}|, |S^{2}_{C}| \ldots |S^{nb}_{C}|\}} \tag{8.14}$$

where *nb* is the total number of buses in the system.

Severity: The severity of overload can be quantified as a percentage of the long-term emergency (LTE) rating of the overloaded line, as follows:

$$Severity = \frac{Load\ \ Flow - LTE}{LTE} \times 100\%, \tag{8.15}$$

where *LTE* rating can be derived from [17], while the *Load Flow* is obtained by carrying out the power flow analysis for the network under an N-1 contingency. The priority of charging/discharging with respect to overloading severity P_{sev} is determined as:

$$P_{sev} = \frac{Sev_C}{Sev_{max}}, \tag{8.16}$$

where Sev_C is the severity of overloading caused by contingency C and Sev_{max} is the severity of the severest overload caused by the severest contingency, which is derived by running the simulation of power flow analysis for all possible N-1 contingencies within the network, calculating the severity of overload they caused using (8.15) and then selecting the maximum.

Potential Consequence: Potential consequences of an N-1 contingency are determined by the number of overloaded lines, assuming that the branch overloaded by the contingency is broken due to the absence of an in-time measure. The priority of charging/discharging in terms of potential consequence P_{PoC} is calculated as follows:

$$P_{PoC} = \frac{Poc_C}{Poc_{max}} \tag{8.17}$$

where Poc_C is the potential consequence of contingency C and Poc_{max} is the severest potential consequence.

Cost to Grid: When charging the EV battery, the cost to the grid is C_c. However, when discharging it, the cost is C_d. From the grid's perspective, the lower the cost the better. Therefore, the priorities of charging and discharging in terms of the cost to grid are evaluated as follows:

Charge:

$$P^c_{CG} = \begin{cases} 1 & C_c \leqslant HCC \\ 0 & C_c > HCC \end{cases} \tag{8.18}$$

Discharge:

$$P^d_{CG} = \begin{cases} 1 & C_d \leqslant LDC \\ \dfrac{HDC - C_d}{HDC - LDC} & LDC < C_d < HDC \\ 0 & C_d \geqslant HDC \end{cases} \tag{8.19}$$

where *HCC* is the high cost to the grid for charging an EV and is set to be zero in this work because the cost to the grid to charge an EV is negative and the cost to discharge an EV is positive in this work. *HDC* and *LDC* are, respectively, the high and low costs to grid for discharging an EV and are set to be £0.02/KWh and £0.01/KWh, respectively [15].

Load Levelling: The priorities of charging and discharging are determined as follows:

Total Load Demand: In order for load levelling (i.e. peak shaving and valley filling) to be effective, it is better for EVs to be charged when the network's original load demand (i.e. the system load without EV integration) is low, so that EVs can store energy for the driving activities and grid operational support that might happen afterwards, while discharged to provide energy to the grid when the network's original load demand is high. Therefore, the priorities of charging and discharging in this case are defined as:

Charge:

$$P^{c}_{LD} = \begin{cases} 1 & d \leqslant LD \\ 1 - \dfrac{d - LD}{0.25(d_{max} - d_{min})} & LD < d < MD \\ 0 & d \geqslant MD \end{cases} \tag{8.20}$$

Discharge:

$$P^{d}_{LD} = \begin{cases} 1 & d \geqslant HD \\ 1 - \dfrac{HD - d}{0.25(d_{max} - d_{min})} & MD < d < HD \\ 0 & d \leqslant MD \end{cases} \tag{8.21}$$

where d_{max} and d_{min} are the maximum and minimum system demand during a day which can be determined using day-ahead load forecasting data or historical data. In this work, real load demand data [15] are used to define these two values. *MD* is the mid-level system demand, calculated as

$$MD = 0.5 \times (d_{max} + d_{min}). \tag{8.22}$$

LD is the low-level system demand, determined by

$$LD = d_{min} + 0.25 \times (d_{max} - d_{min}). \tag{8.23}$$

HD is the high-level system demand, determined by

$$HD = d_{max} - 0.25 \times (d_{max} - d_{min}). \tag{8.24}$$

Potential Consequence: The potential consequence if discharging/not charging the battery is evaluated based on its SOC. If the EV battery is discharged/not charged at a given time, there might not be enough electric energy within the battery to be discharged during high load periods. Therefore, the priorities of charging and discharging are determined as follows:

Charge:

$$P^{c}_{LPoC} = \begin{cases} 1 & SOC \leqslant Sg_{min} \\ 1 - \dfrac{SOC - Sg_{min}}{Sg_{max} - Sg_{min}} & SOC > Sg_{min} \end{cases} \tag{8.25}$$

Discharge:

$$P_{LPoC}^{d} = \begin{cases} 1 & SOC \geqslant Sg_{max} \\ 1 - \dfrac{Sg_{max} - SOC}{Sg_{max} - Sg_{min}} & SOC < Sg_{max} \end{cases} \tag{8.26}$$

where Sg_{min} and Sg_{max} are low and high SOC from the perspective of grid and selected to be 0.4 and 0.8, respectively, for the normal operation of EV batteries.

The final priorities P_f of charging and discharging an EV are calculated in the same way, that is:

$$\begin{aligned} P_f = \ & (P_{SOC} \times w_{SOC} + P_{EP} \times w_{EP}) \times w_{EV} \\ & +((P_{sen} \times w_{sen} + P_{sev} \times w_{sev} + P_{PoC} \\ & \times w_{PoC}) \times w_{NC} + P_{CG} \times w_{CG} + (P_{LD} \\ & \times w_{LD} + P_{LPoC} \times w_{LPoC}) \times w_{LL}) \times w_{G}, \end{aligned} \tag{8.27}$$

where P_{SOC}, P_{EP}, P_{sen}, P_{sev}, P_{PoC}, P_{CG}, P_{LD} and P_{LPoC} are the priorities of charge/discharge with respect to SOC, electricity price, sensitivity, severity, potential consequence of N-1 contingency, cost to grid, total load demand and potential consequence of load levelling, respectively. w_{SOC}, w_{EP}, w_{sen}, w_{sev}, w_{PoC}, w_{CG}, w_{LD} and w_{LPoC} are the corresponding weighting factors. w_{NC} and w_{LL} are the weightings of N-1 contingency and load levelling with respect to grid's concerns. w_{EV} and w_G are, respectively, the weightings of EV users' and grid's concerns.

The above forms the main part of the decision-making process for the dispatch of an EV's battery. It is important to note that the information about the next journey S_n is assumed to be available. The SOC of an EV battery is checked at the beginning of each time interval. If the SOC is less than S_{min}, the battery is charged during the current time interval. Otherwise, if the SOC is larger than S_{max}, it is discharged. The dispatch action is determined as the one that has the highest final priority if the SOC is between S_{min} and S_{max}. How fast the battery is charged/discharged depends on the final priority P_f of charge/discharge. If P_f is no less than 0.9, the EV is charged/discharged at the high current level set here to be 30 A. If P_f is between 0.7 and 0.9, the EV is charged/discharged at the middle-level current of 10 A. If P_f is between 0.4 and 0.7, the EV is charged/discharged at a low-level current of 2 A. Otherwise, when P_f is lower than 0.4, the EV battery is idle during the time interval. The overall AHP-based dispatch strategy is shown in Figure 8.2, which is implemented at the beginning of each time interval for each EV by the PSO using the real-time data gathered a couple of minutes beforehand. The detailed data communication chart is illustrated in Figure 8.3, which describes the procedure of real-time data collection from EVs and system operators, the working process of the PSO and the transmission of dispatch action commands to EVs. The PSO collects data from the EVs, distribution system operator (DSO) and transmission system operator (TSO) and sends dispatch commands to the EVs. For comparison, the flow chart of the rule-based dispatch strategy described in [6] is given in Figure 8.4. The key parameters of the two dispatch strategies are set to be the same for a fair comparison, as shown in Table 8.3.

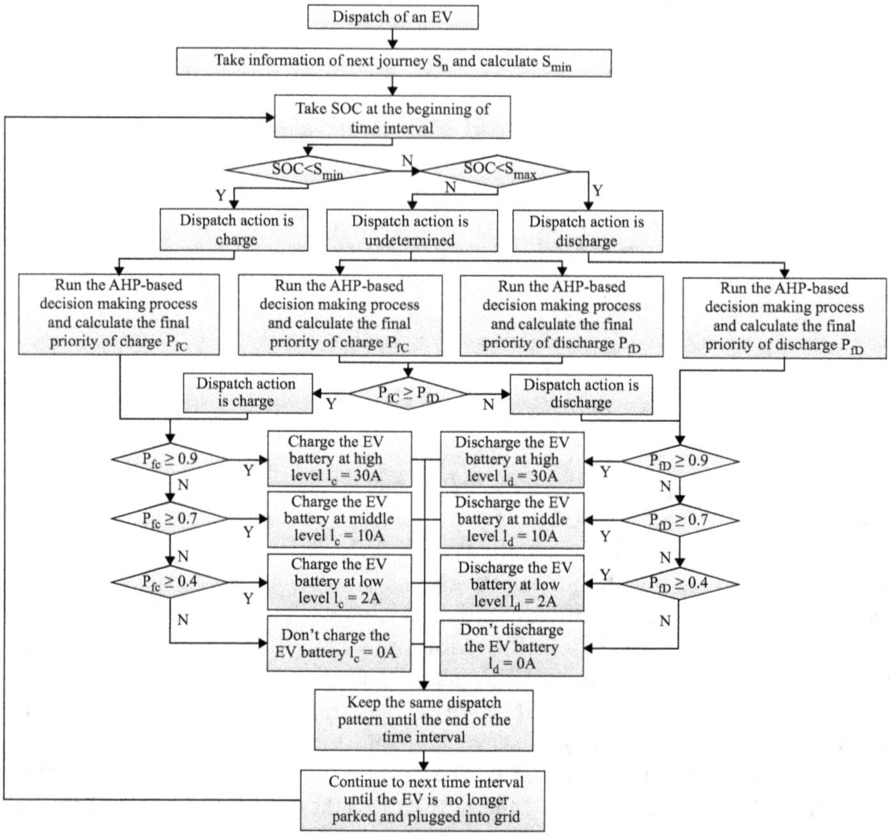

Figure 8.2 AHP-based dispatch strategy for an EV battery

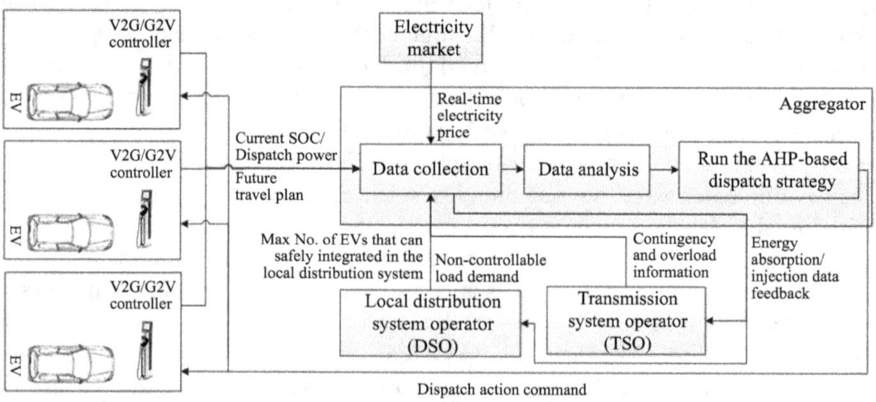

Figure 8.3 Data communication chart

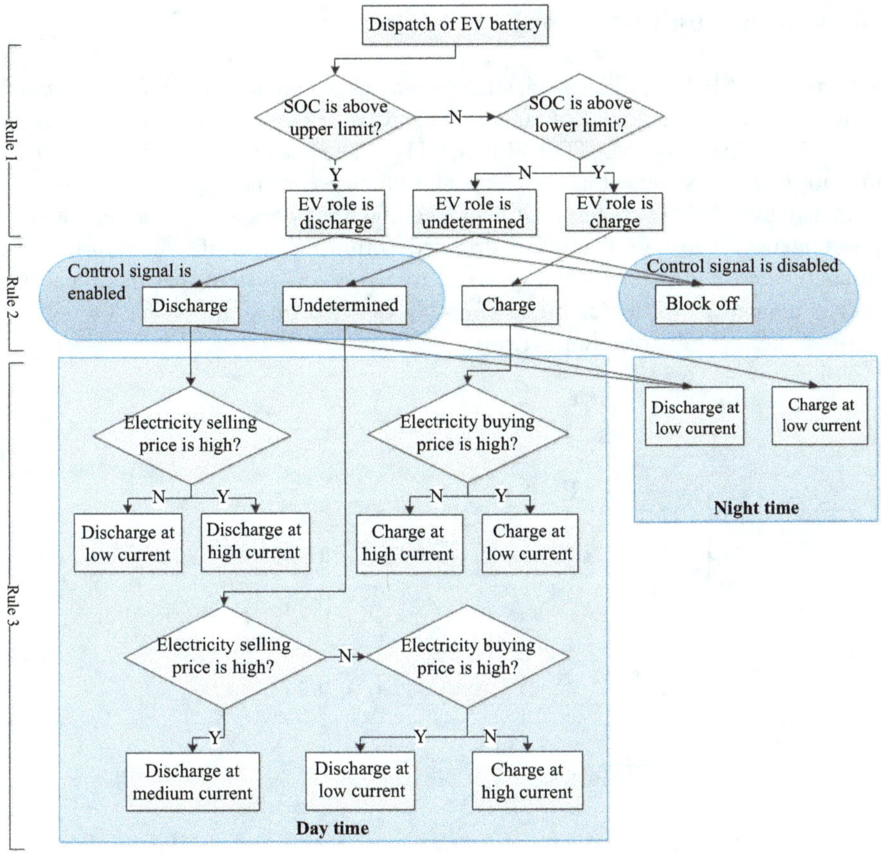

Figure 8.4 Rule-based dispatch strategy described in [6]

Table 8.3 Key parameter settings for AHP-based and rule-based dispatch strategies

Dispatch strategy	SOC_{min}	SOC_{max}	**High buying price** ***hsp*** **(£/KWh)**	**High selling price** ***hbp*** **(£/KWh)**
AHP-based	0.4	0.8	0.15	0.11
Rule-based	0.4	0.8	0.15	0.11
Dispatch Strategy	High Dispatch Current	Medium Dispatch Current	Low Dispatch Current	On-road Discharging Current
AHP-based	30	10	2	20
Rule-based	30	10	2	20

8.4 Simulation test

The proposed AHP-based dispatch strategy was tested on an IEEE RTS [17], which is formed of 24 buses and 38 branches as shown in Figure 8.5, and implemented using MATLAB® and MATPOWER 4.1 [18]. The reason why RTS is used for simulation test is because this chapter mainly focuses on the application of V2G in the transmission system in terms of load levelling and overload alleviation under N-1 contingency. The EVs that are distributed within the local distribution network are aggregated and connected at a bus of transmission system that this local network is connected to, under the assumption that the integration of EVs will not overload the local distribution network.

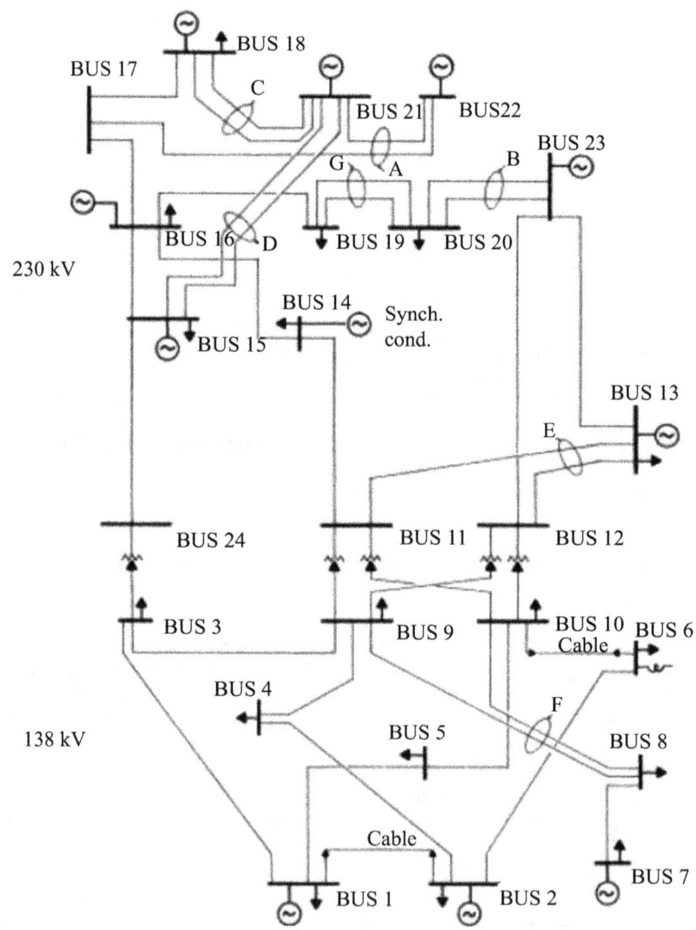

Figure 8.5 Diagram of IEEE reliability test system

The total load demand and system selling/buying price of electricity during a day was taken from [15]. The total load demand is scaled down so that the peak demand during a day is 3,300 MW, which approximately represents the daily load demand in a large UK city, as shown in Figure 8.6. Due to a scarcity of information on the payments by/to EV users when the EVs are charged/discharged, the system selling/buying price in [15] is adjusted based on domestic tariffs using the method proposed in [6] to represent EV users' selling/buying price of electricity. The adjusted system selling/buying price is derived by adding the difference between the average tariff value and the average system selling/buying price to the corresponding system selling/buying price, as shown in Figure 8.7:

$$ASP(t) = SP(t) + \left(\frac{\sum_{t=1}^{T} tariff(t)}{T} - \frac{\sum_{t=1}^{T} SP(t)}{T} \right), \tag{8.28}$$

where a day is divided into T time intervals. $ASP(t)$ and $SP(t)$ are the adjusted system selling/buying price and system selling/buying price at time interval t, respectively, while $tariff(t)$ represents the domestic tariff value at time interval t. The average values of ASSP (i.e. adjusted system selling price) and ASBP (i.e. adjusted system buying price) become the same and equal to the average of domestic tariff; therefore, ASBP can be either higher or lower than ASSP. Therefore, the ASSP and ASBP in Figure 8.7 are used as the EV charging and discharging prices, namely the bp in (8.11) and sp in (8.12), respectively. C_c in (8.18) is defined as the difference between system buying price and system selling price in Figure 8.7, namely $C_c = SSP - SBP$. C_d in (8.19) is set to be the difference between EV discharging price and charging price, that is, $C_d = ASBP - ASSP$. The time step is set to be 30 min. The dispatch actions are determined by the dispatch strategy at the beginning of every time interval and lasts for the entire time interval of 30 min.

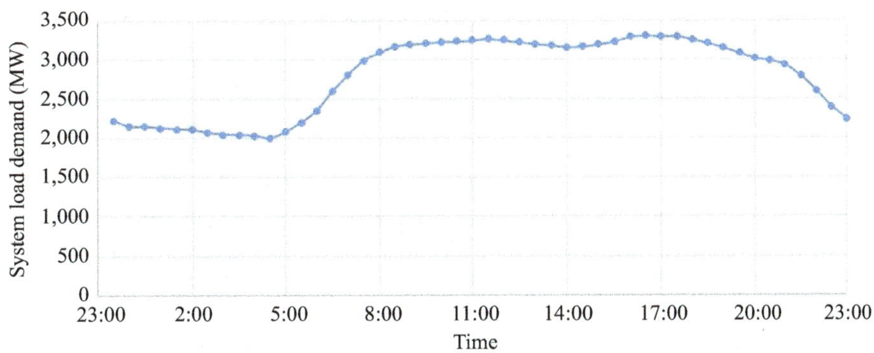

Figure 8.6 Total load demand during a day without EV load

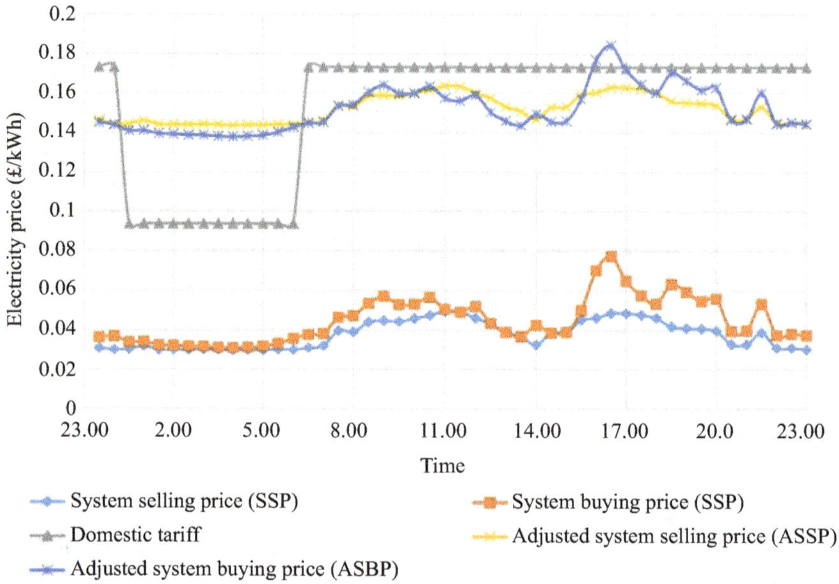

Figure 8.7 Daily electricity price

Table 8.4 EV's travel pattern and S_n during a day

Time	Travel pattern	S_n
23:00–7:00	Parked	0.2
7:00–8:00	on road	n/a
8:00–12:00	Parked	0.1
12:00–12:30	on road	n/a
12:30–13:30	Parked	0.1
13:30–14:00	on road	n/a
14:00–18:00	Parked	0.2
18:00–19:00	on road	n/a
19:00–23:00	Parked	0.2

To test the AHP-based dispatch strategy, an EV is connected to bus 6 when it is parked and plugged into the grid. When the EV is on the road, the battery electric energy is assumed to be consumed at the nominal discharging current (20 A) as assumed in [6]. The EV's travel pattern and electric capacity requirements for the next journey, S_n, during a day are shown in Table 8.4. It is important to note that the travel pattern used here was randomly generated and utilized as an example. The EV can have any kind of travel pattern, which will not affect the feasibility of the proposed dispatch strategy. The simulation results of the EV's SOC and net cost to EV users (i.e. charging cost-discharging payment) during a day are shown in Table 8.5, compared with those generated by applying the rule-based dispatch strategy that has been developed in [6].

In Table 8.5, it is shown that under the same travel pattern the AHP-based dispatch strategy results in better SOC conditions of the EV though costing £0.86 more per day than the rule-based strategy. The improved SOCs produced by the AHP-based dispatch strategy remain in the main within the desired range [0.4, 0.8]. The lowest SOC is just below 0.4, which implies a more reliable driving experience, compared to the lowest SOC of 0.194 when using rule-based strategy. This could justify the higher daily cost of the AHP-based strategy. Furthermore, if the EV battery, without V2G service, is quickly fully charged after every use (i.e. using 30 A charging current) to recover the electricity consumed on the road, the daily cost per vehicle will be £3.08 under the same travel pattern as shown in Table 8.4, which demonstrates the capability of AHP-based dispatch strategy to save costs while ensuring sufficient SOC in the battery for reliable driving.

As for the effectiveness of the AHP-based dispatch strategy in dealing with an N-1 contingency, the EV battery dispatch action can be affected by the occurrence of a severe N-1 contingency and the overloading it causes can be efficiently relieved by power injection/absorption at the bus to which the EV is connected. The simulation results using the AHP-based dispatch strategy show that a car's battery is charged at 479 W from 9:00 a.m. to 9:30 a.m. if no contingency occurred. However, if branch 10 (between bus 6 and 10) is broken (N-1 contingency) at 9:00 a.m., which results in the overloading of branch 5 (between bus 2 and 6), power injection will be required at bus 6 meaning that the connected EV battery is preferably discharged. By applying the AHP-based dispatch strategy, the EV is set to be discharged at 479 W. The rule-based dispatch strategy that was proposed in [6] does not take N-1 contingency into account and thus makes the same decision on the dispatch action of the EV battery regardless of the contingency – using this strategy an EV battery will in fact be charged at 468 W at 9:00 a.m., which will aggravate the overloading caused by the outage of branch 10.

The highest charging/discharging power of an EV is 7,731 W, which is negligible compared to the total load demand of the system. Therefore, in order to test the efficacy of the proposed AHP-based dispatch strategy on load levelling, a total number of 625,000 EVs are assumed to be in the system. This is a predicted number of EVs in a large UK city after 2035 (EVs are predicted to comprise 25% of the total car population after 2035, and there are about 2.5 million cars in a large UK city).

Table 8.5 SOC and daily cost of an EV – Comparisons between the proposed AHP-based dispatch strategy and the rule-based strategy

Dispatch strategy	**Lowest SOC**	**Highest SOC**	**Costs (£)**
AHP-based	0.331	0.694	2.25
Rule-based	0.194	0.657	1.39

These EVs are randomly allocated in the system while power flow analysis is used to make sure that no overload occurs. For the simplicity of simulation, EVs are grouped assuming that EVs within a group have similar SOCs, thus one SOC value is assigned to a group of EVs. Therefore, the initial SOCs of EVs are randomly assigned to the EV groups with a normal probability distribution ($\mu = 0.6, \sigma = 0.1$). A probability curve, illustrated in Figure 8.8 [6], is used to determine the number of parked EVs at each time interval by calculating the product of probability of parked cars at that moment and the total number of EVs available in the power network. When the probability of parked EVs decreases, a corresponding number of EVs are randomly selected to leave the parking lot. When the probability of parked EVs increases, an associated number of EVs arrive at the parking lot with SOCs randomly assigned at the arrival with a normal probability distribution of $\mu = 0.5, \sigma = 0.1$ (assuming that the average driving journey is about 30 min at a constant discharging current of 20 A on road). The newly arrived EVs are then assigned to the EV groups having similar SOCs (the difference is less than 0.05) with them or constitute new EV groups if their SOCs are really different (the difference is larger than 0.05) from the existing EV groups.

In order to measure the total load demand and system transmission loss, the power injected at the slack bus of the test system (bus 13) is evaluated for all the cases including the cases where EVs are charged in an uncontrolled fashion (i.e. EVs are fast charged at a high current level [30 A] after each journey) and where EVs are dispatched by the AHP-based strategy and the rule-based strategy. The power injection at the slack bus represents the power exchange between the power system under investigation and the rest of the grid. Simulation results for the power injections at the slack bus during 7 days are presented in Figure 8.9. The choppy transient response during the first day is due to the random assignment of the SOC of the EVs at the start of the simulation. The following days start with the SOC of EV batteries that have settled to a steady state value, hence the repeatable steady pattern of the power injected at the slack bus during the rest of the days, which confirms the stability of the dispatch strategies. Figure 8.10 shows power injections into the slack bus during the seventh day with the original curve of power injected

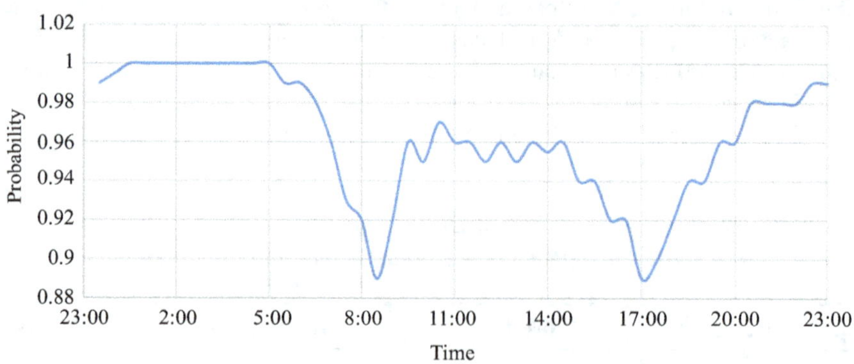

Figure 8.8 Probability of cars that are parked during a weekday

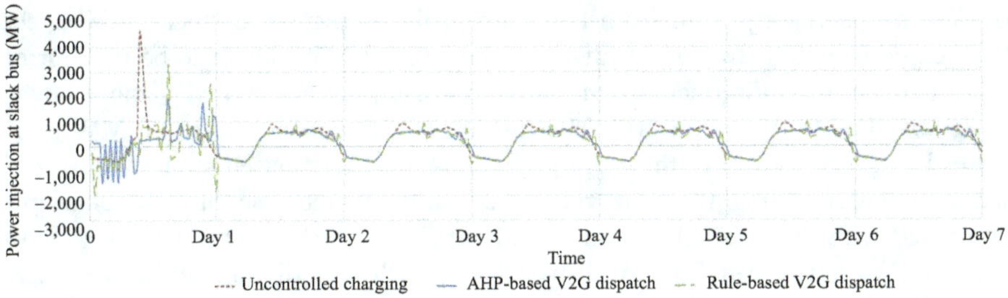

Figure 8.9 Power injections at the slack bus for all three cases during a week

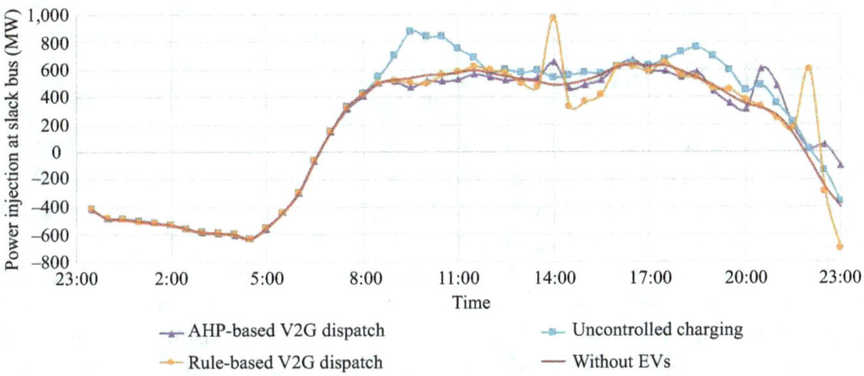

Figure 8.10 Power injections at the slack bus for all three cases during a day

Table 8.6 Numerical analysis of the efficacy of AHP-based dispatch strategy in load levelling compared with that of rule-based dispatch strategy and uncontrolled charging

Dispatch strategy	Lowest power injection (MW)	Highest power injection (MW)
AHP-based	– 626.3	673.4
Rule-based	– 706.5	969.5
Uncontrolled	– 630.5	877.5

at the slack bus when the grid is operated without EVs. Certain key data in Figure 8.10 are given in Table 8.6.

As shown in Table 8.6, the rule-based dispatch strategy and uncontrolled charging result in very high peaks of power injections, which are, respectively, about 200 and 300 MW higher than the peak power injection when using AHP-based strategy. Therefore, the AHP-based dispatch strategy has better performance in peak shaving compared to the two other strategies in the sense of creating much lower peak power injection from the main grid. During the off-peak period, the AHP-based dispatch strategy results in at most 626.3 MW power to be transferred to the main grid, while uncontrolled charging results in about 4 MW more power being absorbed by the main grid. The rule-based strategy results in 80 MW more power being transferred to the main grid. Thus, the AHP-based strategy performs better in valley filling than the rule-based strategy and uncontrolled charging. Compared with the original condition without EVs (631 MW at most being transferred to the main grid), the AHP-based approach actually has effective valley filling, which is not visible in the figure though. Therefore, further development (inclusion of forecast model perhaps) should be made on the dispatch model in the future for better valley filling and peak shaving performance, as discussed in Section 8.6.

8.5 Sensitivity analysis

In order to further investigate and understand the characteristics of the proposed AHP-based dispatch strategy, sensitivity analysis is carried out for many different scenarios. First of all, the effects of changing the weightings of EV users' and grid's concerns (i.e. w_{EV} and w_G, respectively) on the EV daily SOC and cost are investigated, as well as the impacts of the relative importance between SOC w_{SOC} and electricity price w_{EP}.

As shown in Table 8.7, when the weighting of EV users' concerns w_{EV} increases, more attentions are paid to the EV user's benefits; hence, the daily SOC condition of EV improves, as illustrated in Figure 8.11 (the corresponding current levels and final priorities of dispatch decisions are presented in Figures 8.12 and 8.13, respectively), normally leading to an increase of the charging cost. That is not hard to understand: a better SOC condition implies more electric energy to be charged or less to be discharged, which increases the daily cost to EV users. However, when the EV's SOC condition is already good enough within the 40% and 80%, the increase of w_{EV} does not improve the daily SOC condition of the EV but reduces the daily cost for the EV (e.g. when $w_{SOC} = 80\%$, $w_{EP} = 20\%$ with $w_{EV} = 50\%$ and 80%, respectively, as can be seen in Figure 8.11). That is because

Table 8.7 Sensitivity Analysis: the effects of changing the weightings of criteria on the EV daily SOC and cost

$\frac{w_{EV}}{w_G}$	$w_{SOC} = 20\%$, $w_{EP} = 80\%$		$w_{SOC} = 50\%$, $w_{EP} = 50\%$		$w_{SOC} = 80\%$, $w_{EP} = 20\%$	
	SOC	**Cost (£)**	**SOC**	**Cost (£)**	**SOC**	**Cost (£)**
$\frac{20\%}{80\%}$	[0.16, 0.69]	0.66	[0.18, 0.69]	0.82	[0.19, 0.69]	0.89
$\frac{50\%}{50\%}$	[0.19, 0.70]	1.59	[0.33, 0.69]	2.25	[0.47, 0.70]	2.62
$\frac{80\%}{20\%}$	[0.28, 0.75]	1.87	[0.36, 0.71]	2.35	[0.47, 0.70]	2.44

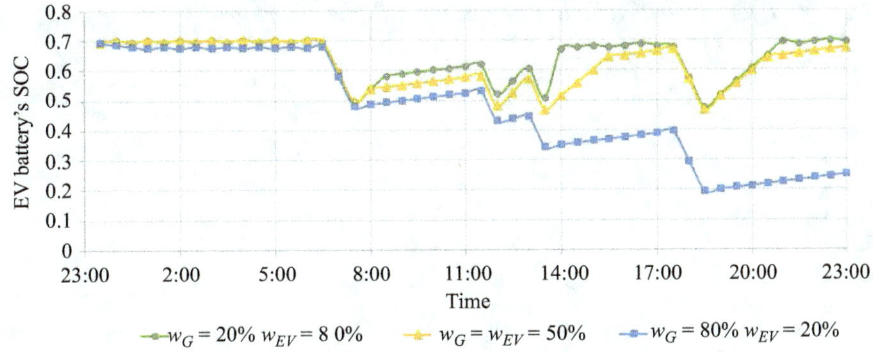

Figure 8.11 Daily SOC conditions of an EV battery under different settings of w_G and w_{EV} but constant w_{SOC} (80%) and w_{EP} (20%)

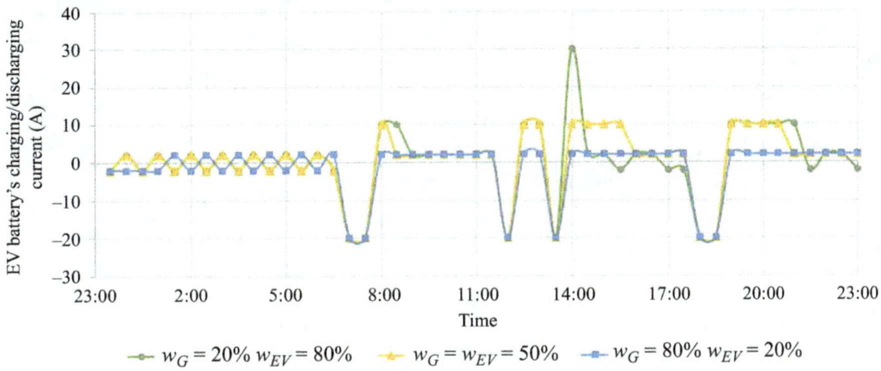

Figure 8.12 Charging/discharging currents of an EV battery under different settings of w_G and w_{EV} but constant w_{SOC} (80%) and w_{EP} (20%)

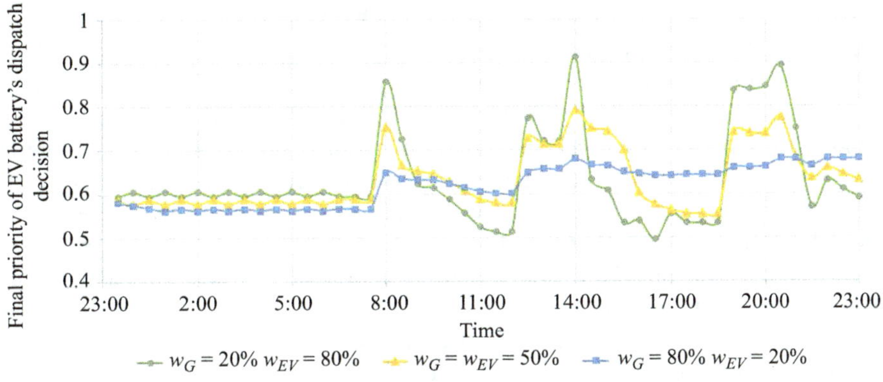

Figure 8.13 Final priorities of dispatch decisions of an EV battery under different settings of w_G and w_{EV} but constant w_{SOC} (80%) and w_{EP} (20%), which decide the dispatch current levels in Figure 8.12 and hence SOC conditions in Figure 8.11

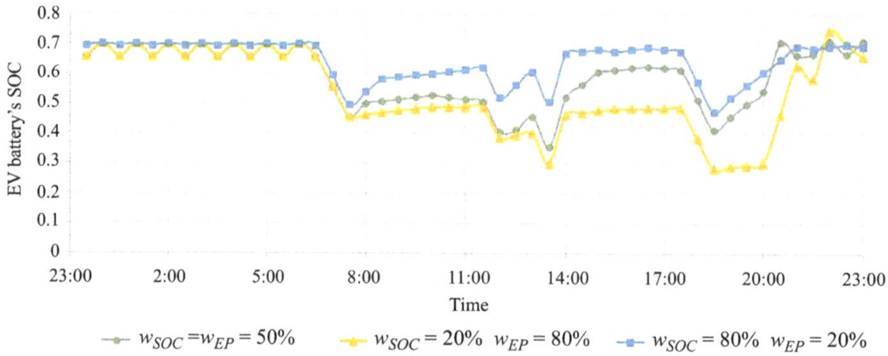

Figure 8.14 Daily SOC conditions of an EV battery under different settings of w_{SOC} and w_{EP} but constant w_G (20%) and w_{EV} (80%)

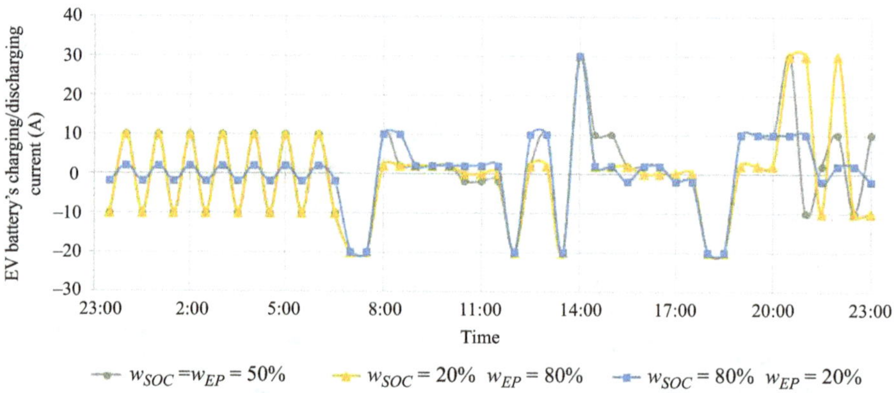

Figure 8.15 Charging/discharging currents of an EV battery under different settings of w_{SOC} and w_{EP} but constant w_G (20%) and w_{EV} (80%)

when the EV's daily SOC condition is already good enough, due to having a high priority, the improvement space becomes very limited. Thus, increasing w_{EV} to continually increase its composite priority will not make significant improvement to the SOC condition. But, the increase of the composite priority of cost due to increasing w_{EV} will tend to charge the EV at the low buying price instead of scattering the charging process over a wide period to avoid increasing the peak load demand. Moreover, when keeping the weightings of EV user and grid concerns unchanged but adjusting the relative importance between the two criteria, SOC and electricity price, the results change as well, as demonstrated in Figure 8.14 (the corresponding current levels and final priorities of dispatch decisions are presented in Figures 8.15 and 8.16, respectively). With a higher weighting given to electricity price (i.e. w_{EP} is higher but w_{SOC} is lower), it will be better to discharge an EV battery due to the high selling price, in order to save costs, than to charge it due to

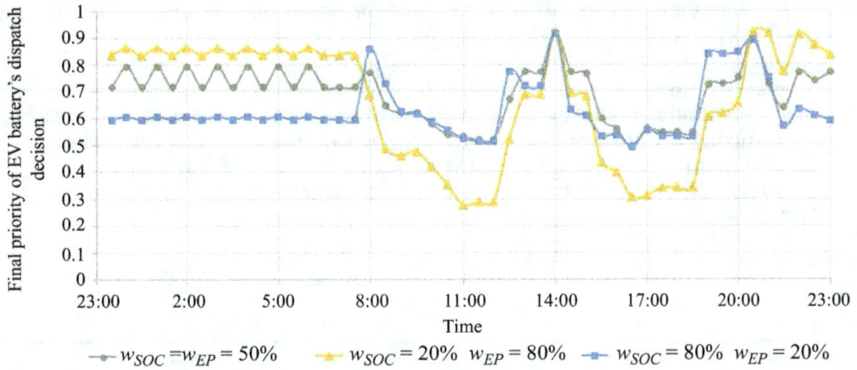

Figure 8.16 Final priorities of dispatch decisions of an EV battery under different settings of w_{SOC} and w_{EP} but constant w_G (20%) and w_{EV} (80%), which decide the dispatch current levels in Figure 8.15 and hence SOC conditions in Figure 8.14

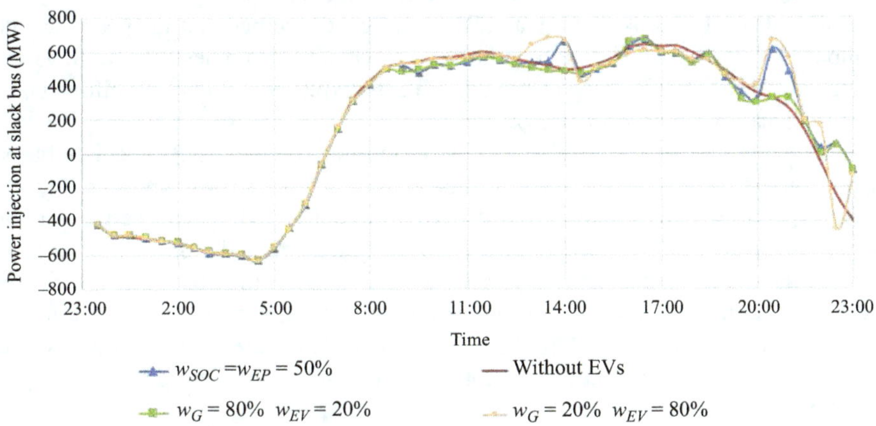

Figure 8.17 Daily power injections at the slack bus under different settings of w_G and w_{EV}

the low SOC. Thus, discharging might be chosen as the current dispatch action and the daily cost to the EV user, unsurprisingly, decreases as SOC condition worsens. However, with higher w_{SOC} but lower w_{EP}, the dispatch action recommended for ensuring that sufficient electricity remains in the battery will have a higher priority to be chosen as the current EV dispatch action than that recommended according to electricity price to save cost. The SOC condition is therefore, as expected, better with increasing daily cost.

As for the impact of different settings of w_G and w_{EV} on the network, similar sensitivity analysis is carried out and the result is shown in Figure 8.17. The corresponding accumulated energy injections are presented in Table 8.8. It is clear in

Table 8.8 Accumulated energy injections at the slack bus under different settings of w_G and w_{EV} in 24-h period

w_G	w_{EV}	Accumulated energy during off-peak period (MWh)	Accumulated energy during peak period (MWh)	Accumulated energy in 24-h period (MWh)
20%	80%	−3,829	7,708	3,879
50%	50%	−3,617	7,413	3,796
80%	20%	−3,613	7,039	3,426
Without EVs		−3,990	7,321	3,331

Figure 8.17 that the network load demand curve is improved, in terms of decreasing peak load demand, with higher weighting assigned to grid's concerns, which is just as expected and easy to understand. It is evident from Table 8.8 that when the weighting of grid concerns (w_G) becomes high at 80% the amount of energy injected at the slack bus reduces significantly with more energy absorbed by the network during off-peak period and less energy consumed during peak period. Compared with original situation without EVs, when w_G becomes 80%, both valley filling and peak shaving are achieved. This is expected as load levelling takes priority as can be seen in Figure 8.17.

As mentioned earlier, the setting of weightings of the AHP model is based mainly on expert experience and common sense. In practice, we envisage that the initial weighting determined based on common sense will be refined based on simulation results at the design stage or over time during the operation stage to produce sensible results. This may require some negotiations between the parties involved who may accept a compromise once they are informed about the implications of their choices and perhaps after their fears are allayed. Indeed, the weighting of the factors and even the factors themselves may need to be changed with time as the system evolves.

Furthermore, the impact of increasing the percentage of V2G-enabled EVs on load levelling is illustrated in Figures 8.18 and 8.19, where V2G-disabled EVs are charged in an uncontrolled way. Figure 8.18 shows the power injected at the slack bus during a 24-h period. As the percentage of V2G-enabled EVs increases, the peak shaving increases. In Figure 8.19, the total accumulated energy injected into the slack bus during a 24-h period is plotted against the percentage of V2G-enabled EVs. This graph was obtained by repeating the simulation 40 times for each percentage of V2G-enabled EVs; each simulation starts with a randomly assigned travel pattern; the mean and standard deviation of injected energy at the slack bus during a 24-h period are calculated for each percentage of V2G-enabled EVs from the results of the 40 simulations. The accumulated energy injection reduces when more EVs participate in the V2G operation. That is just as expected and can be easily understood: increasing the percentage of V2G-enabled EVs raises the available battery capacity that can be used to meet the power system demand.

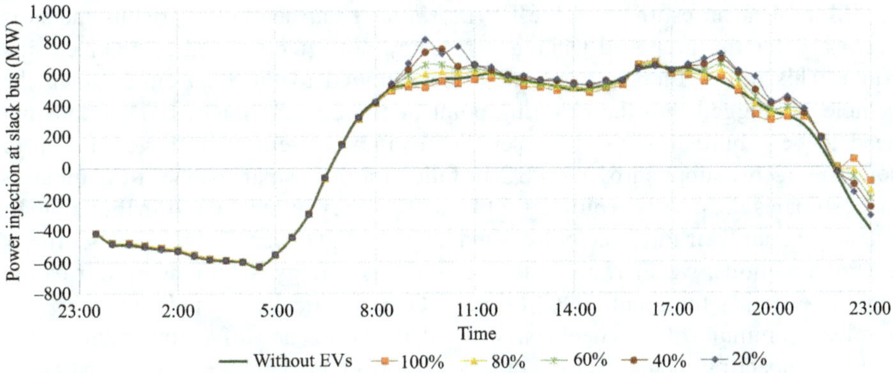

Figure 8.18 Power injections at the slack bus under different percentage of V2G-enabled EVs

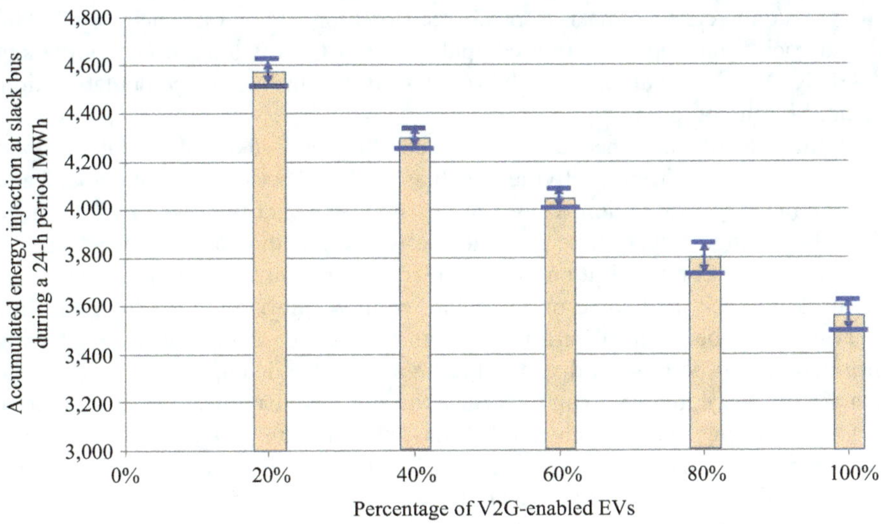

Figure 8.19 Average accumulated energy injections at the slack bus under different percentage of V2G-enabled EVs (Error bars represent ± standard deviation)

It is important to note that this work presents the proposed AHP-based dispatch strategy using an example setting of the criteria weightings in the AHP hierarchy model and the percentage of V2G batteries. However, the settings of weightings can be adjusted by the PSO to accommodate its requirements and the actual situation of the network and EV cluster. The proportion of V2G batteries in practice depends on the EV users' willingness and/or EV chargers' hardware.

Moreover, an estimate of the EV user's next journey energy requirements is necessary for the proposed technique to carry out effective and customized dispatch. This information is expected to be supplied to the PSO either when the vehicle is plugged into the charging point or through the internet. This does not need to be a burden to the EV user. For example, users could have an App (a software agent) supplied by the PSO installed on their smart phones or computers that estimates the energy required for the next journey based on information in their calendar about their future activities (and the time and distance of the next journey), traffic conditions, weather, etc. They can send the energy requirement information (could be simply minimum SOC) to the PSO automatically through the internet or wireless communication channels, without revealing any of their activities and without impacting their privacy. Forecasting tools can also be employed, based on historical records of the EV's previous driving activities, to provide a forecast of EV user's travel pattern including an estimate of the next journey's energy usage. Further research is needed on how this may be implemented in practice.

Furthermore, the AHP-based dispatch strategy is proposed for practical application; therefore, the computational efficiency is an important factor. It takes only 0.3 s on average to make dispatch decisions for the total EV fleet of 625,000 EVs at each time step if using a computer with a Quad Core 3.4 GHz CPU and 16 GB RAM as an example, which provides an insight into the computation time required by the proposed algorithm.

Some information is needed to implement the proposed dispatch strategy, apart from the EV user's driving activities, such as the RTP, network's load demand and N-1 contingency information. These are supposed to be notified a short while ahead of the beginning of every time interval, via wireless communication for example. However, if the communication is suspended due to a sudden accident, for example, hacker attack, and some or all of the required information for the next time interval cannot be obtained in time, then an estimate or a forecast of the missing information will be used to dispatch the EVs. In more detail, if the exact pricing data are unknown, the PSO can either use the available day-ahead electricity price or use forecasting tools, which can be embedded into the proposed model, to predict the price for the next time interval. The uncertain load demand for the next time interval can be dealt with in a similar way, that is, the day-ahead network load demand forecast can be used to implement the proposed dispatch strategy. Besides, if N-1 contingency occurs without being able to inform the PSO, EVs will be dispatched as if no contingency occurred. Due to the inherent redundancy of V2G and other energy storage that are in communication, this may not have a large impact as other EV clusters or stationary energy storage systems can participate in the alleviation of overload caused by the N-1 contingency. Note that it will be better to use exact real-time data in order for the proposed strategy to provide an accurate dispatch of EVs; however, using well-estimated data for a certain time interval (when the communication breaks off) will not affect the dispatch decision of the proposed AHP-based strategy due to its robustness and inherent tolerance of uncertainty. One could envisage that the AHP priority factors may be adjusted depending on the level of confidence in the information, which in turn depends on whether they were based on actual information or estimated one, which could add further robustness. This matter deserves further investigation in the future.

8.6 Conclusions

This chapter has proposed a novel decentralized dispatch strategy for EV batteries using the AHP methodology taking into account the requirements of both EV users and the grid, such that the EV batteries can be dispatched to save cost while ensuring reliable driving experience with sufficient SOC left in the battery and help to support the grid for load levelling or under N-1 contingency. It is important to stress that this work is addressing this particular type of scenario but not limited to this. With slight adjustment, the proposed dispatch strategy can be flexibly applied to other kinds of scenarios that might come to fruition. For example, if home storage batteries are used instead of EV batteries, the dispatch strategy is almost the same except that the electric energy will not be stored for travelling but for home appliance usage during blackouts or peak periods. As the basic/core approach is highly independent of the different scenarios, it can be used as a tool to validate those scenarios and compare their different impacts.

The proposed dispatch strategy was tested on an IEEE RTS. The simulation results demonstrated the feasibility and efficacy of the proposed AHP-based dispatch strategy in satisfying the requirements of both EV users and the grid. Moreover, this strategy has proved to be generally better than the rule-based one proposed in [6]. Compared to the rule-based dispatch approach, it costs £0.86 more a day, per vehicle, but it improves the EV battery's SOC condition and performs better in terms of load levelling. Under the severe N-1 contingency that overloads the branch close to the bus to which the EV is connected, the AHP-based dispatch strategy correctly selects the dispatch action of the EV battery that can help to alleviate the overloading.

In the application of V2G batteries using the proposed AHP-based dispatch strategy, the grid operator gathers the real-time data a short while (e.g. 5 min) prior to the beginning of each time interval, as discussed earlier. RTP data need to be downloaded from the electricity market on a half-hourly basis (i.e. before the start of every time interval of 30 min). RTP can be affected by the real-time demand and supply conditions, which include the dispatch of V2G batteries, that is, the charging as a load or discharging as a source. Moreover, as the electricity price is considered in the V2G battery dispatch, RTP will have an impact on the determination of battery dispatch actions. Hence, there is an interactive effect between V2G battery dispatch and RTP, which will be investigated in the future. Furthermore, the proposed AHP-based dispatch strategy is actually utilized by the grid operator, not by the EV users, as discussed earlier. In order to encourage the V2G operation, incentive payments will be given to the EV users if they agree to let their vehicles participate in the grid operational support when needed. The settings of incentive payments and the relevant policy requires further investigations. Moreover, the stochastic modelling of EV travel patterns in this work is simple to use but not sophisticated enough to reflect the reality.

In the future, EV travel patterns should be modelled in a more realistic way using algorithms such as the Markov process, as discussed in [19], and Copula, as investigated in [20] and the historic data, as utilized in [21], for example. Another improvement that could be made is to include the forecast of the grid's load level and electricity price in the next few hours into the dispatch model. In this way, the dispatch system is aware of the grid's environment in the near future, which might be more or less suitable for charging EVs than the current situation. Therefore, the dispatch system can make a better and more accurate decision in terms of whether to shift the EV charging load to future moments compared with the current situation, in order to achieve better valley-filling results.

References

[1] A. Foley, B. Tyther, P. Calnan, and B. O. Gallachoir, 'Impacts of electric vehicle charging under electricity market operations', *Applied Energy*, vol. 101, pp. 93–102, 2013.

[2] H. Khayyam, J. Abawajy, B. Javadi, A. Goscinski, A. Stojcevski, and A. Bab-Hadiashar, 'Intelligent battery energy management and control for vehicle-to-grid via cloud computing network', *Applied Energy*, vol. 111, pp. 971–981, 2013.

[3] W. Kempton and J. Tomi, 'Vehicle-to-grid power fundamentals: Calculating capacity and net revenue', *Journal of Power Sources*, vol. 144, no. 1, pp. 268–279, 2005.

[4] J. Lopes, F. Soares, and P. Almeida, 'Integration of electric vehicles in the electric power system', *Proceedings of the IEEE*, vol. 99, no. 1, pp. 168–183, 2011.

[5] T. L. Saaty, *The Analytic Hierarchy Process.* New York: McGraw Hill, 1980.

[6] Y. Ma, T. Houghton, A. Cruden, and D. Infield, 'Modeling the benefits of vehicle-to-grid technology to a power system', *IEEE Transactions on Power Systems*, vol. 27, no. 2, pp. 1012–1020, 2012.

[7] L. Wang, S. Sharkh, A. Chipperfield, and A. Cruden, 'Dispatch of vehicle-to-grid battery storage using an analytic hierarchy process', *IEEE Transactions on Vehicular Technology*, vol. 66, no. 4, pp. 2952–2965, 2017.

[8] J. Larminie and J. Lowry, *Electric Vehicle Technology Explained*. England: John Wiley & Sons, 2003.

[9] D. Doerffel and S. M. Sharkh, 'A critical review of using the Peukert equation for determining the remaining capacity of lead-acid and lithium-ion batteries', *Journal of Power Sources*, vol. 155, no. 2, pp. 395–400, 2006.

[10] O. Tremblay, L. Dessaint, and A. Dekkiche, 'A generic battery model for the dynamic simulation of hybrid electric vehicles', in *2007 IEEE Vehicle Power and Propulsion Conference*, Sept. 2007, pp. 284–289.

[11] J. Zhu and J. A. Momoh, 'Optimal VAr pricing and VAr placement using analytic hierarchical process', *Electric Power Systems Research*, vol. 48, no. 1, pp. 11–17, 1998.

[12] J. Zhu and M. Irving, 'Combined active and reactive dispatch with multiple objectives using an analytic hierarchical process', *IEE Proceedings–Generation, Transmission and Distribution*, vol. 143, no. 4, pp. 344–352, 1996.
[13] S. Han, S. Han, and H. Aki, 'A practical battery wear model for electric vehicle charging applications', *Applied Energy*, vol. 113, pp. 1100–1108, 2014.
[14] SAFT, 'Lithium-ion battery life', [Online], Available: http://www.saftbatteries.com/force_download/li_ion_battery_life__TechnicalSheet_en_0514_Protected.pdf.
[15] UK national grid', [Online], Jan. 2014, Available: http://www.bmreports.com/bwx_reporting.htm.
[16] A. Del Rosso and S. Eckroad, 'Energy storage for relief of transmission congestion', *IEEE Transactions on Smart Grid*, vol. 5, no. 2, pp. 1138–1146, 2014.
[17] P. M. Subcommittee, 'IEEE reliability test system', *IEEE Transactions on Power Apparatus and Systems*, vol. PAS-98, no. 6, pp. 2047–2054, 1979.
[18] R. D. Zimmerman, C. E. Murillo-Sanchez, and R. J. Thomas, 'Matpower: Steady-state operations, planning and analysis tools for power systems research and education', *IEEE Transactions on Power Systems*, vol. 26, no. 1, pp. 12–19, 2011.
[19] E. B. Iversen, J. M. Morales, and H. Madsen, 'Optimal charging of an electric vehicle using a markov decision process', *Applied Energy*, vol. 123, pp. 1–12, 2014.
[20] A. Lojowska, D. Kurowicka, G. Papaefthymiou, and L. van der Sluis, 'Stochastic modeling of power demand due to EVs using copula', *IEEE Transactions on Power Systems*, vol. 27, no. 4, pp. 1960–1968, 2012.
[21] Y. Wang, D. Infield, and S. Gill, 'Smart charging for electric vehicles to minimize charging cost', *Proceedings of the Institution of Mechanical Engineers, Part A: Journal of Power and Energy*, pp. 1–15, Dec. 2016.

Chapter 9

Electric vehicles as distributed energy storage for local energy management

Ricardo Morales[1], Jesús Fraile-Ardanuy[2], Álvaro Gutiérrez[2], David Jiménez[2], Benito Artaloytia[2], Roberto Álvaro-Hermana[3,4], Julia Merino[5] and Sandra Castaño-Solis[6]

9.1 Introduction

Electric vehicles (EVs) are relatively new at a commercial scale and are currently dealing with issues like cost, anxiety range and charging capacity [1]; but, they offer enormous possibilities for reducing $C0_2$ and NO_x emissions and carbon footprint if they are integrated by means of renewable sources [2,3].

As the production cost of photovoltaic (PV) energy is closing to the retail prices [4], it is possible to make self-consumption installations profitable, reducing the amount of energy required from the utility grid and decreasing the total electricity cost. In order to achieve this objective, it is necessary to combine the PV generation with electricity storage and demand-response schemes [5–7].

In an office building with PV generation, some of these restrictions can be easily resolved. There is an alignment of working hours with the PV energy production, and there are two different situations that must be analysed. First, if PV generation exceeds the building energy demand during the day, this surplus energy must be stored in a battery and fed back into the building grid when the PV generation is lower than the building energy demand. Thus, PV coverage of energy demand (self-consumption) increases. Second, if PV generation is lower than the building energy demand during all time periods, it is still possible to save money, storing the energy in the batteries when the retail prices are low and powering it back into the building when the utility prices are at the highest value, optimizing

[1]Fronius México SA, Puebla, Mexico
[2]ETSI Telecomunicación, Universidad Politécnica de Madrid, Madrid, Spain
[3]Orkestra – Basque Institute of Competitiveness, Bilbao, Vizcaya, Spain
[4]University of Deusto, Bilbao, Vizcaya, Spain
[5]TECNALIA, Parque Científico y Tecnológico de Bizkaia, Derio, Spain
[6]Escuela Técnica Superior de Ingeniería y Diseño Industrial (ETSIDI), Universidad Politécnica de Madrid, Madrid, Spain

the total electricity cost paid by the building manager. This second situation, where PV energy is the only one used to charge the vehicles, is used in this work.

Even though the Li-ion battery price has dropped significantly in recent years [8], this storage technology is still quite expensive and, in order to avoid overspending, this study proposes using EV's batteries as Battery Energy Storage System. One of the main disadvantages of this scheme is that the total available battery's capacity is variable, but it is dependent on the EV's mobility. In this chapter, an optimization model is developed to minimize the electricity cost purchased from the utility grid by the building manager. This optimization model takes into account several parameters: the availability of the EVs parked at the parking lot (that affects the available storage capacity, which varies at each time period, as EVs come and go along the day), the total energy consumption demanded by the building, the self-PV generation and, finally, the retail energy prices. With this information, the optimization algorithm determines how the EV batteries' charge/discharge of all parked vehicles should be managed to minimize the cost of electricity purchased from the grid.

The optimization system also considers several constraints: the amount of energy that should be stored in each battery at the end of the day, and the maximum energy that can be charged/discharged from these vehicles. We will apply this model to a real building consumption profile.

9.2 System description

9.2.1 Building PV generation and electricity consumption

The *Instituto de Energía Solar*-IES (Solar Energy Institute) is a research centre from Universidad Politécnica de Madrid founded in 1978 and specifically focuses on the fundamentals and practical application of PVs systems. Its headquarters are located at EscuelaTécnica Superior de Ingenieros de Telecomunicación Universidad Politécnica de Madrid (ETSIT-UPM), in Madrid (Spain), and it shares a common parking area with a total capacity of 29 parking decks, as is shown in Figure 9.1.

The total built area of IES is 875 m^2, distributed on three floors, containing several research laboratories, meeting rooms and offices for researchers. There are 13.1 kWp PV cells installed at IES headquarters, integrated into the roofs and the facades, which are shown in Figure 9.2. The PV panel tilt is $\theta_{tilt} = 26°$ and it is oriented directly to the South ($\theta_{orientation} = 0°$). This PV generation is monitored with an hour resolution, along with the total consumption of the building. These measurements are available online at http://monitoring.robolabo.etsit.upm.es/etsit/monitoring.php?tab=ies

Figure 9.3 shows the PV hourly energy generation (in kWh) and the total hourly electricity consumption of this building (in kWh) for three typical working weeks in March, April and May 2016. A blue dotted line represents PV generation, while electric demand is shown with a brown dotted line.

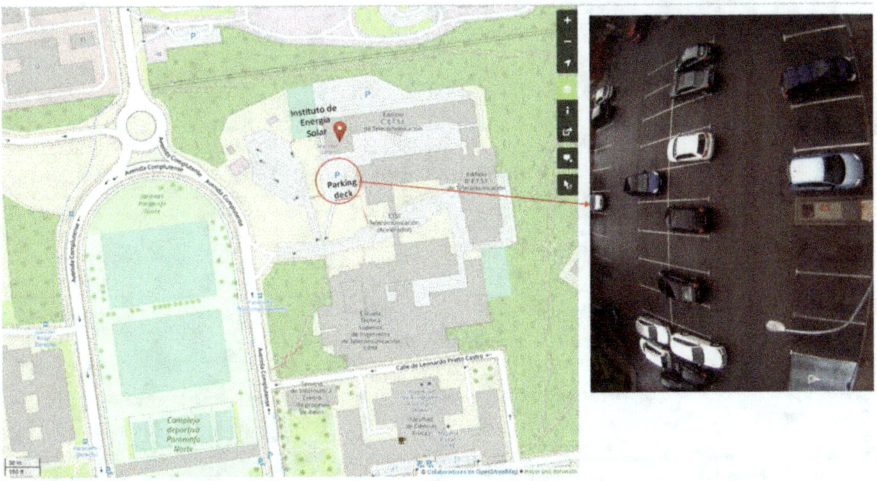

Figure 9.1 Instituto de Energía Solar (Solar Energy Institute) headquarters location with the parking decks highlighted

Figure 9.2 Instituto de Energía Solar (Solar Energy Institute) building

It is observed that, during this 3-month period, the average load demand is around 32 kWh and the peak electric consumption reaches almost 60 kWh, while the maximum PV generation was 8.8 kWh during this period (14%–67% of the peak demand). For that reason, there is not enough PV-generated energy to satisfy all energy load demand of this research centre.

An average week of IES PV hourly energy generation for each season of the year is shown in Figure 9.4. The highest solar production is obtained during spring and autumn seasons, when the outdoor temperature is cooler and the PV panel efficiency is increased. During the winter season, even though the outdoor

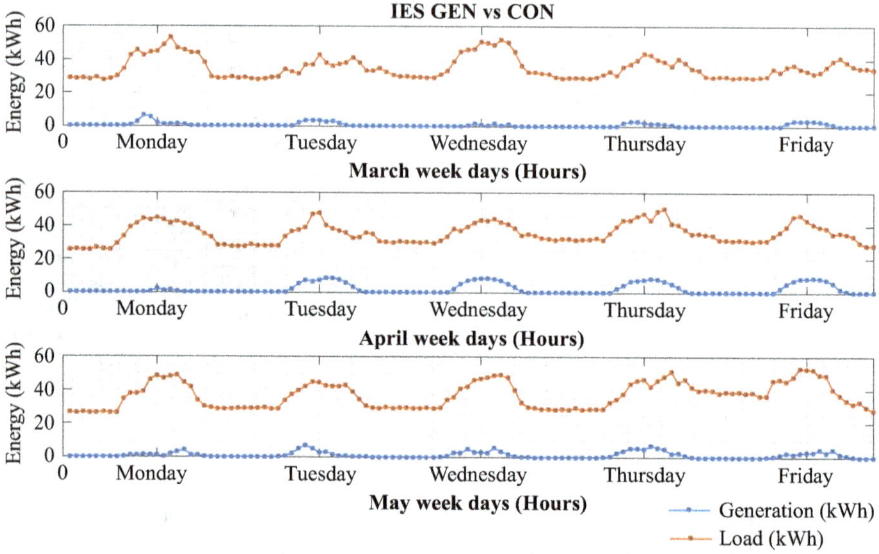

Figure 9.3 IES PV generation (blue line) and electric consumption (brown line) during three different working weeks in March, April and May 2018

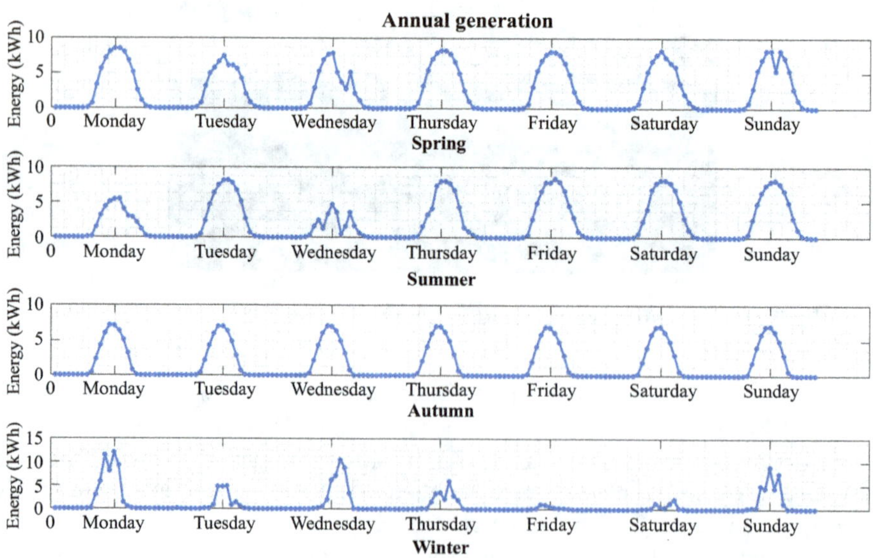

Figure 9.4 IES PV hourly generation (blue line) during an average week in each season of the year

temperature is lower and the panel efficiency is higher, the days are shorter, with less sunlight, and thus the PV generation is significantly reduced.

9.2.2 Grid electricity price

Currently, there are three different electric tariffs in Spain [9]: normal rate 2.0A (which is the default hourly tariff), night rate 2.0 DHA (which is a two periods hourly tariff, with cheaper prices during the night but more expensive prices during the day) and EV rate 2.0 DHS (which is a small variation of the previous 2.0 DHA, with even cheaper prices from 00:00 to 06:00 hours well suited for EVs).

The hourly prices of each of these tariffs are changing every day. Four different representative days, one for each season has been selected for this study. Figure 9.5 shows the three electric tariffs (2.0A, 2.0DHA and 2.0DHS) for these four representative days in spring, summer, autumn and winter.

9.2.3 Mobility information

In order to estimate the aggregated battery capacity available at each hour of the day, the first step was to analyse the parking occupation. With this information, it was possible to know how many vehicles were parked at IES along the day.

Once the parking occupation was calculated, the second step was to estimate the consumption of these vehicles and the energy required to reach the final destination at the end of the day. From this information, the total aggregated battery capacity available for the proposed building energy management was estimated.

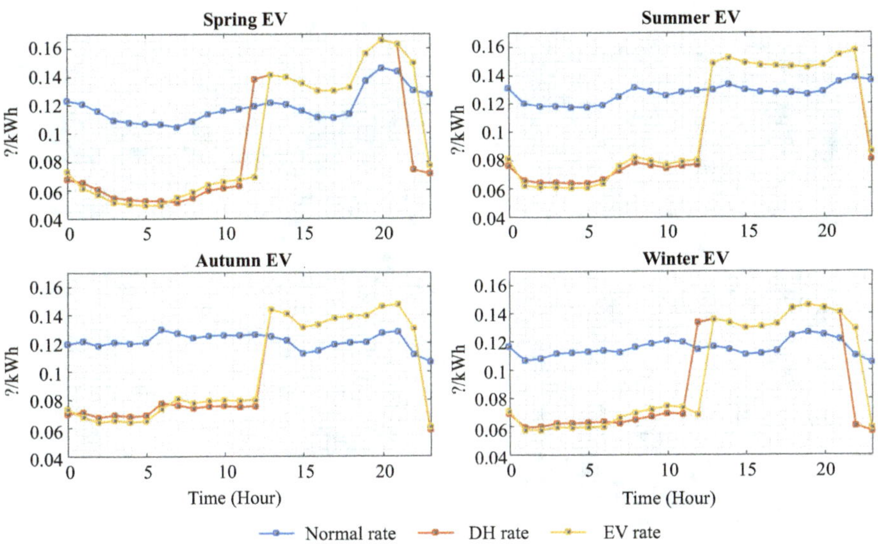

Figure 9.5 Hourly electric tariffs during different seasons

9.2.3.1 Parking occupation

Figure 9.6 shows IES's EV parking deck with 29 places at different time periods. The first (top-left) image in this figure was taken at 10:00 a.m. on Saturday morning and it is observed that there were no vehicles parked at this moment. The second (bottom-left) image from this figure was taken at 8:00 a.m. on Wednesday morning and there were 12 different vehicles parked at this particular hour.

A camera was installed in the parking and the available free parking lots were detected using an artificial intelligence–based algorithm developed in Python and previously trained. The total number of vehicles and free spots are available in real time.

The parking was monitored during five consecutive working days (from Monday to Friday) for 3 months. With this information, an hourly occupancy matrix, such as the one shown in Figure 9.6 (right), was generated for each working day of the week.

In this work, an additional sixth-day scenario was defined to consider a constant static battery capacity. This effect will be simulated in our work, considering that 29 EV were parked all day long.

Figure 9.7 shows the six different hourly occupancy matrix, from Monday to Friday plus the previously defined sixth day. There are 24 different columns in each matrix representing different hours and there are 29 rows, representing each parking deck. Dark blue squares represent free decks, while light blue squares indicate that there is a vehicle parked in this position (defined by the specific row), at this particular time slot (defined by the specific column).

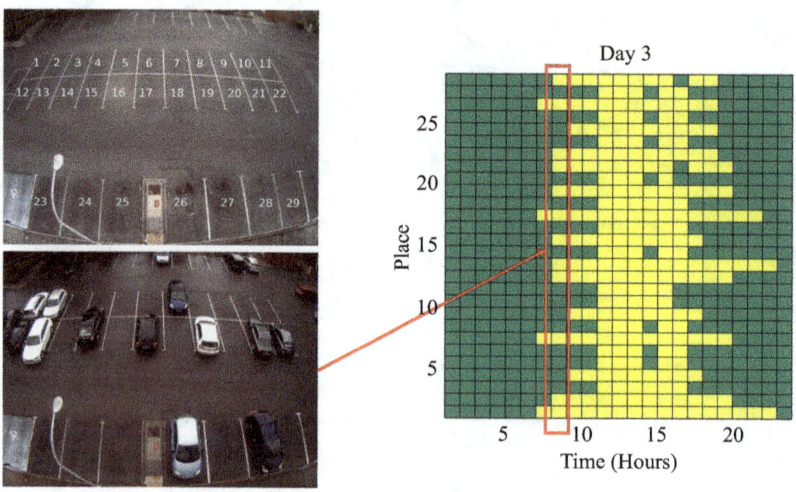

Figure 9.6 EV parking occupation

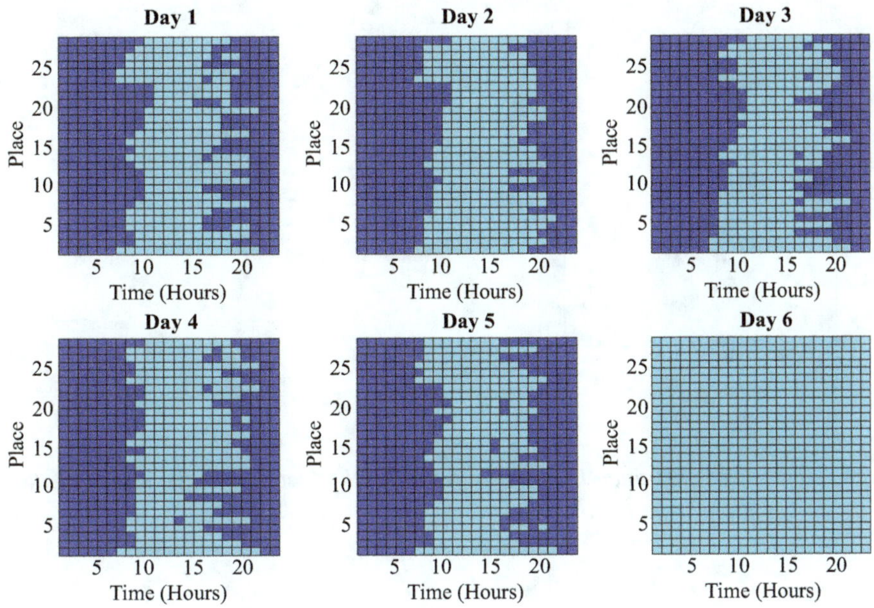

Figure 9.7 Occupation behaviour

9.2.3.2 Initial available capacity

For this research, it is assumed that all vehicles are electric and of the same model (a Nissan Leaf 2016, with a nominal capacity of 30 kWh [10]) and all these cars belong to the IES fleet.

Nissan Leaf has two different charging options, both located at a common charge port at the front of the car: a CHAdeMO inlet, which allows charging to 80% of the nominal battery capacity in 30–40 min (at 50 kW ac) and a conventional Type 2 socket, with a default 3.7 kW onboard charger, which allows charging to 100% of the capacity in 12–14 h (there are some Nissan Leaf models, SV and SL grades, with a 6.7 kW onboard charger available).

In this work, it is assumed that all vehicles can be charged and discharged with maximum power up to 3.7 kW. Additionally, it is also assumed that only 12% of the total capacity available (3.6 kWh) can be used to support the building power grid, in order to reduce battery degradation due to the over-cycling process.

A quantitative online mobility survey was carried out for IES workers to provide insightful information regarding their daily mobility behaviours and 30 employees answered this daily mobility survey. In Figure 9.8, the histogram of the daily average distance travelled by them and the location of the origin of their daily trips are presented. Analysing this figure, it is observed that most of the employees travel less than 30 km per day. The average daily distance is 26.95 km and the maximum distance is 90 km.

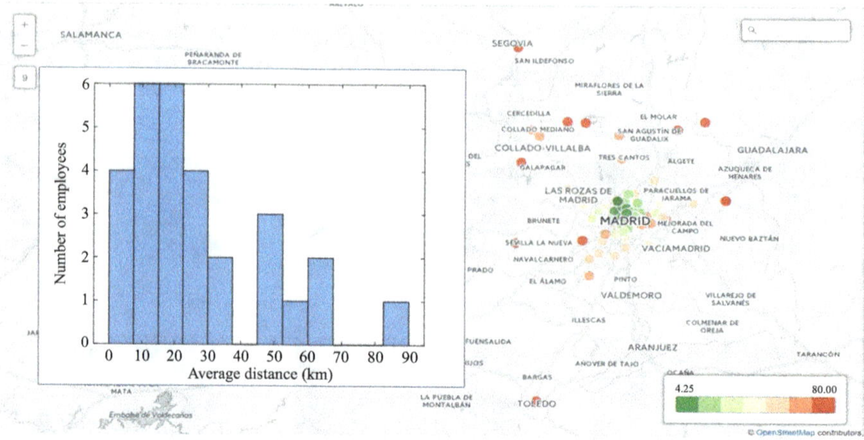

Figure 9.8 Daily distance histogram (left) and the location of the origin of the daily trips (right)

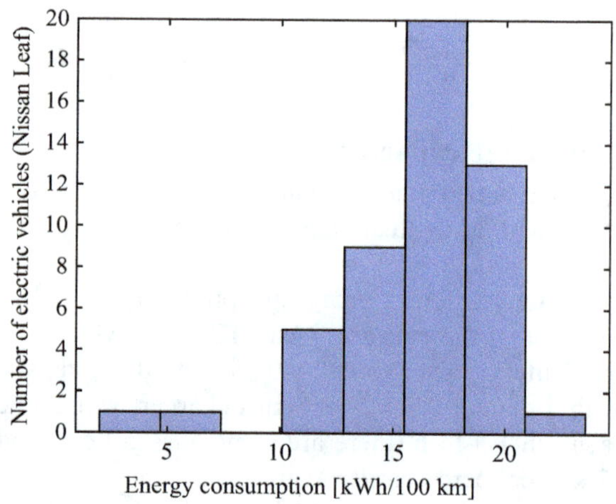

Figure 9.9 Histogram of energy consumption of Nissan Leaf vehicle under real-life conditions

The information about vehicle consumption is obtained from [11], which is a website where users update their fuel consumptions under real driving conditions. The histogram of the electric consumption of 50 different Nissan Leaf owners is presented in Figure 9.9. The average consumption is 16.02 kWh/100 km, with a minimum value of 3.26 kWh/100 km and a maximum value of 22.91 kWh/100 km.

In order to generate different consumptions for each EV, an initial average distance is randomly selected from Figure 9.8. This value is then multiplied by a randomly selected consumption sample extracted from the histogram presented in

Figure 9.9, providing a particular electric consumption value for each EV. All this procedure is highlighted in Figure 9.10. With this data, each particular EV has a slightly different consumption, providing some variability to the model.

Taking into account the mobility constraints of these vehicles during the day, and the available battery capacity of each vehicle in each time period, the aggregated total battery capacity available for the proposed building energy-management system is evaluated, as shown in Figure 9.11.

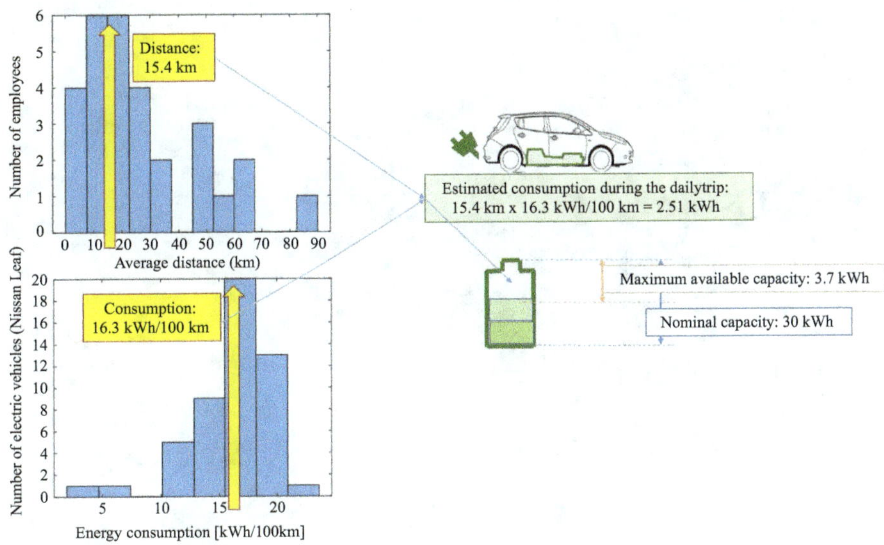

Figure 9.10 Initial battery capacity available per vehicle

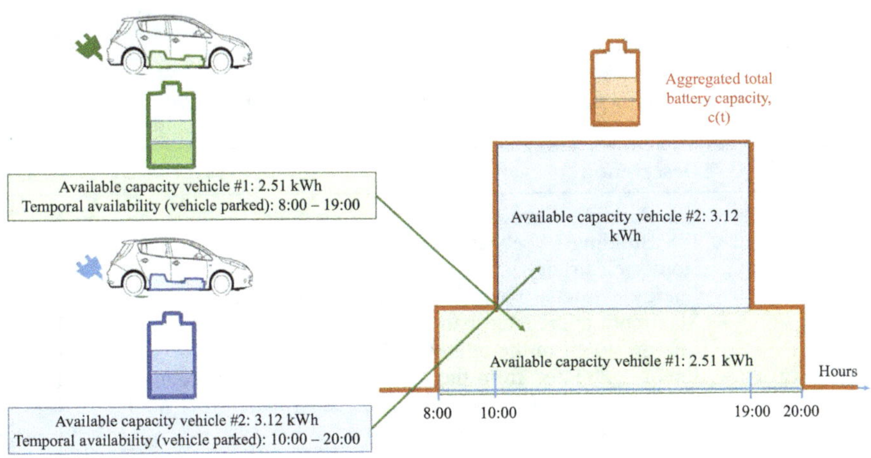

Figure 9.11 Aggregated battery capacity

9.3 Optimization modelling

The optimization problem formulation is defined in this section. Figure 9.12 and Table 9.1 present the main variables and parameters used in this model. The system equations are defined in (9.1)–(9.4).

Equation (9.1) describes the dynamic evolution of the aggregated battery capacity, which depends on the current number of vehicles parked at this period of time, their initial batteries' state of charge and the energy injected/extracted from

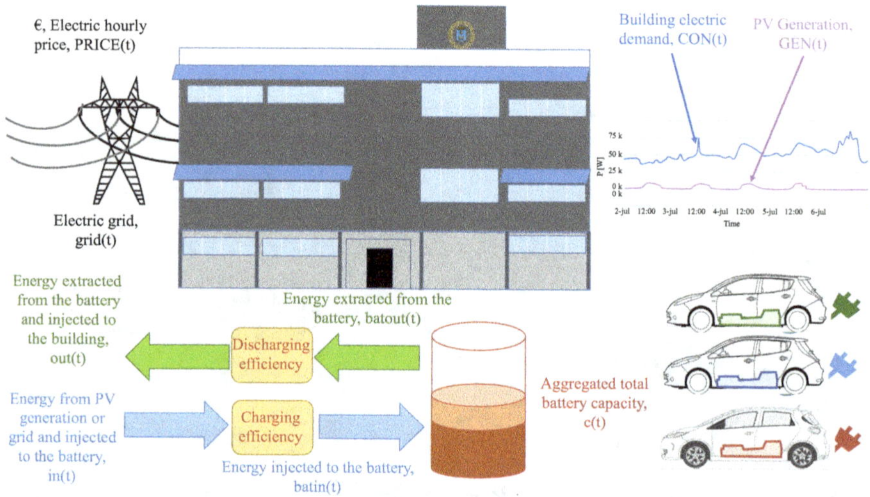

Figure 9.12 Optimization model description

Table 9.1 Variables and parameters from the optimization model

Symbol	Description	Length
$CON(t)$	IES Building electric consumption per hour	1×24
$GEN(t)$	IES Building PV generation per hour	1×24
$PRICE(t)$	Hourly electricity price	1×24
$grid(t)$	Energy extracted from the electric grid per hour	1×24
$c_{max}(t)$	Maximum aggregated battery capacity per hour	1×24
$in(t)$	Energy stored in the aggregated battery per hour	1×24
$out(t)$	Energy extracted from the aggregated battery per hour	1×24
$c(t)$	State of Charge (SOC) of the aggregated battery per hour	1×24
$Occupation(t)$	Number of vehicles parked per hour	1×24
c_{ini}	Initial capacity	1×1
P_{max}	Charging/Discharging nominal power	1×1

this aggregated battery. The parameter c_{ini} is the initial capacity of the aggregated battery. This parameter is zero on the first 5 days (from day 1 to day 5) because there are not EVs parked initially, but this value can change on day 6, when a static battery is assumed to be installed in the proposed system.

Equations (9.2) and (9.3) take into account the charging/discharging efficiency. Finally, (9.4) determines the energy directly extracted from the electric grid to cover the total demand of this system.

$$c(t) = \begin{cases} c(t-1) - batout(t) + batin(t) & \forall t > 1 \\ c_{ini} - batout(t) + batin(t) & t = 1 \end{cases} \tag{9.1}$$

$$batin(t) = \eta_{charging} in(t) \qquad \forall t \tag{9.2}$$

$$batout(t) = \frac{out(t)}{\eta_{discharging}} \forall t \tag{9.3}$$

$$grid(t) = CON(t) - GEN(t) - out(t) + in(t) \forall t \tag{9.4}$$

The objective function to minimize is given by (9.5). This equation evaluates the daily energy cost bought to the electric grid, and the solution of this optimization problem will provide the lowest price to pay at the end of the day.

$$lc = \min \left[\sum_{t=1}^{24} PRICE(t) \cdot grid(t) \right] \tag{9.5}$$

Equations (9.1)–(9.5) must be subjected to the following constraints:

$$c(t) \leq c_{\max}(t) \quad \forall t \tag{9.6}$$

$$out(t) \leq P_{\max} \cdot occupation(t) \quad \forall t \tag{9.7}$$

$$in(t) \leq P_{\max} \cdot occupation(t) \quad \forall t \tag{9.8}$$

where (9.6) indicates that the current capacity of the aggregated battery, *c(t)*, must be lower than $c_{max}(t)$, which is the maximum hourly available capacity. This upper limit is not constant and it will depend on the parking stop dynamics.

The variables, *c(t)*, *out(t)* and *in(t)* are always defined positive.

If all vehicles are parked during all day, this maximum capacity will be 3.7 kWh × 29 = 107 kWh as it is shown in Figure 9.13 (day 6 blue line). In this figure, the maximum hourly available capacity per day is presented, according to the parking occupation shown in Figure 9.7. It is observed that $c_{max}(t)$ is different during each hour and each day of the week.

The following two equations, (9.7) and (9.8), indicate that the energy that can be injected/extracted from this distributed storage system must be lower than the product of the number of available EVs parked at hour *t*, given by the variable *Occupation(t)*, and the maximum charging power, which is limited to 3.7 kW.

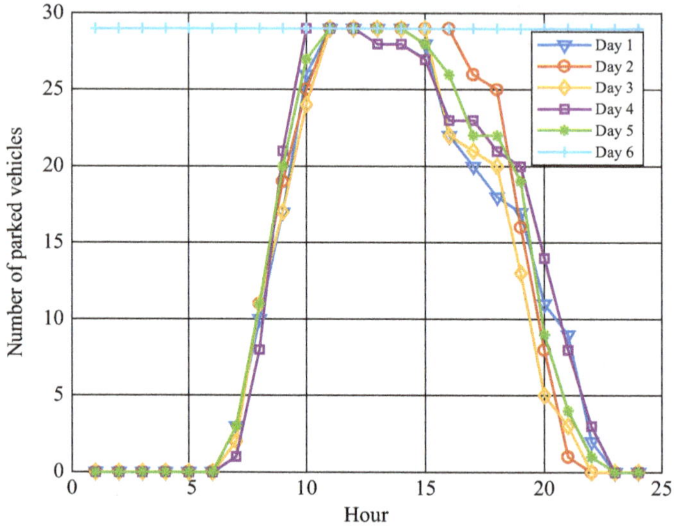

Figure 9.13 Aggregated battery capacity per day

9.4 Scenarios and results

Different scenarios were simulated over 3 months varying EVs parked occupancy, retail electricity price, PV generation and building consumption. These scenarios are defined in Table 9.2.

In Figure 9.14, the first scenario (Day 1) is analysed in detail. Vehicles start arriving at IES headquarters early in the morning (at 7:00 a.m.). Since the energy price is still very low during these hours of the day (denoted by a blue line in the figure), EVs begin to store energy in their batteries (as indicated by the green bars in this figure).

As IES workers continue arriving at their offices, the number of EV available in the parking lot increases (shown by the red line in Figure 9.14) and the optimization building-energy-management system (OBEMS) keeps charging their batteries until 11:00 a.m. At this time, all batteries are completely full and it is not possible to store more energy.

At 13:00 p.m., the EV tariff profile presents a sudden increment and the OBEMS extracts energy from the batteries and injects it into the building grid, which is represented by red bars in the figure, reducing almost to zero the energy purchased to the utility company (depicted by the black line in the figure) during 2 hours (from 14:00 to 16:00 p.m.).

There is a maximum value for the electricity price at 20:00 p.m. For that reason, the OBEMS starts to charge the remaining EVs at 18:00 p.m. and this stored energy is discharged later in the last three available hours (from 19:00 to 22:00 p.m.). Note that IES remains closed from 22:00 p.m. to 7:00 a.m.

Table 9.2 Scenarios under analysis

Season	**EV Tariff**		**DH Tariff**		**Default Tariff**		**Occupation**
Spring	Day 1	Figure 9.17(a)	Day 25	Figure 9.19(a)	Day 49	Figure 9.21(a)	Occupation 1
	Day 2		Day 26		Day 50		Occupation 2
	Day 3		Day 27		Day 51		Occupation 3
	Day 4		Day 28		Day 52		Occupation 4
	Day 5		Day 29		Day 53		Occupation 5
	Day 6		Day 30		Day 54		Constant Occup.
Summer	Day 7	Figure 9.17(b)	Day 31	Figure 9.19(b)	Day 55	Figure 9. 21(b)	Occupation 1
	Day 8		Day 32		Day 56		Occupation 2
	Day 9		Day 33		Day 57		Occupation 3
	Day 10		Day 34		Day 58		Occupation 4
	Day 11		Day 35		Day 59		Occupation 5
	Day 12		Day 36		Day 60		Constant Occup.
Autumn	Day 13	Figure 9.18(a)	Day 37	Figure 9. 20(a)	Day 61	Figure 9. 22(a)	Occupation 1
	Day 14		Day 38		Day 62		Occupation 2
	Day 15		Day 39		Day 63		Occupation 3
	Day 16		Day 40		Day 64		Occupation 4
	Day 17		Day 41		Day 65		Occupation 5
	Day 18		Day 42		Day 66		Constant Occup.
Winter	Day 19	Figure 9.18(b)	Day 43	Figure 9.20(b)	Day 67	Figure 9.22(b)	Occupation 1
	Day 20		Day 44		Day 68		Occupation 2
	Day 21		Day 45		Day 69		Occupation 3
	Day 22		Day 46		Day 70		Occupation 4
	Day 23		Day 47		Day 71		Occupation 5
	Day 24		Day 48		Day 72		Constant Occup.

Figure 9.15 shows the charging/discharging profile from day 34, which represents an average summer day with DH tariff. It is observed that the EVs from early workers were recharged from 7:00 to 8:00 a.m. when the electricity price was the lowest. Then, entry of workers was not as staggered as during the previous scenario and for that reason, the batteries were fully charged at 11:00 a.m., taking advantage of the low electricity price at this particular hour. This energy was progressively discharged until 16:00 p.m. and, at the end of the day, the OBEMS determined to charge (at 20:00 p.m.) and discharge (at 21:00 p.m.) the stored energy, taking advantage of the maximum electricity price at 21:00 p.m. and significantly reducing the amount of energy purchased to the grid.

Figure 9.16 shows the charging/discharging profile from day 62, which represents an average autumn day with normal tariff. The charging/discharging profiles are similar to the previous figures, but in this particular scenario, the electricity price does not have two clearly differentiated periods.

Finally, Figure 9.17 presents the charging/discharging profile from day 72, which represents an average winter day with normal tariff and constant occupation (all vehicles are available during the whole day). These EVs are charged at the lowest electricity price (6:00) and the profile and discharging when the electricity price is the highest.

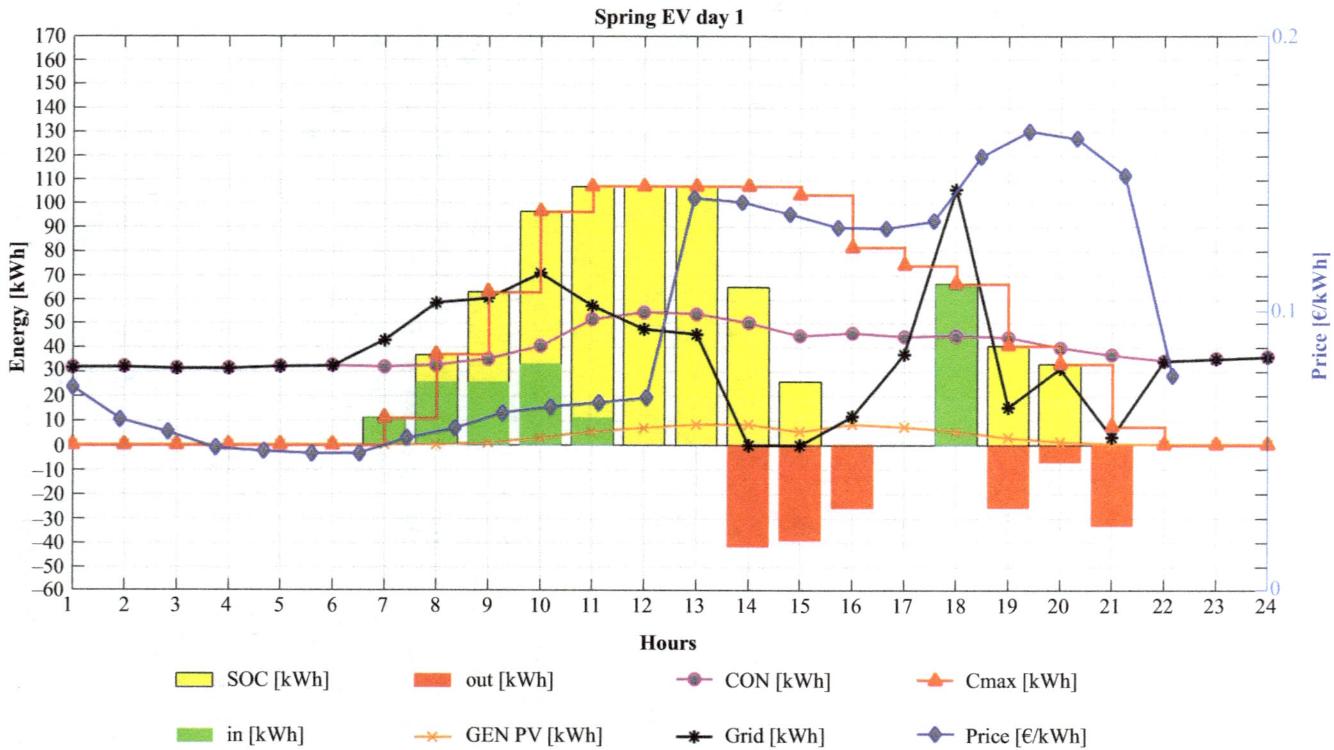

Figure 9.14 Building energy-management system behaviour day 1 (EV tariff, spring, occupation 1)

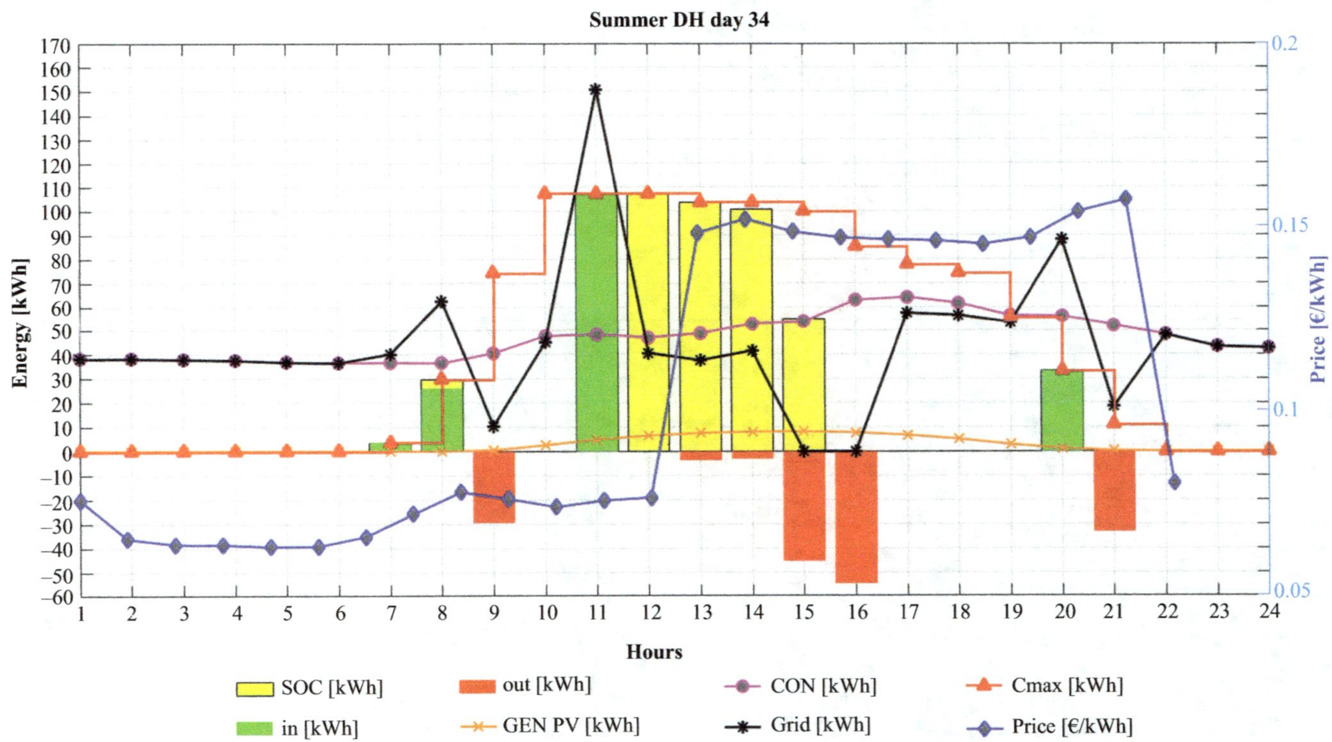

Figure 9.15 Building energy-management system behaviour day 34 (DH tariff, summer, occupation 4)

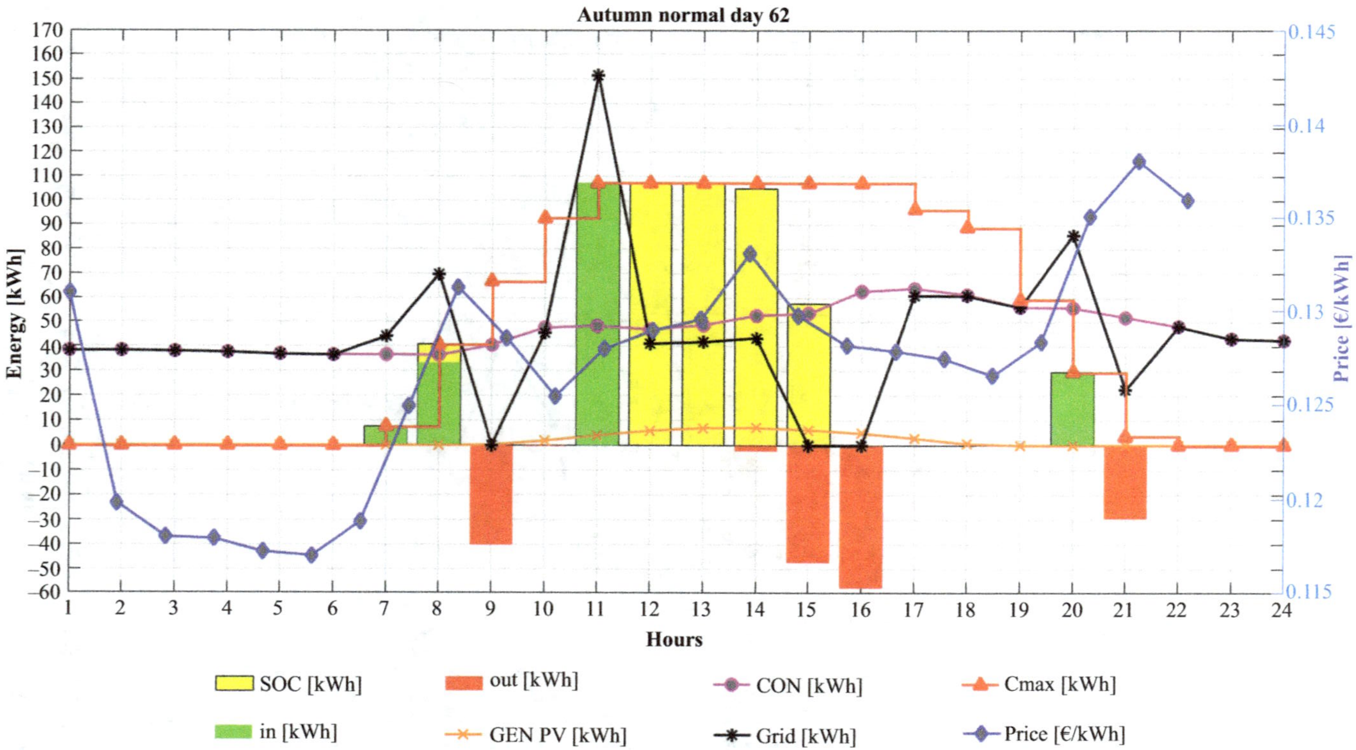

Figure 9.16 Building energy-management system behaviour day 62 (DH tariff, autumn, occupation 2)

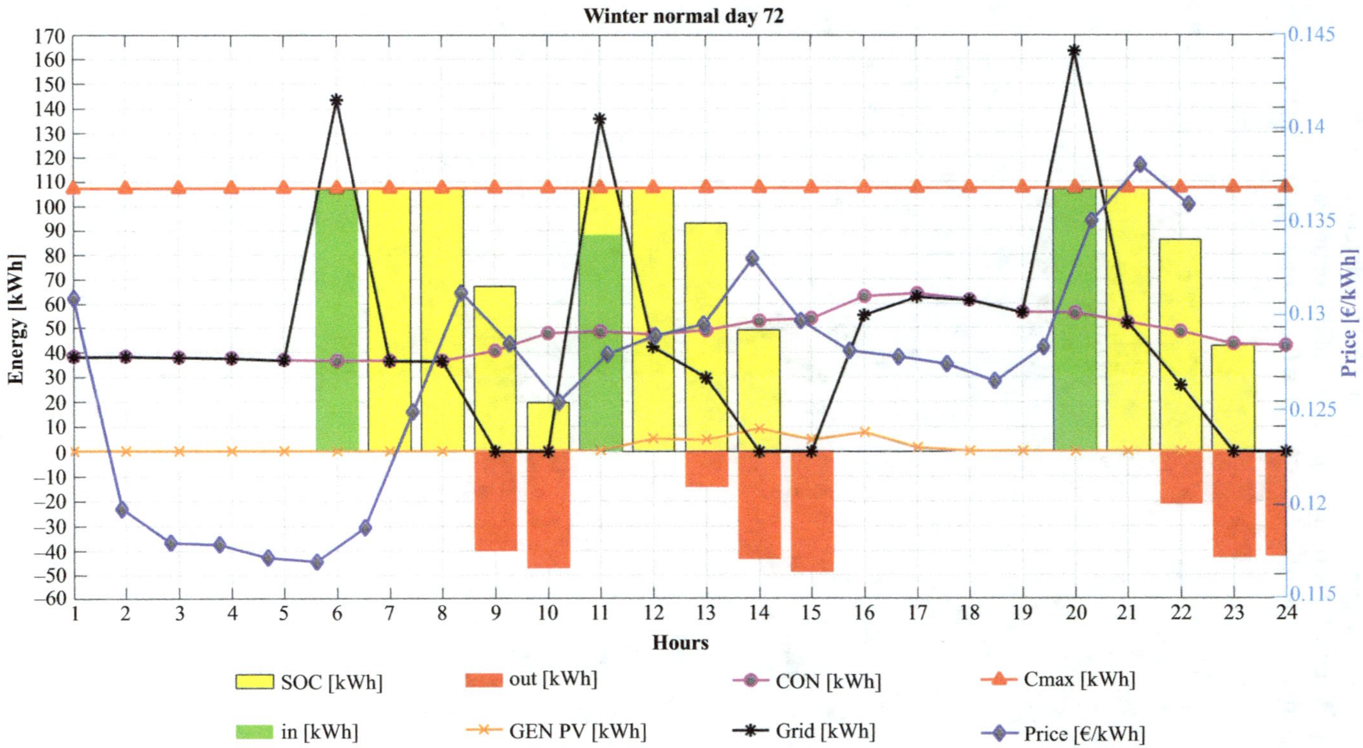

Figure 9.17 Building energy-management system behaviour day 72 (normal tariff, winter, day 72)

The next six figures show the charging/discharging profiles and the SOC evolution of the aggregated battery for the 72 different proposed scenarios.

First, Figure 9.18(a) shows the result of this optimization problem for EV tariff in the spring season (days 1 to 6). This figure presents the time evolution of the

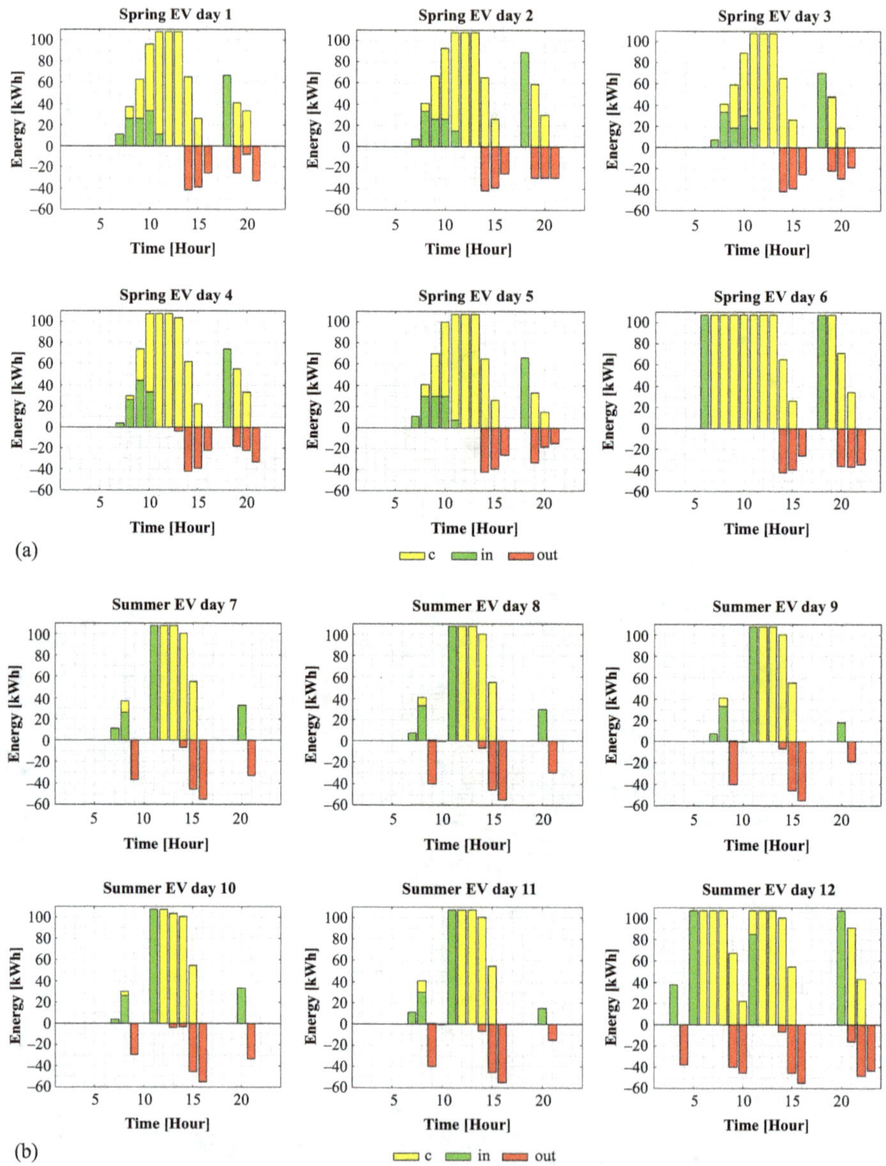

Figure 9.18 (a) SOC evolution and energy extracted/injected in the battery for days 1–6 and (b) SOC evolution and energy extracted/injected in the battery for days 7–12 (EV tariff)

aggregated battery state of charge (yellow bar), the energy extracted from the battery (red bar) and the energy injected (green bar). Figure 9.18(b) shows the same results for the summer season (days 7 to 12).

Figure 9.19(a) shows the result of this optimization problem for EV tariff in the autumn season (days 13 to 18). This figure presents the time evolution of the

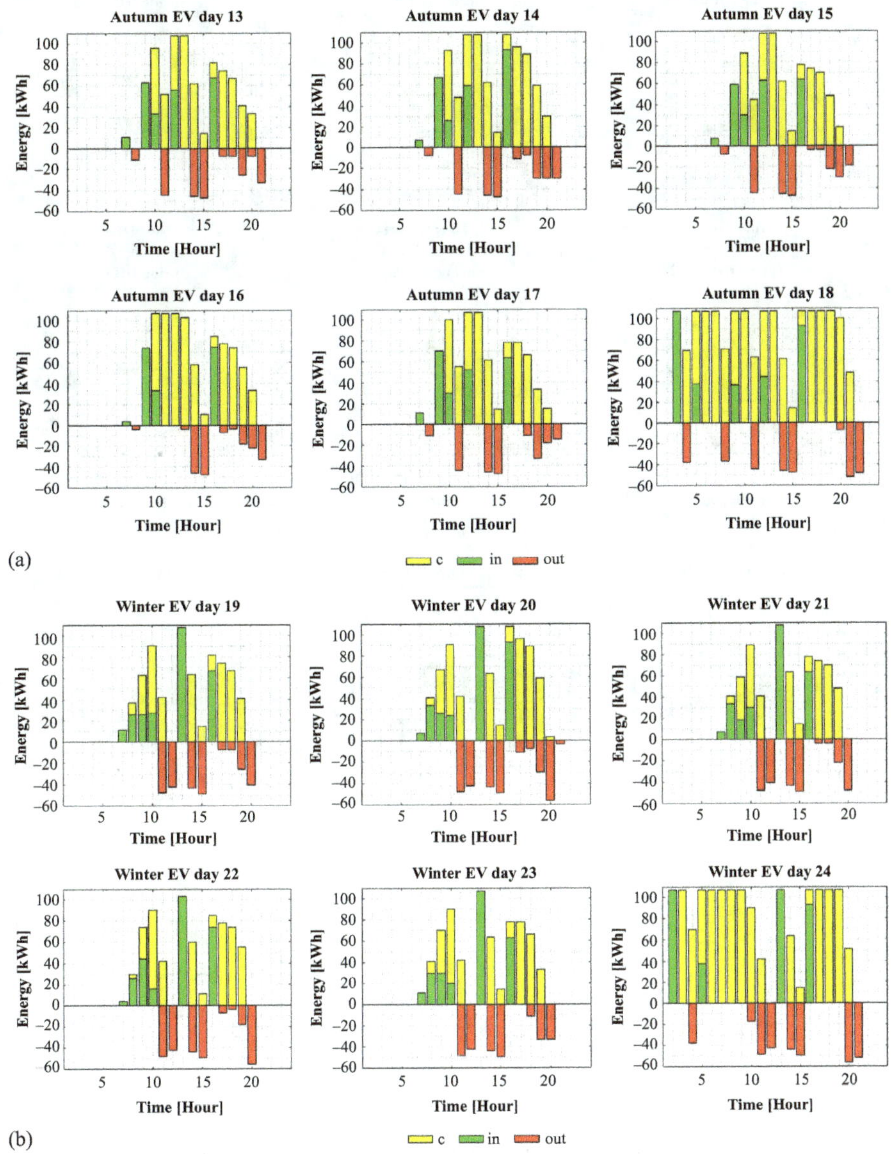

Figure 9.19 (a) SOC evolution and energy extracted/injected in the battery for days 13–18 and (b) SOC evolution and energy extracted/injected in the battery for days 19–24 (EV tariff)

aggregated battery state of charge (yellow bar), the energy extracted from the battery (red bar) and the energy injected (green bar). Figure 9.19(b) shows the same results for the winter season (days 19 to 24).

Figure 9.20(a) shows the result of this optimization problem for DH tariff in the spring season (days 25 to 30). This figure presents the time evolution of the

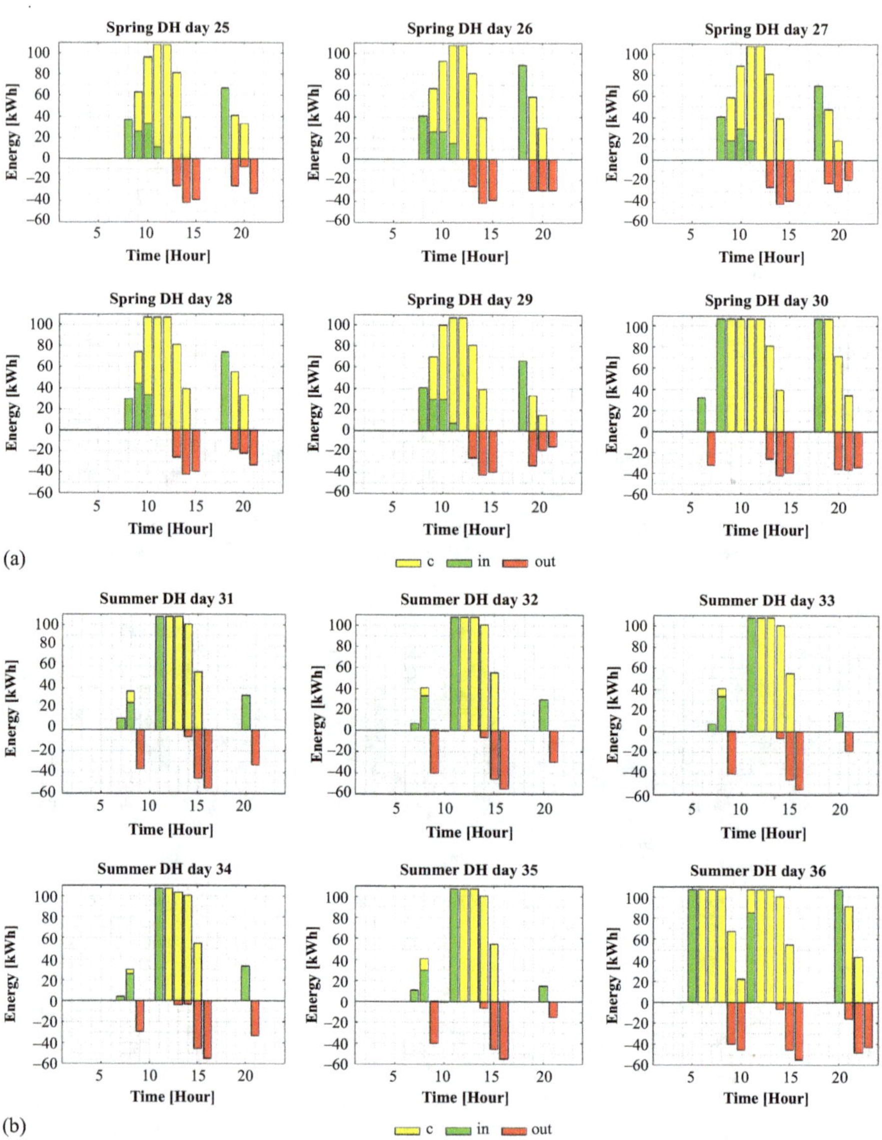

Figure 9.20 (a) SOC evolution and energy extracted/injected in the battery for days 25–30 and (b) SOC evolution and energy extracted/injected in the battery for days 31–36 (DH tariff)

aggregated battery state of charge (yellow bar), the energy extracted from the battery (red bar) and the energy injected (green bar). Figure 9.20(b) shows the same results for summer season (days 31 to 36).

Figure 9.21(a) shows the result of this optimization problem for DH tariff in the autumn season (days 37 to 42). This figure presents the time evolution of the

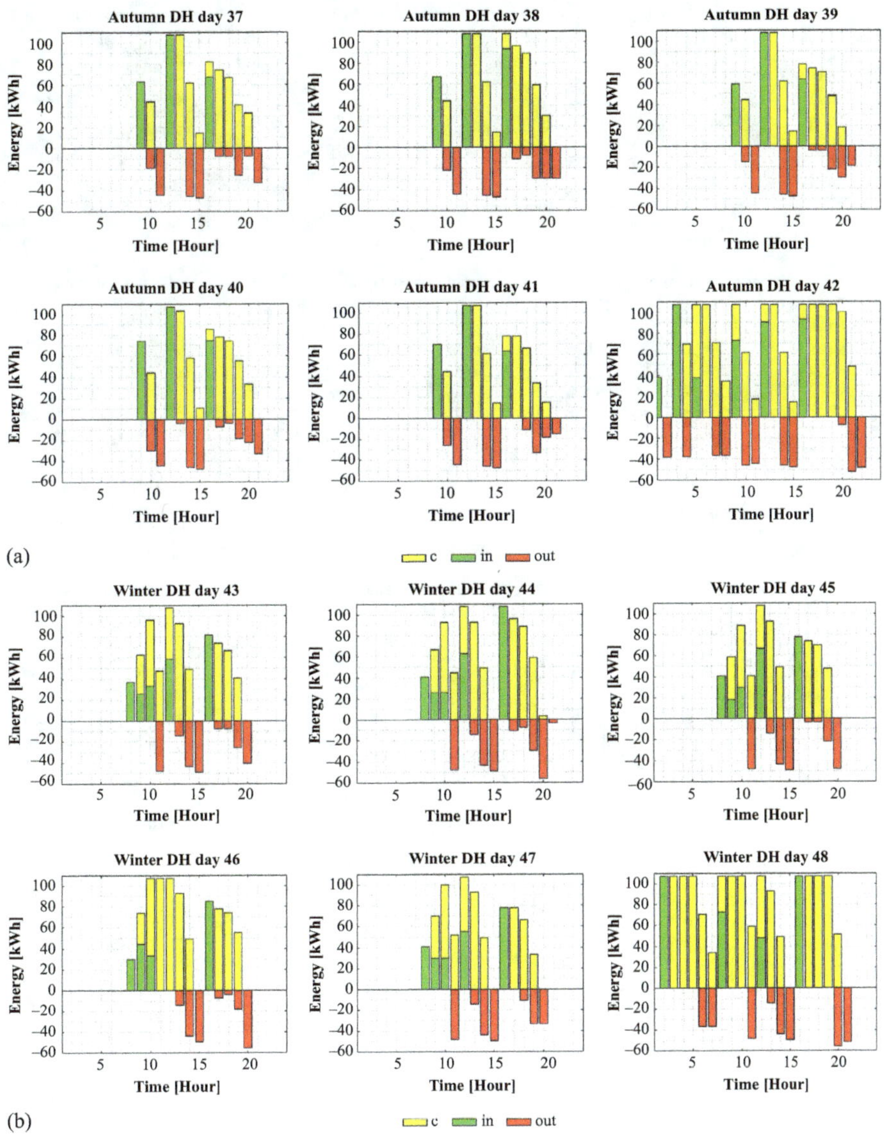

Figure 9.21 (a) SOC evolution and energy extracted/injected in the battery for days 37–42 and (b) SOC evolution and energy extracted/injected in the battery for days 43–48 (DH tariff)

aggregated battery state of charge (yellow bar), the energy extracted from the battery (red bar) and the energy injected (green bar). Figure 9.21(b) shows the same results for the winter season (days 43 to 48).

Figure 9.22(a) shows the result of this optimization problem for default normal tariff in the spring season (days 49 to 54). This figure presents the time evolution of

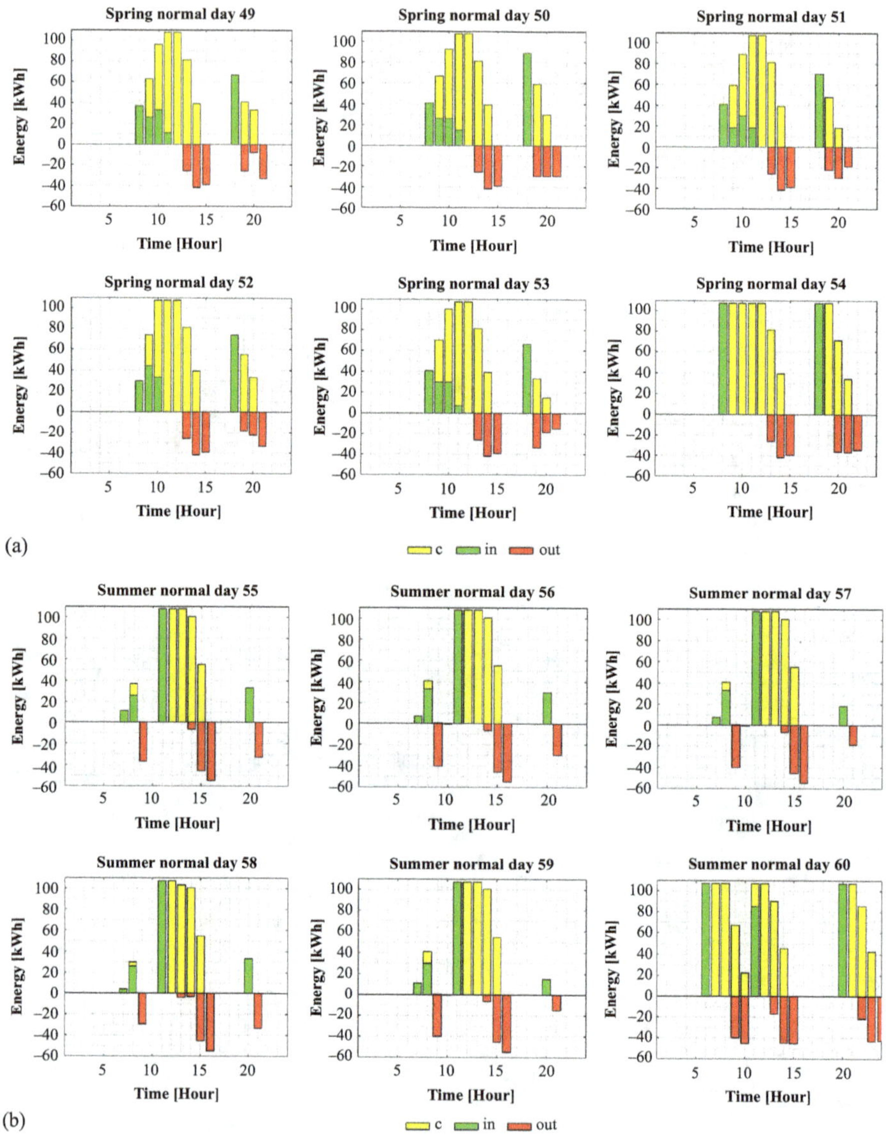

Figure 9.22 (a) SOC evolution and energy extracted/injected in the battery for days 49–54 and (b) SOC evolution and energy extracted/injected in the battery for days 55–60 (normal tariff)

the aggregated battery state of charge (yellow bar), the energy extracted from the battery (red bar) and the energy injected (green bar). Figure 9.22(b) shows the same results for the summer season (days 55 to 60).

Figure 9.23(a) shows the result of this optimization problem for default normal tariff in the autumn season (days 61 to 66). This figure presents the time evolution

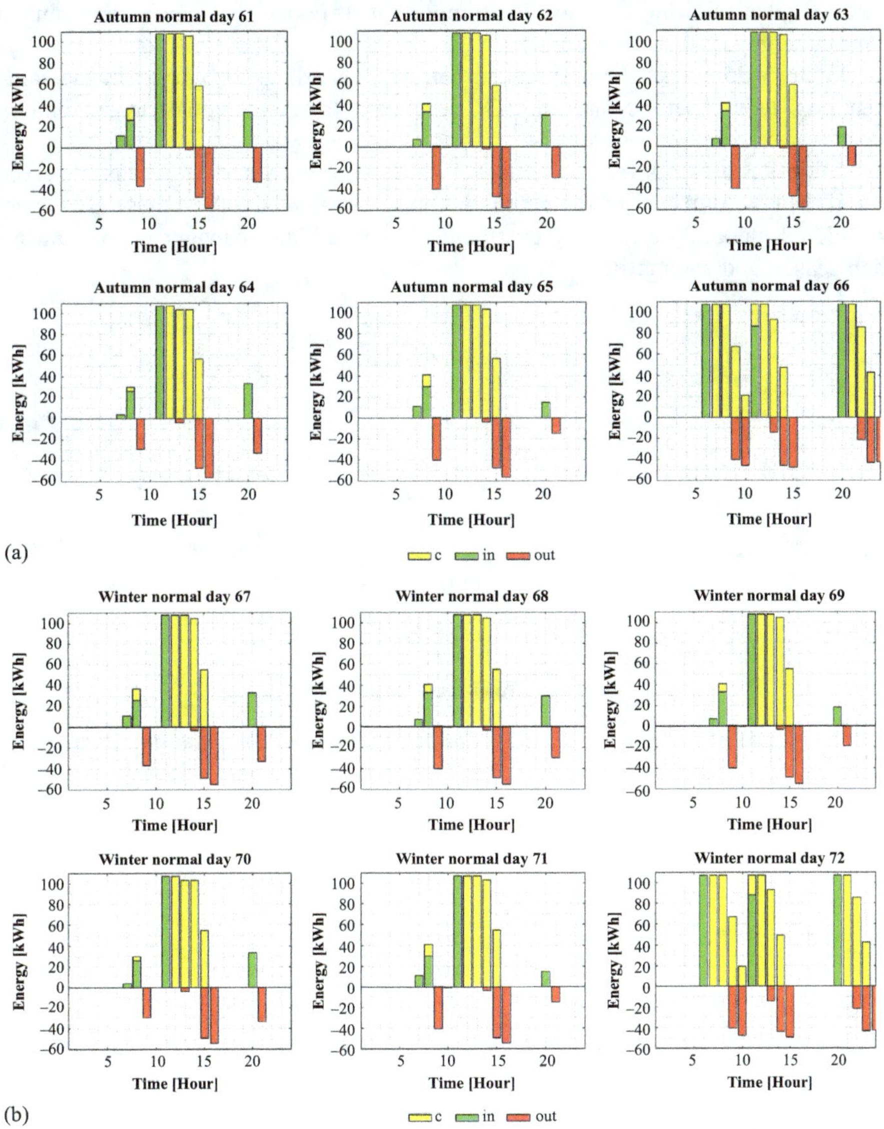

Figure 9.23 (a) SOC evolution and energy extracted/injected in the battery for days 61–66 and (b) SOC evolution and energy extracted/injected in the battery for days 66–72 (normal tariff)

of the aggregated battery state of charge (yellow bar), the energy extracted from the battery (red bar) and the energy injected (green bar). Figure 9.23(b) shows the same results for the winter season (days 67 to 72).

The total aggregated battery capacity, $c_{max}(t)$, available for this energy-management systems varies at each hour. Figure 9.24 represents these variations. Each box plot shows the minimum, 25%, median (red line), 75% and the maximum battery capacity during the day. It is important to notice that day 6 represents a constant aggregated battery capacity.

Figure 9.25 shows the aggregated battery SOC, denoted by $c(t)$, in the same scenarios. It is observed that the SOC is always lower (or equal) to the battery capacity, presented in Figure 9.24.

Table 9.3 presents the daily electricity cost for each scenario. It is observed that, if the possibility of storing energy in the aggregated battery of these EVs is not considered, the daily electricity cost paid by the building manager is constant for each season and each tariff.

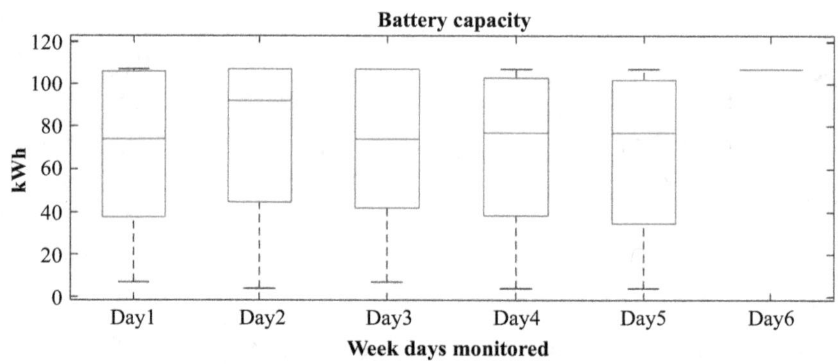

Figure 9.24 Aggregated battery capacity available under each day scenarios

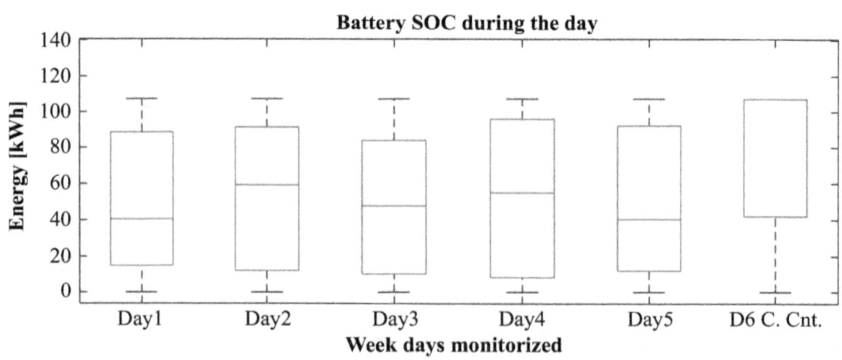

Figure 9.25 Aggregated battery SOC available under each day scenarios

Table 9.3 Electricity cost per day without (and with) the proposed optimization

Season	Payment per day [€]			Payment per day optimization [€]		
	EV tariff	DH tariff	Default tariff	EV tariff	DH tariff	Default tariff
Spring	85.77 €	85.71 €	105.17 €	75.71 €	75.22 €	102.53 €
	85.77 €	85.71 €	105.17 €	75.24 €	74.73 €	102.04 €
	85.77 €	85.71 €	105.17 €	75.70 €	75.19 €	102.50 €
	85.77 €	85.71 €	105.17 €	75.56 €	74.79 €	102.10 €
	85.77 €	85.71 €	105.17 €	75.99 €	75.51 €	102.80 €
	85.77 €	85.71 €	105.17 €	72.72 €	72.81 €	100.08 €
Summer	115.93 €	115.45 €	135.20 €	107.59 €	106.95 €	134.23 €
	115.93 €	115.45 €	135.20 €	107.62 €	106.96 €	134.24 €
	115.93 €	115.45 €	135.20 €	107.64 €	106.98 €	134.26 €
	115.93 €	115.45 €	135.20 €	107.98 €	107.30 €	134.33 €
	115.93 €	115.45 €	135.20 €	107.60 €	106.96 €	134.24 €
	115.93 €	115.45 €	135.20 €	104.96 €	104.92 €	132.25 €
Autumn	112.94 €	112.52 €	137.85 €	105.33 €	104.55 €	136.88 €
	112.94 €	112.52 €	137.85 €	105.18 €	104.38 €	136.89 €
	112.94 €	112.52 €	137.85 €	105.46 €	104.65 €	136.91 €
	112.94 €	112.52 €	137.85 €	105.53 €	104.70 €	136.97 €
	112.94 €	112.52 €	137.85 €	105.43 €	104.65 €	136.89 €
	112.94 €	112.52 €	137.85 €	102.98 €	102.74 €	134.90 €
Winter	108.42 €	107.77 €	138.86 €	99.83 €	99.25 €	137.88 €
	108.42 €	107.77 €	138.86 €	99.49 €	98.90 €	137.89 €
	108.42 €	107.77 €	138.86 €	99.80 €	99.21 €	137.91 €
	108.42 €	107.77 €	138.86 €	99.98 €	99.13 €	137.98 €
	108.42 €	107.77 €	138.86 €	99.81 €	99.23 €	137.89 €
	108.42 €	107.77 €	138.86 €	98.04 €	98.02 €	135.90 €

If the proposed OBEMS is applied, the daily electricity cost is reduced under all considered scenarios. The lowest cost is always reached under day 6 scenario, where all EVs remain parked during the day, as it were a static battery.

If the workers' mobility profiles are also considered in the optimization schedule, the occupation 2 scenario provides the lowest cost for EV tariff (during spring, autumn and winter seasons) and DH tariff (during autumn and winter seasons), while occupation 1 provides the lowest cost for the default tariff (during summer, autumn and winter seasons).

The average daily electricity cost reduction is 8.64€/day in EV tariff, 8.9€/day in DH tariff and 1.4€/day in the default tariff. The estimated annual electricity cost reduction under the proposed OBEMS varies from 2314.13€ for DH tariff (with an annual average reduction of 8.45%) to 364.52€ in the default tariff, which only represents a 1.04% of the annual electricity cost with this particular tariff.

In Figure 9.26, the average cost reduction for each tariff in each season is presented. It is observed that the highest reduction is obtained with DH tariff in spring season (with a reduction of 12.8%) and the lowest reduction is obtained with default (normal) tariff during winters.

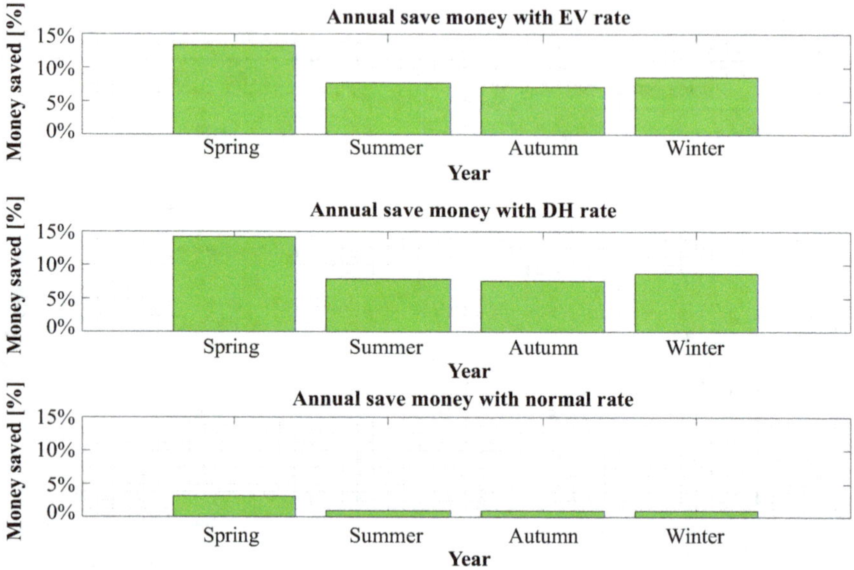

Figure 9.26 Annual saving cost for the different analysed scenarios (different seasons and different tariff)

9.5 Conclusions

The aim of this work was to analyse the integration of a parking deck with EVs with a PV system installed at a research building headquarters in Spain.

The main objective of this study was to take advantage of the distributed storage capacity available in the EVs' batteries to store surplus energy from PV systems (or during off-peak periods, when the grid electricity price is lower) and inject it back to the building electric grid after sunset or during the peak periods, when the electricity price is higher.

Seventy-two different scenarios were analysed, considering three different electricity rates, four different seasons and five different daily mobility profiles. In all studied scenarios, a cost reduction was obtained. Benefits varied mainly depending on the tariff employed. Very low benefits were obtained (less than 1.5% of cost reduction in average) when the default (normal) rate was analysed, since the difference between the maximum and minimum prices along the day is highly reduced. In this case, the investment in new infrastructure to allow this proposed optimized energy-building-management system is not justifiable. Once a larger price variation along the day is introduced with the other dual analysed tariffs (DH and EV), higher annual average cost reductions are obtained, with an average cost reduction of 8.45% in DH tariff and 8.17% in EV tariff. Additionally, DH tariff provides the lowest costs, although its difference with the EV tariff is not remarkable on a day-by-day basis (an average X%) and there are scenarios in which employing an EV tariff has a less cost than a DH tariff.

References

[1] Marc A. Melliger, Oscar P.R. van Vliet, and Heikki Liimatainen, 'Anxiety vs reality – Sufficiency of battery electric vehicle range in Switzerland and Finland', Transportation Research Part D: Transport and Environment, Volume 65, 2018, Pages 101–115, ISSN 1361–9209, https://doi.org/10.1016/j.trd.2018.08.011

[2] KatrinSeddig, Patrick Jochem, and Wolf Fichtner, 'Integrating renewable energy sources by electric vehicle fleets under uncertainty', Energy, Volume 141, 2017, Pages 2145–2153, ISSN 0360–5442, https://doi.org/10.1016/j.energy.2017.11.140.

[3] Antun Pfeifer, Viktorija Dobravec, Luka Pavlinek, Goran Krajačić, and Neven Duić, 'Integration of renewable energy and demand response technologies in interconnected energy systems', Energy, Volume 161, 2018, Pages 447–455, ISSN 0360–5442, https://doi.org/10.1016/j.energy.2018.07.134.

[4] International Renewable Energy Agency, IRENA, 'Renewable Power Generation Cost in 2017', available at: https://www.irena.org/-/media/Files/IRENA/Agency/Publication/2018/Jan/IRENA_2017_Power_Costs_2018.pdf

[5] Nuria Martín-Chivelet, and David Montero-Gómez, 'Optimizing photovoltaic self-consumption in office buildings', Energy and Buildings, Volume 150, 2017, Pages 71–80, ISSN 0378–7788, https://doi.org/10.1016/j.enbuild.2017.05.073

[6] Ghada Merei, Janina Moshövel, Dirk Magnor, and Dirk Uwe Sauer, 'Optimization of self-consumption and techno-economic analysis of PV-battery systems in commercial applications', Applied Energy, Volume 168, 2016, Pages 171–178, ISSN 0306–2619, https://doi.org/10.1016/j.apenergy.2016.01.083

[7] I. Momber *et al.*, 'Plug-in electric vehicle interactions with a small office building: An economic analysis using DER-CAM', IEEE PES General Meeting, Providence, RI, 2010, Pages 1–8, doi: 10.1109/PES.2010.5589485

[8] Claire Curry, 'Lithium-ion battery costs and market', Bloomberg New Energy Finance, 2017, available at: https://data.bloomberglp.com/bnef/sites/14/2017/07/BNEF-Lithium-ion-battery-costs-and-market.pdf

[9] Red Eléctrica de España. 'Active energy invoicing price', available at: https://www.esios.ree.es/en/pvpc

[10] Nissan USA Press Release, 'New 2016 Nissan Leaf now offers best-in-class 107-mile range in affordable, fun-to-drive package', available at: http://nissannews.com/en-US/nissan/usa/releases/new-2016-nissan-leaf-now-offers-best-in-class-107-mile-range-in-affordable-fun-to-drive-package

[11] Spritmonitor Verbrauchswerte real erfahren, available at: http://www.spritmonitor.de/

Chapter 10

Contribution of electric vehicles to power system ancillary services beyond distributed energy storage

Sergio Martinez[1], *Hugo Mendonça*[1], *Rosa M. de Castro*[1] *and Danny Ochoa*[1]

Current electric power systems are experiencing a steady growth in renewable electricity generation and in energy consumption from electric vehicles (EV), involving a number of impacts on the system that increase with the penetration level and the system weakness. As conventional synchronous generation is being displaced by renewable generators, they must take their share in the ancillary services provision and supplementary equipment has to be installed, with its associated cost. This chapter describes how EVs, already connected to the grid with appropriate electronic converters and controls, can be used for contributing to the provision of frequency and voltage control, thus avoiding additional investment in supplementary equipment. The chapter also presents innovative approaches for both ancillary services, based on previous developments in electricity generation from wind and wave energies.

List of abbreviations

AC	Alternating Current
BESS	Battery Energy Storage System
DC	Direct Current
DG	Distributed Generation
EV	Electric Vehicle
LFC	Load Frequency Control
LV	Low Voltage
PQ	Active power and reactive power specified
PV	Active power and voltage specified

[1]Department of Electrical Engineering, ETSI Industriales, Universidad Politecnica de Madrid, Madrid, Spain

rms	Root Mean Square
RoCoF	Rate of Change of Frequency
SAVFI	System Average Voltage Fluctuation Index
SoC	State of Charge

10.1 Introduction

Current electric power systems have experienced an extraordinary growth in electricity generation from clean energy sources during the last years, mainly due to steady cost decline for wind and photovoltaic (PV) technologies [1]. This growth is expected to continue in the coming years along with an increase in the number of electric vehicles (EV) connected to the power system for recharging. Both renewable electricity generation and energy consumption from EVs involve a number of impacts on the power system to which they are connected, which increase with the penetration level and the system weakness [2]. Since this level is already high in the case of renewable generation and it will be steadily increasing in the case of EVs, power systems face the challenge to integrate them in a way so that reliability and efficiency are not affected [3].

To contribute to achieve this goal, a natural approach is to take advantage of the rising number of EVs connected to the grid, with an increasing storage capacity in their batteries, by using them as distributed energy storage for wind or PV plants [4]. Beyond this approach, apart from generating electricity, renewable generators should also be involved in providing ancillary services, such as voltage control, power-frequency control and other grid-support tasks. In older power systems, the bulk electrical energy was generated at conventional power plants, such as hydroelectric or thermal ones, equipped with synchronous electrical generators that, in addition to generating electricity, provided most of the ancillary services. In current power systems, the synchronous generation is being largely displaced by renewable generators, and they must also take their share in the ancillary services provision. As they do not have the same capabilities as the conventional plants they displace, supplementary equipment has to be installed, with its associated cost. EVs seem to be one of the eminent resources for providing various ancillary services due to their defining properties: they are a large load compared with other residential loads, they have quick response with potential bidirectional power flow capabilities and they are available most of the times with a high degree of flexibility [5]. This chapter describes how EVs with appropriate electronic converters and controls, which are already connected to the grid, can be used for contributing to the provision of frequency and voltage control thus avoiding additional investment in supplementary equipment. For example, [6] describes a field-test demonstration in a real Danish distribution grid with an EV providing three ancillary services through unidirectional AC charging.

In relation to the role of EVs in the frequency control of power systems with high levels of renewable generation, it is relevant that the intermittent and non-dispatchable character of wind and PV energy impacts the power system performance in terms of frequency perturbations. Contribution of EVs to secondary and even tertiary frequency controls has been widely described in the literature, for example in [7] and [8]. Lately, some works that deal with taking advantage of EVs connected to the grid for primary power-frequency control have been published. For example, [9] presents a fuzzy controller that, in addition to secondary control, implements a primary frequency control based on perfect and non-delayed information. Section 10.2 of this chapter describes how EVs can contribute to primary power-frequency control and system inertia, and it also presents the authors' proposal for a decentralized control approach, independent of communication links, as an adaptation to connected EVs of the strategy presented in [10] for variable speed-wind turbines.

On the other hand, the reactive power support from distributed generation (DG) opens a range of possibilities to improve the voltage quality. Many approaches address the Volt/var control for voltage regulation, for example [11,12]. In this sense, EVs can be seen as a potential solution to contribute to the voltage regulation of highly renewable energy–penetrated power systems. In fact, [13] proposes a strategy to mitigate over/undervoltage issues by the interaction between electrical vehicles and on load tap-changing transformers. However, the DG penetration does not only affect the voltage level of the grid but also introduces voltage fluctuations caused by the nature of primary sources (solar, wind…) [14] formulates a set of rules that seem interesting to assess the voltage fluctuation in customers that share a distribution service transformer. Although they have been carried out only with PV generation, they could be extrapolated to any other DG. [15] summarizes the impact on the grid of fluctuating power generation and presents a fluctuation power index to relate the voltage fluctuation mitigation with the power fluctuation in the case of storage devices. Reference [16] also uses a storage device to smooth the voltage fluctuation. In this case, the DC-link charges and discharges to smooth the voltage fluctuation in the AC-side. Other mitigating methods can be found in [17] and references therein. Section 10.3 describes how a fluctuating active power injection leads to grid voltage fluctuations and presents the authors' proposal of a reactive power controller for mitigating those fluctuations by using EVs already connected for recharging, without any other additional equipment. This proposal is a simple yet effective adaptation of a control strategy previously developed by the authors for wave energy–conversion systems [18].

10.2 Contribution of electric vehicles to the power system frequency control

As stated in the introduction, this section of the chapter aims at describing how secondary and tertiary frequency controls can be complemented in the short and very short terms with the contribution of EVs to primary power-frequency control and system inertia. The section describes the theoretical basis of the problem and how EVs connected to the grid for recharging can be used for mitigating frequency perturbations by means of an intelligent control on their charging process.

The phenomena involved and the performance of the proposed strategy are illustrated with simulations on a test network.

10.2.1 Theoretical background

In general, the frequency of a power system has to be kept constant, but it depends on the active power balance in the system. A variation of the active power consumed by a load at a given bus leads to a frequency variation in the whole system, reproduced as a speed variation in synchronous rotating generators. As frequencies or speed variations are the consequence of active power imbalance in the system, they are used as control signals for its compensation by acting on the active power injected by generators. Primary (fastest) frequency control is performed by the speed governors of generators and secondary and tertiary (slowest) controls are used for allocating generation.

10.2.2 Case study

In order to illustrate the potential contribution of the EV to the grid-frequency-control tasks when it is plugged into the network, the simple power system depicted in Figure 10.1 will be considered as a case of study. This power system is characterized by two conventional generation technologies: hydroelectric and thermal, which are supplying a single electric load. On the demand side, the aggregated effect of the integration of EVs under recharging has also been considered. According to this approach, the EV is seen as a battery energy storage system (BESS) from the grid side. This BESS is equipped with a grid-connection interface based on a power electronic converter. In this diagram, P_{BESS} is the three-phase active power injected or consumed by the EV-Battery under discharge and charge conditions, respectively. The selection of one or another operation mode will depend on certain control criteria implemented in the power controller that governs the DC/AC converter; although, it is important to mention that, traditionally, the EV is considered as an energy-consumption component of the grid.

Figure 10.2 illustrates the load frequency control (LFC) scheme that represents the short-term dynamics of the power system under study [19]. This representation

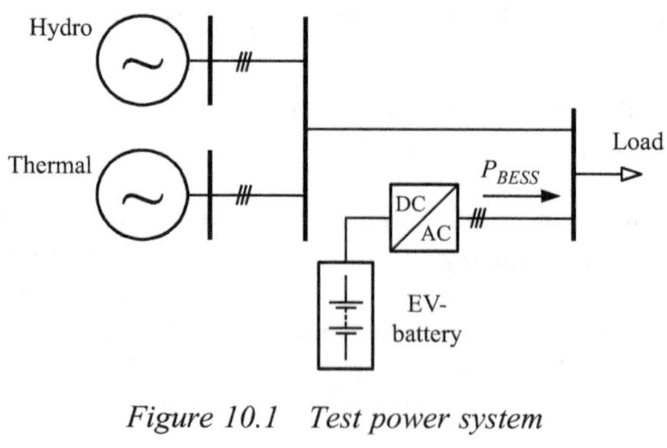

Figure 10.1 Test power system

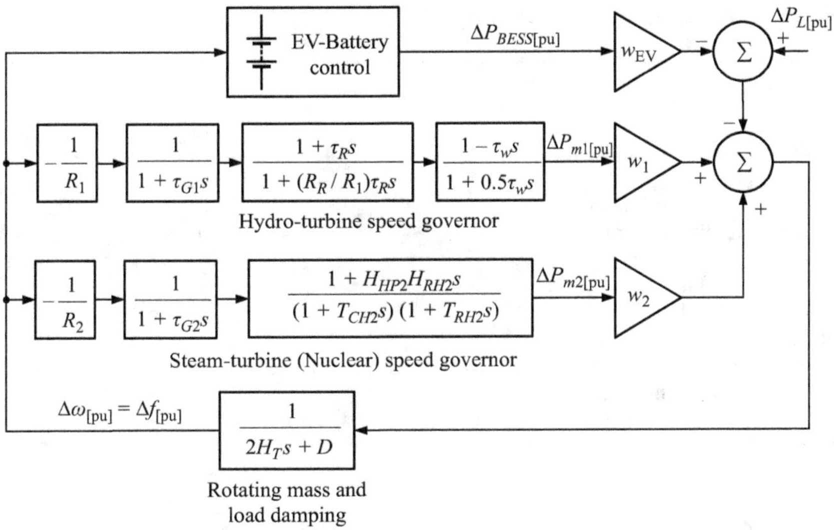

Figure 10.2 LFC scheme of the test power system

Table 10.1 Constant of inertia and LFC-participation weights of the generation technologies

Type of power plant	H_i(s)	w_i (%)
Hydro	3.6	40
Thermal (Nuclear)	7	40
EV-Battery	–	20

is derived from the swing equation of a synchronous generator, linearized by considering small deviations of the variables involved. In this scheme, the constant of inertia, H_T, represents all the available inertial resources and it is constituted by the individual constant of inertia of the operating synchronous generation groups, whose typical values are listed in Table 10.1 [19,20].

On the other hand, the time frame used in this type of analysis (usually, in the order of the seconds) allows focusing the study of the temporal behaviour of conventional generation technologies on the response of the speed governors of their respective mechanical drives. The simplified models of the speed governors that are shown in Figure 10.2 are taken from [19], and the typical values of their variables are summarized in Table 10.2.

ΔP_{m1} and ΔP_{m2} are the deviations of the mechanical power of the mechanical drives of the hydraulic and thermal generator in response to the presence of grid-frequency deviations, $\Delta\omega$. ΔP_L is the electric load deviation and, for this case of study, the only source of disturbance. The damping effect of the load is represented by the constant D (where a typical value of 0.5 p.u. has been taken for this

Table 10.2 Speed governor parameters

Parameter	**Value and units**	
	Hydro	**Thermal**
Droop, R	5%	5%
Main servo time constant, τ_G	0.2 s	0.2 s
Maximum gate opening rate	0.16 pu/s	0.05 pu/s
Maximum gate closing rate	– 0.16 pu/s	– 0.1 pu/s
Water time constant, τ_w	1 s	
Temporary droop, R_T	0.4 pu	
Reset time, τ_R	5 s	
Fraction of total turbine power generated by high pressure, F_{HP}		0.3
Reheater time constant, T_{RH}		5 s
Time constant of main inlet columns and steam chest, T_{CH}		0.3 s

example). ΔP_{BESS} is the injected/consumed EV-Battery active power deviation. Within the considered time scale, this variable will be practically negligible due to the slow charging and discharging process. However, when a grid-frequency-sensitive algorithm is implemented on the active power controller of the EV-Battery, the value of ΔP_{BESS} will begin to gain importance. All the variable deviations introduced before are defined regarding to their respective pre-disturbance values.

Finally, the terms w_1, w_2 and w_{EV} are the participation weights of each generation technology within the LFC scheme, which are defined as the ratio of their respective installed capacity and base power. In this case, the chosen base power corresponds to the total active power of the load. According to this criterion, these factors will reflect the energy share of each type of generation for the electrical load supplying. The assigned values for the participation weights are presented in Table 10.1.

The perturbation applied to the test power system is modelled as a sudden increase of 1% of the electric load. Therefore, it is expected that the frequency control tasks are carried out only by the synchronous generators by default. The time domain simulation results for this case of study are shown in Figure 10.3 as the grey solid lines. These illustrations that show the behaviour of the variables of interest: P_{BESS}, grid-frequency, and the rate of change of frequency (RoCoF), demonstrate that the EV-Battery charger does not contribute to reduce the grid frequency drop under normal operating mode.

However, if an adequate supplementary control scheme is implemented in the active power control of the EV-Battery charger, it is possible to provide this device with an emulated inertial response. The scheme used for this purpose is depicted in Figure 10.4, as an adaptation of the one proposed by [10]. This scheme has been originally devised to be implemented in a variable speed-wind turbine. Hence, after some modifications that scheme could be adapted for allowing its incorporation in the EV-Battery charger used in this study.

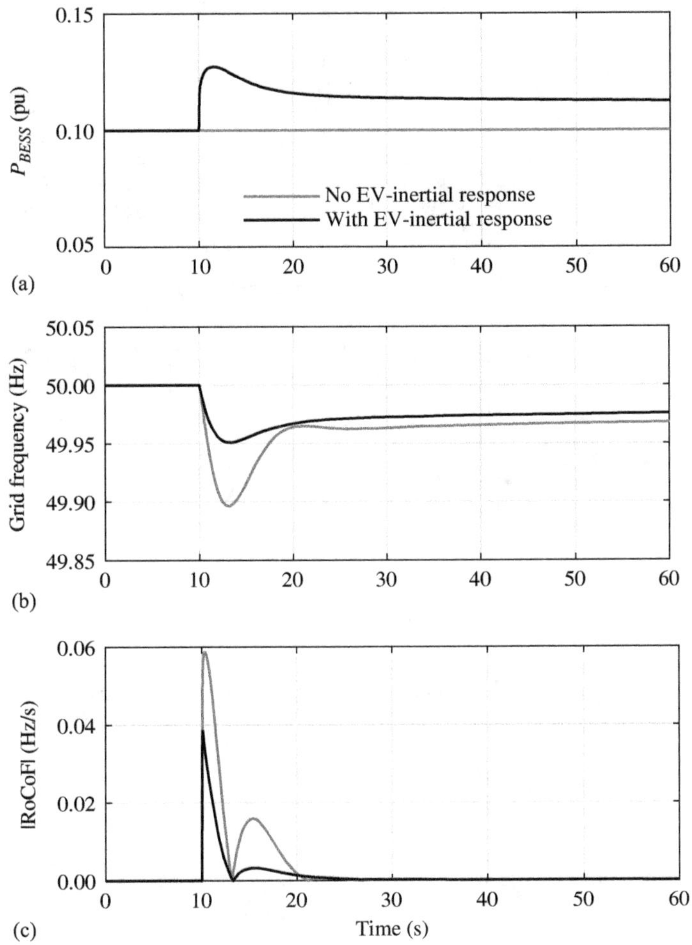

Figure 10.3 Time-domain simulation results: (a) Default EV-Battery charger operation mode (grey solid lines), (b) and (c) Frequency support provided by the EV-Battery charger (black solid lines)

The proposed frequency-control scheme has two control loops: one depends on the grid-frequency deviation (computed as the difference between the nominal frequency, f_{nom}, and the measured frequency, f_s) and the other one depends on the time derivative of the grid frequency. The output signals of both loops (ΔP_1^* and ΔP_2^*) are added to the default active power set-point of the battery charger, P_C^*. These three signals constitute the active power set-point signal of the modified EV-Battery charger, P_{BESS}^*. The dynamic gains that govern the control actions in the supplementary control scheme are defined by (10.1) and (10.2), and depend on the current state of charge (SoC) of the battery. In this example, the initial SoC is chosen as 0.5 p.u., the maximum state of charge (SoC_{max}) as 1 p.u., the initial droop

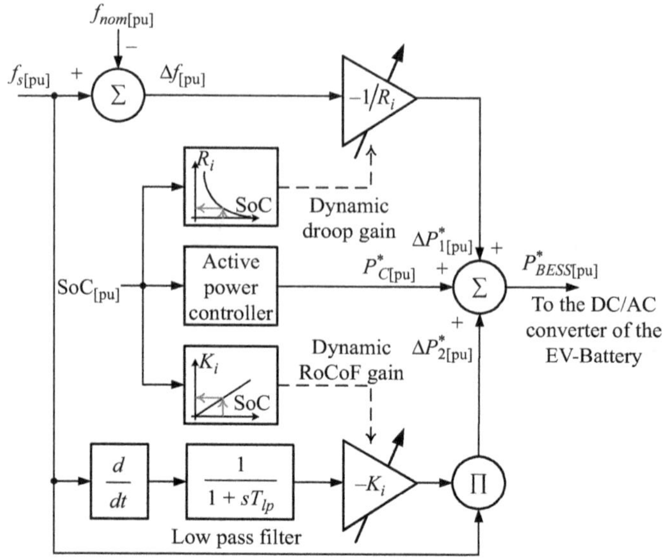

Figure 10.4 Frequency-control scheme implemented in the EV-Battery charger

(R_0) as 2%, the initial RoCoF-constant (K_0) as 50, and the time constant of the low-pass filter, T_{lp}, is set to 100 ms.

$$R_i = R_0\, SoC_{max\ [pu]} / SoC_{[pu]} \tag{10.1}$$

$$K_i = K_0\, SoC_{[pu]} \tag{10.2}$$

Black solid line plots in Figure 10.3 show the benefits of enabling the EV-Battery chargers to provide frequency support to the grid. For this example, the frequency nadir is reduced to 53% and the peak absolute rate of change of frequency is decreased to about 35%.

At this point, the reader should bear in mind that this simple example is intended to demonstrate that power systems can take advantage of the massive integration of EV into the grid for the provision of ancillary services, in this case, the frequency control. However, putting this into practice could constitute a challenging task from a technical, social and economic points of views due to several reasons, such as: the majority existence of single-phase charging stations with the consequent unbalance of the grid, the lack of policies that guarantee the availability of enough number of connected EVs for the provision of ancillary services, control strategies that allow guaranteeing an adequate level of charge when the EV is unplugged, the absence of economic policies for compensating the EV-owner since its battery is used for activities that are responsibility of the system operator, a conclusive research about the effects of the charge and discharge actions on the battery aging, and others.

10.3 Contribution of electric vehicles to voltage control

The fluctuation of the power injected into the grid by wind and PV generators impacts mainly on the power system frequency, and has an impact on the grid voltage. This section of the chapter describes how a fluctuating active power injection leads to grid voltage fluctuations and presents an unbalanced three-phase power flow algorithm that allows to quantitatively analyse the voltage evolution at every phase and bus of a distribution grid driven by this power injection. It also shows how connected EVs can be used for mitigating the grid voltage fluctuations by making use of the reactive capability of the grid-side converter. The performance of the strategy is illustrated with simulations on a test-distribution network based on the European low-voltage (LV) distribution network benchmark with DG from Cigré [21], which presents structural unbalance and unbalanced loads.

10.3.1 Theoretical background

To show how a fluctuating active power injection at a given bus leads to grid voltage fluctuations, let us consider a simplified representation (Figure 10.5) of the grid supplying the bus as a balanced system described by its Thévenin equivalent circuit, with voltage phasor $\underline{E} = E\,(\cos\delta + \mathrm{j}\,\sin\delta)$ and series impedance $\underline{Z} = R + \mathrm{j}\,X$, and a generic bus with:

- A constant power load. Active and reactive power absorbed by the load, P and Q, are independent of bus voltage, $\underline{V}$, that is considered the phase origin, $\underline{V} = V$.
- An electric vehicle load with constant active power consumption, P_{EV}, and controllable reactive power, Q_{EV}.
- A renewable generator with fluctuating active power and constant reactive power generation, P_g and Q_g.

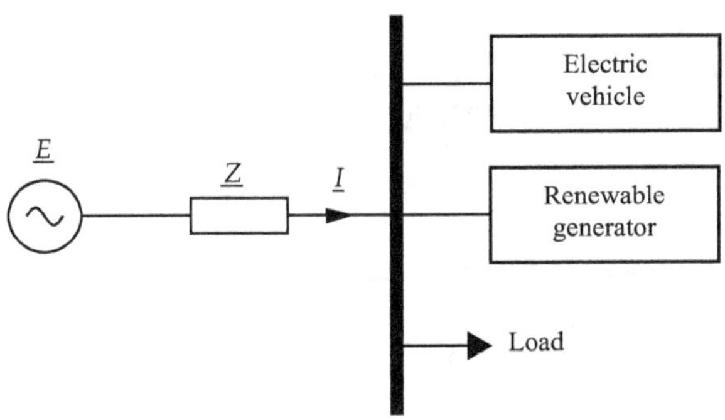

Figure 10.5 Simplified representation of a generic bus

If the current phasor flowing from the grid towards the bus, $\underline{I}$, is expressed in terms of powers:

$$\underline{I} = \left[\left(P + P_{EV} - P_g\right) - \mathrm{j}\left(Q + Q_{EV} - Q_g\right)\right]/V, \tag{10.3}$$

the bus voltage, $\underline{V} = \underline{E} - \underline{ZI}$, can be written as

$$\begin{aligned} V = {} & E(\cos\delta + \mathrm{j}\sin\delta) - \left[R\left(P + P_{EV} - P_g\right) + X\left(Q + Q_{EV} - Q_g\right)\right]/V \\ & + \mathrm{j}\left[X\left(P + P_{EV} - P_g\right) - R\left(Q + Q_{EV} - Q_g\right)\right]/V \end{aligned} \tag{10.4}$$

As δ is small, the imaginary part of (10.4) is negligible and:

$$V^2 - EV + R\left(P + P_{EV} - P_g\right) + X\left(Q + Q_{EV} - Q_g\right) \approx 0 \tag{10.5}$$

Fluctuations of the active power generated at the bus, ΔP_g, along with possible variations of the reactive power consumed by the EV, ΔQ_{EV}, lead to bus voltage variations, ΔV. Taking into account that P, Q, P_{EV} and Q_g are supposed to be independent of voltage, and neglecting second-order terms in (10.5):

$$\Delta V \approx \left(R\Delta P_g - X\Delta Q_{EV}\right)/(2V - E) \tag{10.6}$$

In transmission grids, voltage is mostly sensitive to reactive power variations because R is small in comparison with X. In distribution grids, where R and X are similar, the bus voltage is sensitive to variations in both active and reactive powers. The first term of (10.6) shows that fluctuations of the active power injected in the bus do induce voltage fluctuations. But the second term of (10.6) shows that a controllable reactive power consumption from an EV can compensate these fluctuations.

10.3.2 Unbalanced three-phase power flow

Equation (10.6) gives a first approach to the problem in balanced grids and it is helpful for understanding the general idea. However, it is not suitable for a quantitative assessment in real distribution networks, with unbalanced loads, asymmetry of untransposed lines, and single-phase lines. In this case, more sophisticated analysis tools, like an unbalanced three-phase load flow, are needed [22]. For example, [23] describes a formulation based on current residuals in the sequence frameware, for a N bus network with N_g PV buses (buses in which active power and voltage are specified), the $6N$ unknowns are the real and imaginary parts of zero (0), positive (1) and negative (2) sequence bus voltages. The corresponding $6N$ current mismatch equations can be written as

$$\begin{aligned} \Delta\underline{I}_i^{(j)} &= \underline{I}_{ls,i}^{(j)} + \sum_{k=1}^{N}\sum_{m=0}^{2}\left[\underline{Y}_{ik}^{(j,m)}\underline{V}_k^{(m)}\right], \quad \text{with } i, k \\ &= 1, 2, \ldots N;\ j,\ m = 0, 1, 2 \end{aligned} \tag{10.7}$$

where: j, m are the indices for sequence components; i, k are the subscripts for buses i and k, respectively; $\underline{I}_{ls,i}{}^{(j)}$ is the current demanded at bus i for sequence j due to the loads and generators connected to the bus; $\underline{V}_k{}^{(m)} = V_{r,k}{}^{(m)} + \mathrm{j}\ V_{x,k}{}^{(m)}$ represents the complex voltage at bus k and sequence component m and $\underline{Y}_{ik}{}^{(jm)} = G_{ik}{}^{(j,m)} + \mathrm{j}B_{ik}{}^{(j,m)}$ is the sequence component admittance matrix element that connects bus i with bus k.

The system of 6N nonlinear complex equations (10.7) can be solved by a Newton-Raphson recursive algorithm by splitting it into 12N nonlinear real equations for the real and imaginary parts of current mismatches:

$$\Delta I_{r,i}^{(j)} = I_{lsr,i}^{(j)} + \sum_{k=1}^{N}\sum_{m=0}^{2}\left[G_{ik}^{(j,m)} V_{r,k}^{(m)} - B_{ik}^{(j,m)} V_{x,k}^{(m)}\right] \tag{10.8}$$

$$\Delta I_{x,i}^{(j)} = I_{lsx,i}^{(j)} + \sum_{k=1}^{N}\sum_{m=0}^{2}\left[B_{ik}^{(j,m)} V_{r,k}^{(m)} + G_{ik}^{(j,m)} V_{x,k}^{(m)}\right] \tag{10.9}$$

10.3.3 Contribution of electric vehicles to voltage control

While an EV is connected to the grid through a power electronic converter for recharging, or even for injecting power into the grid in case of a bidirectional converter, the main function of the grid-side part of the converter is to keep the voltage of its DC link constant by transferring the active power from the grid to the vehicle, or vice versa. But advanced converters are also able to interchange reactive power with the grid, within given limits that are defined by a current, I_{max}, or a volt-ampere (VA) constraint, S_{max}, sometimes complemented with other limits related to stability or other practical issues. According to (10.6), for a balanced system and for a given active power generation ($\Delta P_g = 0$), if the converter is set to increase its reactive power consumption or decrease its reactive power injection ($\Delta Q_{EV} > 0$), the bus voltage will drop. Conversely, if a decrease in the reactive power absorption of the converter, or an injection increase, is imposed ($\Delta Q_{EV} < 0$), the bus voltage will rise. Thus, the reactive capabilities of the converter can be used for the additional objective of controlling the bus voltage.

Real power systems, particularly at the distribution level, are more complicated, as they are unbalanced, that is, the bus voltages are different for each phase. Additionally, the grid-side part of an electric vehicle power converter can adopt different topologies. A common one is a three-phase inverter, only able to interchange three-phase balanced reactive power with the grid, from just one control variable, for example, the mean of the rms voltage in each phase of a bus, as in [18]. As the electrical dynamics is much faster than that of the oscillations induced by the active power variations of renewable generators, the distribution grid can be seen as quasi-static, and the voltage phasors in each bus and phase can be computed in steady state as the response to a given time-variable input, $P_g(t)$.

10.3.4 Case study

The three-phase power flow formulation described in Section 10.3.2 allows assessing in detail the impact on the voltages at different buses on the grid of a fluctuating active power injection, $P_g(t)$, at a given bus. To illustrate its application, let us consider the European LV distribution network with DG [21] shown in Figure 10.6. This distribution grid is unbalanced due to line asymmetries and the unbalanced characteristic of the loads. In this case, the loads are modelled as constant power,

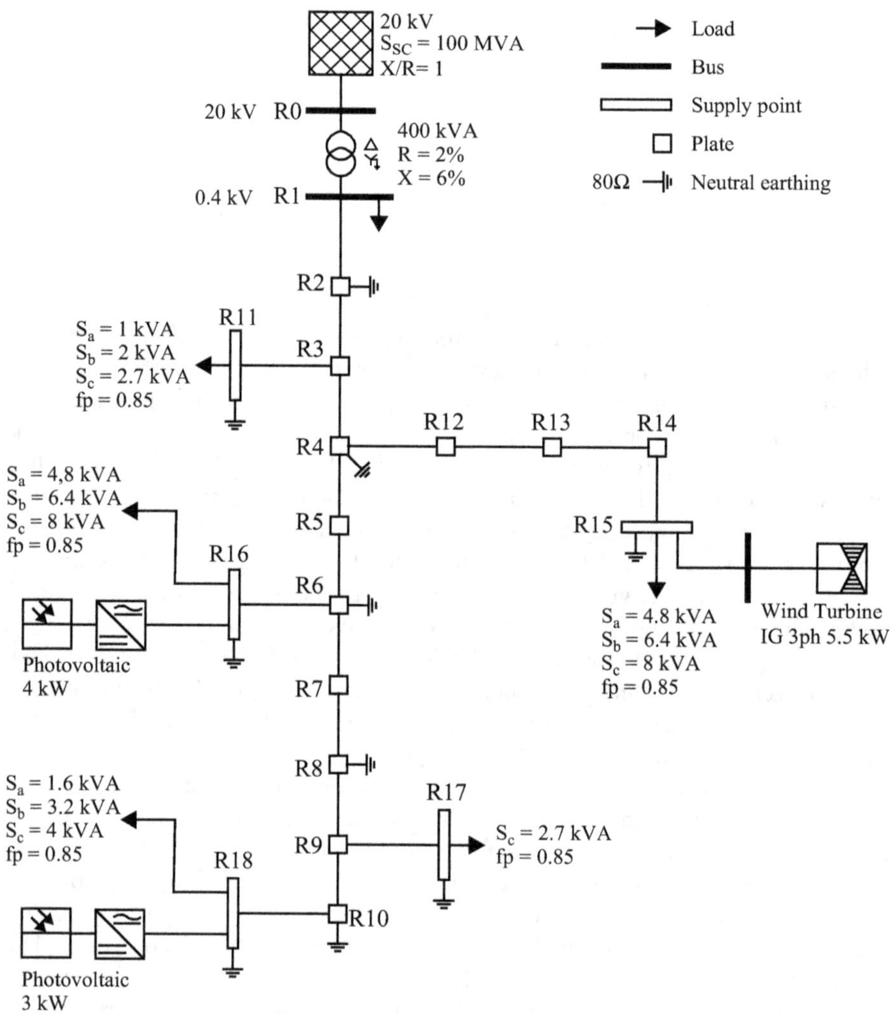

Figure 10.6 Modified European LV distribution network with distributed generation [21]

although they could be replaced by constant impedance or constant current or a combination of any of these three models.

As a case study, the original test grid is modified so as to include three EVs as part of the loads in buses R15, R16 and R18. They are modelled as constant PQ balanced loads (balanced loads in which active and reactive powers are specified) and, once connected to the grid, they absorb a constant active power of 7 kW and can absorb or generate reactive power with a limit of 5 kvar.

Figure 10.7 shows the details of each phase voltage at every bus of the modified test grid, computed by using the load-flow formulation of Section 10.3.2, without DG and electric vehicles.

To show the impact of a fluctuating active power injection from the DG at buses R15, R16 and R18, the profiles of $P_g(t)$ shown in Figure 10.8 have been considered. The active power time series from the generator at bus R15 has been obtained from the per unit profile of the base case of Figure 10.13 in [24], scaled to a rated power of 5.5 kW. Similarly, the active power time series from both PV generators, at buses R16 and R18, have been obtained from [25] and scaled to their rated power of 4 and 3 kW, respectively.

Time evolution of the grid magnitudes can be computed as successive quasi-steady states, obtained from the power flow formulation of Section 10.3.2, via a continuation load flow [26]. The black solid lines of Figure 10.9 show the thus computed time evolution of the three-phase voltages at buses R15, R16 and R18, revealing the variations induced by the fluctuating active power injection from the wind generator. Although the voltage variations are stronger at bus R15, they are also propagated to the other buses of the grid.

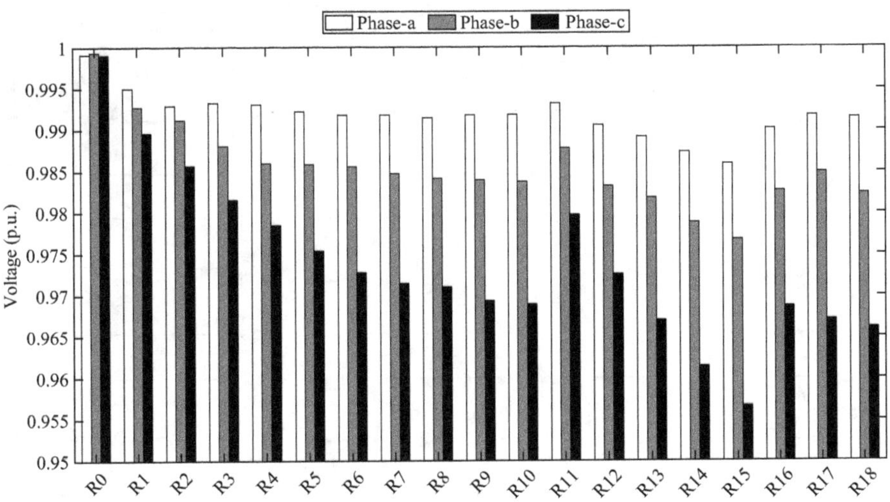

Figure 10.7 Phase voltages of the modified test network

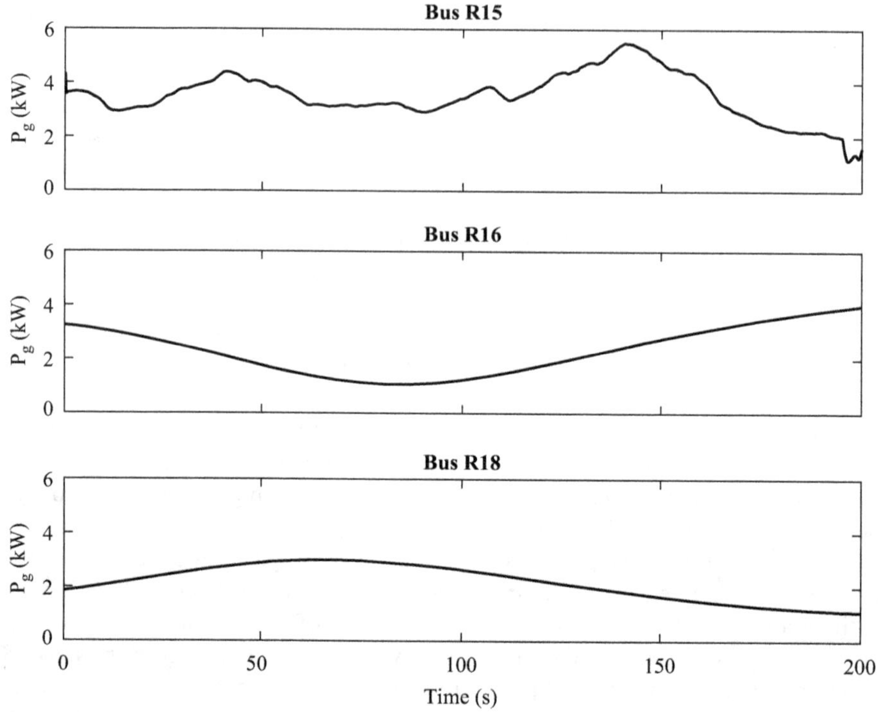

Figure 10.8 Active power injected by the distributed generation at buses R15, R16 and R18

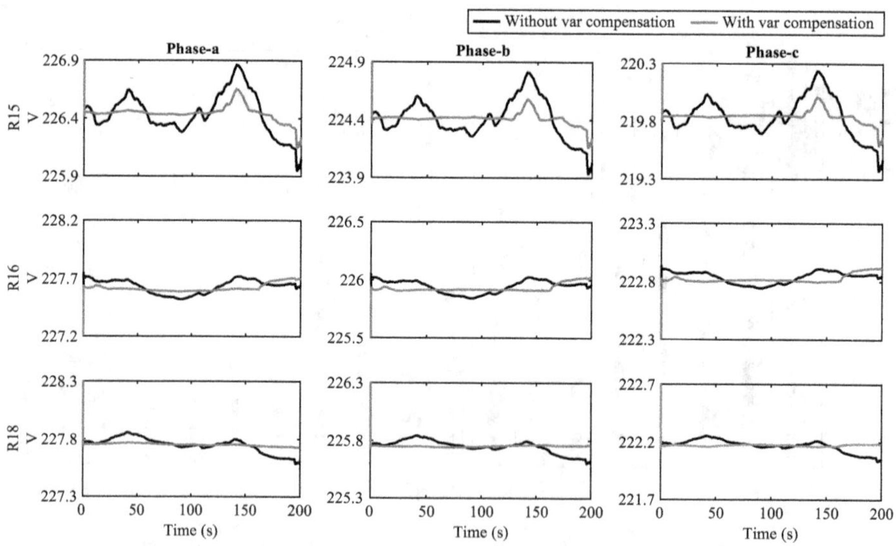

Figure 10.9 Voltage variations at buses R15, R16 and R18 induced by a fluctuating active power injection at bus R15

As stated in Section 10.3.3, the EVs connected for recharging can be used for mitigating the grid voltage fluctuations by using the reactive capability of their charger converter. For example, Figure 10.10 shows a simple implementation of a reactive power controller at a generic bus, *i*, where the plant is the power flow formulation of Section 10.3.2 and the outputs are the three-phase rms voltages. As the bus voltages are different for each phase and the grid-side converter is only able to generate three-phase balanced reactive power, the controller uses the mean bus voltage, defined as the mean of the rms voltage in each phase.

The grey lines of Figure 10.9 show the computed time evolution of the three-phase voltages at buses R15, R16 and R18 when the var compensation from the EVs is implemented. Generally, the voltage fluctuations are clearly smoothed.

Note, however, that there are some periods in which the reactive capability limits of the EV converters (±5 kvar) are reached and, thus, the smoothing capability is also limited. To gain a deeper insight into this issue, Figure 10.11 shows the evolution at buses R15, R16 and R18 of: the mean voltage (black solid lines), the mean voltage set point of the controller (black dotted lines) and the reactive power absorbed by the electric vehicle converters (grey solid lines).

Figure 10.12 shows the effect of using the reactive power compensation from the EVs in all the buses of the grid as the difference between the maximum and minimum voltages of the period at each phase.

A more systematic analysis of the reduction of voltage fluctuations can be derived by extending the [27] definition for balanced networks to an unbalanced power system as

$$VF_{k\Phi} = 1/T \sum_{t=1}^{T} |V_{k\Phi}(t) - V_{k\Phi}(t-1)| \tag{10.10}$$

where T is the number of time samples of the period under study and $VF_{k\Phi}$ is the voltage fluctuation index at bus k and phase Φ. From this, the system average

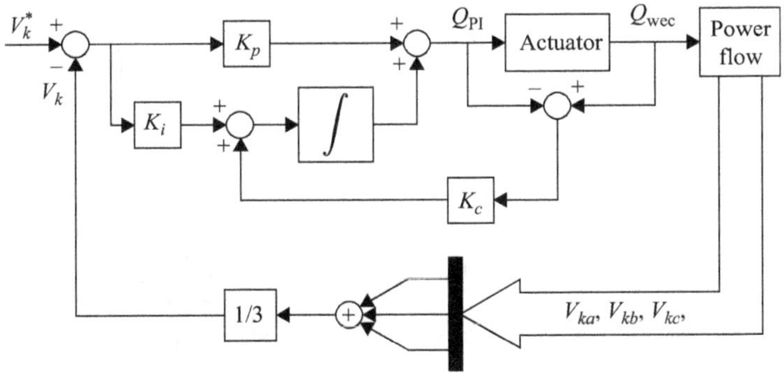

Figure 10.10 Simple reactive power controller for the electric vehicle grid-side converter

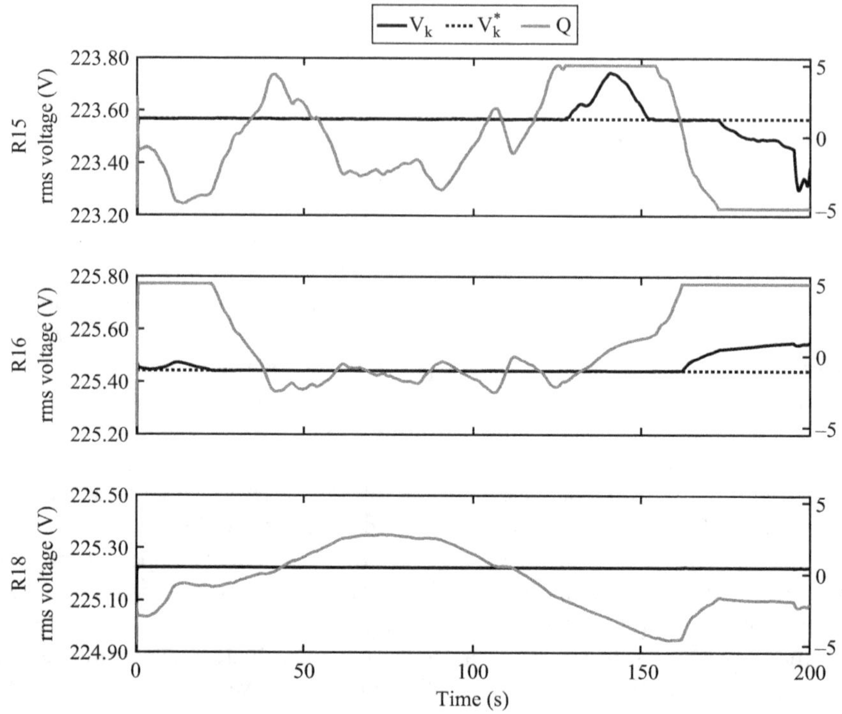

Figure 10.11 Evolution of voltage and reactive power

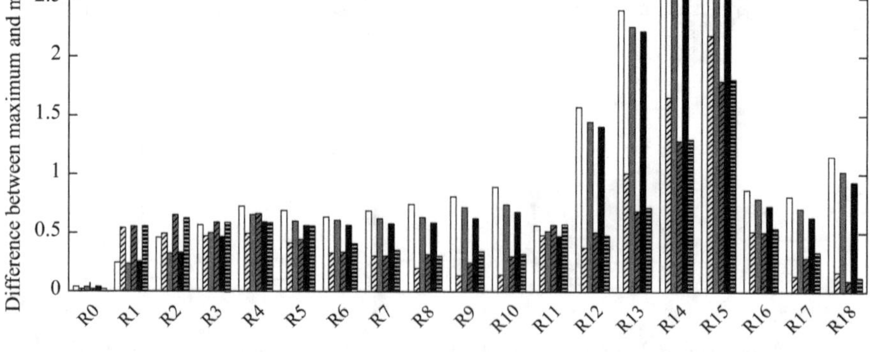

Figure 10.12 Difference between maximum and minimum voltage per bus and phase, without and with reactive compensation from electric vehicles

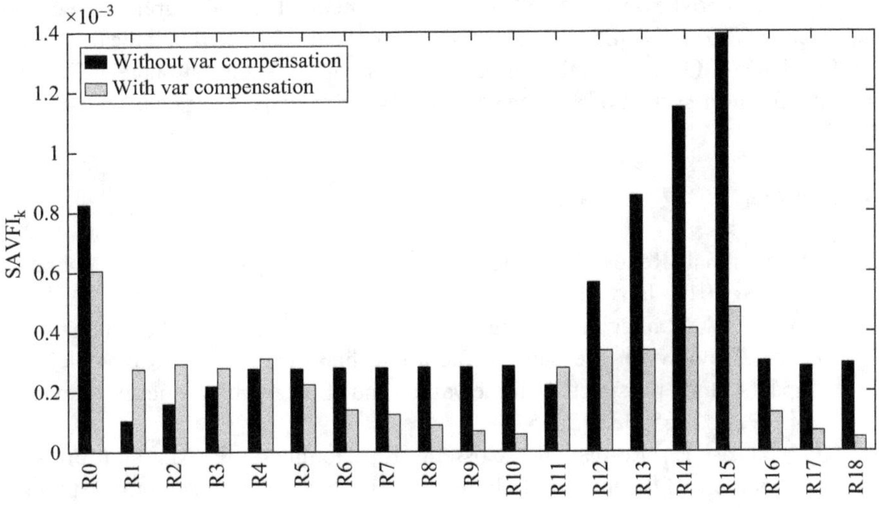

Figure 10.13 System average voltage fluctuation index per bus, with and without reactive power compensation from electric vehicles

voltage fluctuation index (SAVFI) per bus [28] can be computed as

$$\mathrm{SAVFI}_k = 1/n_\Phi \sum\nolimits_{\Phi \in} \kappa VF_{k\Phi} \tag{10.11}$$

where n_Φ and κ are the number and set of phases at bus k, respectively. Figure 10.13 shows the SAVFI for each bus of the test network for the 200 s period under consideration, without and with reactive power compensation.

10.4 Conclusion

This chapter has shown how EVs connected to a power system for recharging can be used as providers of the ancillary services of primary frequency control, inertia and voltage control. Appropriate control strategies have been presented for these services and their effectivity has been tested for some simulation case studies. Further research will include a wider variety of cases and practical implementation at a microgrid.

Acknowledgements

This work was supported in part by the European Commission under project OPS Master Plan for Spanish Ports (grant number: 2015-EU-TM-0417-S) that is co-financed by the Connecting Europe Facility (CEF) for the building of the European Union's TEN-T. It was also supported in part by the Secretaria de Educacion Superior, Ciencia,

Tecnologia e Innovacion (SENESCYT), Government of the Republic of Ecuador (grant number: 2015-AR6C5141). The authors would also like to thank the companies ELECGALAPAGOS S.A. and EOLICSA for providing information related to the wind-diesel power system of the San Cristobal Island (Galapagos, Ecuador).

References

[1] International Renewable Energy Agency, Renewable Power Generation Costs in 2017, Jan. 2018.
[2] Yong, J. Y., Ramachandaramurthy, V. K., Tan, K. M., and Mithulananthan, N. A. 'Review on the state-of-the-art technologies of electric vehicle, its impacts and prospects', Renewable and Sustainable Energy Reviews, vol. 49, pp. 365–385, 2015.
[3] Fernandez, L., Roman, T., Cossent, R., Domingo, C. M., and Frias, P. 'Assessment of the impact of plug-in electric vehicles on distribution networks', IEEE Transactions on Power Systems, vol. 26, no. 1, pp. 206–213, 2011.
[4] Kempton, K., and Tomic, J. 'Vehicle-to-grid power implementation: From stabilizing the grid to supporting large-scale renewable energy', Journal of Power Sources, vol. 144, no. 1, pp. 280–294, 2005.
[5] Andersen, P. B., Marinelli, M., Olesen, O. J., *et al.* 'The Nikola project intelligent electric vehicle integration', In IEEE PES Innovative Smart Grid Technologies, Europe, Istanbul, pp. 1–6, 2014.
[6] Knezovic, K., Martinenas, S., Andersen, P. B., Zecchino, A., Marinelli, M. 'Enhancing the role of electric vehicles in the power grid: Field validation of multiple ancillary services', IEEE Transactions on Transportation Electrification, vol. 3, no. 1, pp. 201–209, 2017.
[7] Galus, M. D., Koch, S., and Andersson, G. 'Provision of load frequency control by PHEVs controllable loads and a cogeneration unit', IEEE Transactions on Industrial Electronics, vol. 58, no. 10, pp. 4568–4582, 2011.
[8] Masuta, T., and Yokoyama, A. 'Supplementary load frequency control by use of a number of both electric vehicles and heat pump water heaters', IEEE Transactions on Smart Grid, vol. 3, no. 3, pp. 1253–1262, 2012.
[9] Falahati, S., Taher, S. A., and Shahidehpour, M. 'Smart deregulated grid frequency control in presence of renewable energy resources by EVs charging control', IEEE Transactions on Smart Grid, vol. 9, no. 2, pp. 1073–1085, 2018.
[10] Lee, J., Muljadi, E., Srensen, P., and Kang,Y. C. 'Releasable kinetic energy-based inertial control of a DFIG wind power plant', IEEE Transactions on Sustainable Energy, vol. 7, no. 1, pp. 279–288, 2016.
[11] Barr, J., and Majumder, R. 'Integration of distributed generation in the Volt/var management system for active distribution networks', IEEE Transactions Smart Grid, vol. 6, no. 2, pp. 576–586, 2015.
[12] Wang, L., Yan, R., and Saha, T. K. 'Voltage management for large scale PV integration into weak distribution systems', IEEE Transactions on Smart Grid, vol. 9, no. 5, pp. 4128–4139, 2018.

[13] Cheng, L., Chang, Y., and Huang, R. 'Mitigating voltage problem in distribution system with distributed solar generation using electric vehicles', IEEE Transactions on Sustainable Energy, vol. 6, no. 4, pp. 1475–1484, 2015.

[14] Parchure, A., *et al.* 'Investigating PV generation induced voltage volatility for customers sharing a distribution service transformer', IEEE Transactions on Industry Applications, vol. 53, no. 1, pp. 71–79, 2017.

[15] Woyte, A., *et al.* 'Voltage fluctuations on distribution level introduced by photovoltaic systems', IEEE Transactions on Energy Conversion, vol. 21, no. 1, pp. 202–209, 2006.

[16] Tong, X., *et al.* 'DC voltage follow-up control for mitigating the power fluctuation of distributed generation', En Industrial Electronics Society, IECON 2017-43rd Annual Conference of the IEEE. IEEE, pp. 6380–6383, 2017.

[17] Shivashankar, S., *et al.* 'Mitigating methods of power fluctuation of photovoltaic (PV) sources–A review', Renewable and Sustainable Energy Reviews, vol. 59, pp. 1170–1184, 2016.

[18] Mendonça, H., de Castro, R., Martinez, S., and Montalban, D. 'Voltage impact of a wave energy converter on an unbalanced distribution grid and corrective actions', Sustainability, vol. 9, no. 10, pp. 1–16, 2017.

[19] Kundur, P. '*Power system stability and control*', 1st ed. New York: McGraw-Hill, 1994.

[20] Anderson, P. M., and Fouad, A. A. '*Power system control and stability*', 2nd ed. Piscataway, N.J.: IEEE Press, Wiley-Interscience, 2003.

[21] Strunz, K., Abbasi, E., Fletcher, R., and Joos, G. 'TF C6. 04.02: TB 575–Benchmark systems for network integration of renewables and distributed energy resources', CIGRE, Technical Report, 2014.

[22] Balamurugan, K., and Srinivasan, D. 'Review of power flow studies on distribution network with distributed generation', In Proceedings of the IEEE 2011 IEEE Ninth International Conference on Power Electronics and Drive Systems (PEDS), Singapore, 5–8, pp. 411–417, Dec. 2011.

[23] Mayordomo, J., Izzeddine, M., Martinez, S., Asensi, R., Exposito, A. G., and Xu, W. 'Compact and flexible three-phase power flow based on a full Newton formulation', IEE Proceedings on Generation, Transmission and Distribution, vol. 149, no. 2, pp. 225–232, 2002.

[24] Ochoa, D., and Martinez, S. 'Frequency dependent strategy for mitigating wind power fluctuations of a doubly-fed induction generator wind turbine based on virtual inertia control and blade pitch angle regulation', Renewable Energy, vol. 128, part A, pp. 108–124, 2018.

[25] Bureau of Metrology of the Australian Government. 'About one minute solar data', 29 August 2012. Retrieved from http://www.bom.gov.au/climate/data/oneminsolar/about-IDCJAC0022.shtml

[26] Ajjarapu, V., and Christy, C. 'The continuation power flow: A tool for steady state voltage stability analysis', IEEE Transactions in Power Systems, vol. 7, no. 1, pp. 416–423, 1992.

[27] Viawan, F. A., and Karlsson, D. 'Combined local and remote voltage and reactive power control in the presence of induction machine distributed generation', IEEE Transactions on Power Systems, vol. 22, no. 4, pp. 2003–2012, 2007.

[28] Ding, F., Nagarajan, A., Baggu, M., *et al.* 'Application of autonomous smart inverter Volt-VAR function for voltage reduction energy savings and power quality in electric distribution systems: Preprint', Technical Report, National Renewable Energy Laboratory(NREL): Golden, CO, USA, 2017.

Chapter 11

Electric vehicles for renewable energy integration in isolated power systems

Yaofang Wang[1], *Sandra Castaño-Solis*[2], *Jesús Fraile-Ardanuy*[3], *David Jiménez*[3], *Benito Artaloytia*[3], *Roberto Álvaro-Hermana*[4,5] *and Julia Merino*[6]

11.1 Introduction

Global warming is increasing the probability of extreme weather episodes such as storms, floods, heat waves, and droughts worldwide. The primary cause of this increment in temperature is due to the effect of greenhouse gas (GHG) emissions generated by human activities such as burning fossil fuels for electricity generation or transportation.

A recent report from the Intergovernmental Panel on Climate Change (IPCC) highlights the importance of limiting global warming temperature to 1.5 °C instead of 2 °C (or more) in the following years, but this will require unprecedented transitions in different sectors [1].

According to the United States Environmental Protection Agency (EPA), electric generation and heat production were responsible for 25% of the global GHG emissions in 2010, while another 14% of global GHG emissions were generated by the transportation sector [2]. Therefore, it is necessary to promote different strategies in order to sharply reduce the current global dependence on fossil fuels in both the sectors, through different actions such as integration of renewable energy in the generation mix and electrification of the transportation sector.

Renewable energies such as hydropower, wind, and solar contribute to reducing global GHG emissions, energy dependence, and the cost of electricity, but they have a main drawback: the availability and intermittency of its generation, which creates several technical and nontechnical issues such as power, frequency,

[1]École NationaleSupérieure de l'Électronique et de ses Applications (ENSEA), Cergy-Pontoise, France
[2]Escuela Técnica Superior de Ingeniería y Diseño Industrial (ETSIDI), Universidad Politécnica de Madrid, Madrid, Spain
[3]ETSI Telecomunicación-Universidad Politécnica de Madrid, Madrid, Spain
[4]Orkestra – Basque Institute of Competitiveness, Bilbao, Vizcaya, Spain
[5]University of Deusto, Bilbao, Vizcaya, Spain
[6]TECNALIA, Parque Científico y Tecnológico de Bizkaia, Derio, Spain

and voltage fluctuations, long-term dynamic stability, increment in operating and control costs, and so on, which affect the electrical grids [3,4].

These problems are more significant in the islands. Most of the generating power installed in these isolated grids are based on diesel groups, having an extremely high dependence on fossil fuels, with high operation and transportation costs and a high level of GHG emissions. Additionally, these systems have a limited capacity to accommodate renewable energy, due to the weakness of the power grid, which presents lower inertia, due to the reduced number of generators, and poorly meshed transmission networks [5–7]. In order to ensure the electric supply, generation units are oversized and the ratio between the peak and valley demand is higher than in the continental grids.

Despite these limitations, introducing renewable sources into electrical grids to reduce the dependence on fuel importation for electricity generation has become an important task to guarantee the energy and economic sustainability of many islands around the world. To improve the controllability and reliability of renewable plants, different energy storage technologies have been proposed in the literature, such as electrochemical batteries, flywheels, or pumped hydro-energy storage (PHES) [8,9]. In particular, PHES plants are very cost-effective installations, but they are necessary both at the appropriate place and geographical resources, and the setup demands high capital costs.

With the current storage technology available, it is now possible to become a virtually energy-independent electric system. For example, in 2016, Tesla Company deployed a 1.4 MW solar plant and a 6 MWh lithium-ion battery energy storage system in the island of Ta'u in American Samoa, avoiding burning over 400,000 L of diesel per year (plus shipping) in order to supply the electricity required by its nearly 600 inhabitants [10].

The introduction of electric vehicles (EVs) is another alternative to reduce GHG emissions in these isolated systems, because these vehicles are more suitable for the short average daily distances on the islands. EVs can also contribute to integrating renewable energies in these isolated systems, becoming flexible loads and helping to reduce the peak–valley ratio and flattering the energy demand curve [11–13]. Additionally, the mobility characteristics in islands make them appropriate to promote the electrification of the transportation sector.

The vehicle-to-grid (V2G) concept was proposed by Kempton and Letendre [14]. The main idea is to use battery storage available from EVs for providing power to the grid and using the surplus renewable energy for charging them. Introducing V2G as an energy storage system presents advantages such as making use of large-scale distributed storage systems; the batteries of EVs present very low cost for grid operators, easy implementation, and compatibility with the environmental policy.

This work analyses the option of charging an EV fleet as a distributed energy storage system to increase the participation of renewable energy in the particular isolated power system of an island from the Canary Archipelago (Spain). A comparative study will be performed to quantify the amount of diesel generation needed and CO_2 emission after the integration of EVs with different hydro-electric power

Figure 11.1 Location of the Spanish island of El Hierro, at the northwest of the African continent

plant conditions at the beginning, in order to study how the integration of EVs could influence the grid.

11.2 El Hierro's electrical system description

El Hierro is one of the seven islands that constitute the Canary Islands Archipelago, with an average population of around 10,000 people. It is the most western and southern island of the archipelago, and is located on the northwest of the African continent, as shown in Figure 11.1. After its nomination as a biosphere reservoir by UNESCO in 2000, the economic activity of the island changed from agriculture and fishing activities to tourism and small-scale industries, producing a progressive increment in the energy demanded in the island. With the aim to reduce fossil fuel importation and to introduce renewable energy sources to supply this electricity demand, the Council of El Hierro boosts the development of feasibility studies of renewable plants. As a result, the hydro-wind plant project called Gorona del Viento was developed by the Council of El Hierro, Unelco S.A. and the Canary Islands Technological Institute (I.T.C.) [15].

The renewable plant Gorona del Viento is composed of the following subsystems [15]:

- *Wind farm*: Five wind turbine models, Enercon E-70 [16], with a rated power of 2.3 MW each, producing a total power of 11.5 MW.
- *Pumping plant*: Two pumping groups of 1.5 MW each and six pumping groups of 0.5 MW each, with a total capacity of 6 MW.

- *Upper reservoir*: This is a lake with a capacity of 380.000 m^3 located at 698 m above sea level in the La Caldera crater, which is an extinct volcano.
- *Lower reservoir*: This is a dam built near the diesel power plant of Llanos Blancos, with a usable capacity of 150.000 m^3.
- *Hydroelectric power station*: Four Pelton turbine groups with a power of 2.8 MW each for a total power of 11.3 MW. The maximum flow rate is 2 m^3/s with a gross head of 653 m.
- *Electrical substation*: This is the interconnection point of the hydroelectric plant, wind farm, pumping plant, and the electrical grid.
- *Pipeline system*: This system is formed by two aerial pipes to communicate to the reservoirs. The penstock is 2350 m long with a diameter of 1 m while the pumping pipeline is slightly longer at 3015 m, and narrower with a diameter of 0.8 m.

The schematic setup of the plant is shown in Figure 11.2. The hybrid wind-pumped hydro-storage plant stores surplus wind power by pumping water up to 698 m to the crater La Caldera. If the winds are calm or the demand exceeds electrical supply, water is released from the upper reservoir through the penstock to the

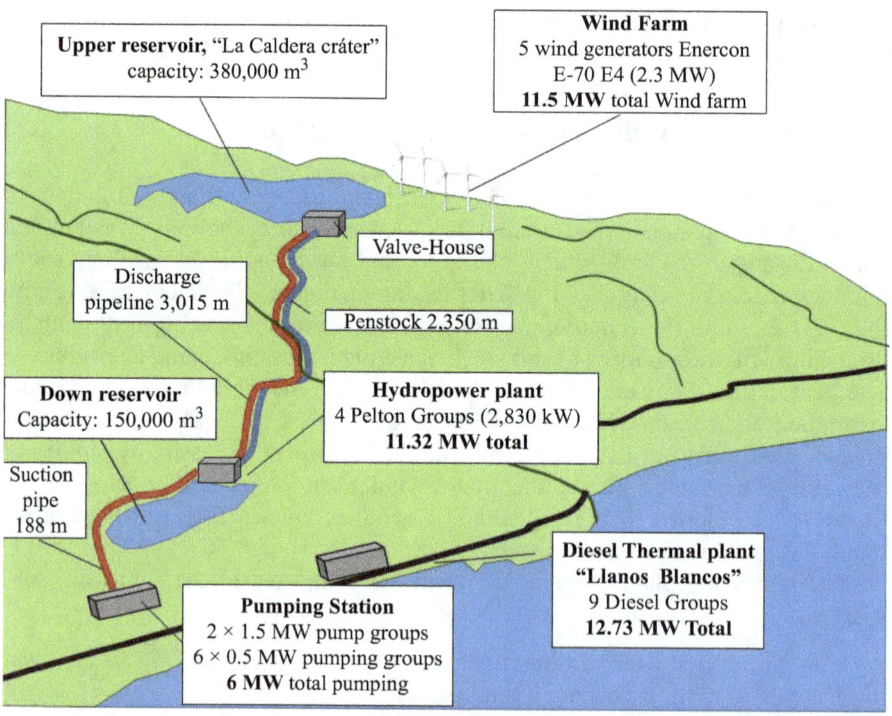

Figure 11.2 Schematic setup of the hydro-wind plant

Table 11.1 Variables and parameters from the optimization model

Concept	Before (2011)	After (2015)
Thermal power plant		
Rated power	13 MW	15 MW
Annual energy production	44.6 GWh	13.7 GWh
Annual diesel consumption	9.810 Ton	3.014 Ton
Total electricity generation cost	10.79 M€	3.24 M€
Wind farm		
Rated power	0 MW	11.5 MW
Annual energy production	0 GWh	35.9 GWh
Total electricity generation cost	0 M€	2.58 M€
Pumped hydropower plant		
Rated power	0 MW	10 MW
Annual energy production	0 GWh	6.4 GWh
Total energy generated	**44.6 GWh**	**56 GWh**

hydroelectric power station to generate electricity, refilling the lower basin. When there is an excess of wind power or a reduction of the electricity demand (typically from midnight to sunrise), water stored in the lower reservoir is pumped back to the upper reservoir.

This project has allowed supplying the electrical demand of the island for 8 consecutive days in 2017 and during 18 days in January–February 2018, with 100% renewable energy [17]. Also, more than 10,000 tons of fossil fuels have been saved and almost 30,000 tons of CO_2 have not been emitted into the atmosphere [18].

Table 11.1 shows the comparison of the El Hierro's electrical system before and after the startup of Gorona del Viento wind–hydro power plant [18].

11.3 Data description

The description of the main data required in this work is presented in this section.

11.3.1 Electric power system data

The 2018 annual electricity demand (Figure 11.3) and annual wind generation (Figure 11.4) with an hourly resolution for El Hierro Island was directly obtained from Red Electrica de España (REE), a partly state-owned and public limited Spanish Corporation that operates the national transmission grid in Spain [19].

The minimum electricity consumption was 2.17 MW, reached in March 1, 2018, at 3:00 and the maximum value was reached in August 24, 2018, at 21:00 (7.044 MW). The average electricity consumption during this year was 4.786 MW.

The wind generation with hourly resolution in the same period is presented in Figure 11.4. The minimum wind generation was 0 MW and the maximum value was reached on June 4, 2018, at 23:00 (10.73 MW). The average electricity

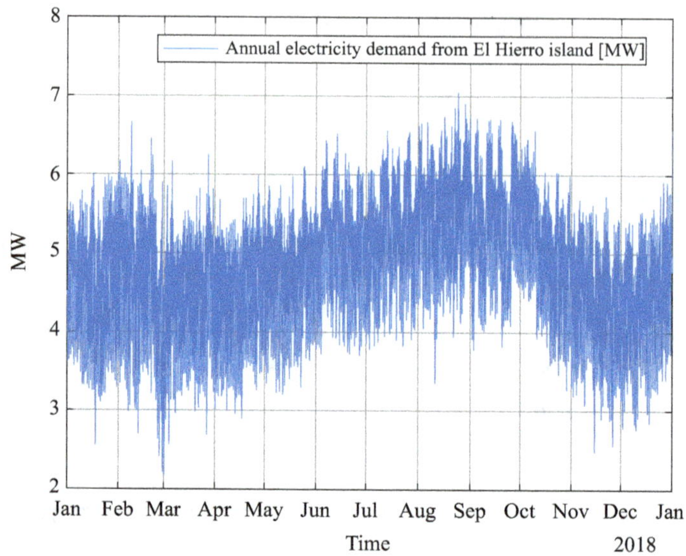

Figure 11.3 Annual power demand in El Hierro Island (January–December 2018)

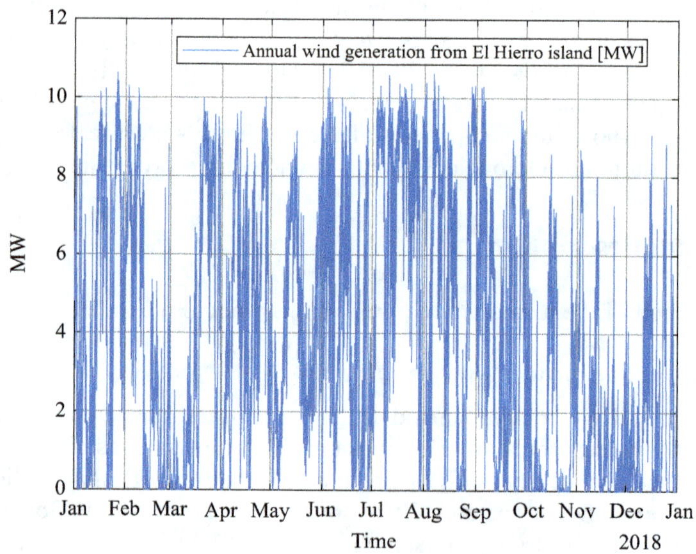

Figure 11.4 Annual wind generation in El Hierro Island (January–December 2018)

consumption during this year was 3.985 MW, very close to the average electricity consumption.

The electric consumption data histogram and the wind generation data histogram is presented in Figure 11.5. From this figure, it is observed that for more than 18 h, the wind farm production is almost 0 (20.55% of the annual hours), but there are hours when the wind production significantly exceeds the electric demand on the island, generating an energy surplus that can be stored and used in the future.

According to the 2017 environmental impact assessment in Llanos Blancos diesel-fueled power plant (DFPP), the CO_2 emissions per generated MWh is around 0.67–0.68 tCO2/MWh [20]. This value will be used in our study.

Despite living in the Canary Islands, where the electricity generation cost is significantly higher than in mainland Spain, Canarian inhabitants pay the same electricity tariff as the rest of the Spaniards. The hour-by-hour electricity price, which depends on the spot price, is also variable with time. There are three different domestic tariffs named as 2.0A (general tariff), 2.0DHA (night rate tariff), and 2.0DHS (EV tariff, with three periods: peak, off-peak, and super off-peak). This last tariff, 2.0DHS, is shown in Figure 11.6 [21]. As it can be seen in this figure, the electricity price is lower from 22:00 to 21:59 hours. Therefore, it is assumed that all-EVs will be charged in from 22:00 to 05:00 in order to get the vehicle ready early in the morning, when the first trips begin.

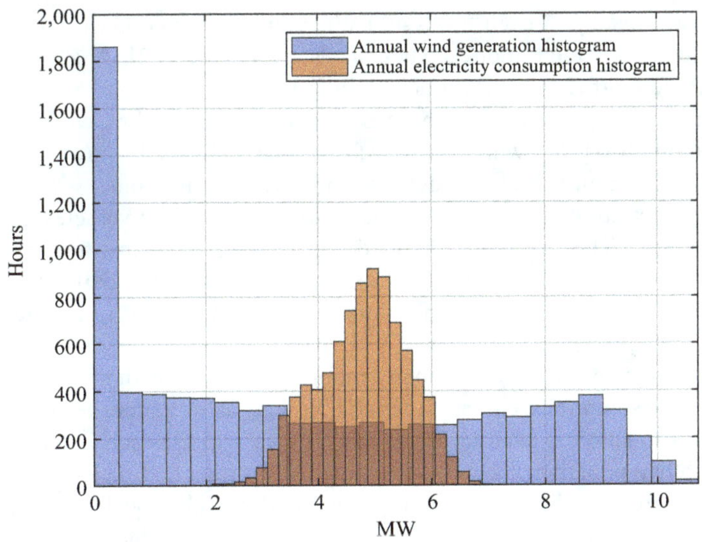

Figure 11.5 Annual wind generation and annual electricity consumption histograms in El Hierro Island (January–December 2018)

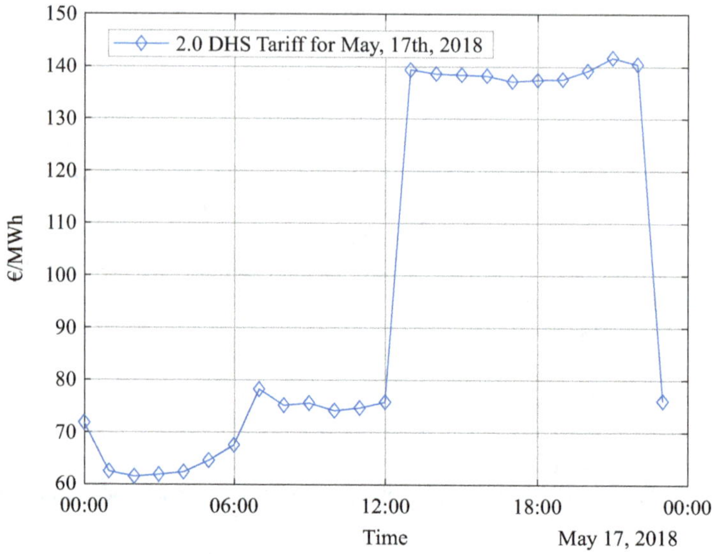

Figure 11.6 Hourly 2.0DHS tariff (May 17, 2018)

11.3.2 Mobility data

The nominal capacity of each EV is 40 kWh, which is currently the common value in different commercial models such as Renault Zoe [22], Nissan Leaf [23], Volkswagen e-Golf [24], or Hyundai Ioniq [25].

The average consumption of these EVs is 0.17 kWh/100 km, which is obtained from Spritmonitor [26], which is a website that collects fuel (or energy) consumption of different vehicles under real-life driving conditions.

The current number of internal combustion engine (ICE) vehicles in El Hierro Island is 4,848. This value is obtained from the Canary Institute of Statistics (*Instituto Canario de Estadística*, ISTAC) [27]. The average daily travelled distance, 25 km, and the hourly distribution of trips (represented in Figure 11.7) are obtained from Colmenar-Santos et al. [28,29].

In order to provide a more realistic mobility behavior with some variability in the electric consumption demand for each EV, a probability density function based on a gamma function is used to select the distance trip for each individual driver. The probability is higher around the average daily travelled distance (25 km), as shown in Figure 11.8. Similar profiles, based on Rayleigh probability functions are also used to model the arriving and departure time for each individual driver, as it was proposed in Ardanuy [30].

Finally, in order to evaluate the CO_2 emissions due to conventional ICE vehicles and according to the European Environment Agency (EEA) [31], average CO_2 emissions of 130 g/km is chosen in this work.

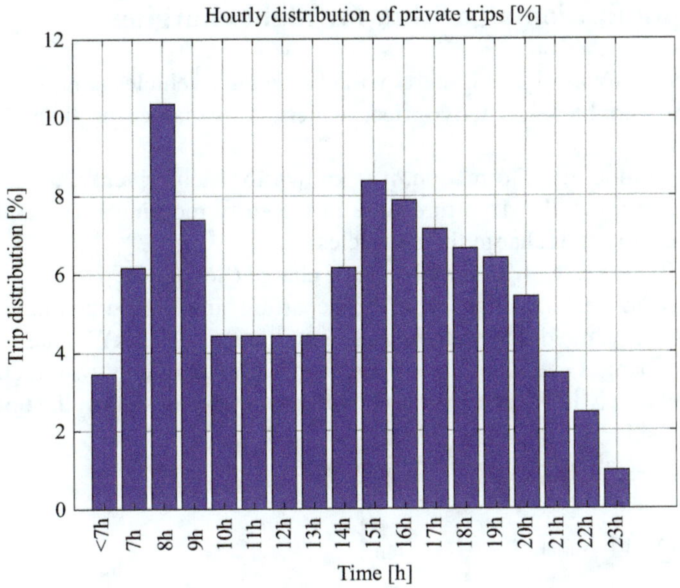

Figure 11.7 Hourly distribution of private trips in Tenerife [30]

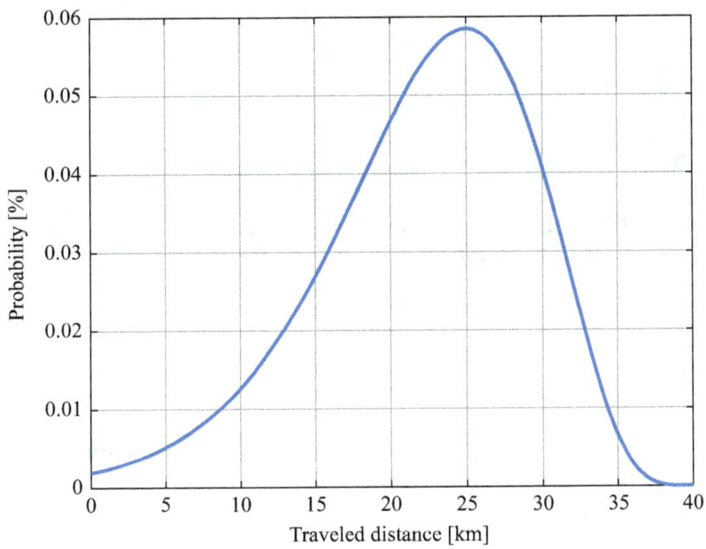

Figure 11.8 Hourly distribution of private trips in Tenerife [30]

11.4 Optimization algorithm for night charging

All EV drivers involved in this study will charge their vehicles at night, taking into account the variable electricity price, which is available for them on a day-ahead basis.

From the mobility information, it is known for each driver, v, the arrival time in the evening, $t_a(v)$, the departure time in the early morning, $t_d(v)$, and the total demanded energy to recharge their vehicles.

Each VE arrives with an initial state of charge (SoC), $SOC(t_a(v)) = SOC_{ini}$ and it is required to be fully charged at the departure time. The maximum power at which EVs are charged during the night period is $CR = 3.3$ kW. Due to the electricity price changes every hour and the required amount of energy to recharge each EV being relatively low, each hour is discretized in 15-min slots in the optimization algorithm (Table 11.2).

The optimization problem is defined as

$$\min\left[\sum_t \mathrm{PEX}_{\mathrm{supply}}(t)\cdot i(t,v)/\gamma_{\mathrm{eff}}\right] \tag{11.1}$$

Subject to the following restrictions:

$$soc(\mathrm{t_a}(v),v)=\mathrm{SOC_{ini}}(v) \quad ; \quad soc(\mathrm{t_d}(v),v)=\mathrm{SOC_{max}} \tag{11.2}$$

$$\mathrm{SOC_{ini}}(\mathrm{v})\leq soc(t,v)\leq \mathrm{SOC_{max}} \tag{11.3}$$

$$i(t,v)\leq \gamma_{\mathrm{eff}}\cdot \mathrm{CR}\cdot \mathrm{BCT}(t,v)\cdot \Delta\mathrm{t} \tag{11.4}$$

$$soc(t,v)=soc(t-1,v)+i(t,v)/\mathrm{C} \tag{11.5}$$

$$soc(t,v), i(t,v)\geq 0 \tag{11.6}$$

Table 11.2 Definition of indexes, variables, and parameters from the optimization model

Description	Symbol	Value	Unit	
Vehicle	v	[1,V]	–	Index
Time slot	t	$[T_a,T_d]$	–	Index
Energy hourly price	$PEX_{supply}(t)$	$[PEX_t]$	€/kWh	Parameter
Battery capacity	C	40	kWh	Parameter
State of charge	$soc(t,v)$		%	Variable
Number of time period t per hour	Δt	4	–	Parameter
Energy extracted from the grid	$i(t,v)$		kWh	Variable
Charge rate	CR	3.3	kW	Parameter
Charge efficiency rate	γ_{eff}	0.95	–	Parameter
Maximum allowed SOC	SOC_{max}	100	%	Parameter
Initial SOC	$SOC(t_a(v))$	SOC_{ini}	%	Parameter
Conn./Discon. Vector	$BCT(t,v)$	{0,1}	–	Parameter

The objective function (11.1) is to minimize the cost of charging a particular EV, defined by the index *v*, during the night, considering a charging efficiency of 95%. Constraint (11.2) defines the initial and final SoC of each vehicle *v* at the beginning and end of the optimization period. Constraint (11.3) limits that the SoC must be located between both extreme SoC values during the optimization period.

Equation (11.5) presents the time evolution of the battery SoC, assuming that the charging process is linear. The charging process will be available depending on the vector BCT*(t,v)*, which is equal to 1 from the arriving time to the departure time and 0 out of this period.

Finally, variables are defined as positive to guarantee that energy input is always positive.

GAMS® and Cplex Solver optimizer were used for solving the individual night charging optimization problem for each driver.

11.5 Scenarios

11.5.1 Scenario 1. Base scenario. No VE-No PHEP

In all analyzed scenarios, it is assumed that the wind generators are the main energy source in El Hierro Island. Wind energy is intermittent and the DFPP is used as an auxiliary generation, providing the complementary energy to cover the current demand, after evaluating the diesel-generated energy required, the CO_2 emission associated with the electric production is calculated, as shown in Figure 11.9.

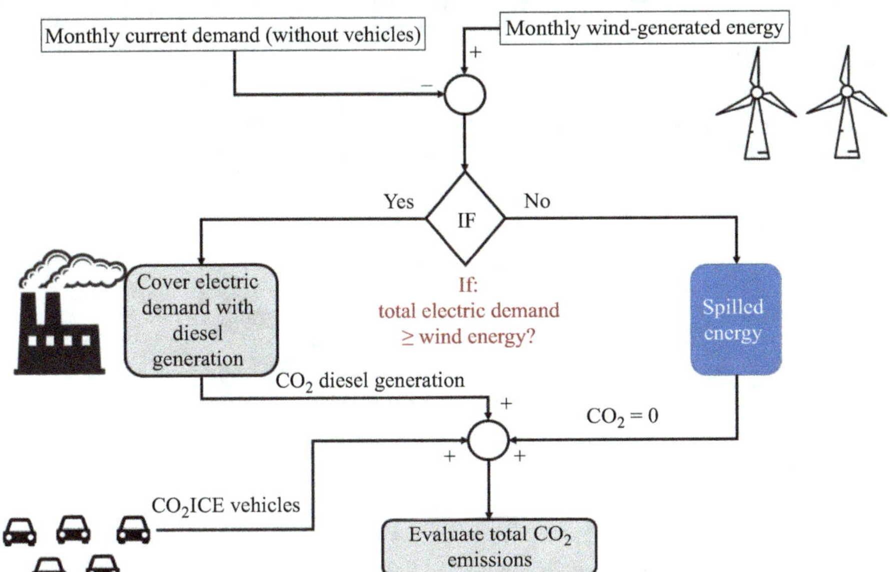

Figure 11.9 Scenario 1: Algorithm description

Figure 11.10(a) presents the monthly wind generation and demand from El Hierro Island during one month (December 2018 which is the third month with the lowest wind production in the registered time series). The monthly wind-generated energy in this period is 1.3895 GWh and the current electric demand is 3.271 GWh; therefore, this wind-generated energy is insufficient to cover the current demand and it is necessary to provide energy from an alternative source. The diesel energy required to complement the demand is 2.07 GWh. From these values, it is observed that the sum of the wind-generated energy plus the diesel-generated energy is greater than the current demand. This is because wind generation often does not coincide with demand and there are some periods of low demand, with an excess of wind generation, when the energy cannot be stored and it is spilled. The monthly amount of spilled energy is 188.5793 MWh during this month, representing the 13.5% of the generated wind energy in the same period.

Figure 11.10(b) highlights the following situation: when wind generation is higher than the current demand (from December 23 to 25), the surplus energy is spilled, but when the current demand is higher than the current wind generation (during December 22, for example), the complementary energy is provided by the DFPP.

Figure 11.11 shows the CO_2 emissions in this scenario. These emissions will have two different components, the first is due to the periods of time which the DFPP is operating, burning fuel, and emitting CO_2. The second component is due to the emissions generated by passenger cars in their daily journeys. In this case, it is assumed that the vehicles are moving with the mobility profile presented in Figure 11.7, and they are emitting CO_2 in these periods. In order to simplify the study, it is assumed that the mobility pattern is repeated every day, even on weekends. The CO_2 emissions in this situation are 2143.9 tons of CO_2 during December 2018.

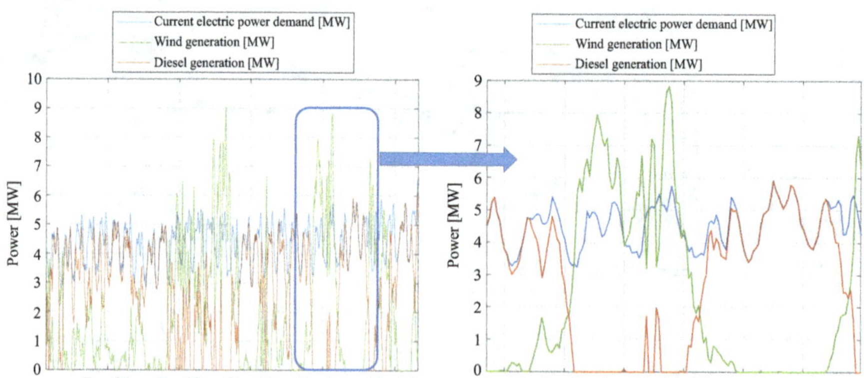

Figure 11.10 Scenario 1: Electric current demand, wind generation, and diesel generation in El Hierro Island (a) during December 2018 and (b) during December 22–27, 2018

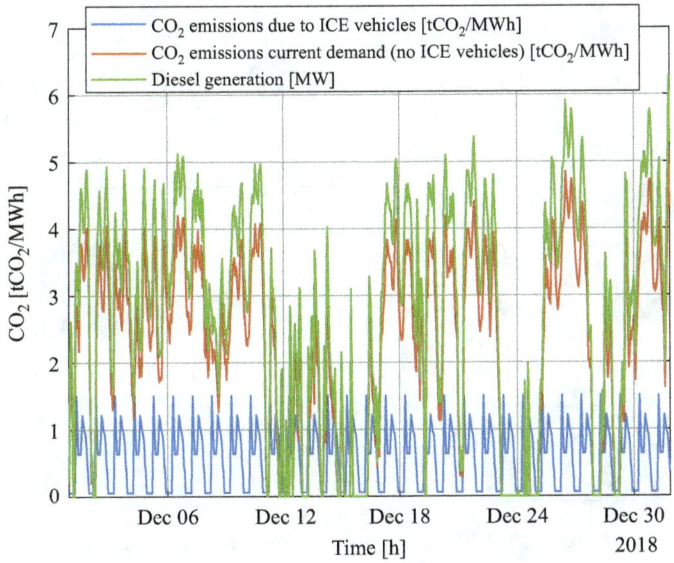

Figure 11.11 Scenario 1: Diesel emissions due to electric generation and ICE vehicles mobility in El Hierro Island during December 2018

11.5.2 Scenario 2. Base scenario+VEs-No PHEP

In this second scenario, all ICE passenger vehicles (4,848) in El Hierro Island are replaced by EVs. It is assumed that the mobility pattern does not change and all EVs are recharged during the night period, from 22:00 to 5:00, when the electric tariff 2.0DHS reaches the lowest values, using the optimization algorithm described in Section 4. The wind energy surplus can be used to charge EVs during this night period, reducing the diesel consumption and CO2 emissions. This algorithm is shown in Figure 11.12.

Figure 11.13 shows the monthly wind generation, the increased electric demand (the current electric demand from El Hierro Island and the new demand due to recharging EVs), and the required diesel generation during December 2018. The new electric demand (3.8984 GWh) will be the previous demand, 1.3895 GWh, plus the electric demand due to EVs (20.24 MWh per day, which represents 627.4 MWh per month). Under this assumption, the required diesel energy to complement this new demand is 2.6213 GWh, which represents an increment of 26.63% from Scenario 1. The wind energy spilled in this situation is 112.3331 MWh, representing a significant reduction of 40.43% from the previous scenario.

The total amount of CO_2 emitted is 2149.4 tons CO_2/month, which only represents an increment of 0.26% from the previous scenario.

Table 11.3 presents all these results, extending the previous analysis to 1 year (2018). It is observed that the total wind-generated energy is lower than the electric demand in both the scenarios, therefore, a complementary energy source will be always required. The transition from ICE vehicles to VE ones will increase the total

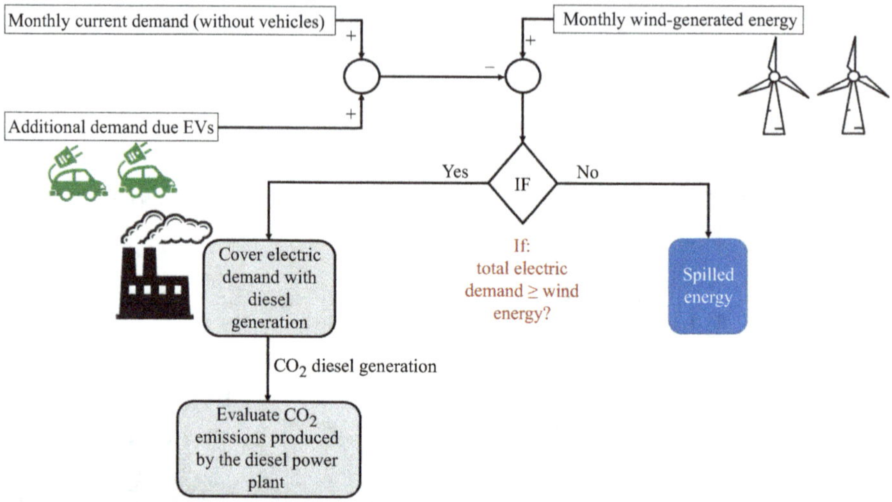

Figure 11.12 Scenario 2: Algorithm description

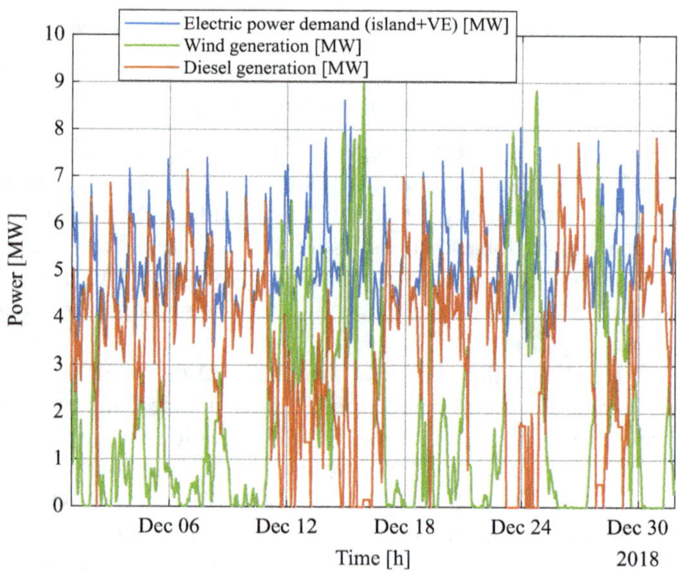

Figure 11.13 Scenario 2. Electric demand, wind generation, and diesel generation in El Hierro Island during December 2018

electric demand on the island, causing an increment of the diesel generation, but this solution optimizes the wind generation, reducing the amount of spilled energy from 9.51 GWh, which is 27.2% of the total energy generated by the wind, to 6.39 GWh (representing a reduction of almost 10%).

Table 11.3 Total energy demand, wind-generated energy, diesel-generated energy, and spilled energy under Scenarios 1 and 2

2018	**Energy demand [GWh]**			**Diesel energy [GWh]**		**Spilled energy [GWh]**	
Month	**Sc. 2 (VE)**	**Sc. 1**	**Wind Gen.**	**Sc. 2 (VE)**	**Sc. 1**	**Sc. 2 (VE)**	**Sc. 1**
Jan	4.034	3.406	3.510	1.313	1.042	0.789	1.1452
Feb	3.6343	3.0676	2.495	1.5924	1.2485	0.4531	0.676
Mar	3.8897	3.2622	2.556	2.0075	1.6248	0.6739	0.9187
Apr	3.7976	3.1904	3.3618	1.0549	0.7821	0.6191	0.9534
May	4.0878	3.4603	3.0048	1.4665	1.0639	0.3836	0.6084
Jun	4.2338	3.6266	3.2581	1.4743	1.1644	0.4986	0.796
Jul	4.4948	3.8674	5.4525	0.51	0.4066	1.4677	1.9917
Aug	4.6764	4.0489	4.1711	1.3207	1.091	0.8155	1.2132
Sep	4.458	3.8507	3.0798	1.7392	1.4213	0.361	0.6503
Oct	4.3563	3.7288	1.2808	3.159	2.6017	0.0835	0.1537
Nov	3.7535	3.1463	1.3748	2.5114	1.9869	0.1328	0.2155
Dec	3.8984	3.2709	1.3895	2.6213	2.0701	0.1123	0.188
Total	**49.314**	**41.926**	**34.934**	**20.770**	**16.503**	**6.390**	**9.510**

Table 11.4 Total CO_2 emissions under Scenarios 1 and 2

2018	**10^3 tons of CO_2 emitted**			
Month	**Sc. 2 (VE)**	**Sc. 1 (ICE)**	**Sc.1 (current demand. No ICE)**	**Variation**
Jan	1.077	1.300	0.854	−17.2%
Feb	1.3057	1.427	1.0238	−8.5%
Mar	1.6462	1.7788	1.3324	−7.5%
Apr	0.865	1.0733	0.6413	−19.4%
May	1.2026	1.3188	0.8724	−8.8%
Jun	1.2089	1.3869	0.9548	−12.8%
Jul	0.4182	0.7798	0.3334	−46.4%
Aug	1.083	1.3411	0.8947	−19.2%
Sep	1.4261	1.5975	1.1655	−10.7%
Oct	2.5903	2.5798	2.1334	0.4%
Nov	2.0594	2.0613	1.6293	−0.1%
Dec	2.1494	2.1439	1.6975	0.3%
Total	**17.031**	**18.789**	**13.533**	−9.4%

Table 11.4 shows the CO_2 emissions under Scenarios 1 and 2 for a whole year (2018). The second column presents the monthly CO_2 emissions produced by the diesel power plant to cover the increment in the electricity demand due to the EV deployment (Scenario 2). The fourth column represents the CO_2 emissions produced by the diesel power plant to cover the current demand (Scenario 1) and the

third column is the sum of the values of the fourth column plus the emissions due to the ICE vehicles. It is observed that, in general, promoting the transition from ICE vehicles to VEs reduces these polluting emissions in almost all months, between −0.1% (November) and −46.5% in July, the month with the highest wind generation. For only 2 months (October and December), there is an insignificant CO_2 emissions increase of less than 0.5%, between Scenarios 1 and 2.

11.5.3 Scenario 3. Base scenario+VEs-PHES

In this third scenario, the available PHES in El Hierro Island is used to store the excess of wind-generated energy by pumping the water from the lower basin to the upper one. This stored energy is converting back to electricity when it is demanded, as proposed in Figure 11.14.

According to the El Hierro PHES specifications, the lower reservoir can hold 150.000 m^3 with a gross head of 653 m; therefore the maximum energy storage capacity is 267 MWh. PHES are very efficient, with round-trip efficiencies greater to 80%–87% and this value will be considered in this work [32].

As it was considered in the previous scenario, all ICE passenger vehicles in El Hierro Island are replaced by EVs and these vehicles are recharged during the night period, from 22:00 to 5:00, when the electric tariff 2.0DHS is the lowest.

Three different subscenarios are considered depending on the initial storage capacity available in the upper reservoir.

11.5.3.1 Subscenario 3.1. Base scenario+VEs+PHES at 0%

In this first subscenario, it is assumed that the upper reservoir is completely empty at the beginning of the study (January 1, 2018). Figure 11.15 presents the monthly

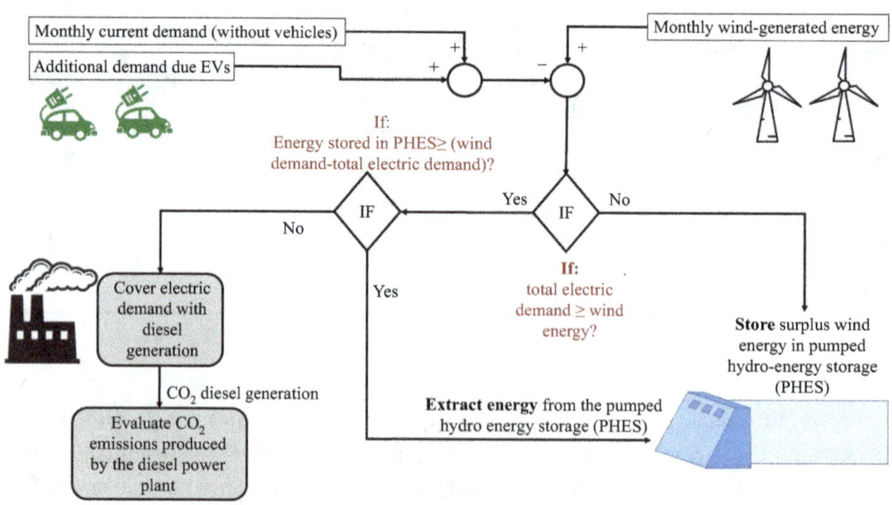

Figure 11.14 Scenario 3: Algorithm description

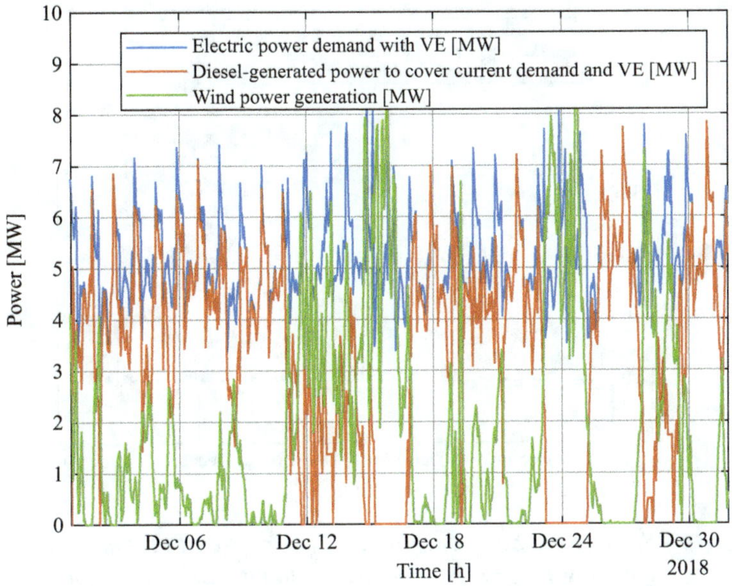

Figure 11.15 *Scenario 3. Subscenario 3.1. Electric demand, wind generation, and diesel generation in El Hierro Island during December 2018 (upper reservoir = 0%)*

wind generation and demand from El Hierro Island during the same month that has been previously analyzed in Scenarios 1 and 2 (December 2018).

The monthly wind-generated energy, current demand (only to cover the island demand, without vehicles) and current demand including the energy required to charge all EVs are the same as in previous scenarios (1.3895 GWh of wind-generated energy, 3.271 GWh of current demand, and 3.898 GWh of current demand plus energy for the EVs).

The diesel energy required to cover the current demand, taking into account that now it is possible to store the surplus of wind energy in PHES, is 1.927 GWh (compared to 2.07 GWh in Scenario 1, which represents a reduction of −6.9%).

The diesel energy required to cover the increment of demand due to the EV recharging process is 2.529 GWh (compared to 2.6213 GWh in Scenario 2, which represents a reduction of −3.5%).

In both the situations, the amount of wind-generated energy spilled is zero, which indicates that all wind energy has been used in the system, and therefore, optimizing the amount of energy generated by the diesel power plant.

Figure 11.16 shows the time evolution of the PHES during December 2018 to cover the current demand of 3.271 GWh, highlighting the moments when the energy is stored (in blue) or used (in red). It is observed that the upper reservoir is not completely full at any time and the final level of this reservoir at the end of this month is zero.

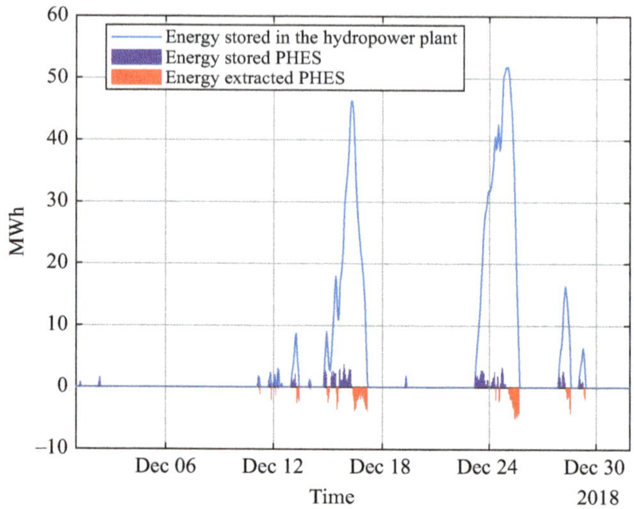

Figure 11.16 Scenario 3. Subscenario 3.1. Energy stored at the PHES when covering the current demand (without VEs) during December 2018 (upper reservoir = 0%)

Figure 11.17 presents the time evolution of the same PHES during the same period when covering the additional demand of 3.898 GWh (considering the current demand plus energy for the EVs). It is observed that, the energy stored in the upper basin in this situation is lower than the previous one. The reason is that a significant amount of wind energy is directly used to recharge EVs, avoiding storing the energy in the PHES.

11.5.3.2 Subscenario 3.2. Base scenario+VEs+PHES at 50%

In this second subscenario, it is assumed that the upper reservoir is of half-capacity at the beginning of the study (January 1, 2018). The diesel energy required to cover the current demand (without VEs) is lower, 1.8112 GWh, because the upper reservoir is used to cover the initial electricity demand, as it is shown in Figure 11.18. In the previous subscenario, it was necessary to cover this demand with diesel generation. The maximum energy stored at PHES is 51.9 MWh on December 25, 2018 at 01:00.

The diesel energy required to cover the increment of demand due to EV recharging process, is 2.4201 GWh. The time evolution of the PHES during the same period when covering the additional demand is shown in Figure 11.19. Again, it is observed that, the energy stored in the upper basin during this month never exceed 40 MWh (it reaches a maximum value of 28.16 MWh on December 23, 2018 at 21:00), due to the amount of wind energy which is directly used to recharge EVs, avoiding storing the energy in the PHES.

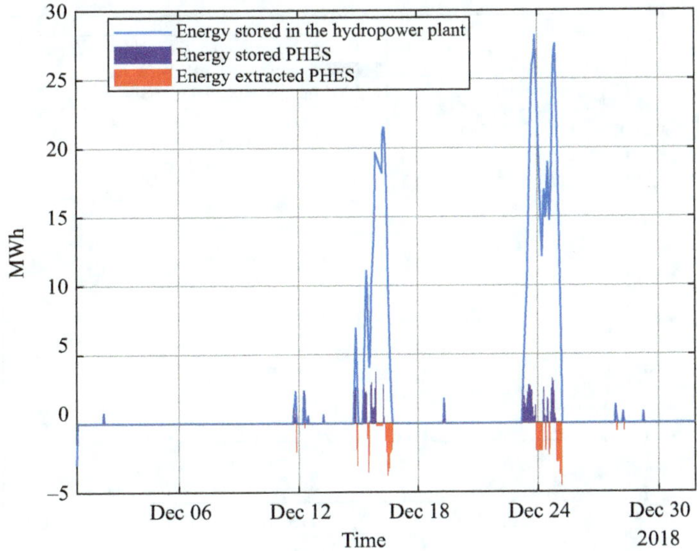

Figure 11.17 Scenario 3. Subscenario 3.1. Energy stored at the PHES when covering the additional demand (including VEs) during December 2018 (upper reservoir = 0%)

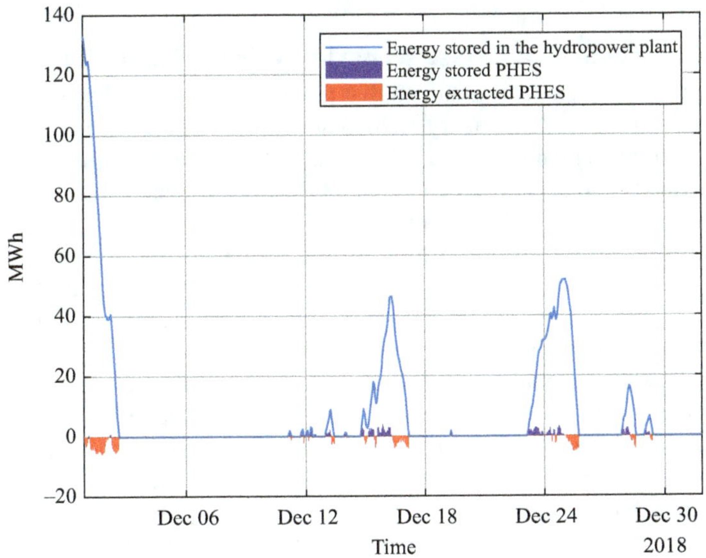

Figure 11.18 Scenario 3. Subscenario 3.2. Energy stored at the PHES when covering the current demand (without VEs) during December 2018 (upper reservoir = 50%)

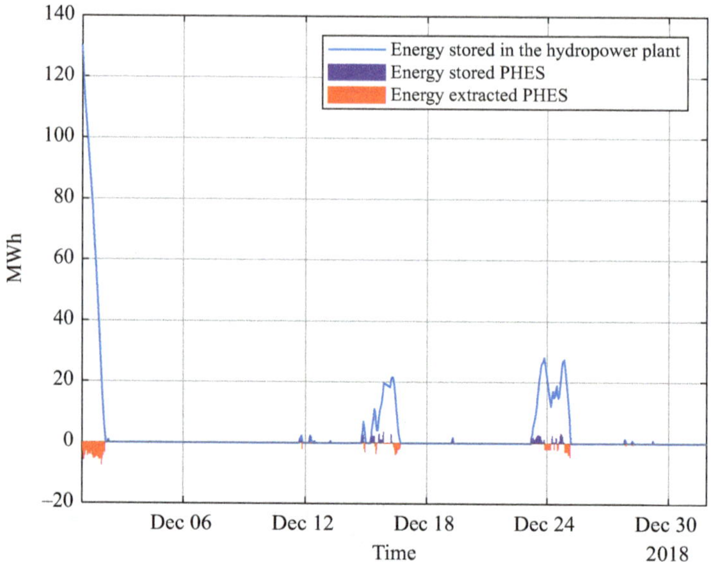

Figure 11.19 Scenario 3. Subscenario 3.2. Energy stored at the PHES when covering the additional demand (considering VEs) during December 2018 (upper reservoir = 50%)

11.5.3.3 Subscenario 3.3. Base scenario+VEs+PHES at 100%

In this third subscenario, it is assumed that the upper reservoir is completely full at the beginning of the study (January 1, 2018 at 00:00). The diesel energy required to cover the current demand (without VEs) reaches the lowest value (1.695 GWh). Again, the maximum energy stored at PHES is 52 MWh reached on December 25, 2018 at 01:00 as shown in Figure 11.20.

The diesel energy required to cover the increment of demand due to EV recharging process is 2.304 GWh. As it was shown in the previous scenario, the energy stored in the upper basin during this month reaches a maximum value of 28.16 MWh on December 23, 2018, at 21:00 (see Figure 11.21).

To analyze the reduction of the annual diesel consumption and CO_2 emissions during a whole year (2018), the Scenario 3 algorithm is applied for the first month (January 2018), with the previous initial conditions in the upper reservoir (0%, 50%, or 100%). After the execution of this algorithm, it is applied again to the rest of the months sequentially, initializing the level of the upper reservoir with the level obtained in the last hour of the last day from the previous month.

Figure 11.22 presents the time evolution of the energy stored at the PHES, covering the additional demand (including VEs) during January 2018, when there is no water stored in the upper reservoir. The energy stored at the end of this month (January 31, 2018, at 23:59) is 188.009 MWh.

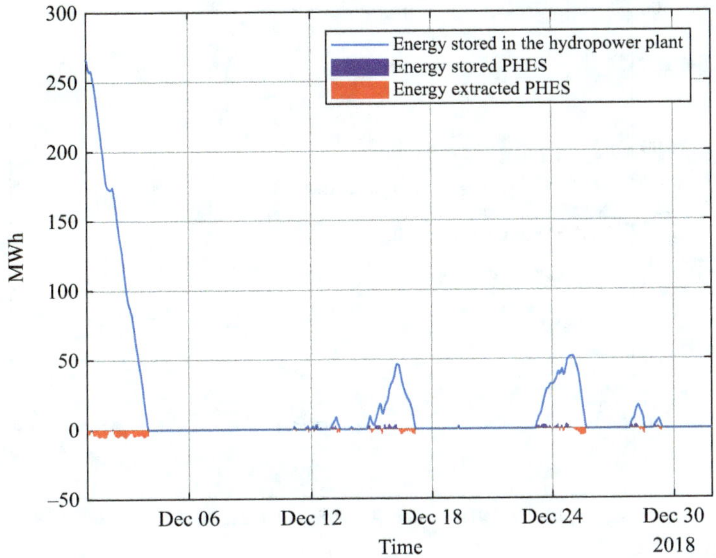

Figure 11.20 *Scenario 3. Subscenario 3.3. Energy stored at the PHES when covering the current demand (without VEs) during December 2018 (upper reservoir = 100%)*

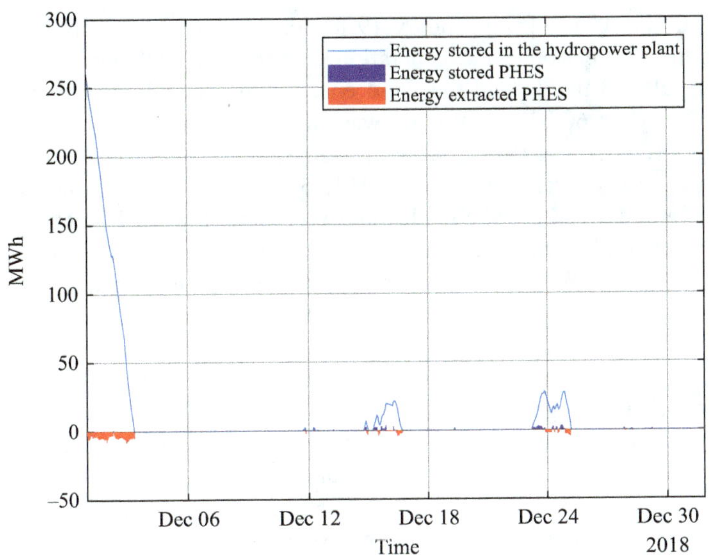

Figure 11.21 *Scenario 3. Subscenario 3.2. Energy stored at the PHES when covering the additional demand (considering VEs) during December 2018 (upper reservoir = 100%)*

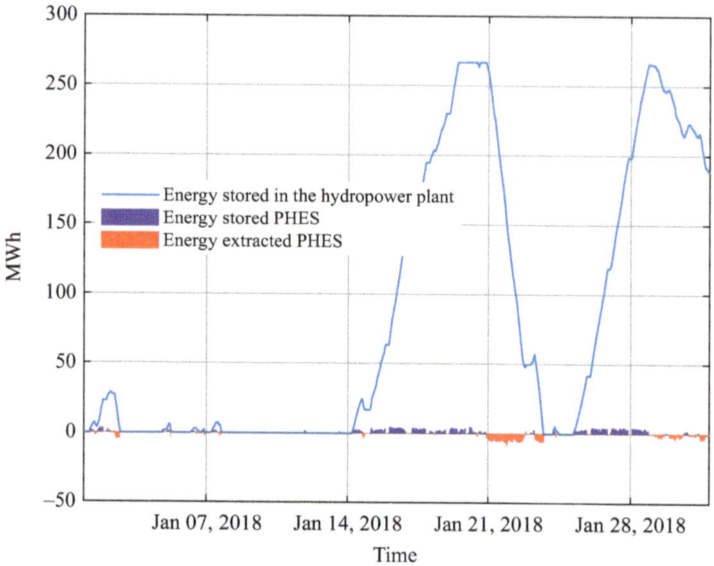

Figure 11.22 Scenario 3. Subscenario 3.1. Energy stored at the PHES when covering the additional demand (including VEs) during January 2018 (upper reservoir = 0%)

In is observed that, from January 19 to 20, the upper reservoir is completely full. If during this period, wind generation exceeds the total demand, there is no storage capacity available and the wind energy will be spilled. This is shown in Figure 11.23, when 60.3373 MWh are wasted.

Figure 11.24(a) and (b) presents the same analysis during Subscenarios 3.2 (initial upper reservoir capacity equal to 133.5 MWh) and 3.3 (initial upper reservoir capacity equal to 267 MWh). In both scenarios, the final level of the upper reservoir at the end of the month reaches the same value as in the previous subscenario, 188.009 MWh. Therefore, this value will be the initial upper reservoir capacity for February month analysis for the following three subscenarios.

In Subscenario 3.3, the upper basin is kept completely full during two time periods in this month (between January 1 and 2, and 19 and 20), and 88.3626 MWh of wind-generated energy is wasted (see Figure 11.25).

Figure 11.26 presents the energy stored at the PHES when covering the additional demand (including VEs) during February 2018 (Figure 11.26(a)), March (Figure 11.26(b)), April (Figure 11.26(c)), and May (Figure 11.26(d)).

The upper reservoir capacity at the end of February is 0 MWh, 76.0424 MWh at the end of March, 115.4252 MWh at the end of April and 49.1219 MWh at the end of May.

Figure 11.27 presents the energy stored at the PHES when covering the additional demand (including VEs) during June 2018 (Figure 11.27(a)), July (Figure 11.27(b)), August (Figure 11.27(c)), and September (Figure 11.27(d)).

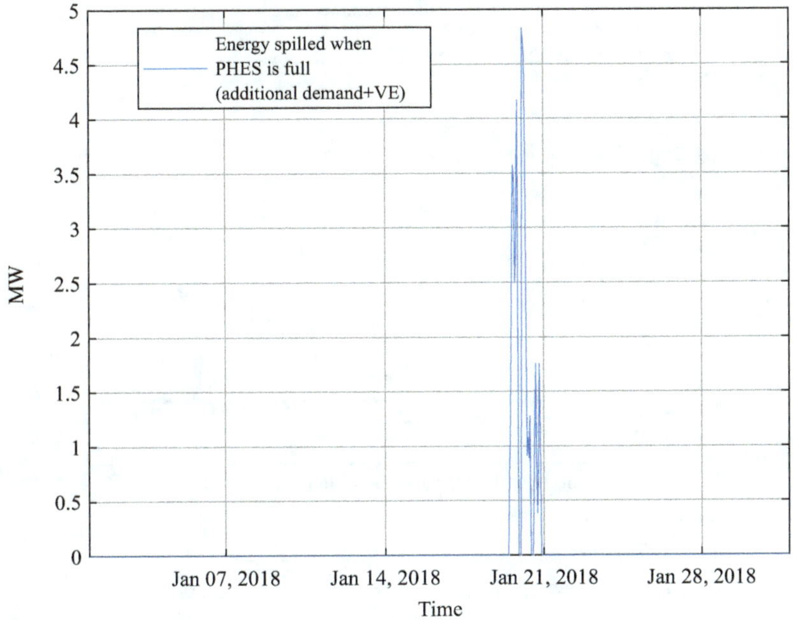

Figure 11.23 Scenario 3. Subscenario 3.1. Wind energy spilled during January 2018 (upper reservoir = 0%)

The upper reservoir capacity at the end of June is 0 MWh, 144.3453 MWh at the end of July, 212.3038 MWh at the end of August and 0 MWh at the end of September.

Finally, Figure 11.28 shows the energy stored at the PHES when covering the additional demand (including VEs) during October 2018 (Figure 11.28(a)) and November (Figure 11.28(b)). The energy stored at the PHES during December was previously presented in Figure 11.21.

The upper reservoir capacity at the end of October is 0 MWh and 0.4928 MWh at the end of November.

Table 11.5 presents the total diesel-generated energy and wind-generated spilled energy during the first subscenario (3.1), when the upper reservoir is empty. Comparing these values with those shown in Table 11.3, it is observed that the annual diesel-generated energy has reduced from 20.77 to 17.011 GWh (−18.1%), reducing also the CO_2 emissions (from 17.031 to 13.949 Mtons CO_2) and the amount of wind energy wasted, from 6.39 to 1.289 GWh, which represents a reduction of −79.8%.

Table 11.6 presents the total diesel-generated energy and wind-generated spilled energy during the first subscenario (3.2), when the upper reservoir is at 50% of its nominal capacity. Comparing these values with those shown in Table 11.3, it is observed that the annual diesel-generated energy has reduced from 20.77 MWh

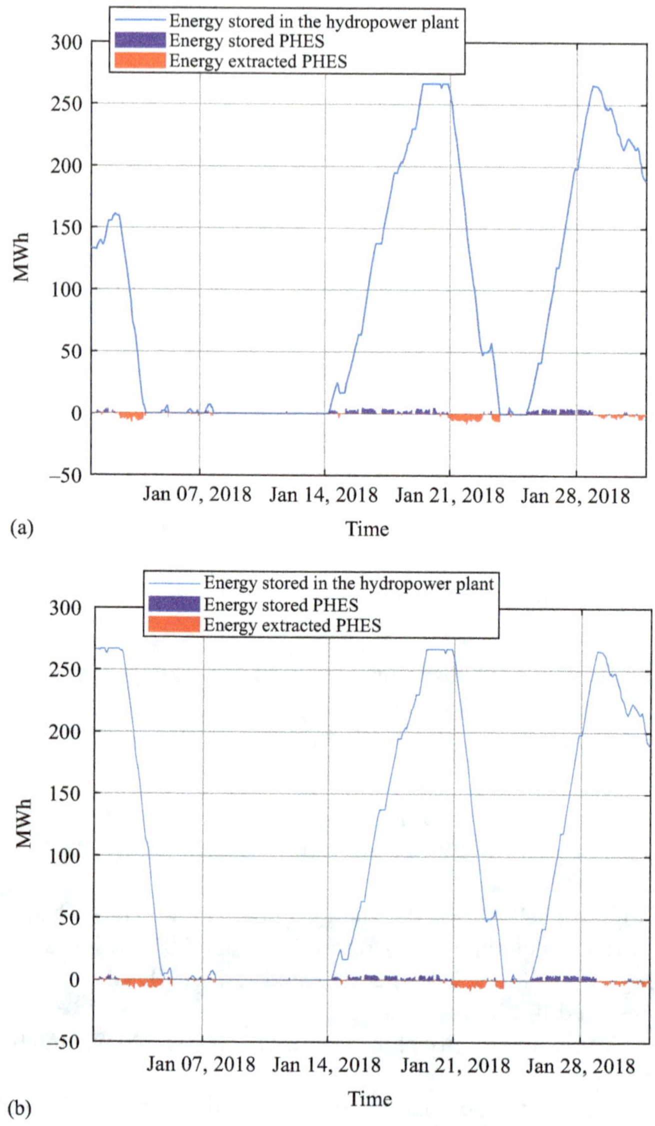

Figure 11.24 Scenario 3. (a) Subscenario 3.2. Energy stored at the PHES when covering the additional demand (including VEs) during January 2018. (upper reservoir = 50%). (b) Sub-scenario 3.3. Energy stored at the PHES when covering the additional demand (including VEs) during January 2018 (upper reservoir = 100%)

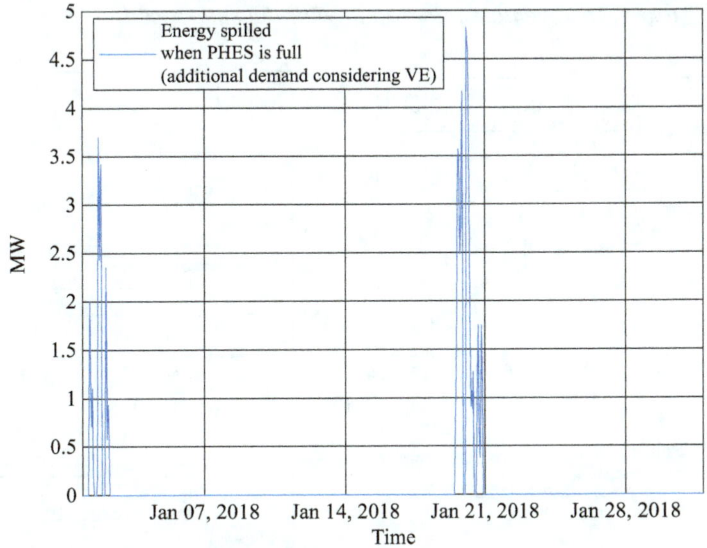

Figure 11.25 Scenario 3. Subscenario 3.3. Wind energy spilled during January 2018 (upper reservoir = 100%)

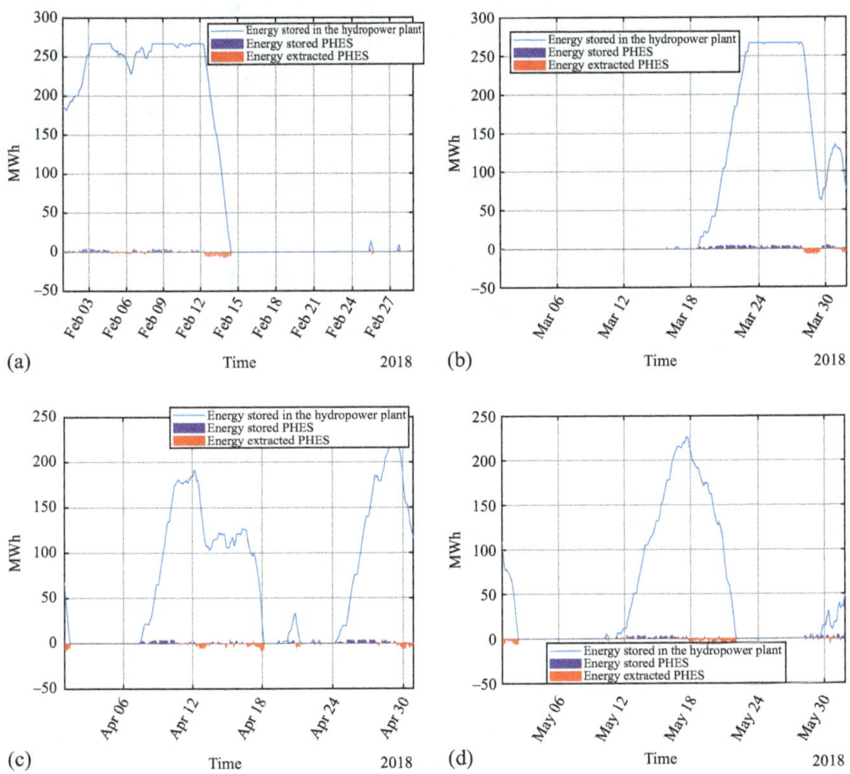

Figure 11.26 Scenario 3. Subscenario 3.3. Wind energy spilled during (a) February 2018, (b) March 2018, (c) April 2018, and (d) May 2018 (upper reservoir = 100%)

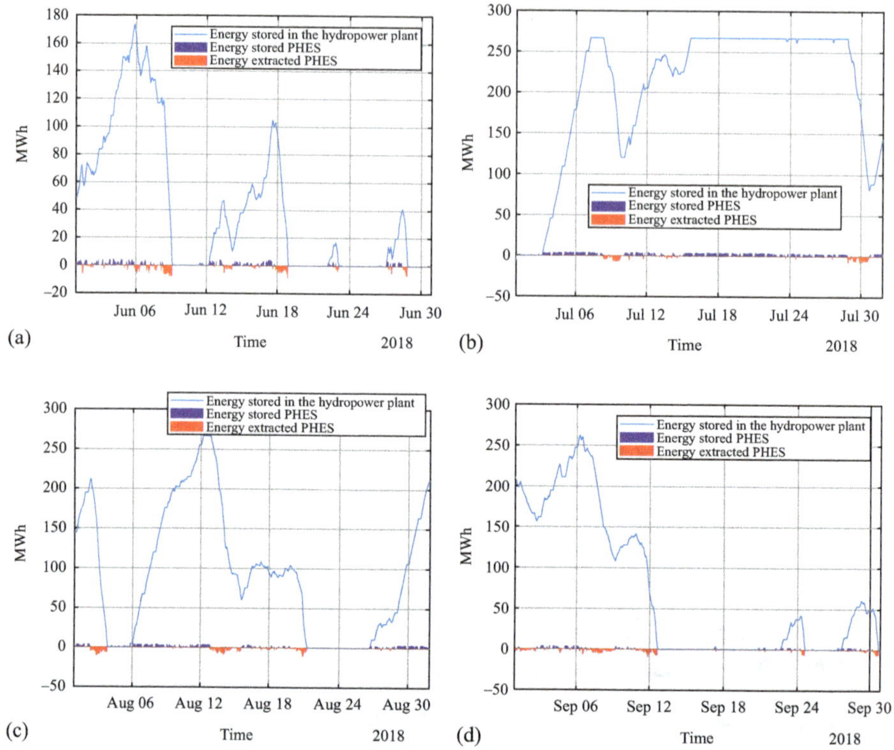

Figure 11.27 Scenario 3. Subscenario 3.3. Wind energy spilled during (a) June 2018, (b) July 2018, (c) August 2018, and (d) September 2018 (upper reservoir = 100%)

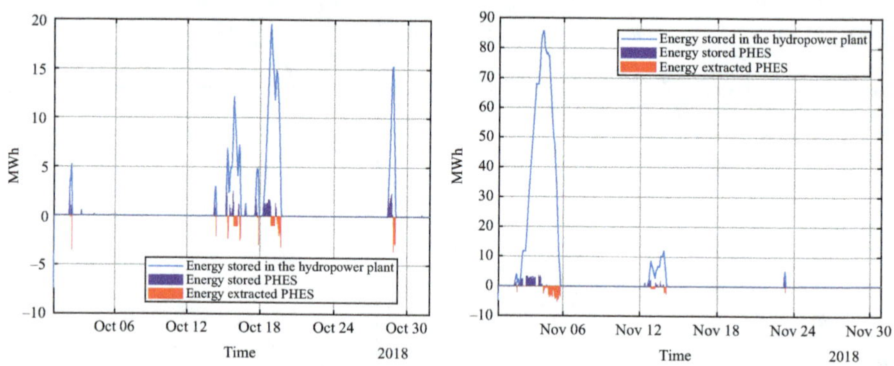

Figure 11.28 Scenario 3. Subscenario 3.3. Wind energy spilled during (a) October 2018 and (b) November 2018 (upper reservoir = 100%)

Table 11.5 Total energy demand, wind-generated energy, diesel-generated energy, and spilled energy under Scenario 3. Subscenario 3.1 (upper reservoir = 0%)

2018	Energy demand [GWh]			Diesel energy [GWh]	Spilled energy [GWh]
Month	**Sc. 2 (VE)**	**Sc. 1**	**Wind Gen.**	**Sc. 2 (VE)**	**Sc. 2 (VE)**
Jan	4.034	3.406	3.510	0.931	0.06
Feb	36.343	30.676	2.495	1.275	0.218
Mar	38.897	32.622	2.556	1.76	0.233
Apr	37.976	31.904	33.618	0.621	0
May	40.878	34.603	30.048	1.119	0
Jun	42.338	36.266	32.581	1.054	0
Jul	44.948	38.674	54.525	0.143	0.723
Aug	46.764	40.489	41.711	0.811	0.055
Sep	4.458	38.507	30.798	1.281	0
Oct	43.563	37.288	12.808	3.084	0
Nov	37.535	31.463	13.748	2.403	0
Dec	38.984	32.709	13.895	2.529	0
Total	**49.314**	**41.926**	**34.934**	**17.011**	**1.289**

Table 11.6 Total energy demand, wind-generated energy, diesel-generated energy, and spilled energy under Scenario 3. Subscenario 3.1 (upper reservoir = 50%)

2018	Energy Demand [GWh]			Diesel energy [GWh]	Spilled energy [GWh]
Month	**Sc. 2 (VE)**	**Sc. 1**	**Wind gen.**	**Sc. 2 (VE)**	**Sc. 2 (VE)**
Jan	4.034	3.406	3.510	0.815	0.06
Feb	36.343	30.676	2.495	1.275	0.218
Mar	38.897	32.622	2.556	1.76	0.233
Apr	37.976	31.904	33.618	0.621	0
May	40.878	34.603	30.048	1.119	0
Jun	42.338	36.266	32.581	1.054	0
Jul	44.948	38.674	54.525	0.143	0.723
Aug	46.764	40.489	41.711	0.811	0.055
Sep	4.458	38.507	30.798	1.281	0
Oct	43.563	37.288	12.808	3.084	0
Nov	37.535	31.463	13.748	2.403	0
Dec	38.984	32.709	13.895	2.529	0
Total	**49.314**	**41.926**	**34.934**	**16.895**	**1.289**

to 16,895 MWh (−18.7%), reaching 13.854 Mtons of CO_2 emissions and the same reduction in the spilled energy as the previous scenario (1.289 GWh).

Finally, Table 11.7 presents the total diesel-generated energy and wind-generated spilled energy during the first subscenario (3.3), when the upper reservoir is completely full. Comparing these values with those shown in Table 11.3, it is

Table 11.7 Total energy demand, wind-generated energy, diesel-generated energy, and spilled energy under Scenario 3. Subscenario 3.1 (upper reservoir = 100%)

2018	**Energy demand [GWh]**			**Diesel energy [GWh]**	**Spilled energy [GWh]**
Month	**Sc. 2 (VE)**	**Sc. 1**	**Wind gen.**	**Sc. 2 (VE)**	**Sc. 2 (VE)**
Jan	4.034	3.406	3.510	0.593	0.088
Feb	36.343	30.676	2.495	1.275	0.218
Mar	38.897	32.622	2.556	1.76	0.233
Apr	37.976	31.904	33.618	0.621	0
May	40.878	34.603	30.048	1.119	0
Jun	42.338	36.266	32.581	1.054	0
Jul	44.948	38.674	54.525	0.143	0.723
Aug	46.764	40.489	41.711	0.811	0.055
Sep	4.458	38.507	30.798	1.281	0
Oct	43.563	37.288	12.808	3.084	0
Nov	37.535	31.463	13.748	2.403	0
Dec	38.984	32.709	13.895	2.529	0
TOTAL	**49.314**	**41.926**	**34.934**	**16.673**	**1.317**

observed that the annual diesel-generated energy has reduced from 20.77 MWh to 16,673 MWh (−19.7%), reaching 13.779 Mtons of CO_2 emissions. The wind spilled energy during the year has slightly increased (1.317 GWh), because during the first days of January, there is an excess of wind generation and the upper basin is completely full, so it is not possible to store the surplus energy and it is wasted.

11.6 Conclusion

In this work the option of charging an EV fleet as a distributed energy storage system to increase the participation of renewable energy in an isolated power system has been presented. A comparative study has been performed quantifying the amount of diesel generation needed and CO_2 emission emitted, after the integration of EVs.

It has been proven how the electrification of the transportation sector will increase the utilization factor of the wind power plant, from 72.77% to 81.71%. Additionally, diesel generation and CO2 are significantly reduced.

If the isolated electric system has an additional storage system such as PHES, it is possible to increase the utilization factor significantly, reaching 96.31% in this study.

The energy storage capacity determines the amount of wind energy wasted in the process. In this work, the limitation factor is the small capacity of the lower reservoir of the PHES.

In the future, the use of other storage system previously used in isolated power system such as Li-ion battery packs or the use of V2G schemes will be considered.

References

[1] IPCC, "Global Warming of 1.5 °C. An IPCC special report on the impacts of global warming of 1.5 °C above pre-industrial levels and related global greenhouse gas emission pathways, in the context of strengthening the global response to the threat of climate change, sustainable development, and efforts to eradicate poverty," October 2018. http://ipcc.ch/report/sr15/

[2] EPA, "Global Greenhouse Gas Emissions Data," October 2018. https://www.epa.gov/ghgemissions/global-greenhouse-gas-emissions-data#Sector

[3] American Physical Society (APS), "Integrating Renewable Electricity on the Grid," https://www.aps.org/policy/reports/popa-reports/upload/integratingelec.pdf Accessed 23 August 2019.

[4] D. Eltigani, and S. Masri. "Challenges of integrating renewable energy sources to smart grids: A review," *Renewable and Sustainable Energy Reviews*, vol. 52, no. C, pp. 770–780, 2015.

[5] H. Vasconcelos, C. Moreira, A. Madureira, J. P. Lopes and V. Miranda, "Advanced Control Solutions for Operating Isolated Power Systems: Examining the Portuguese islands," *IEEE Electrification Magazine*, vol. 3, no. 1, pp. 25–35, 2015.

[6] O. Erdinc, N.G. Paterakis, and J. Catalão, "Overview of insular power systems under increasing penetration of renewable energy sources: opportunities and challenges," *Renew Sustain Energy Rev*, vol. 52, pp. 333–346, 2015.

[7] Y. Kuang, Y. Zhang, B. Zhou, C. Li, Y. Cao, L. Li, *et al.* "A review of renewable energy utilization in islands," *Renew Sustain Energy Rev*, vol. 59, pp. 504–513, 2016.

[8] International Renewable Energy Agency (IRENA), "Electricity Storage and Renewables for Island Power. A Guide for Decision Makers," May 2012. https://www.irena.org/-/media/Files/IRENA/Agency/Publication/2012/Electricity-Storage-and-RE-for-Island-Power.pdf

[9] Yueyu Tong, Ji Liang, Hua Kun Liu, Shi Xue Dou, "Energy storage in Oceania," *Energy Storage Materials*, 2019, ISSN 2405-8297, https://doi.org/10.1016/j.ensm.2019.04.031

[10] A. Peters. "The Island of Samoa Is Going 100% Renewable - with Tesla's Help," *Fast Company*, 2018, https://www.fastcompany.com/90209339/the-island-of-samoa-is-going-100-renewable-with-teslas-help, Accessed Sep 2018.

[11] L. Zhang, F. Jabbari, T. Brown, and S. Samuelsen, "Coordinating plug-in electric vehicle charging with electric grid: valley filling and target load following," *J. Power Sources*, vol. 267, pp. 584–597, 2014.

[12] K. Zhan, Z. Hu, Y. Song, N. Lu, Z. Xu, and L. Jia, "A probability transition matrix based decentralized electric vehicle charging method for load valley filling," *Electr Power Syst Res*, vol. 12, pp. 1–7, 2015.

[13] Camus, C., and Farias, T. "The electric vehicles as a mean to reduce co_2 emissions and energy cost in isolated regions. The Sao Miguel (Azores) case study," *Energy Policy*, vol. 43, pp. 153–165, 2012.

[14] W. Kempton, and S. E. Letendre, "Electric vehicles as a new power source for electric utilities," *Transportation Research Part D: Transport and Environment*, vol. 2, no. 3, pp. 157–175, 1997.

[15] Francisco Javier Garcia Latorre, Jose Juan Quintana, and Ignacio de la Nuez, "Technical and economic evaluation of the integration of a wind-hydro system in El Hierro island," *Renewable Energy*, vol. 134, 2019, pp. 186–193, ISSN 0960-1481, https://doi.org/10.1016/j.renene.2018.11.047.

[16] EnerconGmbJ. "ENERCON product overview," Enercom GmbH, Aurich, Germany, 2016.

[17] Red Eléctrica de España (REE), "For 18 consecutive days, the island of El Hierro covers 100% of its electricity demand with renewable energy," Press release, February 2018. https://www.ree.es/en/press-office/press-release/2018/02/18-consecutive-days-island-el-hierro-covers-100-its-electricity-demand-with-renewable-energy

[18] Gobierno de Canarias. "Anuario Energético de Canarias 2016." http://www.gobiernodecanarias.org/ceic/energia/doc/Publicaciones/AnuarioEnergeticoCanarias/ANUARIO-ENERGETICO-CANARIAS-2016.pdf Accessed March 2019.

[19] Red Eléctrica de España, "El Hierro electricdemand," https://demanda.ree.es/visiona/canarias/el_hierro/total Accessed May 2019.

[20] Endesa. "Declaración ambiental. CD Llanos Blancos," https://www.endesa.com/content/dam/enel-es/home/sostenibilidad/medioambiente/gestionambiental/documentos/2017/declaracion-medioambiental-llanos-blancos.pdf Accessed May 2019.

[21] Endesa. "Electricity Access Rates," https://www.endesaclientes.com/electricity-access-rates Accessed May 2019.

[22] Renault. Renault Zoe Spec. https://www.renault.co.uk/vehicles/new-vehicles/zoe.html Accessed May 2019.

[23] Nissan. Nissan Leaf Spec. https://www.nissan.co.uk/vehicles/new-vehicles/leaf.html Accessed May 2019.

[24] Volkswagen. Volkswagen Golf Spec. https://www.volkswagen.co.uk/new/golf/explore/e-golf Accessed May 2019.

[25] Hyundai. Hyundai Ioniq Spec. https://www.hyundai.co.uk/new-cars/ioniq/electric Accessed May 2019.

[26] Sprinmonitor. https://www.spritmonitor.de/en/ Accessed 23 August 2019.

[27] ISTAC. Instituto Canario de Estadística. Estadística del Parque de Vehículos. 2017. http://www.gobiernodecanarias.org/istac/jaxi-istac/tabla.do?uripx=urn:uuid:31730d2f-86a8-4f0b-a706-5e5942702b7b&uripub=urn:uuid:172cc83a-4789-4f72-bf57-a4d0147c0656 Accessed May 2019.

[28] Antonio Colmenar-Santos, Ana-Rosa Linares-Mena, David Borge-Diez, Carlos-Domingo Quinto-Alemany, "Impact assessment of electric vehicles on islands grids: A case study for Tenerife (Spain)," *Energy*, vol. 120, pp. 385–396, 2017.

[29] A. Colmenar-Santos, Carlos de Palacio-Rodriguez, Enrique Rosales-Asensio, and David Borge-Diez, "Estimating the benefits of vehicle-to-home in islands: The case of the Canary Islands," *Energy*, vol. 134, pp. 311–322, 2017.

[30] Jesús Fraile Ardanuy, "Analysis of the impact of charging of Plug-in and EV in Spain," International Conference on Renewable Energies and Power Quality, ICREPQ'12, Santiago de Compostela (Spain), March 28–30, 2012.

[31] European Environment Agency. "Average CO2 emissions from new passenger cars," https://www.eea.europa.eu/data-and-maps/daviz/average-emissions-for-new-cars-4#tab-chart_1 Accessed May 2019.

[32] Ioannis Kougias, and Sándor Szabó, "Pumped hydroelectric storage utilization assessment: Forerunner of renewable energy integration or Trojan horse?" *Energy*, vol. 140, part 1, pp. 318–329, 2017.

Chapter 12

A solar- and wind-powered charging station for electric buses based on a backup batteries concept

Jakub Jurasz[1,2] *and Bartłomiej Ciapała*[3]

This chapter focuses on a solar- and wind-powered charging station for electric buses, which is equipped with a set of backup batteries that serve as energy storage. Additional batteries are used to compensate for the variable and stochastic energy generation of the climate-driven solar and wind sources. The research presented here focuses on analysing the charging station's reliability in terms of covering observed load, its economic performance, impact on the power system and the potential to reduce greenhouse gas emissions (in CO_2 equivalent). The results show that such a station can not only minimise electric buses' impact on the environment and ensure satisfactory reliability levels but also be a profitable investment.

12.1 Introduction

CAPEX	capital expenditures (investment costs)
EB(s)	electric bus(es)
EV(s)	electric vehicle(s)
NPV	net present value
O&M	operation and maintenance
PV	photovoltaics
SOC	state of charge
V2G	vehicle to grid
(v)RES	(variable) renewable energy sources
WT	wind turbine

[1]Department of Engineering Management, AGH University of Science and Technology, Cracow, Poland
[2]School of Business, Society and Engineering, Mälardalen University, Västerås, Sweden
[3]Department of Fossil Fuels, AGH University of Science and Technology, Cracow, Poland

The first electric vehicles (EVs) came into existence in the mid-nineteenth century, but were soon replaced by more cost-efficient and effective ones propelled by internal combustion engines. However, technological progress, as well as environmental concerns, led to EVs' resurgence at the beginning of the twenty-first century. Although EVs are perceived as a viable option to reduce human impact on the natural environment and to limit the amount of fumes inhaled by citizens, they also create a multitude of threats (and, simultaneously, opportunities) to modern power systems, which have been deeply penetrated by variable renewable energy sources (vRES). Obviously, in order to minimise the impact of transportation on the environment, the electrical energy used in EVs has to come from environmentally friendly sources such as wind turbines (WTs) or photovoltaics (PV). The research presented here is motivated by a recent declaration by the Polish government to support the development of electro mobility and achieve a goal of 1 million EVs (about 5% of the current fleet) on Polish roads by the end of 2025. However, achieving this ambitious goal will not significantly reduce transport-related CO_2 emissions, as the EV batteries will be charged from a power system dominated by hard and brown coal. So, in general, minimising emissions from burning petrol in cars will result in increasing emissions from the power system. Moreover, if charging were synchronised and simultaneous (on purpose or because of commonality in the population's daily schedules), it may lead to instability in the grid.

Along with the government's efforts to popularise EV, local authorities are investing significant amounts of money in order to replace the oldest municipal buses with electric buses (EBs). This activity, which aims to limit fumes and particulate matter emissions, especially in city centres, also causes a problem by creating a high and variable energy demand. Although the impact of a single EB, or several EBs, is negligible, an entire fleet may lead to power supply problems, especially if charging stations were spread across non-industrial parts of the city or placed in rural areas, where the electricity grid is not very resistant to high variations in demand.

Considering the above, the objective of this chapter is 2-fold: first, we aim to present the concept of a solar- and wind-powered charging station for EBs, which uses supplementary backup batteries to overcome the problem of the intermittent nature of vRES. Second, we analyse the financial, environmental and grid-related aspects of such a solution. To obtain the answers to the questions asked, we have applied mathematical modelling and available meteorological and technical data.

12.1.1 Related works

There is a growing body of literature on the problem of vRES integration into the power grid [1]. This issue has to be addressed, because intermittent energy sources introduce into the power system operation a stochastic element on the supply side. Therefore, the need to simultaneously bear in mind the multidimensionality of the decision process makes its management a more complicated task.

Among the several approaches [2] proposed to facilitate vRES integration (including, interconnecting geographically distributed variable sources to smooth their energy supply by benefiting from the temporal and spatial complementarity of resources, using demand side management, storing energy, oversizing renewable generation, and forecasting vRES energy generation), the use of EV batteries in the 'vehicle to grid' (V2G) concept looks like an especially promising option. On the one hand, switching to electric-powered transportation will exorbitantly increase the world demand of electrical energy, while on the other hand, it provides some flexibility to the owners of the vRES generators, as they have access to a form of mobile battery storage and can provide various services to the energy sector, such as peak shaving.

From the perspective of this chapter, of particular interest are papers that dealt with the problem of managing the process of charging single – or whole fleets of – EVs. Álvaro and Fraile-Ardanuy [3] proposed different charging strategies for managing an EV fleet. The results showed that their algorithms can be used to ensure the optimal deployment of a charging station. In the follow-up paper, Alvaro-Hermana *et al.* [4] suggested that it is possible to reduce the impact of EVs during peak periods by not only taking energy from the grid but trading it with other EVs parked in the same place. The algorithm ensures that no discontinuities will occur in the charging process and the personal constraints of the EVs owners will be maintained.

Another paper that investigated the impact of charging EVs on the local distribution grid [5,6] proposed coordinated charging of EVs to minimise distribution losses. The problem of an optimal scheduling for charging and discharging of EVs has been analysed by He *et al.* [7]. Mehta *et al.* [8] proposed smart charging strategies for optimal integration of plug-in EVs within the existing distribution system infrastructure.

Naturally, the possibility of using renewables in the charging process has also been addressed in the literature. Fuentes *et al.* [9] analysed the model of a solar-powered vehicle-sharing system based on building integrated PVs, which seems to be especially interesting for urban areas where the development of wind generation is strongly limited. Fathabadi [10] proposed a novel grid-connected PV-powered charging station. The author focused on the process of effectively converting solar energy into electricity based on a precise maximal-power-point tracking technique. Fathabadi developed his model [11] and equipped the charging station with a wind generator. According to the author, this was the first study to ever consider such a hybrid as a viable power source for EV charging stations. The concept of Fathabadi [10] utilising solar and wind energy in plug-in EVs was proposed. What seems as a follow-up of recent studies [11,12], the authors equipped the EV charging station with a battery and fuel cell in order to increase the power station reliability. Kaldellis *et al.* [13] analysed the operation of a stand-alone solar EV charging station. According to the authors, the proposed station is one of the most environmentally friendly solutions capable of decarbonising the European transport system.

All the above referenced works constitute a significant methodological basis for this chapter.

12.2 Methods and data

12.2.1 Methods and problem formulation

The schematic overview of the proposed EB charging station is presented in Figure 12.1. As can be seen, it consists of two variable energy sources in the form of PV and wind generation. The charging station has some storage potential in the form of backup batteries, which can be used when the demand does not match the supply from solar–wind sources. Apart from this, it is connected to the local energy network, which is used both as a source of energy when solar–wind generation and storage is not sufficient to cover the demand, and as a surplus energy acceptor when energy surpluses from solar–wind generation occur.

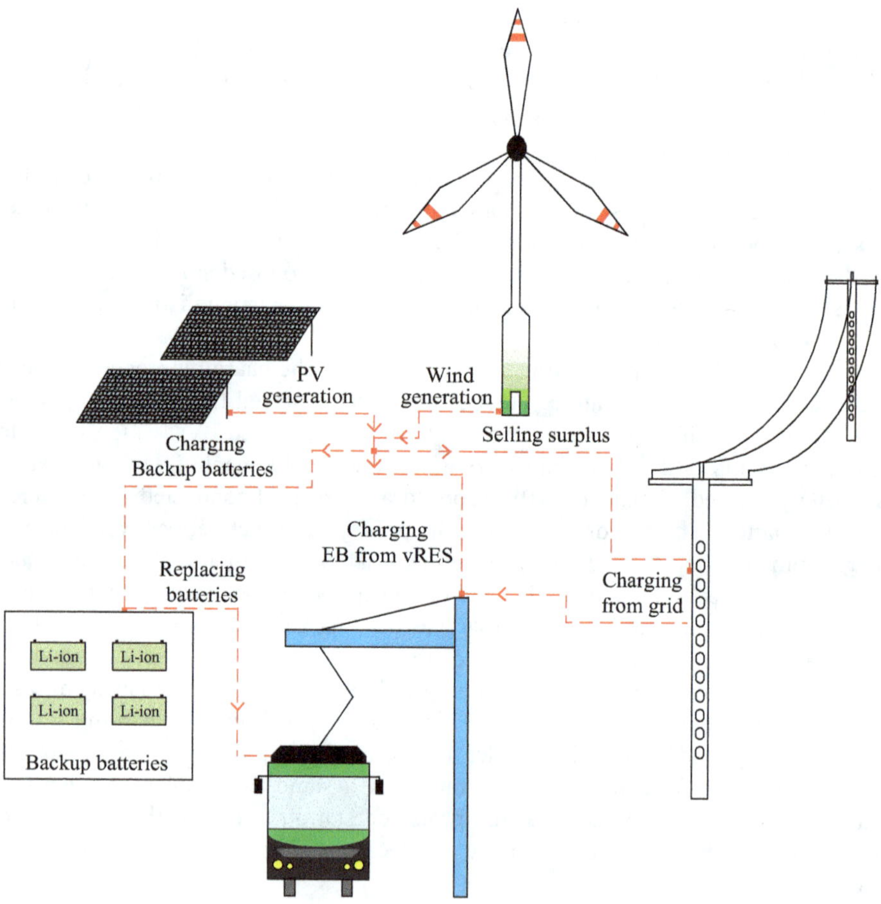

Figure 12.1 Schematic overview of electric-bus charging station, with energy-flow directions

There are several parameters that describe the time during which the bus is at the charging station. The first is 'duty time' – how much time may be spent charging the bus to a given level. This given level is the minimum accumulated energy sufficient for performing tasks. 'Reliability' describes the probability of successfully charging the batteries to the required level in time. The longer the procedure duration, the lower the level of battery charge, and a lower acceptable level of reliability facilitates the use of vRES, as it makes the unstable energy flux inherent in vRES more tolerable.

In the example, 100% reliability (excluding power grid malfunctions), a 45-min duty time and a 100% charge level have been applied. The charging procedure study is depicted in Figure 12.2. Naturally, it can be adapted to different timescales and underlying assumptions, that is, an additional diesel generator may be used to increase reliability. The charging procedure in Figure 12.2 is presented in a simplified version, which is of course subject to the constraints of the charging station capacity.

In general, the first 30 min of the 45-min charging process are based entirely on the vRES sources, whereas the last 15 min can be covered by the remaining load from the vRES, or by using the available backup batteries, or from the grid.

One may rightly ask, on what basis will the decision to use grid power be made? Please bear in mind that although this is a deterministic model and full data is provided to the model on a small temporal scale (15 min), the difference between a deterministic and probabilistic model may be very small (as observed for short-term forecasts of PV energy generation) [14]. The input data may be provided by an individual meteorological station or by a professional governing body of private meteorological agencies.

Several methods of PV yield assessment are available, from analytical recalculations to artificial neural networks. In our analysis we have used formulas used *inter alia* by François *et al.* [15] to calculate the energy yield from PVs. To simulate the energy generation from wind generation, we have applied the power curve (as in Table 12.1) representative for a Vestas V39 500-kW WT. The simulation of the charging process of batteries has been simulated by Campana *et al.* [16]. The remaining constraints and relations have been described by means of mixed integer mathematical programming.

12.2.2 Input data

It is important to note the general purpose of coupling energy sources. Hybridised energy sources are used with the aim of providing a power curve to fit a given energy demand curve. When two vRESs placed in the same place produce energy in time periods that do not overlap, we call this 'time complementarity'. When two installations (usually of the same type) placed in two different areas produce energy in different time periods, we call this 'spatial complementarity'. For example, PV installations in Poland and in Hawaii are perfectly spatially complementary. On the other hand, the temporal complementarity between wind and solar power is also clearly visible (mainly at the monthly scale) and has already been investigated

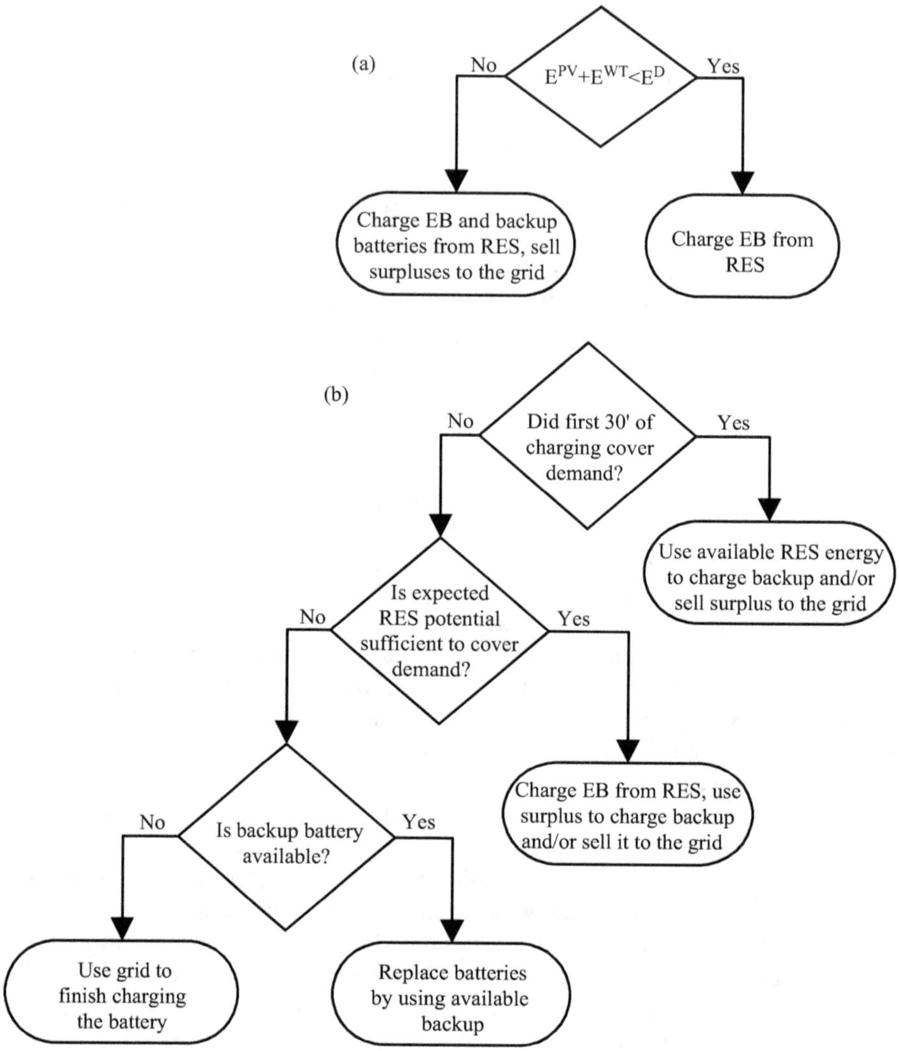

Figure 12.2 Charging station operation strategy: (a) first 30 min of charging period and (b) last 15 min

in-depth [17,18]. Of course, a perfect match is impossible and, thus, adding some sort of buffer significantly improves the fit between curves. It may be noted that wind and solar power production takes place at different times over the year, which means these sources have a high time complementarity.

When the production is defined and is relatively predictable, there is an energy demand to be described. One of the important factors that must be considered while designing a sustained EB network is the variability in energy consumption when the heating and cooling of bus interiors is required. Some manufacturers partly

Table 12.1 Values of parameters used in the analysis

Parameter	Value
PV CAPEX [20]	885.7 [Euro/kW]
WT CAPEX [20]	1,364 [Euro/kW]
Li-ion CAPEX [20]	250 [Euro/kWh]
Li-ion CAPEX [20]	261 [Euro/kW]
WT emissions [20]	10 [kg CO_2/MWh]
PV emissions [20]	45 [kg CO_2/MWh]
Grid emissions	717 [kg CO_2/MWh]
Green certificates	20 [Euro/MWh]
PV O&M [20]	2.5% of CAPEX
WT O&M [20]	2.7% of CAPEX
Li-ion O&M [20]	1.4 % of CAPEX
Li-ion replacement	Every 10 [years]
Grid cost change	3 [%/year] increase
Carbon trading change	1 [%/year] increase
Emissions reduction [21]	0.8 [%/year]
Grid electricity cost [21]	1.5 [%/year] increase
Li-ion emissions [22]	175 [kg CO_2/kWh]
Li-ion capacity	100 [kWh]
Discount rate [23]	5 [%]
Carbon trading	20 [Euro/tonne CO_2]
WT power curve. Equation for wind speeds ranging from 3 to 13 [m/s]	$y = -0.0041x^5 + 0.0091x^4 + 1.7962x^3 - 18.43x^2 + 80.85x - 121.07$

mitigate this problem with petroleum-fuelled space heaters. Avoiding significant energy expenditures by weak cooling and heating may also be a solution. Otherwise, the depletion time of a fully charged battery may vary over the year. The effective range may vary not only due to temperature, but also because of the number of passengers and traffic congestion. On the other hand, the technical specifications of the bus and the battery itself are very important. Thus, specific equations should be addressed individually to each investment.

In the example, for simplicity, the load has been assumed to be constant over the whole year. The first bus arrives at the charging station at 5.30 in the morning and it is expected to be fully charged within 45 min. The next bus arrives exactly 1 h later and the last one is at the station at 23.30 in the evening. In total there are 19 charging events per day. We did not differentiate between working days or holidays.

For the purpose of our analysis we have assumed that each charging event will require the station to provide 100 kWh of electric energy in the 45 min after the bus arrives. The need to provide 100 kWh of electricity results from the considered

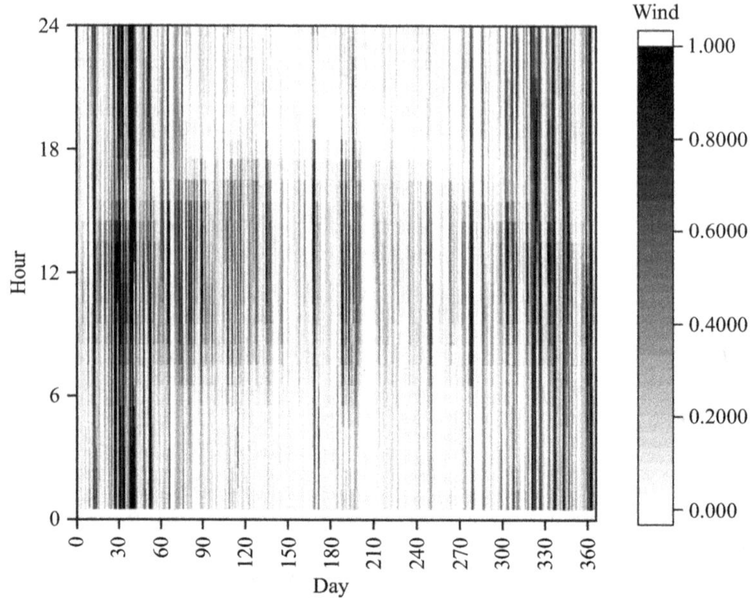

Figure 12.3 Annual spectrum of wind generation. Each hour of the year is represented by a rectangle. The darker it is, the more the energy produced (normalised)

energy consumption in EB rounded to 2 kWh/km (literature provides 1.91 kWh/km [19]) and an assumed bus journey length from point A to B and back to A of 50 km.

Figures 12.3 and 12.4 visualise the hourly variability of solar and wind generation from a WT and PV installation. Data has been gathered on a 15' time step covering the whole year for 2016 and representative for southern Poland.

Table 12.1 summarises the main assumptions regarding efficiencies, costs, environmental impact and the remaining relevant parameters

12.3 Discussion and results

This section is divided into five parts. The first part discusses the ability of a PV–WT-based charging station to cover the observed demand and ensure satisfactory reliability. The second analyses the financial aspects of such an investment. The third focuses on the environmental impact, the fourth investigates the impact on the power grid, while the last section concludes with some potential future research directions.

12.3.1 Self-sufficiency and reliability of supply

Designing a power source using climate-driven RES and designated to cover any form of load is a challenging task. Not only does the energy availability tend not to match the demand, but it also varies a lot on the temporal and spatial scales.

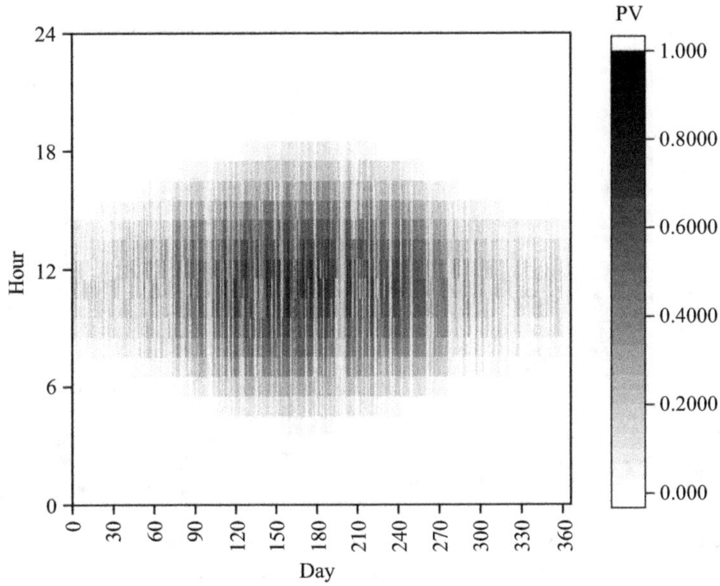

Figure 12.4 Annual spectrum of PV generation. Each hour of the year is represented by a rectangle. The darker it is, the more the energy produced (normalised)

We have analysed two structures (Case I and Case II) of installed capacity in PV and WT sources. In both, we have assumed that the solar generation should be equal to the wind generation. In the first case, the aggregated annual energy yield was equal to the annual demand. In the second case, the energy yield was twice as high. The possibility of setting such PV, wind and storage capacity may provide power to EBs and simultaneously be a party of the renewable energy stock (i.e. sell a declared amount of energy to the system grid). Case I had 361 kW in PVs and 179 kW in WTs, while Case II had 722 kW and 358 kW, respectively. Naturally, the smaller capacity in wind generation resulted from its greater capacity factor in that region. These proportions should be optimised, taking into account all circumstances, and especially the local climatic conditions.

The results of our calculations are shown in Figure 12.5. The continuous and dotted black lines represent a charging station's capability to cover the demand using vRES sources. In the process, we aimed to examine how varying the number of backup batteries (each of 100 kWh capacity) impacts the charging station's self-sufficiency. Naturally, the greatest increase in covering the demand from renewables happens when the first battery is added to the system. This clearly shows the important role of energy storage in cases of vRES-based power sources. Each subsequent battery does not significantly impact the system's self-sufficiency, but it can be as high as 95% (Case II, ten batteries). On average, each 1 kWh of storage increases the share of the demand covered from renewables by 0.024% and 0.26%

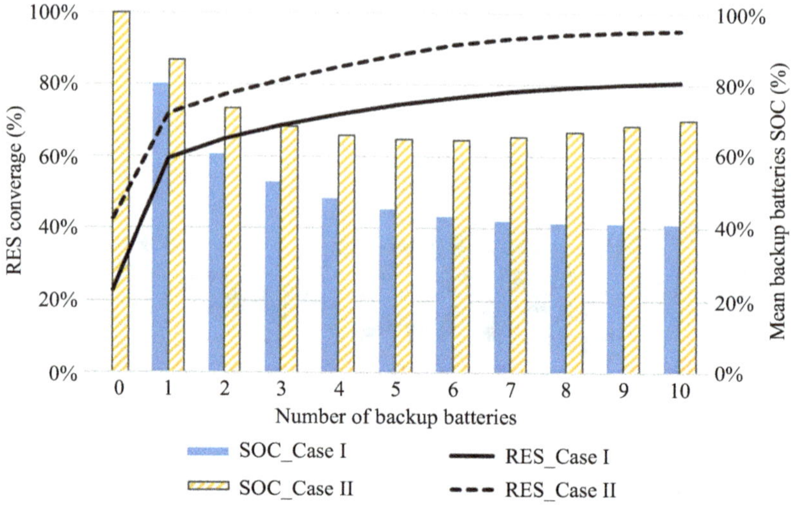

Figure 12.5 Charging station self-sufficiency depending on the number of backup batteries as well as their average state of charge

in Cases I and II, respectively. Increasing the storage capacity of the systems tends to lower the mean stage of charge (SOC) of batteries (Case I), but when the energy generation is oversized (Case II), we can observe that after exceeding six batteries their SOC starts to pick up. This is due to the fact that with greater storage capacity more energy surpluses can be stored and the mean SOC does not so often reach the lowest values. It can be said that a higher mean SOC increases the batteries' longevity as the exploitation of a single battery is lower, but on the other hand it entails greater capital expenditures.

12.3.2 Economic analysis

To verify whether the proposed charging station scheme in given wind and solar conditions and an observed energy market situation is economically justified, we have used the net present value (NPV) measurement. All the considered parameters necessary for calculations provided in Table 12.1 were used. The interpretation of the NPV indicator is quite straightforward. If it is lower than 0, the project will be a financial sinkhole, if it is greater than 0, it will bring a profit, and if it is equal to 0, then it will bring neither profit nor loss, and other (non-monetisable) criteria should sometimes be decisive.

12.3.2.1 On economic justification

The calculation of NPV requires several major assumptions to be made with regard to the development of the market situation. Therefore, the results should be handled with caution and subject to thorough sensitivity analysis. In our analysis, we have used the NPV index as an objective function for the optimisation problem formulated as follows: *Maximise the NPV of an investment in the EB charging station by changing the installed capacity in wind and solar generation as well as the capacity of backup batteries.* Such optimisation was subject to constraints with regard to the system operation (resulting from the algorithm in Figure 12.4.) and the assumption that the volume of energy surpluses should not be greater than double the energy demand. Depending on country regulations, other assumptions are possible. For example, when selling energy to the grid is impossible, the power generation will have to be cut off. Although seemingly complex, these optimisation procedures may be conducted with the commonly available software, for example, implemented in MS Excel and solved by means of the in-built Solver (Evolutionary method with default settings). This was the case with the authors' example.

12.3.2.2 On energy balance

The optimisation results show that the charging station should consist of 800 kW installed in wind generation, 490 kW in PVs, a set of five backup batteries of 500 kWh cumulative capacity and a charging station with a 400-kW capacity. Such a station, according to our calculations, should cost roughly €1.9 million, and its cost structure would be as presented in Figure 12.6. Such a station would be able to cover 92.5% of the observed energy demand, but on an annual scale it would generate almost 1.4 GWh of energy surpluses (sold to the grid). Clearly, the station strongly benefits from selling the energy to the grid, as the cost of renewable generation starts to become cost competitive with conventional generation. Naturally, the energy surpluses could also have been sent to another charging station located some distance away. Their location would enable them to benefit from the effect of spatial vRES complementarity as well as (probably) different charging patterns. The effect of an obvious system oversizing is that the backup batteries' mean SOC is 74.5% and can be as high as 85% in February or as low as 66% in

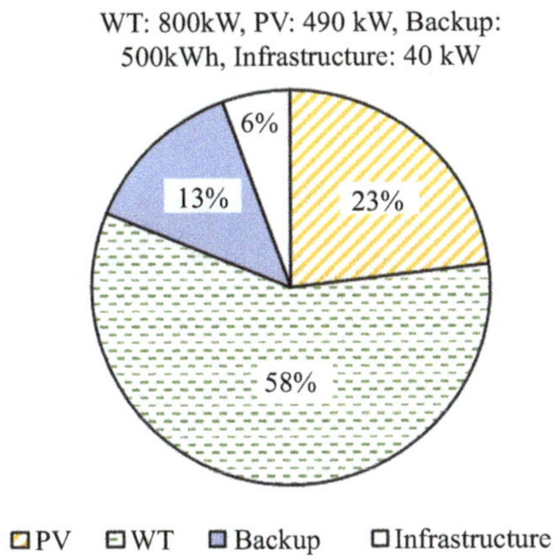

Figure 12.6 Cost structure of the optimised charging station

September. Such a station would directly cover 57% of the observed demand, while 35.5% would be covered by means of backup batteries and the remaining part from the grid.

The resulting NPV index for such an investment (after 30 years) is positive and amounts to €417,000. This means that in the long run this investment would be beneficial to the investor. However, such an analysis is subject to a threat resulting from making a multitude of assumptions about future cost development scenarios and other parameters. In Figure 12.7, we have presented the impact of varying values of the selected parameters on the NPV index. For each, we have changed it in the range from 0.75 to 1.25 of its reference value as presented in Table 12.1. The red line indicates the NPV resulting from the previous optimisation.

12.3.2.3 Once again on financial balance

As can be seen, the highest impact on the NPV comes from wind generation investment costs (WT CAPEX); energy cost on the market – the price at which it is bought from the grid to cover the demand (energy price); the discount rate and the price at which the energy is being sold to the grid (energy selling price). The role of wind generation investment cost is understandable as it heavily contributes to the overall cost of the system (58% as shown in Figure 12.6). The energy price at which it is being bought impacts the potential value of savings, which on the other hand result from the system's self-sufficiency index. The higher it is, the lower the volume of energy that has to be bought from the grid and the greater the savings, and larger expenditures on energy storage capacity are thus reasonable. Greater energy market prices lead to increasing profitability of energy storage, making it economically justified.

An oversized system, such as the one considered here, is prone to varying energy selling prices. A 25% price decrease (from €40/MWh to €30/MWh) may reduce the NPV value by over 60%. However, in a coal-dominated power system

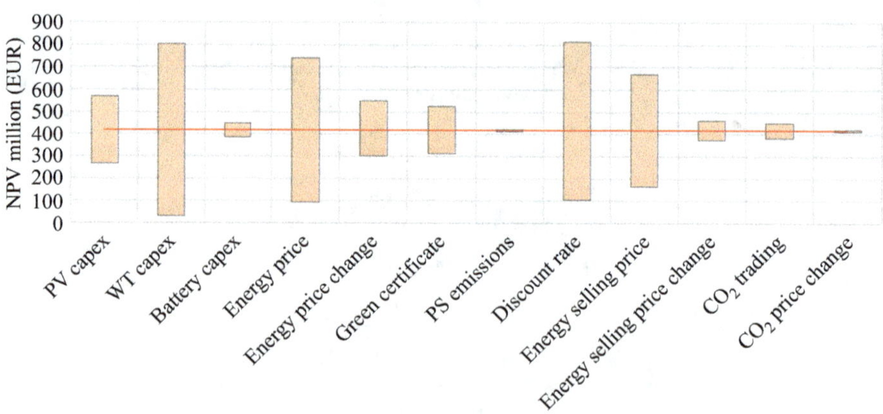

Figure 12.7 Impact of changing input parameters on the NPV. Upper range of each bar indicates NPV after +25% value increase; lower range of each bar indicates NPV after −25% value decrease

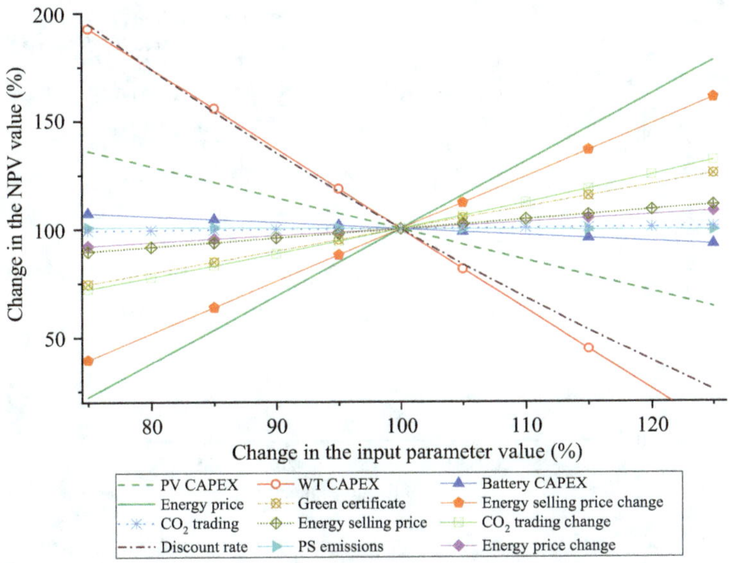

Figure 12.8 Impact of input parameters on the NPV. The central point (100% on both axes) is a reference point

this is a highly unlikely scenario, as the cost of such conventional generation tends to exhibit an upward trend. Although the green certificate price variation (base of €20/MWh) does not very strongly impact the NPV, it is crucial to the project's positive economic performance. Further analysis has shown that the elimination of green certificates leads to a negative NPV. However, it has to be stressed that such outcomes are only valid for the system considered here, as it highly relies on energy surpluses.

The remaining parameters exhibit significantly lower impact on the system's economic performance. Neither battery investment costs, nor the power system's annual reductions in CO_2 emissions (PS emissions), nor CO_2 trading prices and any future changes therein, nor changes in energy selling price have more than a 10% impact on NPV. In the considered scenario, even removing CO_2 trading from the analysis does not lead to a negative NPV. It is clear that green certificates will promote selling energy surpluses (oversizing and generating them), whereas CO_2 trading will support self-consumption.

For better understanding, the impact of input parameters on the NPV is shown in Figure 12.8. The figure clearly shows how a percentage change in one parameter affects the economic side of the investment.

12.3.3 Environmental impact

For EVs, in general, to minimise their impact on the natural environment, it is not only vital to use a non-polluting fuel but also to ensure that it comes from a sustainable and renewable source. Currently, in energy systems dominated by

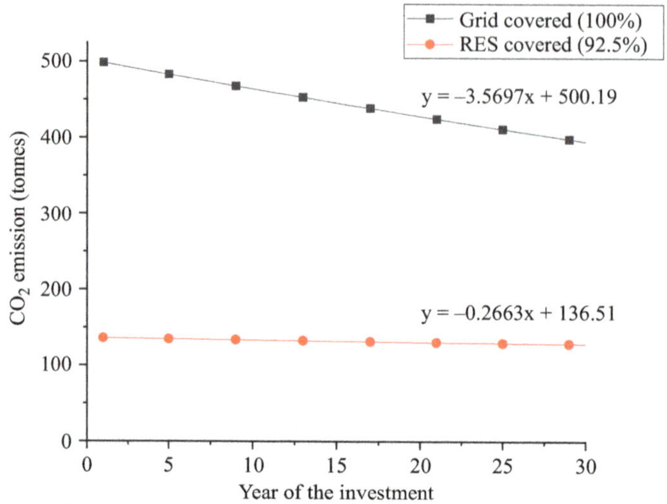

Figure 12.9 *CO_2 emissions if EBs were charged only from the grid and a 92.5% self-sufficient RES-based charging station*

conventional generation (especially coal) each kWh used in EVs will have a significant carbon footprint. In the case of the Polish power system, on average, generating 1 MWh of electricity results in emitting 717 kg of CO_2.

Figure 12.9. shows the cumulative emissions of charging batteries from the grid and a 92.5% self-sufficient solar- and wind-powered charging station. Emissions are given in tonnes of CO_2 and relate to an annual demand of 695.4 MWh. During the first year of our analysis the emissions would amount to 499 tonnes in the 'grid' case and 136 tonnes of CO_2 in the 'RES' case. Although the grid covered case shows a significant downward trend (CO_2 emissions tend to decrease due to the growth in renewables) the cumulative emissions over the next 30 years would be 336% higher than in the case of a 92.5% self-sufficient RES-based charging station. The very low downward trend for the renewables-powered charging station results from the fact that the share of demand covered from the grid (which exhibits lowering emissions) accounts for only 7.5%.

12.3.4 Impact on the grid

A charging station powered by intermittent renewable energy sources (RESs) has to ensure a certain level of reliability. This usually comes with adding storage capacity or connecting it to a dispatchable power source. As already shown in Section 12.3.1, additional storage capacity strongly increases the system reliability without the need to oversize its generating potential. If one does aim to maximise self-sufficiency while minimising the impact on the grid's operation (by a bi-directional flow of energy – selling and purchasing) one solution is to increase the storage capacity – in this case, the number of backup batteries.

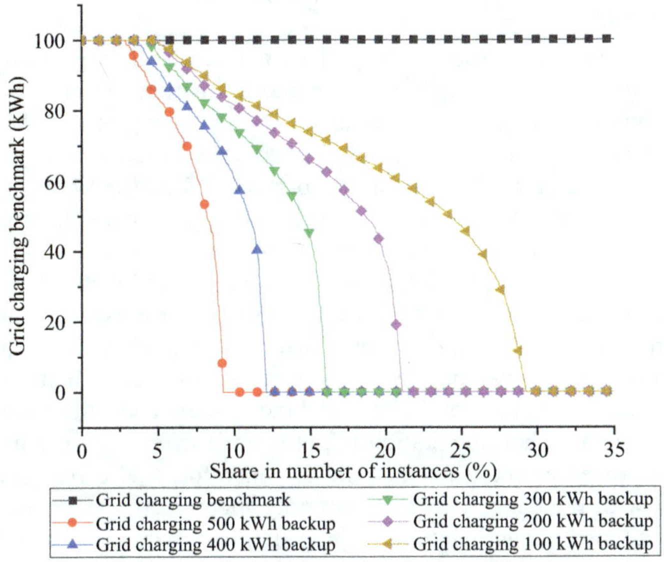

Figure 12.10 Ordered chart of EB charging events. Horizontal axis limited to 35% as in all cases charging station covered 70% or more of charge requests

When a charging station is covering the demand of buses randomly arriving at the depot, it heavily burdens the power grid with sudden peaks in requested power. If in the future the EB, or EVs in general, are about to dominate transportation modes, the demand on electricity would undoubtedly sky-rocket. This would have an immense impact on the operation of the power system. In this analysis, we have shown that a vRES-powered charging station is capable of covering 92.5% of the demand while still being economically justified (although huge volumes of energy are being sold to the grid).

Over the considered year, there were in total almost 7,000 charging requests. Our analysis has shown (Figure 12.10.) that a charging station with parameters as in Figure 12.6 can completely cover slightly over 90% of them, while 4% will be covered partially and the remaining part, entirely from the grid. Reducing the storage capacity by one battery increases the share of not-completely-covered charging requests by 3%. Analysing the lines in Figure 12.9, it is clear that such a charging station has huge potential to reduce the overall volume of energy covered by the grid, but still requires the grid to be available to deliver power.

Further analysis shows that in such a station, the use of the charging infrastructure's full potential will be significantly reduced but it will be still necessary to provide the requested installed power, just as it is for an entirely grid-powered station.

12.3.5 Future works

Obviously, the analysis presented here does not cover all aspects of a vRES-powered charging station operating on the energy market but encapsulates only a small section of it. The conducted analysis allows us to conclude that, in future works, a stochastic model should be applied to assess the reliability and profitability of such a solution. The stochastic part should address both the variable generation (decisions made based on forecasted energy generation from PVs and WTs) and the fact that buses would not always arrive as scheduled. Random arrival times, as well as changing time (duration), required to fully charge the battery may significantly impact the efficacy of the proposed charging station operation modes. It seems quite likely that the variability in frequency and duration of charging events may require a larger set of backup batteries. To deal with the climate-driven nature of vRES and the randomness of bus arrival times, we think that such a charging station might benefit from coupling it with some dispatchable power source, such as a diesel generator, fuel cell or run-off-river power plant with pondage. This may not only increase the reliability and self-sufficiency but also strengthen the charging station's market position by making it resistant to changing energy prices or energy policies. However, it is obvious that using hydropower will be limited to only selected locations, as already shown in [24].

12.4 Conclusion

This chapter presented the concept of a grid-contacted EB charging station powered by solar and wind generators. The unavoidable impact of the intermittent nature of both energy sources is minimised by using additional batteries that are charged when energy surplus from vRES occurs and then exploited when vRES potential is not sufficient to cover the current demand. From the presented methodology, data, results and analysis, the following conclusions can be drawn:

- By equipping a charging station with additional backup batteries (used to store energy surpluses or to replace empty ones in EBs), one is capable of achieving relatively high levels of energetic self-sufficiency;
- Oversizing the charging station energy sources in such a manner that they provide on an annual basis double the required energy increases of the station energy self-sufficiency by 14% on average;
- The charging station which is operating on the energy market tends to be at once (on an annual scale) both an energy consumer (when solar-wind generation and storage is not sufficient to cover the load) and a producer (when energy generation is greater than available storage capacity and demand);
- Owing to the above economic aspects, the charging station will be strongly dependent on two market instruments: green certificates (as a way of supporting RES development) and carbon trading (as a tool effectively advocating

for self-consumption and avoiding purchasing energy from the grid the price of which will increase);

- RES sources usually have a low potential in terms of energy yield per unit of occupied surface. This is especially visible in the case of solar and wind generation in areas with moderate resources. Therefore, such a station should be located on the city outskirts or in rural areas, where sufficient area is available for PV and/or WT project development. Naturally, one can locate it in the city centre and use the roof potential of surrounding buildings or invest in an addition grid connection to a remote vRES;
- A charging station using vRES-based generators is prone to the same dangers as any energy source that relies on intermittent and climate-driven sources, and therefore it usually cannot provide 100% reliability without significant oversizing. So although we were able to significantly reduce the volume of energy bought from the grid it was not possible to reduce the range of charging volumes, which remain at 0–100 kWh.

References

[1] Jones LE. Renewable energy integration: practical management of variability, uncertainty and flexibility in power grids [Internet]. [cited 14 March 2019]. Available from: https://books.google.pl/books?hl=pl&lr=&id=b_6pDQAAQBAJ&oi=fnd&pg=PP1&dq=Renewable+energy+integration:+practical+management+of+variability+uncertainty,+and+flexibility+in+power+grids&ots=uDRhMXCUKz&sig=HN-T3iP-bnBDC4dElMb7ZuIRQzM&redir_esc=y#v=onepage&q=Renewable energy integration%3A practical management of variability%2C uncertainty%2C and flexibility in power grids&f=false

[2] Delucchi MA, and Jacobson MZ. Providing all global energy with wind, water, and solar power, Part II: reliability, system and transmission costs, and policies. Energy Policy [Internet]. 2011 [cited 14 March 2019]; 39(3):1170–90. Available from: https://www.sciencedirect.com/science/article/pii/S0301421510008694

[3] Alvaro R, and Fraile-Ardanuy J. Charge scheduling strategies for managing an electric vehicle fleet parking. In: IEEE EUROCON 2015 – *International Conference on Computer as a Tool* (*EUROCON*) [Internet]. *IEEE*; 2015 [cited 14 March 2019]. p. 1–6. Available from: http://ieeexplore.ieee.org/document/7313732/

[4] Alvaro-Hermana R, Fraile-Ardanuy J, Merino J, and Knapen L. Optimization of night electric vehicle fleet charging at regional level. *Int J Therm Environ Eng* [Internet]. 2016;15(1):43–9. Available from: https://uhdspace.uhasselt.be/dspace/handle/1942/25045

[5] Clement-Nyns K, Haesen E, and Driesen J. The Impact of Charging Plug-In Hybrid Electric Vehicles on a Residential Distribution Grid. *IEEE Trans Power Syst* [Internet]. 2010 [cited 2019 Mar 14];25(1):371–80. Available from: http://ieeexplore.ieee.org/document/5356176/

[6] Sortomme E, Hindi MM, MacPherson SDJ, and Venkata SS. Coordinated Charging of Plug-In Hybrid Electric Vehicles to Minimize Distribution System Losses. *IEEE Trans Smart Grid* [Internet]. 2011 [cited 14 March 2019];2(1):198–205. Available from: http://ieeexplore.ieee.org/document/5664815/

[7] He Y, Venkatesh B, and Guan L. Optimal Scheduling for Charging and Discharging of Electric Vehicles. *IEEE Trans Smart Grid* [Internet]. 2012 [cited 14 March 2019];3(3):1095–105. Available from: http://ieeexplore.ieee.org/document/6244822/

[8] Mehta R, Srinivasan D, Khambadkone AM, Yang J, and Trivedi A. Smart Charging Strategies for Optimal Integration of Plug-In Electric Vehicles Within Existing Distribution System Infrastructure. *IEEE Trans Smart Grid* [Internet]. 2018 [cited 14 March 2019];9(1):299–312. Available from: http://ieeexplore.ieee.org/document/7448958/

[9] Fuentes M, Fraile-Ardanuy J, Risco-Martín JL, and Moya JM. Feasibility Study of a Building-Integrated PV Manager to Power a Last-Mile Electric Vehicle Sharing System. *Int J Photoenergy* [Internet]. 26 April 2017 [cited 14 March 2019];2017:1–12. Available from: https://www.hindawi.com/journals/ijp/2017/8679183/

[10] Fathabadi H. Utilizing solar and wind energy in plug-in hybrid electric vehicles. Energy Convers Manag [Internet]. 2018 [cited 14 March 2019];156:317–28. Available from: https://www.sciencedirect.com/science/article/pii/S0196890417310543

[11] Fathabadi H. Novel battery/photovoltaic hybrid power source for plug-in hybrid electric vehicles. Sol Energy [Internet]. 2018 [cited 14 March 2019];159:243–50. Available from: https://www.sciencedirect.com/science/article/pii/S0038092X17309519

[12] Fathabadi H. Fuel cell hybrid electric vehicle (FCHEV): Novel fuel cell/SC hybrid power generation system. Energy Convers Manag [Internet]. 2018 [14 March 2019];156:192–201. Available from: https://www.sciencedirect.com/science/article/pii/S019689041731035X

[13] Kaldellis JK, Spyropoulos G, and Liaros S. Supporting Electromobility in Smart Cities Using Solar Electric Vehicle Charging Stations. In: *Mediterranean Green Buildings & Renewable Energy* [Internet]. Cham: Springer International Publishing; 2017 [cited 14 March 2019]. p. 501–13. Available from: http://link.springer.com/10.1007/978-3-319-30746-637

[14] Kleissl J. (ed) *Solar energy forecasting and resource assessment*. Academic Press; 2013.

[15] François B, Zoccatelli D, and Borga M. Assessing small hydro/solar power complementarity in ungauged mountainous areas: A crash test study for hydrological prediction methods. Energy [Internet]. 2017 [cited 14 March 2019];127:716–29. Available from: https://www.sciencedirect.com/science/article/abs/pii/S0360544217304656

[16] Campana PE, Yang Z, Anders L, Hailong L, and Jinyue Y. An Open-source Platform for Simulation and Optimization of Clean Energy Technologies. Energy Procedia [Internet]. 2017 [cited 14 March 2019];105:946–52. Available from: https://www.sciencedirect.com/science/article/pii/S1876610217304642

[17] Hoicka CE, and Rowlands IH. Solar and wind resource complementarity: Advancing options for renewable electricity integration in Ontario, Canada. Renew Energy [Internet]. 2011 [cited 14 March 2019];36(1):97–107. Available from: https://www.sciencedirect.com/science/article/pii/S0960148110002600

[18] Jurasz J, Wdowikowski M, Kaźmierczak B, and Dąbek P. *Temporal and spatial complementarity of wind and solar resources in Lower Silesia (Poland).* Kaźmierczak B, Kutyłowska M, Piekarska K, Jouhara H, and Danielewicz J, (eds). E3S Web Conf [Internet]. 7 November 2017 [cited 14 March 2019];22:00074. Available from: http://www.e3s-conferences.org/10.1051/e3sconf/20172200074

[19] van Goethem S. Smart choices for cities – clean buses for your city. Cleaner bus options for your municipality [Internet]. 2014 [cited 14 March 2019]. Available from: https://civitas.eu/sites/default/files/documents/sam_van_goethem_presentation_civitas_webinar_clean_buses.pdf

[20] Carlsson J, Fortes M del MP, Marco G de, Giuntoli J, Jakubcionis M, Jäger-Waldau A, *et al.* ETRI 2014 – *Energy Technology Reference Indicator projections for 2010–2050* [Internet]. JRC Science and Policy Reports. 2014. p. 1–108. Available from: https://setis.ec.europa.eu/system/files/ETRI 2014.pdf

[21] Eurostat statistics [Internet]. Available from: http://ec.europa.eu/eurostat

[22] Romare M, and Dahllöf L. The Life Cycle Energy Consumption and Greenhouse Gas Emissions from Lithium-Ion Batteries A Study with Focus on Current Technology and Batteries for light-duty vehicles [Internet]. 2017 [cited 14 March 2019]. Available from: www.ivl.se

[23] Ligus M. Efektywność inwestycji w odnawialne źródła energii: analiza kosztów i korzyści. CeDeWu. Wydawnictwa Fachowe; 2011.

[24] Jurasz J, and Ciapała B. Solar–hydro hybrid power station as a way to smooth power output and increase water retention. Sol Energy [Internet]. 2018 [cited 14 March 2019];173:675–90. Available from: https://www.sciencedirect.com/science/article/pii/S0038092X18307527

Chapter 13

Deploying stochastic coordination of electric vehicles for V2G services with wind

Sulabh Sachan[1] *and Nand Kishor*[1]

In this work, we evaluate the effect of charging and discharging plan on integration of electric vehicles (EVs) with wind penetrated distribution grid. A probabilistic model is established by considering the stochastic environment of wind and the accessibility of network constraints. Next, a methodology based on fuzzy approach is applied to model the load profile of EVs with the stochastic availability of EVs, that is, arrival and departure times. Based on this stochastic load profile of EVs, particle swarm optimization with evolution strategy (ESPSO) procedure is applied to incorporate EVs into the distribution grid. Peak load shaving is discussed with vehicle to grid (V2G), having considered the EV charging cost, degradation cost of EV battery, and frequency regulation incomes along with spinning reserve earnings in single objective function. The study is further extended with a varied range of EV penetration in the distribution network.

Notation

EVs	electric vehicles
ESPSO	particle swarm optimization with evolution strategy
V2G	vehicle to grid
LPMF	load profile modeling framework
PDF	probability distribution function
SSE	squares of error
DT	departure time
AT	arrival time
DM	daily mileage
SOC	state of charge
TOU	time of use

[1]Department of Electrical Engineering, Motilal Nehru National Institute of Technology Allahabad, Prayagraj, India

DoD	discharge rate of EV
PSO	particle swarm optimization
DNO	distribution network operator
v	wind speed
m	contour coefficient
c	scale coefficient
W	discrete wind speeds
$v(l)$	discrete wind speed for speed range $v(l) \pm \frac{\varphi}{2}$
φ	width of the speed range
v_{in}, v_r, v_{out}	Cut-in, rated, cut out wind speed
P_r	maximum wind power output
d	travel distance
μ	mean deviation
σ	standard deviation
$reg(t)$	regulation service price
$SR(t)$	spinning reserve service price
$v^k_{is^k_{id}}, x^k_{is^k_{id}}$	velocity and position of the particle
w	inertia weight
k	iteration
i, d	particle and dimension number

13.1 Introduction

Nowadays, electric vehicles (EVs) are playing a significant role in the transportation sector. EVs can decrease carbon contamination from transportation, and aid to recover the air quality. They have the potential to be substituted for combustion-engine vehicles for a wide range of transportation.

However, EV integration into the grid raises several challenges. Due to their considerable base loads, charging of EVs can certainly affect the distribution network. For example, EV integration amplifies the peak load demand [1], increases power losses, causes voltage variations [2], and so on. EVs require more power to get charged than commonly used household appliances in the charging scenario. Owing to this, voltage and power flow violates the critical limit on the distribution network [3]. In the dumb charging scheme, the peak load of distribution network may be met while EV charging and thus loading impact arises on the distribution network [4].

Some impacts of EVs on the distribution level are deliberated in [5]. The voltage will fall substantially and losses in the network could also be on the rise [6]. The problem of voltage violation along with power loss minimization has been discussed in-depth via numerous charging patterns in [7] and [8].

In the literature, different assumptions have been considered by authors to model the driving pattern of EVs. For instance, in [9], arrival of EVs at home till late night is assumed. In [10] and [11], a probabilistic methodology has been suggested to show the arrival, departure, and day-by-day mileage using copula [12]. A Markov-chain-based strategy is anticipated in [13] and [14] to find the link among EV arrival, departure, and daily mileage (DM). Although these works [12–14] are successful in showing the EV driving configurations, additional study on the vehicle features is still desired. Additionally, the policies for endorsing EVs should be planned by taking into account the capacity of the prevailing grid and the implementation of data and correspondence innovation in forthcoming smart grids.

In recent trends, vehicle to grid (V2G) technology is adopted frequently to reshape the traditional view of the smart grid. EVs are encouraged as energy storage devices because diffusion of recurrent generation elements and loads are increasing. V2G can assist the power network by cutting the peak demand and functioning as auxiliary amenities in the form of frequency regulation along with spinning reserves.

Execution of V2G needs an actual commercial model tied with a more innovative battery technology. In [15] and [16], numerous optimization techniques are used to lessen the effects of EVs on the distribution system. Furthermore, certain works have been commended for exploring the V2G benefits and feasibility [17,18]. There is a scope to associate these methodologies in order to address this integrated problem.

In [19], the authors discussed the scheduling of V2G in the context of a centralized unit commitment toward renewable energy sources. However in their study, the uncertainties related to the stochastic nature of power availability from these power sources are not considered; that is, the model is absolutely deterministic.

The above discussion suggests that few works have been carried out assuming the stochasticity in driving pattern and various features of vehicle, including renewable penetration. This work herewith presents probabilistic model of wind power generation along with load profile modeling framework (LPMF) of EVs. LPMF considers the features of driving configuration and vehicle constraints to evaluate the correlation among EV arrival, departure, and DM. In the study, we present a stochastic adaptive fuzzy model to produce the driving pattern. With adaptive fuzzy model of EVs, this chapter has established an innovative commercial model for EVs to offer frequency regulation service and spinning reserve service along with contribution in peak demand saving in a distribution network. The V2G services are considered in [17–19]. In this work, V2G is managed in the grid with wind generation along with price and load ambiguity. Real-time pricing is modeled on the availability of wind power to inspire EVs owners to contribute to peak demand saving, by offering economic incentives. A smart charging strategy is suggested to coordinate EVs in a residential distribution grid toward flattening the load demand, and providing V2G services like frequency regulation service and spinning reserve service by minimizing the total cost as objective function.

The objective function is solved by the particle swarm optimization-based evolution strategy (ESPSO) algorithm.

The major contributions of this work are as follow: (1) probabilistic modeling of wind power generation, (2) stochastic modeling of EVs as load demand, (3) coordinating the effect of wind power availability and stochastic charging of EVs with real-time pricing, and (4) building a commercial model to apply the V2G services.

13.2 Test system

The proposed study is based on the IEEE-34 distribution network [20]. This distribution network is transformed to 12 kV level from the original system, with different segments, including loads and line impedances scaled appropriately. This network is coupled to the grid at bus 800. In the distribution network, there are 75 houses and each house has one EV. The base demand of the network is scaled from [21], and the power flow calculation is performed using power system toolbox (PST) [22]. The wind generation of 80 kW is included at bus 840, as shown in Figure 13.1.

13.3 Probabilistic model of wind power

In [23] and [24], wind speed (v) distribution is modeled using Weibull distribution function. Equation (13.1) presents the modeling of wind speed variation via probability distribution function (PDF) according to Weibull distribution as

$$f(v) = \left({}^{m}/_{c}\right)\left({}^{v}/_{c}\right)^{m-1} exp\left[-\left({}^{v}/_{c}\right)^{m}\right] \tag{13.1}$$

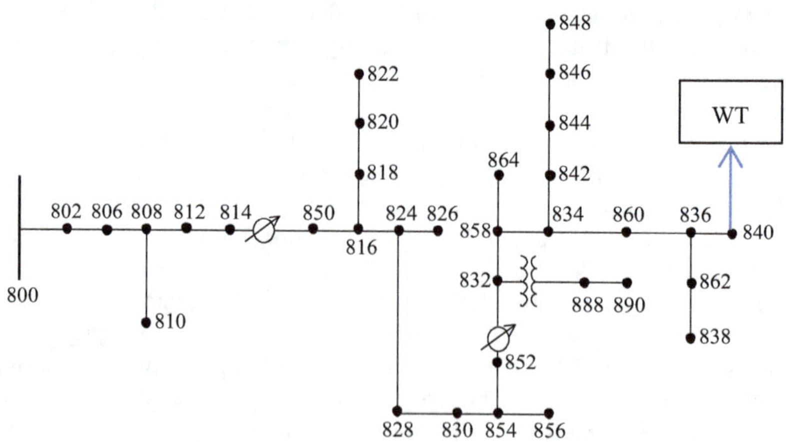

Figure 13.1 Modified IEEE 34 distribution network

The probability distribution of wind speed in (13.1) can be considered as a set of discrete data shown in Figure 13.2(a) as

$$F_v = \{v(l), F_v(l): l = 0, \quad 1, \dots W\} \tag{13.2}$$

The probability $F_v(l)$ of the discrete wind speed is given as

$$F_v(l) = \int_{v(l)-\frac{\varphi}{2}}^{v(l)+\frac{\varphi}{2}} f(v)dv = exp\left\{-\left[\frac{\mathbf{v}(\boldsymbol{l})-\frac{\varphi}{2}}{\mathbf{c}}\right]^m\right\} - exp\left\{-\left[\frac{v(l)+\frac{\varphi}{2}}{c}\right]^m\right\} \tag{13.3}$$

The output power associated with wind speed is given as [25]

$$P = \left\{\begin{matrix} 0 & 0 \le v < v_{in} or v > v_{out,} \\ a + bv^3 & v_{in} \le v < v_r \\ P_r & v_r \le v \le v_{out} \end{matrix}\right\} \tag{13.4}$$

In the study, $v_{in} = 4m/s$, $v_r = 16m/s,\ and\ v_{out=}25m/s$. The constants a and b are expressed as $a = -\frac{P_r v_{in}^3}{v_r^3 - v_{in}^3},\ and\ b = P_r/(v_r^3 - v_{in}^3)$. The relation between discrete wind speed distribution $F_v(l)$, wind power output $P(i)$, and wind speed v along with the discrete probabilistic distribution of the power output $F_p(i)$ is depicted in Figure 13.2. The discrete probabilistic distribution of the power output

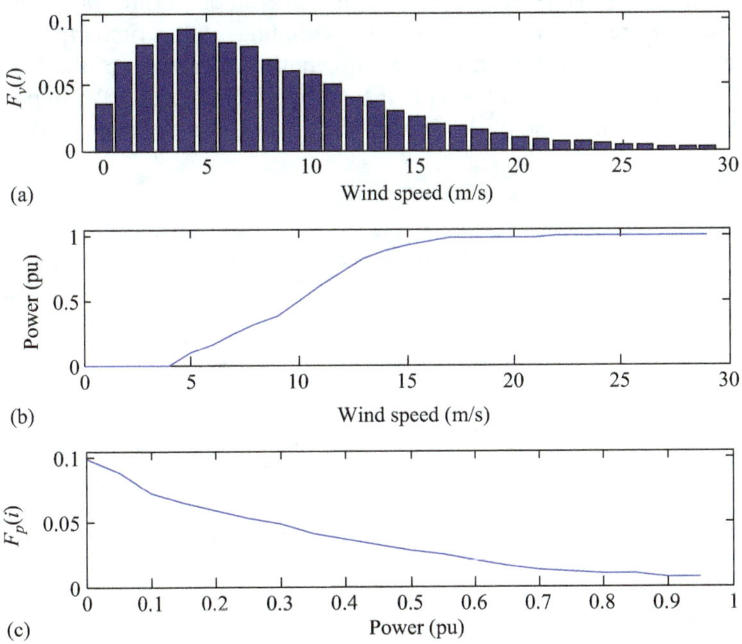

Figure 13.2 Probabilistic distribution of wind output: (a) discrete wind speed distribution, (b) wind power output, and (c) discrete probabilistic distribution of power output

$F_p(i)$ can be now given as the probability set:

$$F_p = \{P(i), F_p(i){:}i = 0, 1, 2, \ldots. M\} \tag{13.5}$$

Here, M depicts the states of wind power.

The probabilistic distribution of wind generation at every moment of time is essential to explore the disparity in wind speed at every instant of a day. The probability distribution at the nth time step $F_{pn}(i)$ can be given by the probability set:

$$F_{pn} = \{P_n(i), F_{pn}(i){:}i = 0, 1, 2, \ldots. M\} \tag{13.6}$$

The relationship between variations in v wind speed and height can be as [26]

$$v_1 = v_2 \frac{ln\left(\frac{h_1}{z}\right)}{ln\left(\frac{h_2}{z}\right)} \tag{13.7}$$

Here, v_1 depicts the measured wind at height h_1, v_2 is noted as the wind speed at height h_2. $z = 0.05$ is the roughness length.

13.4 Stochastic load modeling of EV

In this study, the arrival time (AT) and departure time (DT) data are referred from the US Travel data (NHTS) [27]. The data are centered on the analysis of American vehicle users. Here, DT is defined as the beginning of the first trip, and AT is defined as ending of the final trip. The computation procedure of the EV driving pattern is shown in Figure 13.3. As per the AT and DT data, the EVs in percentage versus the DT/AT is depicted in Figure 13.4. Based on the computed optimum

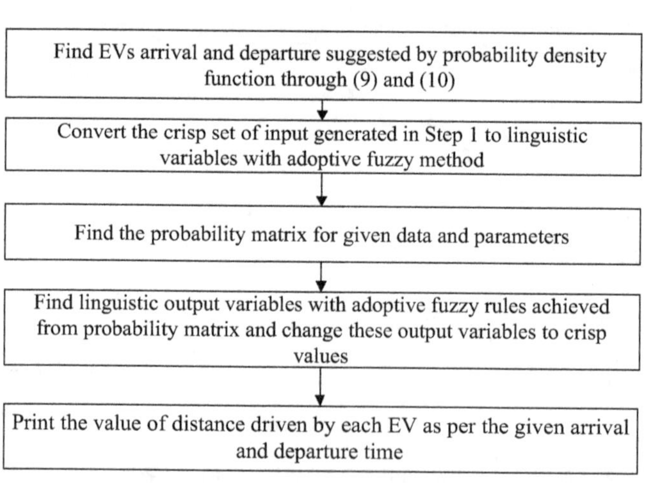

Figure 13.3 Computational method for driving pattern of EVs

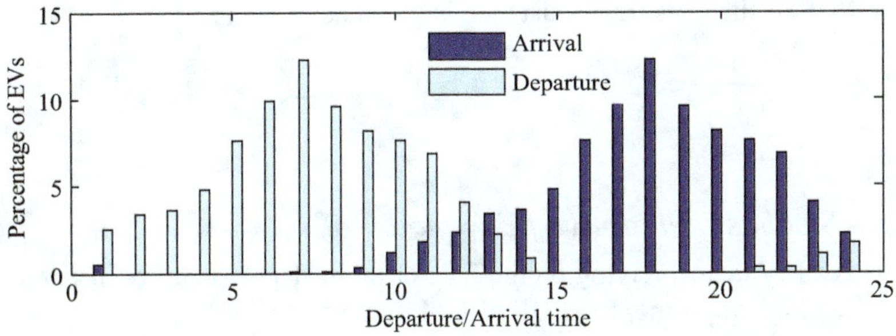

Figure 13.4 Distribution of departure/arrival time

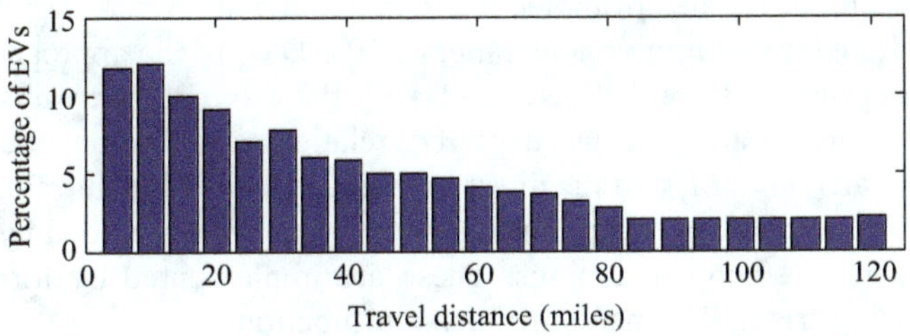

Figure 13.5 Stochastic pattern of driving distance

parameters of membership function, the stochastic adaptive model can forecast the EV driving pattern.

The PDF of arrival, departure, and DM of EVs is calculated. The real curve fit is assessed as the sum of squares of error (SSE) given as

$$SSE = \sum_{i=1}^{n} w_i (y_{oi} - y_{pi})^2 \tag{13.8}$$

where y_{oi} is the observed value, y_{pi} is the expected value by curve fit, and the value of w_i is selected as 1.

As illustrated in Figure 13.5, the PDF of DT is a normal distribution, given as

$$F_{dep}(t) = \frac{1}{\sigma\sqrt{2\pi}} e^{-(t-\mu)/2\sigma^2}, \quad 0 < t < 24 \tag{13.9}$$

where $\mu = 9.97$ and $\sigma = 2.2$.

The PDF of the AT of EVs can be described as

$$F_{arr}(t) = \frac{1}{\sigma\sqrt{2\pi}} e^{-(t-\mu)/2\sigma^2}, \quad 0 < t < 24 \tag{13.10}$$

where $\mu = 17.01$ and $\sigma = 3.2$.

The PDF of the EVs travel distance is given as

$$F_d(d) = \frac{1}{d\sigma\sqrt{2\pi}} e^{-(lnd-\mu)/2\sigma^2}, \quad d > 0 \tag{13.11}$$

Here, $\mu = 3.2,\ \sigma = 0.9$.

13.4.1 Stochastic adaptive fuzzy model of EVs

The DM is a key factor, relying on the EV AT and DT [11–13]. For dissimilar sets of arrival and leaving time, the PDF of DM may vary.

In this subsection, a stochastic adaptive fuzzy model is described to establish the driving patterns of EVs. As cited earlier, the main assignment of demonstrating the charging load of EVs is to determine the time when EVs are meant to be connected to grid and their initial state of charge (*SOC*). The *SOC* can be categorized in various steps, and it deviates from one phase to another after charging. The varying *SOC* can be transformed into various ranges of the DM. In the approach applied in this study, fuzzy logic is applied here to classify the driving pattern.

There is a direct, although not explored, relation between the stochasticity of EV departure, arrival, DM, and fuzziness expressed by membership functions. The stochastic distribution changes the discontinuous mappings provided by crisp logic systems into more smooth mappings. These are implemented by formulation of fuzzy rules using triangular shape membership functions.

As illustrated in Figure 13.6(a) and (b), input variables, DT and AT, are defined by symmetrical five-segmented triangular membership functions. The inputs are converted into linguistic variables as very early (VE), little early (LE), normal (N), little late (LL), and very late (VL).

Similarly, the output DM as illustrated in Figure 13.6(c), associates the corresponding linguistic variables as small (S), small-medium (SM), medium (M), medium-large (ML), and large (L). The maximum and minimum limits along with the mean value of each parameter can be computed from its PDF. The values of these parameters are considered as $Min_{dep} = 4,\ \mu_{dep} = 9.97,\ Max_{dep} = 21$, $Min_{arr} = 7$, $\mu_{arr} = 17.01$, $Max_{arr} = 29$, $Min_d = 0$, $\mu_d = 35.64$, $Max_d = 100$.

Now, the probability matrix P_M is given below as

$$P_M = \begin{bmatrix} p^d_{VE,VE} & p^d_{VE,LE} & p^d_{VE,N} & p^d_{VE,LL} & p^d_{VE,VL} \\ p^d_{LE,VE} & p^d_{LE,LE} & p^d_{LE,N} & p^d_{LE,LL} & p^d_{LE,VL} \\ p^d_{N,VE} & p^d_{N,LE} & p^d_{N,N} & p^d_{N,LL} & p^d_{N,VL} \\ p^d_{LL,VE} & p^d_{LL,LE} & p^d_{LL,N} & p^d_{LL,LL} & p^d_{LL,VL} \\ p^d_{VL,VE} & p^d_{VL,LE} & p^d_{VL,N} & p^d_{VL,LL} & p^d_{VL,VL} \end{bmatrix}$$

$$\forall d \in [S, SM, M, ML, L] \tag{13.12}$$

where every set of probability matrix P_M is a row vector. This indicates the probabilistic distribution over the membership functions of DM.

For example, one element of probability matrix is obtained as follows:

$$p^d_{VE,LL} = \left[p^S_{VE,LL},\ p^{SM}_{VE,LL},\ p^M_{VE,LL},\ p^{ML}_{VE,LL},\ p^L_{VE,LL}\right] \tag{13.13}$$

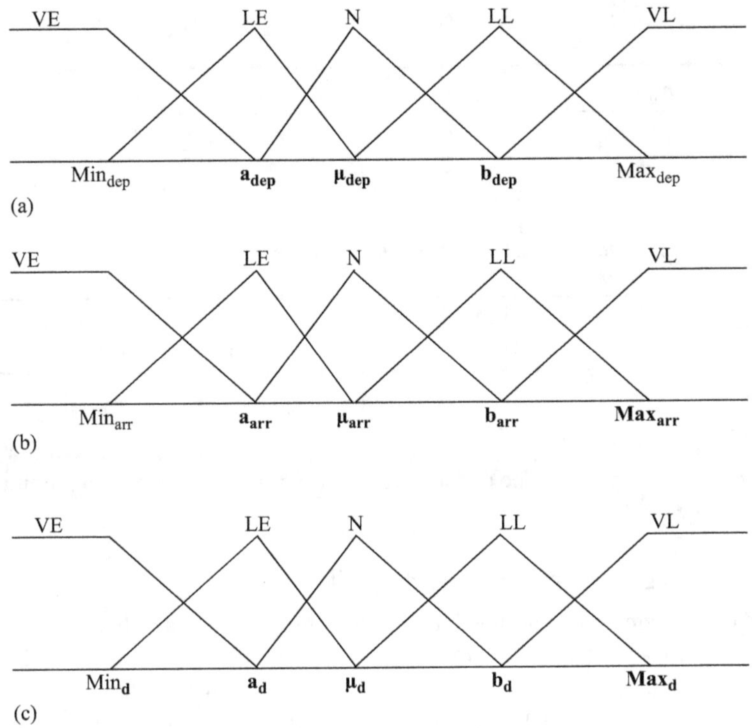

Figure 13.6 Membership functions: (a) pattern of DT, (b) pattern of AT, and (c) pattern of travel mileage

The anticipated rules are defined according to the if-then rule declarations.

To check the validity of driving pattern, the constraints related to membership functions must be appropriately defined. The particle swarm optimization (PSO) algorithm is used to select the optimal constraints of the membership functions.

PSO is essentially created through recreation of a rush of winged creatures in two-measurement space [28]. In PSO, a flock of particles optimize an assured objective function. Every particle identifies its best value (P_{best}) and its *XY* location. Additionally, each particle selects the best value in the group among (G_{best}) and (P_{best}). The best site of the *i*th particle is kept as $Pbest_i^k$ and the best position of all the particles is kept as global best , $Gbest_i^k$. The improved velocity of every element can be premeditated by the present velocity and the distance from (P_{best}) and (G_{best}) as given below:

$$V_i^{k+1} = wV_i^k + C_1 * r_1 * \left(Pbest_i^k - X_i^k\right) + C_2 * r_2 * \left(Gbest_i^k - X_i^k\right) \quad (13.14)$$

$$S_i^{k+1} = S_i^k + V_i^{k+1} \quad (13.15)$$

$$w_i = w_{max} - k \times \frac{w_{max} - w_{min}}{k_{max}} \quad (13.16)$$

Table 13.1 Optimal values of membership function parameters

a_{dep}	b_{dep}	a_{arr}	b_{arr}	a_d	b_d
9.13	13.27	11.93	19.22	13.68	87.74

Table 13.2 Percentage of EVs with various ranges

Percentage	EV-30	EV-40	EV-60
	27%	50%	23%

where v_i^k and S_i^k are the velocity and position of particle i at iteration k and w_i is the inertia weight of particle i. The optimal value of membership function parameters is given in Table 13.1.

13.4.2 Charging level and type of EVs

In order to accommodate the EV integration into the grid, the charging level of 4 kW at 230V/46A [8] and 120V/15A and 240V/50A [9] have been introduced. In this work, the charging level of 3.6 kW is considered with equal probability for each EV. The break-up of each type of EV is given in Table 13.2 [29]. The number 'x' associated with each EV denotes the total distance covered in miles with full charged capacity. For example, EV-40 suggests that it has a range of 30 miles.

13.4.3 Initial SOC and EVs load profile

SOC for EV is termed as the fraction of energy enduring in the battery. In this study, the minimum *SOC* is taken as 20% to achieve enhanced working life of the battery. Assume that the EV has a range of d_R, and energy consumption of EVs is comparative to the travel distance d. The initial *SOC* of EV according to daily travel distance d is

$$SOC_{initial} = \left\{ \begin{matrix} \left(1 - \dfrac{\aleph \times d}{d_R}\right) \times 100\%, & 0 < \aleph d < 0.8 d_R \\ 20\%, & \aleph d > 0.8 d_R \end{matrix} \right\} \tag{13.17}$$

The energy required for EVs battery is

$$E_{req} = \frac{(1 - SOC_{initial}).C}{\tau} \tag{13.18}$$

where C is the capacity of EVs battery, τ is the charging efficiency factor, and $\aleph$ is the mileage factor.

On the basis of the above discussion on anticipated modeling framework, the load profile of EVs is determined as illustrated in Figure 13.7. Initially, the driving

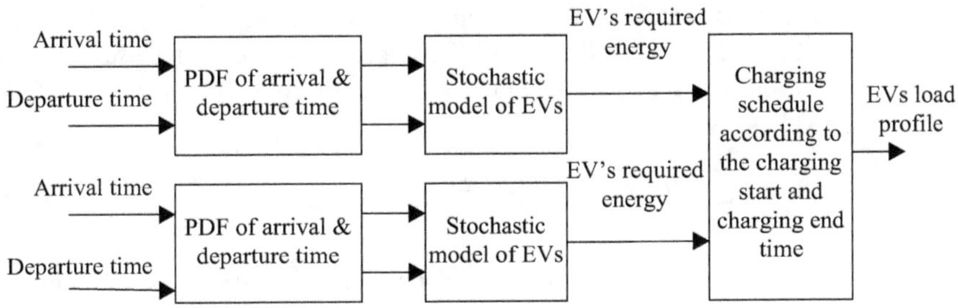

Figure 13.7 EV load modeling framework

pattern is created through planned stochastic technique as stated above. The required energy is calculated from (13.17) and (13.18). Thus, the EV load demand is estimated on the basis of required energy and available charging schedule at the charging location.

13.5 Financial and operational modeling

13.5.1 Real-time pricing policy

The real-time price depends significantly on the load demand, rising sharply as the load demand increases. In order to decrease the peak load of the network, a time of use (TOU) rate strategy is settled in order to schedule the charging of EVs according to the real-time price. The price is given as

$$r(t) = \beta_1 + \beta_2 \times \alpha^{\frac{P_{sys}^t + P^{wind} - P_{avg}}{P_{avg}}} \tag{13.19}$$

where β_1, β_2, α are the real-time price parameters, P_{sys}^t is demand from grid, P^{wind} is the wind output power, and P_{avg} is the average load demand from grid.

Due to wind power availability and the varying charging demand of EVs, P_{sys}^t and P^{wind} are time-varying quantities. The value of $\beta_1 = 0.1\$/\text{kWh}$, $\beta_2 = 0.2\$/\text{kWh}$, and $\alpha = 10$ are assumed in this study.

13.5.2 Degradation cost of EVs battery

Battery degradation is one of the challenging factors in V2G operation [30]. The cost associated with degradation is estimated as wear in V2G operation due to added cycling of battery. It is reported in [30] that the life cycle of the EV battery is governed through the discharge rate of EV (*DoD*) defined via the Wohler curve. Wohler curve [31] suggests that *DoD* is reciprocal to the failures of battery cycles. Precisely, Wohler curve is given as [31]

$$N_c(DoD) = \theta \times (DoD)^{\varnothing} \tag{13.20}$$

where constraints θ and $\varnothing$ control the category of battery. N_c is the total number of battery cycles. In practice, lithium-ion batteries are dominantly used in EVs [31].

The Wohler curve constraints for the above said battery are $a = 1331$ and $b = -1.825$ at 80% discharge. Precisely, the degradation cost of battery for one discharge from totally charged state $(DoD = 0)$ to a definite state $(DoD = DoD_S)$ is defined as [31]

$$C_d(0, DoD_s) = \frac{Cost_{EVbat} \times DoD_s \times E_{bat}}{N_c(DoD_s)} \tag{13.21}$$

where N_c is the number of cycles, $Cost_{EVbat}$ is battery investment cost (\$), and E_{bat} is usable energy of battery (kWh). The investment cost for EVs battery is anticipated to be \$315 [31].

Using (13.21), the degradation cost for discharge cycle from DoD_i to DoD_f is calculated as

$$C_d(DoD_i, DoD_f) = C_d(0, DoD_f) - C_d(0, DoD_i) \tag{13.22}$$

The overall degradation cost is the sum of degradation costs with respect to the total discharges given as

$$Cost_{deg} = \sum_{k=1}^{N_{dis}} C_d^k(DoD_i, DoD_f) \tag{13.23}$$

where N_{dis} is the number of discharge cycles.

13.5.3 Frequency regulation

In V2G operation, EVs can be used to deliver frequency regulation service for the distribution network. The EV aggregators share a contract with distribution network operator (DNO), which offers incentives to EVs owners for contribution in the regulation facility [32].

13.5.4 Spinning reserves

EVs with V2G proficiencies are highly appropriate for providing spinning reserves [32] because (1) EVs can respond rapidly to emergencies when required, and (2) since spinning reserves are hardly used for operation, it does not distress the battery lifetime so much as in other markets. Similarly, reserve incentive is framed between EV aggregators and DNO, increasing the reliability and performance of the distribution grid. In this study, V2G operation is assumed to be benefitted via regulation and spinning reserve incentives.

13.6 Formulation of charging/discharging strategy

In this section, we describe the charging/discharging scheme based on the above discussion in a concise mathematical formulation of optimization function (represented in (13.39)).

13.6.1 Charging/discharging energy of EVs

The prospect of charging during a day is given as a vector $T = [1, \ldots\ldots, t, \ldots\ldots.., T]$; here T represents the total time slots, while the EV is a vector $P = [1, \ldots., n, \ldots\ldots, N]$. For the nth EV, the required charging time and energy can be generated through the procedure given in Figure 13.7.

As verified in [9], EVs charging up to 100% of SOC makes more profit for the EV owner when it is used for V2G services as frequency regulation or spinning reserve service. The charging plan can be given by the following vector C_S as

$$C_{S,n} = \left[C_{S,n}^{t_{in,n}}, \ldots\ldots, C_{S,n}^{t}, \ldots\ldots., C_{S,n}^{t_{out,n}}\right] \forall n \in N$$

$C_{S,n}^t = 1$ means the charging state of nth EV at time slot t; $C_{S,n}^t = -1$ refers to the discharging state of nth EV at time slot t, and $C_{S,n}^t = 0$ represents the idle state of EV.

The energy constraint is defined as

$$E_{req,d} = \sum_{t=t_{in,n}}^{t_{out,n}} C_{S,n}^t \times P_{rate}^n \forall n \in N \tag{13.24}$$

where P_{rate}^n is the maximum charging power of the nth EV. It is considered that the EVs can respond to V2G services in idle state only. Accordingly, the following three vectors are defined to represent the charging, discharging, and idle state of EVs for every time instant as

$$C_n = \left[C_n^{t_{in,n}}, \ldots\ldots, C_n^t, \ldots\ldots., C_n^{t_{out,n}}\right] \forall n \in N \tag{13.25}$$

$$C_n^t = \left\{ \begin{matrix} 1 & if\ C_{S,n}^t = 1 \\ 0, & otherwise \end{matrix} \forall t \in T; \quad \forall n \in N \right\} \tag{13.26}$$

$$D_n = \left[D_n^{t_{in,n}}, \ldots\ldots, D_n^t, \ldots\ldots., D_n^{t_{out,n}}\right] \forall n \in N \tag{13.27}$$

$$D_n^t = \left\{ \begin{matrix} 1 & if C_{S,n}^t = -1 \\ 0, & otherwise \end{matrix} \forall t \in T; \quad \forall n \in N \right\} \tag{13.28}$$

$$I_n = \left[I_n^{t_{in,n}}, \ldots\ldots, I_n^t, \ldots\ldots., I_n^{t_{out,n}}\right] \forall n \in N \tag{13.29}$$

$$I_n^t = \left\{ \begin{matrix} 1 & if C_{S,n}^t = 0 \\ 0, & otherwise \end{matrix} \forall t \in T; \quad \forall n \in N \right\} \tag{13.30}$$

where n is the nth EV and t depicts the tth time slot.

The EVs capacity for frequency regulation can be termed as

$$P_{Reg}^t = \sum_{n=1}^{N} I_n^t \times P_{rate}^n, \quad \forall t \in T \tag{13.31}$$

Similarly, the spinning reserve capacity of EVs can be reflected in the form:

$$P_{SR}^{t} = \sum_{n=1}^{N} I_{n}^{t} \times P_{rate}^{n}, \quad \forall t \in T \tag{13.32}$$

The load demand due to charging of EVs is demonstrated as in (13.33) and the average demand of the network is calculated using (13.34).

$$P_{EV}^{t} = \sum_{n=1}^{N} C_{S,n}^{t} \times P_{rate}^{n}, \quad \forall t \in T \tag{13.33}$$

$$D_{n} = \frac{1}{T} \sum_{t=1}^{T} \left(P_{base}^{t} + P_{EV}^{t} \right) \tag{13.34}$$

where P_{base}^{t} is the base demand of the network.

The discharge energy (V2G) delivered by EVs is determined from:

$$E_{dis} = \sum_{n=1}^{N} \sum_{t=t_{in,n}}^{t_{out,n}} D_{n}^{t} \times \mathrm{P}_{\mathrm{rate}}^{\mathrm{n}} \tag{13.35}$$

13.6.2 *Optimizing strategy and function*

The optimizing strategy may be applied either to flatten the load demand, reducing charging cost based on real-time price, or even defined in terms of benefit earned by EV owners for incentive services. In the liberalized electricity market operation, one should realize that objectives may be defined for charging strategy in the interest of different groups, whether they are energy producers, DSO, EV owners, or aggregators. As a consequence, objectives set for charging optimization gets misaligned. The situation in the market gets worse with a high amount of wind penetration.

We, therefore, frame the objective function for charging cost of EVs considering different financial factors of wind-driven electricity prices. The objective function is represented in terms of charging cost, degradation cost during *V2G* operation, and the revenue earned by offering *V2G* services. Here, the charging cost comprises of the charging cost as well as the profit of EVs via V2G operation. The charging cost is as below:

$$Cost_{chg} = \sum_{t=1}^{T} P_{EV}^{t} \times r(t) \tag{13.36}$$

The degradation cost during V2G services is given in (13.23), and the profit earned through regulation service is as below:

$$Earn_{reg} = \sum_{t=1}^{T} P_{Reg}^{t} \times reg(t) \tag{13.37}$$

The revenue earned by spinning reserve service is given as

$$Earn_{SR} = \sum_{t=1}^{T} P^{t}_{Reg} \times SR(t) \tag{13.38}$$

The following optimization function expressed in terms of total cost for charging of EVs is minimized as

$$Cost = Cos_{chg} + Cost_{deg} - Earn_{reg} - Earn_{SR} \tag{13.39}$$

13.6.3 ESPSO

This subsection briefly explains an algorithm of PSO based on evolution strategy ESPSO. This two-layer space search algorithm explains the objective function. Initially, an optimal V2G plan is selected with the help of an evolution strategy in the upper layer; and on behalf of this V2G scheme, a PSO-based optimization process is anticipated to get the charging procedure at the lower layer.

The conventional methods for optimization are not efficient to solve the charging process described through the three states (charging, discharging, and idle) of EVs. The anticipated ESPSO methodology explains this problem by separating it into two layers, which significantly contracts the search space.

Here, each EV is observed as a dimension in the search space, and the initial point of sequence x_{sd} is the coordinate in this precise dimension. The V2G approach is developed according to (13.40) and (13.41).

The novel and developed particles keep modifying their flying trajectories as expressed by (13.42)–(13.44) as

$$es^{k}_{id} = s^{k}_{id} + w \times N(0, \sigma_d) \tag{13.40}$$

$$s^{k+1}_{id} = \begin{Bmatrix} es^{k}_{id}, & if\ Cost_i < pBest_i \\ s^{k}_{id}, & otherwise \end{Bmatrix} \tag{13.41}$$

$$v^{k+1}_{is^{k+1}_{id}} = wv^{k}_{is^{k}_{id}} + C_1 \times rand_1 \times \left(pBest_i - x^{k}_{is^{k}_{id}}\right) + C_2 \times rand_2 \times \left(gBest_i - x^{k}_{is^{k}_{id}}\right) \tag{13.42}$$

$$x^{k+1}_{is^{k+1}_{id}} = x^{k}_{is^{k}_{id}} + v^{k+1}_{is^{k+1}_{id}} \tag{13.43}$$

$$\mathrm{w} = \mathrm{w_{max}} - \mathrm{k} \times \frac{\mathrm{w_{max}} - \mathrm{w_{min}}}{\mathrm{k_{max}}} \tag{13.44}$$

Here s^{k}_{id} is the basic V2G plan of element, es^{s}_{id} is the developed V2G strategy.

The steps evolved in the proposed ESPSO algorithm are as follows:

1. Initialize all the particles in the search space;
2. Randomly select the position and velocity of each particle;
3. Generate the particles as suggested in (13.40);

4. Assess the wellness of every unique particle and its corresponding developed particle concerning the goal;
5. For best fitness value store the position of each particle;
6. Update the position and velocity of each particle according to (13.42) and (13.43);
7. In the event of maximum constraints value, go to Step 8; otherwise, return to Step 2;
8. Output the optimal result.

13.7 Case studies and discussion

13.7.1 Comparison of load profiles of EVs on different modeling schemes

In this study, the modeling of EV charging load demand is analyzed under the following cases:

- Case 1 (Proposed strategy): EV charging load profile modeled by the adaptive fuzzy model and LPMF.
- Case 2: EV load model considers the driving patterns, without assuming the EV constraints.
- Case 3: EVs load model considers the EV parameters, without driving pattern.

The simulation results obtained for the above cases using different charging schemes, dumb charging, PSO optimized charging, and proposed strategy, are illustrated in Figure 13.8. It can be noticed that the load demand of the above three cases are reasonably dissimilar. As indicated in the demand profiles for Cases 2 and 3, it can be inferred that imprecise modeling of the driving pattern will bring error in estimation of the EV load demand. It is worth noting that EVs charging for Case 3 (i.e. without driving pattern) dictate the maximum variation in the EV load profile. EV charging load profile modeled by the adaptive fuzzy model and LPMF result in more accurate estimation. Note further that the total load (base + EVs) with the proposed ESPSO strategy does not increase the expected network peak load, unlike dumb charging strategy. The valley period of base load is utilized via charging of EVs and thus ensuring a flat load demand profile.

13.7.2 Impact of wind and EV penetration

In this subsection, charging strategy is formulated corresponding to the charging costs and peak load demand minimization. This formulation is based on the available knowledge of real-time price, driving pattern, and base load. The effect of wind power availability according to wind speed on the variation in real-time price is taken into account, as described in the above section.

Figure 13.9 shows the real-time price with wind power using different charging schemes. The simulation is performed with 40% EV penetration level. As it can be observed, the maximum price corresponds to the peak load demand duration using dumb and PSO smart charging. On the other hand, the proposed ESPSO keeps a modest variation in the real-time price.

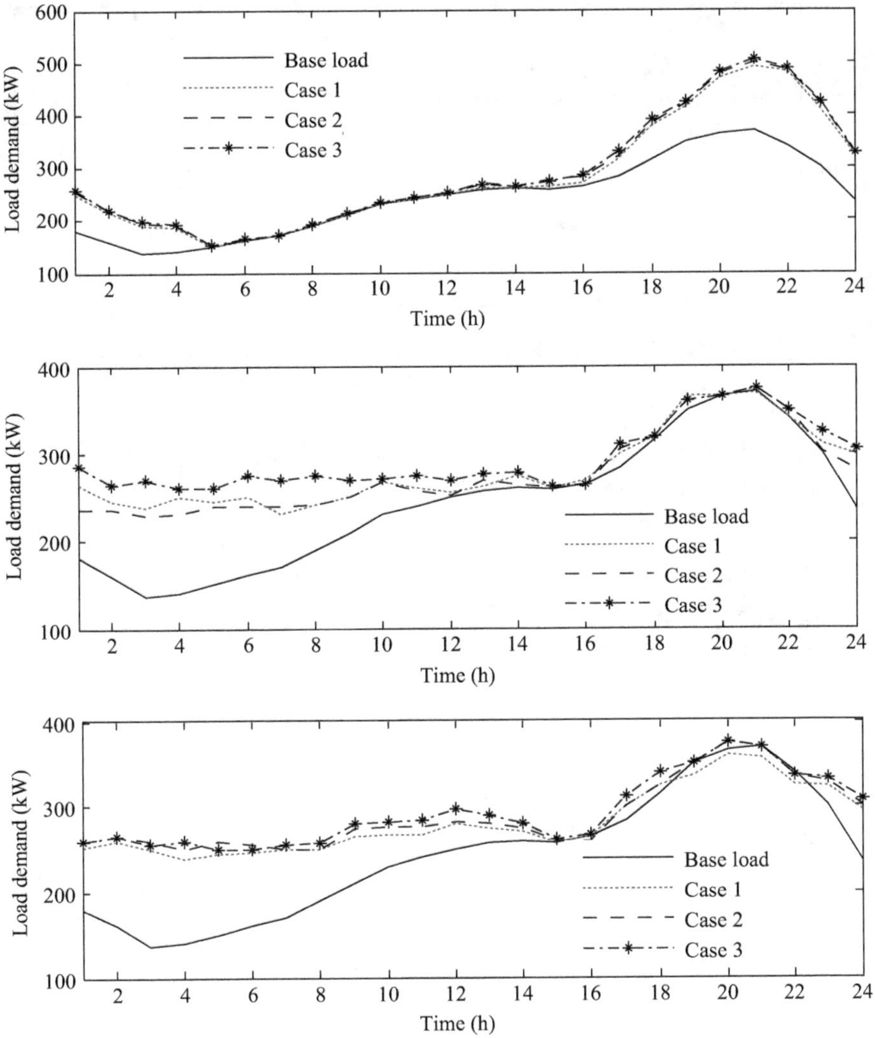

Figure 13.8 Load–demand curve with different EV modeling methods and charging schemes: (a) dumb charging, (b) PSO-centered smart charging, and (c) proposed ESPSO strategy

Figure 13.10 depicts the load demand (base load + EVs) of the distribution network with respect to the proposed charging schemes. It is indicated that in the dumb charging strategy, EVs impose additional congestion on the distribution grid during the peak base load demand. On the other hand, using PSO optimized smart charging as shown in Figure 13.10(a), a flat profile of EVs load demand is achieved, during the valley period of base load demand. It is worth noting that dumb charging dictates the maximum variation in load profile. Also, additional

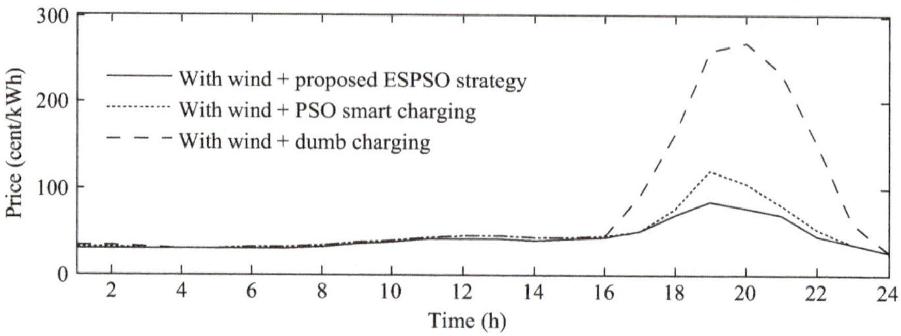

Figure 13.9 Real-time pricing for different charging strategies

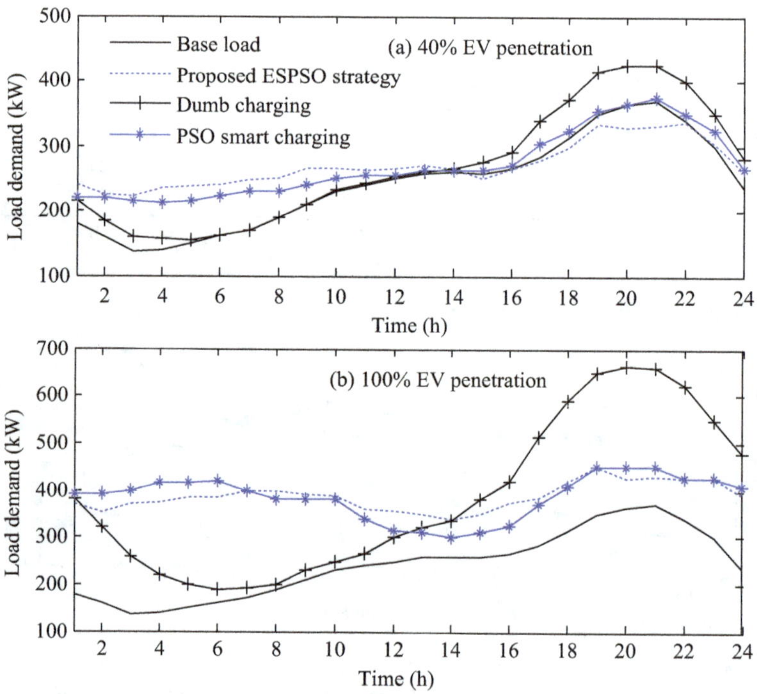

Figure 13.10 Load–demand curves for different charging algorithms: (a) 40% EV penetration and (b) 100% EV penetration

load is not demanded at the time of base load peak period. On the other hand, the load demand associated with the proposed charging scheme suggests least variation. In a broader perspective, the proposed charging scheme having considered the driving pattern and battery attributes, not only contributes to flat demand from EVs (valley period of base load) on the distribution grid, but also marginally leads to reduced responsive demand on the grid during peak period.

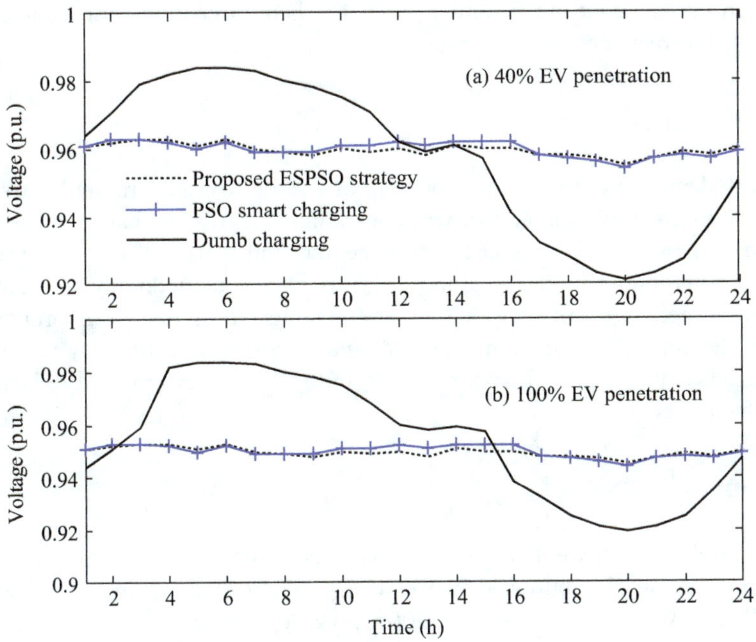

Figure 13.11 Voltage profile for different charging algorithms: (a) 40% EV penetration and (b) 100% EV penetration

Table 13.3 Cost of different charging strategies at 40% EV penetration

Charging strategy	Charging cost ($)	Degradation cost ($)	Earnings by V2G services ($)	Total benefit cost ($)
Dumb charging	782.25	N/A	57.32	724.93
PSO smart charging	122.34	N/A	57.32	65.02
Proposed ESPSO strategy	99.73	17.66	48.84	64.81

Furthermore, Figure 13.10(b) suggests that the proposed ESPSO control scheme distributes less V2G power for peak load saving at the maximum EV infiltration status. For higher EV integration, V2G capacity is more supported either for frequency regulation or spinning reserve service. In this situation, the load demand is flattened by shifting the high demand in low demand hours.

Figure 13.11 indicates the voltage profile for various charging algorithms at different EV penetration levels. The anticipated ESPSO procedure can decrease the voltage deviation successfully.

Table 13.3 depicts the charging cost by the different schemes, indicating the lowest cost being achieved by the ESPSO charging scheme. The degradation cost is also taken into account in the said scheme. The benefit cost lies in the same range as that of the smart charging scheme.

13.8 Conclusion

This work depicts a probabilistic modeling of wind power generation by assuming the stochastic idea of wind and system constraints. Next, a stochastic modeling for the load demand of EVs has been implemented. On behalf of the stochastic EV model, a commercial model is proposed to use V2G methodology in a domestic distribution network. In addition, an ESPSO algorithm has been proposed to enhance the control schemes of EVs and attain various objectives by refining the power quality, decreasing the peak load, offering V2G services, and reducing the total effective cost.

References

[1] L. Kelly, A. Rowe and P. Wild, "Analyzing the impacts of plug-in electric vehicles on distribution networks in British Columbia," in Proc. *Electrical Power & Energy Conference*, 2009, pp. 1–6.

[2] S. Deilami, A. S. Masoum, P. S. Moses, and M. A. S. Masoum, "Real-time coordination of plug-in electric vehicle charging in smart grids to minimize power losses and improve voltage profile," *IEEE Trans. Smart Grid*, 2011, vol. 2, no. 3, pp. 456–467.

[3] S. Babaei, D. Steen, T. A. Le, O. Carlson, and L. Bertling, "Effects of plug-in electric vehicles on distribution systems: A real case of Gothenburg," in *Proc. IEEE Innov. Smart Grid Technol. Conf.Eur. (ISGT)*, Gothenburg, Sweden, Oct. 2010, pp. 1–8.

[4] R. C. Green, II, L. Wang, and M. Alam, "The impact of plug-in hybrid electric vehicles on distribution networks: A review and outlook," Renew. Sustain. Energy Rev., 2011 vol. 15, no. 1, pp. 544–553.

[5] E. Sortomme and M. A. El-Sharkawi, "Optimal charging strategies for uni-directional vehicle-to-grid," *IEEE Trans. Smart Grid*, 2011, vol. 2, no. 1, pp. 131–138.

[6] K. C. Nyns, E. Haesen, and J. Driesen, "The impact of charging plug-in hybrid electric vehicles on a residential distribution grid," *IEEE Trans. Power Syst.*, 2010, vol. 25, no. 1, pp. 371–380.

[7] K. Clement, E. Haesen, and J. Driesen, "Coordinated charging of multiple plug-in hybrid electric vehicles in residential distribution grids," in *Proc. Power Syst. Conf. Expo.*, Seattle, WA, USA, Mar. 2009, pp. 1–7.

[8] S. Shafiee, M. Firuzabad, and M. Rastegar, "Investigating the impacts of plug-in hybrid electric vehicles on power distribution systems," *IEEE Trans. Smart Grid*, 2013, vol. 4, no. 3, pp. 1351–1360.

[9] Jayakrishnan R. Pillai, P. Thogersen, J. Moller, and B. Bak-Jensen, "Integration of electric vehicles in low voltage Danish distribution grids," Power and Energy Society General Meeting, *IEEE*, July 2012, no. 978-1-4673-2727-5.

[10] A. Lojowska, D. Kurowicka, G. Papaehthymiou, and L. Sluis, "Stochastic modeling of power demand due to EVs using copula," *IEEE Trans. Power Syst.*, 2012, vol. 27, no. 4, pp. 1960–1968.

[11] N. G. Paterakis, M. Gibescu, "A methodology to generate power profiles of electric vehicle parking lots under different operational strategies," *Applied Energy*, 2016, vol. 173, pp. 111–123.

[12] T. Lee, B. Adornato, and Z. Filipi, "Stochastic modeling for studies of real-world PHEV usage: Driving schedule and daily temporal distributions," *IEEE Trans. Veh. Technol.*, 2012, vol. 61, no. 4, pp. 4193–1502.

[13] Amini, M. Hadi, Amin Kargarian, and OrkunKarabasoglu. "ARIMA-based decoupled time series forecasting of electric vehicle charging demand for stochastic power system operation." *Electric Power Systems Research* 140 (2016), pp. 378–390.

[14] ReinhildeD'hulst, Fjo De Ridder, Bert Claessens, LukKnapen and Davy Janssens. "Decentralized coordinated charging of electric vehicles considering locational and temporal flexibility," *Int. Trans on Electrical Energy Systems*, 2015, vol. 25, no. 10, pp. 2562–2575.

[15] Kempton, Willett, and Steven E. Letendre. "Electric vehicles as a new power source for electric utilities." *Transportation Research Part D: Transport and Environment*, 1997, vol. 2, no. 3, pp. 157–175.

[16] S. Han, S. Han, and K. Sezaki, "Economic assessment on V2G frequency regulation regarding the battery degradation," in Proc. *IEEE PES Innovative Smart Grid Technol. (ISGT),* Jan. 2012, pp. 1–6.

[17] A. Y. Saber and G. K. Venayagamoorthy, "Plug-in vehicles and renewable energy sources for cost and emission reductions," *IEEE Trans. Ind.Electron.*, 2011, vol. 58, no. 4, pp. 1229–1238.

[18] Jun Tan and L. Wang, "Integration of Plug-in Hybrid Electric Vehicles into Residential Distribution Grid Based on Two-Layer Intelligent Optimization," *IEEE Trans. Smart Grid*, 2014, vol. 5, no. 4, pp. 11774–1784.

[19] IEEE 34 Node Test Feeder—IEEE Distribution System Analysis Subcommittee [Online]. Available:http://www.ewh.ieee.org/soc/pes/dsacom/testfeeders.html

[20] J.A. Jardini, C.M. Tahan, and M.R. Gouvea, "Daily load profiles for residential, commercial and industrial low voltage consumers," *IEEE Trans. Power Del.*, 2000, vol. 15, no. 1, pp. 375–380.

[21] J.H. Chow and K.W. Cheung, A toolbox for power system dynamics and control engineering education and research, *IEEE Transactions on Power Systems*, 1992, vol. 7, no. 4, pp. 1559–1564.

[22] S. Zhang *et al.*, "Statistical voltage quality assessment method for grids with wind power generation," *IET Renew. Power Gener.*, 2010, vol. 4, no. 1, pp. 43–54.

[23] I. Y. F. Lun and J. C. Lam, "A study of Weibull parameters uses long term wind observations," *Renew. Energy*, 2000, vol. 20, no. 2, pp. 145–153.

[24] T. Ackermann, "Wind power in power systems," *Wind Eng.*, 2006, vol. 30, no. 5, pp. 447–449.

[25] R. Meir *et al.*, DTI Noise Working Group, ETSU, U.K., The Assessment and Rating of Noise from Wind Farms, Final Report ETSU-R-97, Sep. 1996.

[26] National Household Travel Survey, 2009 [Online]. Available at: http://nhts.ornl.gov

[27] J. Kennedy and R. C. Eberhart, "Particle swarm optimization," in *IEEE Int. Conf. Neutral Netw.*, 1995, pp. 1942–1948.

[28] S. Sachan and N. Kishor, "Charging of electric vehicles under contingent conditions in smart distribution grids," *2016 IEEE International Conference on Power Electronics, Drives and Energy Systems (PEDES)*, Trivandrum, 2016, pp. 1–5.

[29] YugongLuo, Tao Zhu, Shuang Wan, Shuwei Zhang, and Keqiang L, "Optimal charging scheduling for large-scale EV deployment based on the interaction of the smart-grid and intelligent-transport systems," *Energy*, 2016, vol. 97, pp. 359–368.

[30] D. Dallinger, J. Link, and M. Büttner, "Smart grid agent: Plug-in electric vehicle," *IEEE Trans. Sustain. Energy*, 2014, vol. 5, no. 3, pp. 710–717.

[31] *Electric Vehicles–The Benefits and Barriers*, S. Soylu Ed. Rijeka, Croatia: InTech, 6 Sep 2011, p. 252, ISBN 978-953-307-287-6.

[32] Junjie Hu *et al.*, "A multi-agent system for distribution grid congestion management with electric vehicles," in *Engineering Applications of Artificial Intelligence, Elsevier*, 2015, vol. 38, no. 1, pp. 45–58.

Chapter 14

Optimal location and charging of electric vehicle with wind penetration

Sulabh Sachan[1] *and Nand Kishor*[1]

Large number of electric vehicles influences the aggregate power request significantly. Demand estimation is typically intended for the occasionally changing load designs. Though, with the fleet of electric vehicles, daily charging request makes conventional strategies less precise. In this work, prime site for centralized charging of electric vehicles and its charging scheme in a modified IEEE-34 distribution network is planned. To determine the optimum site, an hourly load demand is changed at specified junction nodes and the equivalent voltage sensitivity indexes are calculated. These indexes are derived from the Newton–Raphson load flow analysis in proposed distribution network and aid in choosing appropriate site for charging. Next, in the study, optimum charging of electric vehicles at selected node is conferred based on real-time pricing and wind power output data. The charging based on real-time price would lead to shift in the situations of peak load occurrence in the network. Similarly, charging based on wind power availability would aid in optimizing the customer's benefit in terms of cost. The peak period is shifted to valley period wherein real-time price is low or wind output is high.

Nomenclature

EV	Electric Vehicle
NRLF	Newton–Raphson Load Flow
CDF	Congestion distribution factor
DCE	Profit in charging/discharging shifting
MC	Marginal cost
P_{ij}, Q_{ij}	Real and reactive power flow respectively between bus i and j
δ_i, δ_j	Voltage angle at bus i and bus j

[1]Department of Electrical Engineering, Motilal Nehru National Institute of Technology Allahabad, Prayagraj, India

n	Number of nodes in distribution network
[J]	Jacobian matrix
U1, U2, U3, U4	Voltage change (angle and magnitude) against power (active and reactive)
ΔP_m	Required real power to remove congestion
P_{ij}^c	Real power obtained by load flow
a_{ij}^m	Congestion distribution factor (CDF) of active power for line ij and node m
S_{ij}^{max}	Maximum line apparent power
r_{ij}^m	Congestion distribution factor (CDF) of reactive power for line ij and node m
ΔF	Real and reactive power flows matrix
$[CDF]$	Congestion distribution factors matrix
ΔI	Real and reactive power nodal injections matrix
φ	Mileage factor of electric vehicle
d_R	Electric vehicles range
SOC	Final state-of-charge of each electric vehicle in parking lot
C	Electric vehicles battery capacity
τ	Electric vehicles charging efficiency factor
E_{req}	Energy requires for electric vehicles battery
$O_{f,l}$	Optimized load of charging EVs
L[t]	Real-time load curve
N	Total number of vehicles of each type i
X_m	Starting time of charging for m – th vehicle
$T_{C_{m,l}}$	Time required for full charging to $m - th$ vehicle
$SoC_{m,l}$	Final state-of-charge of $m - th$ vehicle
$CS_{m,l}$	Rate of charge for $m - th$ vehicle
$O_{f,r}$	Minimum charging cost of EV
P[t]	Real-time price curve
$T_{C_{m,r}}$	Time required for full charging to $m - th$ vehicle in pricing scenario
$SoC_{m,r}$	Final state-of-charge of $m - th$ vehicle in pricing scenario
$CS_{m,r}$	Rate of charge for $m - th$ vehicle in pricing scenario
$O_{f,w}$	Optimized load for EVs charging according to wind power output and the real-time price data
$W[t]$	Wind power output
$T_{C_{m,w}}$	Time required for full charging to $m - th$ vehicle in wind power and pricing scenario
$SoC_{m,w}$	Final state-of-charge of $m - th$ vehicle in wind power and pricing scenario

$CS_{m,w}$	Rate of charge for $m-th$ vehicle in wind power and pricing scenario
C_w	Marginal cost of electric energy
P	Net demand supplied to feeder
P_{max}	Maximum demand supplied to feeder
P_{min}	Minimum demand supplied to feeder
A_1, A_2	Area represents charging and discharging respectively in load curve of Figure 14.13
W_{EVC}	Energy transfer to the electric vehicle during charging
W_{EVD}	Discharged energy from electric vehicle
n	Round trip efficiency of EVs battery
C_C	Extra cost of electric vehicles charging energy
C_D	Extra cost of electric vehicles discharging energy
T_B	Discharging time of electric vehicle during peak demand
W_min	Minimum wind power output
W_avg	Average wind power output
W_max	Maximum wind power output

14.1 Introduction

Plug in electric vehicles (EVs) will play a central role in decarbonising road transport in the near future. But this new type of electricity load needs strategic management, there has been a rising concern regarding the large scale penetration of EVs as energy storage units in the power systems. Several studies focused on the siting and sizing of energy storage systems [1] in distribution network when breakdown occurs. Unscheduled placement of energy storage increases the vulnerabilities of voltage fluctuations that may damage other devices in the distribution network [2]. Charging and discharging of battery is not feasible at some instances due to voltage variation constraints. Furthermore, the proficiency of the energy storage to maintain the voltage and frequency within limit in the distribution network can be deteriorated. The optimal storage site and size also determines the verge for renewable energy penetration, and the so-called stable operation (voltages and frequency) of a system [3,4]. Having an oversized storage connected at a nonoptimal bus within a considerably large-sized distribution grid will not intrinsically have the best impact in terms of stability on the rest of the buses. The energy storage system supplies to the utility grid at high demand period. The appropriate location of energy storage in distribution network also empowers to relieve the voltage problems at various buses [5].

Electric vehicles (EVs) are receiving much attention recently because of their ability as environmentally friendly power source. Hence, public charging places are anticipated to increase at distribution network level in future [6]. EVs will be charged via charging stations in distribution network. EVs charging demand [7] in residential area will be supported by distribution network. Due to the regular long-time usage of the cars, EV requires more power than common household appliance in charging scenario [8]. Consequently, voltage drop and power flow violation may arise on distribution grid [8]. Luo *et al.* in [9] proposed a charging scheme to tackle the problem of voltage violation and power loss minimization. Kempton *et al.* in [10] suggested a smart charging technique is to shift peak load into valley load periods. According to Amini *et al.* in [11], EV battery operates as V2G to decrease adverse effects at peak load.

Soares *et al.* in [12] proposed that inappropriate strategy in location of battery storage and size brings the issue of low/high voltage in micro grids. Distribution system comprises of feeders and each of them having provision for connection of EVs. Due to either breakdown or congestion problem at any feeder, the EVs charging may not be allowed at home. In such situation, a centralized charging station at an optimally selected feeder with proper scheduling is a viable option. Montoya *et al.* [13] have proposed the location of centralized charging station with the objective of minimizing system costs including power loss and voltage deviation. Neyestani *et al.* [14] discussed the expansion planning for distribution network considering site location and sizing of EVs charging stations. The objective function is formulated in terms of investment cost and the functioning cost. Similarly, according to [15], the objective function defined as combined cost of charging stations and EVs consumers is minimized to solve the proposed problem. The optimal planning of EVs charging stations is defined as a two-step optimization problem in [16]. Initially, the optimal places of EVs charging stations are identified based on screening factors, such as environmental concerns and service radius of charging stations. In the second step that follows, optimal size while minimizing the total cost is determined. In the framework toward optimal planning development, charging strategies with battery swap stations (BSSs) is presented in [17]. A combined agenda for scheduling of public charging spots and roadside fast charging stations in a metropolitan area is proposed in the previous studies [18,19]. The objective function minimizes the social budgets of the complete charging structure. In addition to the mentioned issues, there is a major concern regarding the effect of EV integration on the stochastic operation of power systems [10]. This issue is tackled in [11]. The results have been later used for solving chance-constrained unit commitment problem.

Although several researchers (see for example [14–17]) have discussed the study related to allocation of centralized charging location in the distribution grid; in their studies, authors have not performed sensitivity analysis to find optimal location. In other words, the proposed solutions mostly evaluates the effect of different allocations of the power distribution network.

Sachan *et al.* [20] have performed sensitivity analysis for centralized charging location inside distribution network but they did not perform congestion management for EVs. Congestion management along with stochastic load modeling of EVs to shift charging/discharging is added significantly in this paper.

It is worth noting that on integration of EVs for charging/discharging, especially in the distribution grid, the network loading would be severely affected.

Thus, it is important that voltage sensitivity analysis is performed so as to decide the best location for charging/discharging of EV according to the existing base load conditions. Also, it is important to determine the number of EVs for charging/discharging. On considering these issues related to EVs integration in the distribution, the problem has been formulated in the chapter and results have been discussed.

The significant highlights of the paper can be summarized as follows:

- A novel algorithm for placement of public charging station for EVs on distribution grids.
- Advantage of this methodology is that the situations of over voltage/under voltage in the distribution network can be overcome by performing charging/discharging of EVs at suggested optimal location. This methodology also ensures the stability of distribution network.
- EVs charging according to the variation in real-time pricing and wind power availability in the distribution grid. The charging based on real time price would lead to shift in the situations of peak load occurrence in the network. Similarly, charging based on wind power availability would aid optimizing the customer's benefit in terms of cost.

14.2 Test system

The proposed study is engrossed on IEEE-34 distribution network [21]. The proposed distribution network is transformed to 12 kV level from original system [21]. This network is coupled to the main grid at bus 800. A renewable source (wind system) is included in the distribution network as depicted in Figure 14.1.

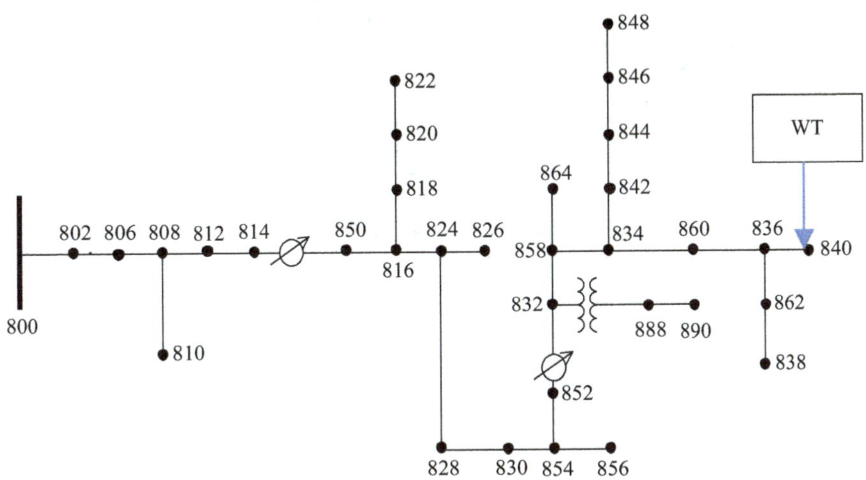

Figure 14.1 Modified IEEE 34 distribution network [21]

In this work, base demand is taken as fixed demand, whereas EVs charging load is considered as controllable demand. It is supposed that utility offers forecasted electricity price and base demand of a day. The objective of this paper is to obtain the best optimal location for centralized charging (G2V) and discharging (V2G) of EVs in distribution system. The study also depicts load shift optimization of EV's charging pattern based on real time pricing and wind power output data.

14.3 Sensitivity analysis

The presented methodology considers the voltage sensitivity analysis, which is usually implemented in load flow analysis to find the connection between voltage magnitude and active/reactive power at a precise bus. These correlations are commonly described in $P-V$ and $Q-V$ curves. In a transmission network, a $Q-V$ curve is studied to evaluate the impact of the reactive power on voltage magnitude in terms of dQ/dV [13]. In distribution grid, the association between voltage and power is not as straightforward as that in the transmission network, as the length of cable opted for distribution network is much shorter than that of transmission system. Due to this inductance of cable has less value than that of transmission network. This needs to have both, the active as well as reactive power supply installed in the system. The flowchart of the proposed methodology is depicted in Figure 14.2.

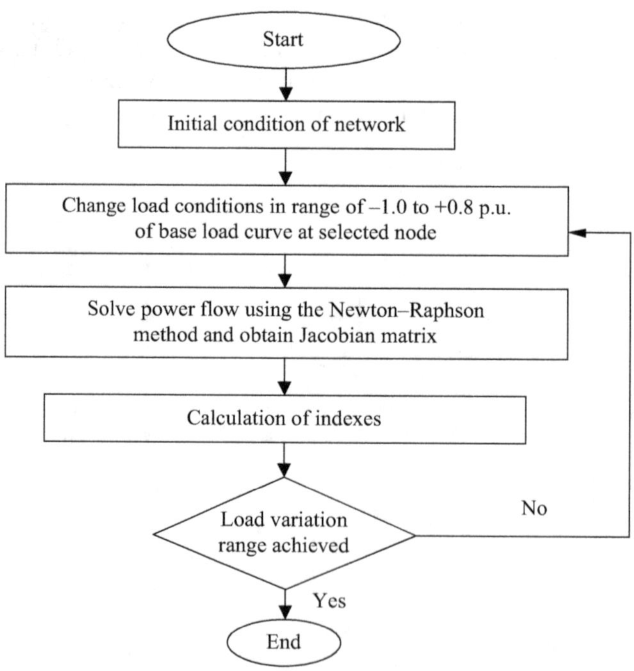

Figure 14.2 Flow chart of proposed methodology

In voltage sensitivity study, four indexes are derived from Newton–Raphson Load Flow (NRLF) analysis to estimate the sensitivity of load buses in the proposed distribution network.

The real and reactive power calculations for bus i and j are calculated from:

$$P_{ij} = V_i^2 G_{ii} + \sum_{\substack{j=1 \\ j \neq i}}^{N} \left| V_i V_j Y_{ij} \right| \cos\left(\theta_{ij} - \delta_i + \delta_j\right) \tag{14.1}$$

$$Q_{ij} = -V_i^2 B_{ii} - \sum_{\substack{j=1 \\ j \neq i}}^{N} \left| V_i V_j Y_{ij} \right| \sin\left(\theta_{ij} - \delta_i + \delta_j\right) \tag{14.2}$$

By differentiating (14.1) and (14.2) with respect to voltage angles and magnitudes, the NRLF equation is depicted as (14.3).

$$\begin{bmatrix} \Delta P \\ \Delta Q \end{bmatrix} = \begin{bmatrix} J_1 & J_2 \\ J_3 & J_4 \end{bmatrix} \begin{bmatrix} \Delta \delta \\ \Delta V \end{bmatrix} \tag{14.3}$$

Equation (14.3) depicts sensitivity indexes in form of voltage angles and magnitudes. The change observed in voltage angle and magnitude with respect to supplied real and reactive power can be attained by describing inverse Jacobian matrix [U] as in the form:

$$\begin{bmatrix} \Delta \delta \\ \Delta V \end{bmatrix} = \begin{bmatrix} \mathrm{U}_1 & \mathrm{U}_2 \\ \mathrm{U}_3 & \mathrm{U}_4 \end{bmatrix} \begin{bmatrix} \Delta P \\ \Delta Q \end{bmatrix} \tag{14.4}$$

$$U_1 = \begin{bmatrix} \frac{\partial \delta_2}{\partial P_2} & \frac{\partial \delta_2}{\partial P_3} \cdots & \frac{\partial \delta_2}{\partial P_n} \\ \vdots & \vdots & \vdots \\ \frac{\partial \delta_n}{\partial P_2} & \frac{\partial \delta_n}{\partial P_3} \cdots & \frac{\partial \delta_n}{\partial P_n} \end{bmatrix}; \; U_2 = \begin{bmatrix} \frac{\partial \delta_2}{\partial Q_2} & \frac{\partial \delta_2}{\partial Q_3} \cdots & \frac{\partial \delta_2}{\partial Q_n} \\ \vdots & \vdots & \vdots \\ \frac{\partial \delta_n}{\partial Q_2} & \frac{\partial \delta_n}{\partial Q_3} \cdots & \frac{\partial \delta_n}{\partial Q_n} \end{bmatrix}$$

$$U_3 = \begin{bmatrix} \frac{\partial V_2}{\partial P_2} & \frac{\partial V_2}{\partial P_3} \cdots & \frac{\partial V_2}{\partial P_n} \\ \vdots & \vdots & \vdots \\ \frac{\partial V_n}{\partial P_2} & \frac{\partial V_n}{\partial P_3} \cdots & \frac{\partial V_n}{\partial P_n} \end{bmatrix}; \; U_4 = \begin{bmatrix} \frac{\partial V_2}{\partial Q_2} & \frac{\partial V_2}{\partial Q_3} \cdots & \frac{\partial V_2}{\partial Q_n} \\ \vdots & \vdots & \vdots \\ \frac{\partial V_n}{\partial Q_2} & \frac{\partial V_n}{\partial Q_3} \cdots & \frac{\partial V_n}{\partial Q_n} \end{bmatrix}$$

U_1, U_2, U_3, and U_4 depict the impact of voltage variation (angle and magnitude) with respect to power (active and reactive) on each bus of distribution network. They comprise all the data about the line structure and configuration. Here, matrix elements start from the partial derivatives of the second element rather than the first element because first elements belong to the slack bus and rest elements correspond to the load bus of distribution network. Next, subsequent four sensitivity indexes are depicted [13].

$$\text{Index } I_1 = \frac{\sum_{k=2,i=2}^{n} \frac{\partial \delta_k}{\partial P_i}}{n};$$

$$\text{Index } I_2 = \frac{\sum_{k=2,i=2}^{n} \frac{\partial V_k}{\partial P_i}}{n}$$

$$\text{Index } I_3 = \frac{\sum_{k=2,i=2}^{n} \frac{\partial \delta_k}{\partial Q_i}}{n};$$

$$\text{Index } I_4 = \frac{\sum_{k=2,i=2}^{n} \frac{\partial V_k}{\partial Q_i}}{n}$$

14.3.1 Node selection

In optimal location for centralized charging, the sensitivity indexes are estimated when the Newton–Raphson method converges to meet the necessities of the load flow. The integration of EVs to the system is expected to cause congestion of line segments. Therefore, in a meshed system having several nodes, changing pattern of EVs integration may cause line flows to vary. Subsequently, line segments that are not even essentially closer to EVs charging location may be driven to their limits as their number changes during the 24 h.

The rated base demand is applied at junction nodes 816, 834, and 836 in the distribution network and load flow solution is obtained followed by sensitivity indexes. The variation of base demand is depicted in Figure 14.3.

Further, the four indexes are calculated for every node by varying base demand to decide appropriate locations for EVs charging. Here, the rated base demand is changed in the range of −1.0 to +0.8 p.u. in the network and the proposed voltage sensitivity indexes are estimated. Figure 14.4 represents the variation of sensitivity indexes at all the nodes to explore the possibility of charging/discharging of EVs. It is important that these indexes do not violate the constraint limits of the network so as to maintain stability. Any node showing most sensitivity toward the connected base load demand is most suitable for EVs integration. It is obvious that the pattern of indexes variation remain similar since, the variables, voltage, angle, real power, and reactive power are related to each other in nonlinear characteristic form of the dynamic behavior.

Here, it can be concluded that the voltage sensitivity indexes prominently describe the best node to install the EVs charging station, with regulation of real and reactive power, having substantial effect on voltage magnitudes of network.

As depicted above, the base demand is varied in the range of −1.0 to +0.8 p.u. in the network. Here negative range (−1.0 to 0) represents that power is taken from

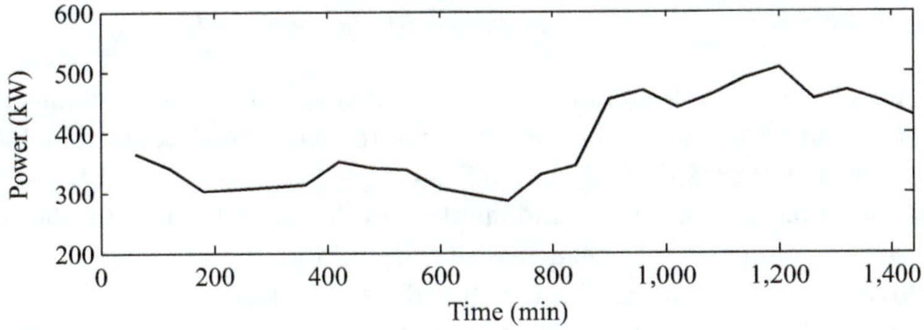

Figure 14.3 Base load demand

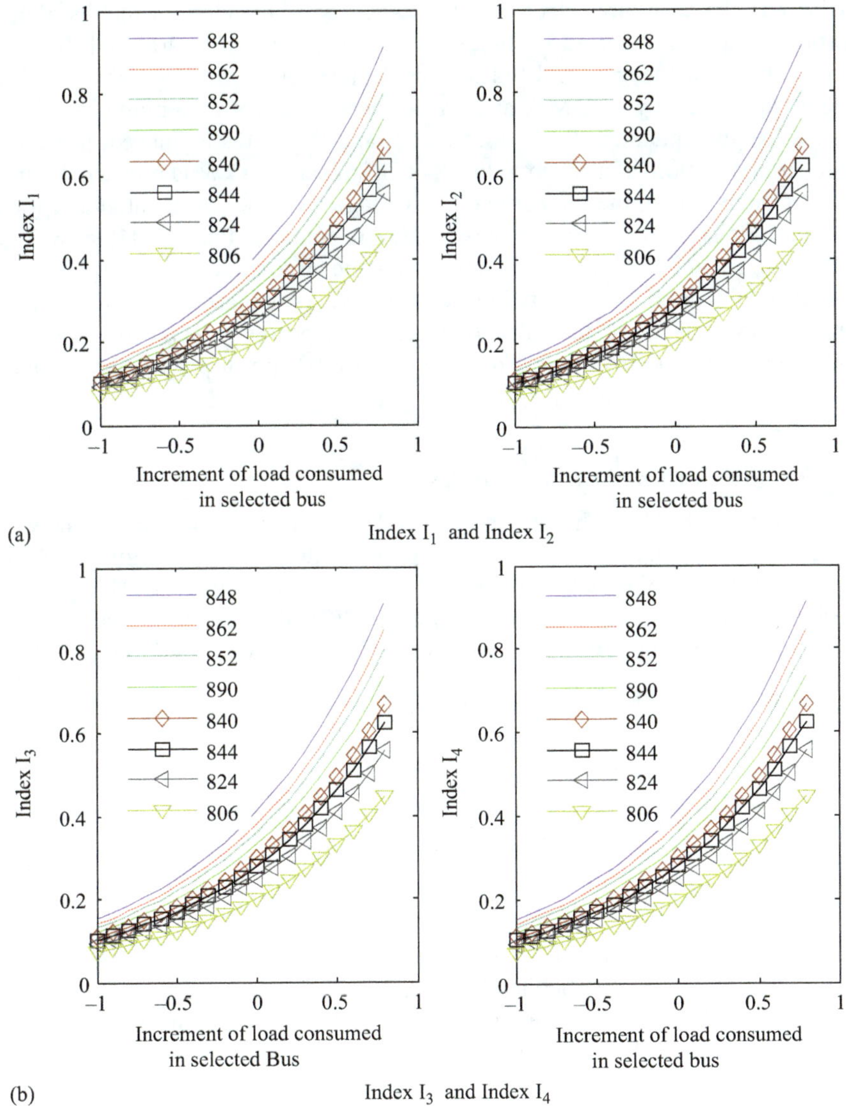

Figure 14.4 Variation of indices at different nodes with loading conditions

grid and positive varying load range (0 to 0.8) represents that power is injected into grid. From Figure 14.4, it can be seen that the indexes for selected four locations (bus 890, bus 848, bus 852, and bus 862) are very large, which indicates that the variation of voltage magnitude and angles are more sensitive to the varying injecting power range (0–0.8) than that of varying load power range (−1.0 to 0). From above discussions it can be concluded that overall effect of charging and discharging of EVs at most sensitive bus make the distribution network more stable than that of at least sensitive bus.

As a consequence, indexes I_2 and I_4 have substantial effect. Four sites; bus 890, bus 848, bus 852, and bus 862, denoted as sites I–IV, correspondingly may be considered. The buses are selected on their sensitivity intended for the distribution network, compared indicated by Indexes I_2 and I_4 in Figure 14.4(a) and (b).

According to Index I_2 and Index I_4 as depicted in Figure 14.4(a) and (b) the average voltage change per MW at selected locations screen the graph in different network loading situations. Thus, locations II and IV are distinguished. It means if EVs charging station is installed at bus 848 or 862, and 10 kW is injected at one of these locations. Then, the largest average voltage increase can be attained, for all buses in the distribution network. Also, if EV is connected at bus 848 or 862 and 10 kW is added at one of these locations. Then, the largest average voltage decrease per node is obtained, for all buses in the network. As depicted above the effect of injecting power is more than that of consuming power at same bus thus both locations are suitable to install a public charging station in proposed distribution network.

14.3.2 System operation algorithm

As the location of public charging station is decided inside a distribution network, the next issue is to tackle the congestion problem due to EVs charging and other loads. The EVs can remove congestion from distribution grid. The proposed algorithm for congestion management through *EV* is depicted in Figure 14.5. The algorithm to avoid congestion using EV is described in detail by following steps:

1. The real power needed to remove congestion in a line can be obtained by (14.5).

$$\Delta P_m = \frac{\left(P_{ij}^c - \sqrt{\left(S_{ij}^{max}\right)^2 - Q_{ij}^2}\right)}{a_{ij}^m} \tag{14.5}$$

ΔP_m = Required real power to remove congestion
P_{ij}^c = Real power obtained by load flow
a_{ij}^m = Congestion distribution factor (CDF) of active power for line ij and node m, S_{ij}^{max} = maximum line apparent power and Q_{ij} = reactive power flow

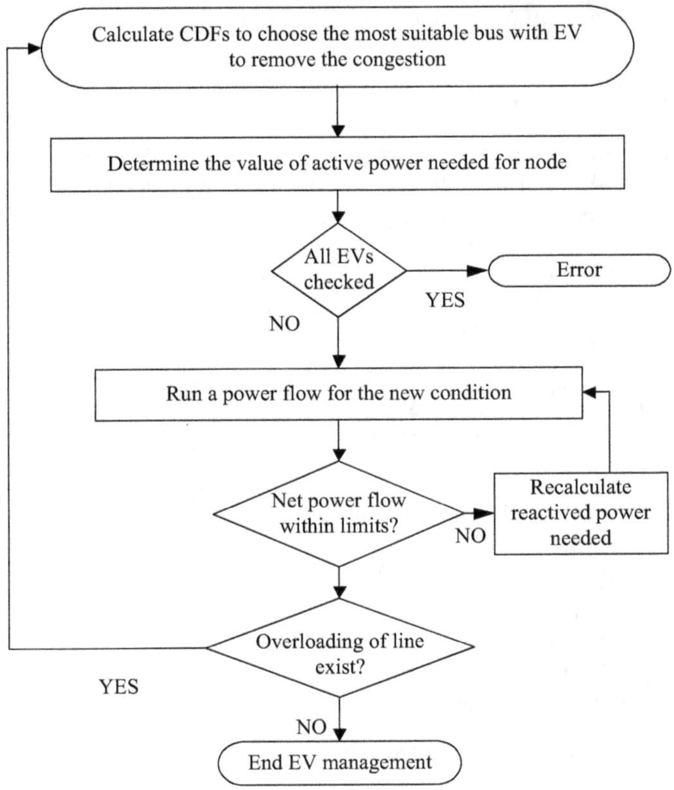

Figure 14.5 EVs management algorithm block diagram

2. Through power flow, it is checked if total power flow is within a limit, according to (14.6).

$$(1-\varepsilon)\cdot S_{ij}^{max} < S_{ij}^{c} < S_{ij}^{max} \tag{14.6}$$

where ε defines a limit for net power flow.
3. The change in reactive power flow for the line is defined by reactive power CDFs, as depicted in (14.7).

$$Q_{ij}^{w} = Q_{ij}^{w-1} + r_{ij}^{m}\cdot \Delta P_{m}^{w-1} \tag{14.7}$$

w = iteration number
r_{ij}^{m} = CDF of reactive power for line ij and node m
Q_{ij}^{w} is used to calculate EVs contribution for other node (14.8).

$$\Delta P_{m}^{w} = \Delta P_{m}^{w-1} + n^{w}\cdot \Delta x^{w} \tag{14.8}$$

n^{w} = weight lower than 1

Δx^w = correction term calculated from (14.9)

$$\Delta x^w = \pm\left(\Delta P_m^{up} - \Delta P_m^0\right) \tag{14.9}$$

where plus/minus sign is selected if net power flow is above/below the limit defined in (14.6). ΔP_m^{up} is the amount of real power calculated in (14.5) with updated reactive power. The algorithm is repeated until congestion is alleviated from each line of distribution grid. Congestion distribution factors (CDFs) can be obtained using Jacobian as defined earlier in (14.3). It is clearly shown that these matrices are linked with magnitudes and angle in each node except slack bus.

$$\begin{bmatrix} \Delta P_{ij} \\ \Delta Q_{ij} \end{bmatrix} = \begin{bmatrix} J_1^f & J_2^f \\ J_3^f & J_4^f \end{bmatrix} \begin{bmatrix} \Delta\delta_m \\ \Delta V_m \end{bmatrix} \tag{14.10}$$

It is possible to obtain active and reactive CDFs as depicted in (14.11).

$$\Delta F = J^f.J^{-1}.\Delta I = [CDF].\Delta I \tag{14.11}$$

ΔF = real and reactive power flows matrix
$[CDF]$ = CDFs matrix
ΔI = real and reactive power nodal injections matrix

The CDFs matrix can be divided into four submatrices:

$$[CDF] = \begin{bmatrix} CDF_1 & CDF_2 \\ CDF_3 & CDF_4 \end{bmatrix} \tag{14.12}$$

where CDF_1 and CDF_3 contain the real and reactive power CDFs a_{ij}^m and r_{ij}^m.

In conclusion, EVs that assisted the network drawing power from their batteries, which is preventing charging or execution of V2G; recover the power used in added time periods. Such kind of charging is included as a demand in the power flow calculations and charging is allowed if network constraints are within limits, otherwise it is deferred to other time periods.

14.4 Stochastic modeling for EVs load demand

In practice, the charging/discharging pattern of EVs is dynamic, depending on the arrival and departure rates. For illustration, the driving distance of each EV varies from others; departure/arrival at charging station is not the similar; some owners may have requirements to drive a lot during weekends, etc. These random behaviors can be predicted based on certain patterns. The patterns can be attained by large number of surveys. A methodology can be formulated in a stochastic form, which makes the analyses of the network more realistic. The arriving/leaving time

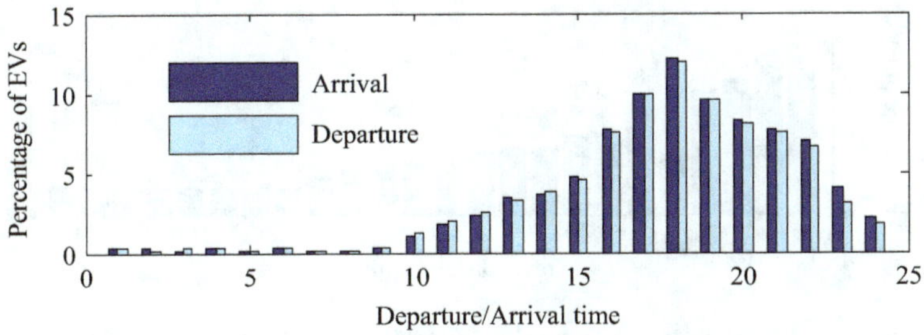

Figure 14.6 Distribution of departure/arrival time

data in this chapter is from National Household Travel Survey (NHTS) [22]. It is based on the survey of American vehicle users. The arrival and departure of EVs is plotted in Figure 14.6. It can be seen that maximum cars arrive at charging station around 18:00 h. To get the departure time of EVs, comparable distribution profile is anticipated. The peak of departure time distribution is represented during 17:00–18:00 h, which is same as arrival time. As depicted in Figure 14.6, the departure time of vehicles at charging station follows a normal distribution which can be expressed as follows:

$$f_{dep}(t) = \frac{1}{\sigma\sqrt{2\pi}} e^{-(t-\mu)/2\sigma^2}, \quad 0 < t < 24 \tag{14.13}$$

where $\mu = 16.97, \sigma = 3.1$.

Also, the pdf of EVs arrival at charging station is a normal distribution and can be depicted as follows:

$$f_{arr}(t) = \frac{1}{\sigma\sqrt{2\pi}} e^{-(t-\mu)/2\sigma^2}, \quad 0 < t < 24 \tag{14.14}$$

where $\mu = 17.01, \sigma = 3.2$.

The pdf of the EVs travel distance (Figure 14.7) can be depicted as follows:

$$f_d(d) = \frac{1}{d\sigma\sqrt{2\pi}} e^{-(\ln d-\mu)/2\sigma^2}, \quad d > 0 \tag{14.15}$$

In this case $\mu = 3.2, \sigma = 0.9$.

Where d is the distance traveled by EV, μ is the mean of lnd, and σ is the standard deviation.

14.4.1 Charging level and type of EVs analysis

The charging rate of EVs could be diverse due to several charging amenities. For illustration, in [7] charging rate is considered of 4 kW at 230 V/46A outlet in Belgium. In [23] outlets of 120 V/15 A and 240 V/50 A are considered. As the

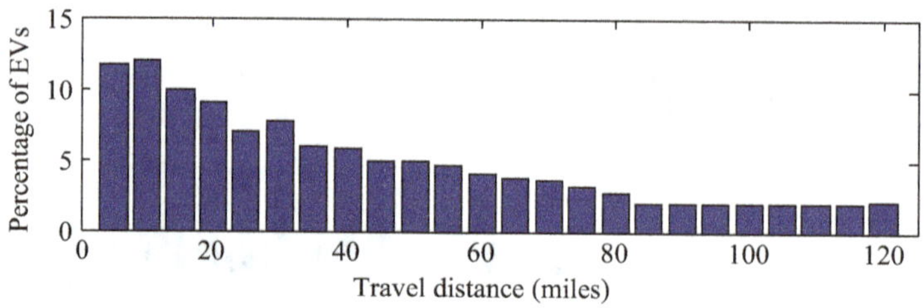

Figure 14.7 Stochastic pattern of driving distance

Table 14.1 Percentage of EVs with various ranges [23]

EVs Category Percentage	EV-30 27%	EV-40 50%	EV-60 23%

Table 14.2 Storage capacity [23]

EVs Category Vehicle Type	EV-30	EV-40	EV-60
Compact Sedan	7.8 kWh	10.4 kWh	15.6 kWh
Mid-size Sedan	9 kWh	12 kWh	18 kWh
Mid-size SUV	11.4 kWh	15.2 kWh	22.8 kWh
Full-size SUV	13.8 kWh	18.4 kWh	27.6 kWh

study of this chapter is engrossed on residential level impact of EVs, the charging level of 3.6 kW is considered with equal probability for each EV.

EVs are categorized by its range and the available percentage of EV-x is depicted in Table 14.1 [23]. For example, EV-40 depicts that it can travel 40 miles. Various categories of EV-x have different energy consumption per mile and storage capacities, and they are depicted in Table 14.2 [23].

14.4.2 Stochastic fuzzy modeling

By exploring the data of NHTS, it is depicted that the arrival time and departure time of a EVs at centralize charging station are two autonomous actions. The probability density functions of the arrival and departure time do not have a reliant configuration when their combination depict the activity of EVs. But it becomes a dissimilar case when data of daily mileage is considered. The daily mileage is very reliant on the arrival time and departure time [6]. For altered groupings of arrival time and departure time, the probability density of the daily mileage may be changed. Here, a stochastic modeling based on fuzzy logic is suggested to create driving configurations of EVs. The plugged in time, plugged out time and initial state of charge (SOC) of EVs are the key factors to model the charging pattern of

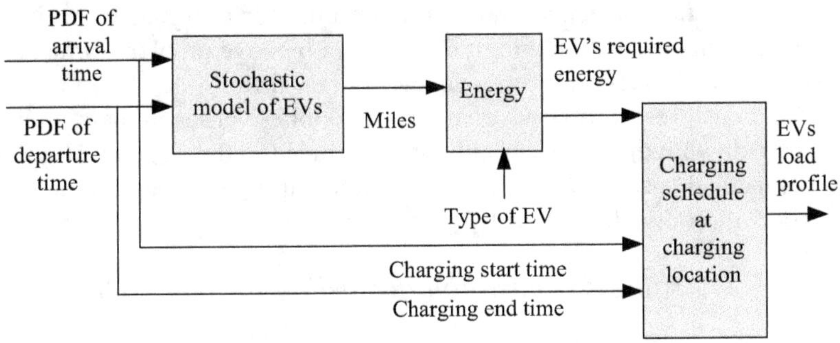

Figure 14.8 EV load modeling framework

EVs. These three factors can be controlled splendidly using the concept of fuzzy logic. Here, a similar study on fuzzy logic-based stochastic model [24,25] is erected to investigate the connection between the departure time, arrival time, and daily mileage of EVs. Similar study has been performed in [25] for home charging of EVs.

14.4.3 Initial SOC and EVs load profile

State of charge for EV is termed as the percentage of energy remaining in EVs battery. In this chapter minimum SOC is fixed to 20% to increase the life of EVs battery. A mileage factor φ is considered in this paper. EVs range is considered as d_R, and consumption of EVs energy is comparative to the travel distance d.

The initial SOC of EV according to daily travel distance d is

$$SOC_{initial} = \begin{Bmatrix} \left(1 - \dfrac{\varphi \times d}{d_R}\right) \times 100\%, 0 < \varphi d < 0.8 d_R \\ 20\%, \varphi d > 0.8 d_R \end{Bmatrix} \tag{14.16}$$

The energy anticipated to EVs battery is

$$E_{req} = \frac{(1 - SOC_{initial}) \cdot C}{\tau} \tag{14.17}$$

The load profile of EVs is attained by anticipated load profile modeling framework (LPMF) as depicted in Figure 14.8. First, the driving pattern is created by the planned stochastic technique as stated in this section. Then required energy is obtained according to (14.16) and (14.17). Finally, the EVs load profile is achieved by the required energy and available charging schedule at charging location.

14.5 Peak load shifting optimization

The EVs can be charged from the grid in an uncoordinated form [26]. Though, implementation of coordinated charging scheme would decrease the effect on the network. This section presents a host of algorithms for charging EVs via public charging station in distribution grid. It depicts the objective problem formulation of EVs charging in different forms; load curve, real-time price (RTP), and wind power. In the past, researchers have not addressed these cases in their studies and thus comparison with their study could not be included in this chapter. However, in this study practical cases have been considered taking into account in the scenario of future operation distribution grid.

14.5.1 EV charging optimization according to load curve

According to the load curve as described (Figure 14.3), there is a peak variation from 1,200 to 1,260 min during evening hours. Hence, an objective function (14.18) is formed based on real load data as

$$O_{f,l} = \sum_{m=1}^{m=N} \sum_{t=X_m}^{t=X_m+T_{C_{m,l}}} L[t] \tag{14.18}$$

$$T_{C_{m,l}} = \left(1 - SoC_{m,l}\right)/CS_{m,l} \tag{14.19}$$

with constraints as

$$X_m \geq A_{t_m}$$

$$0.2 \leq SoC_{m,l} \leq 1$$

where, $O_{f,l}$ is the optimized load of charging EVs, L[t] is the real-time load curve, X_m (controllable variable) is the starting time of charging for $m-th$ vehicle, $T_{C_{m,l}}$ is the time required for full charging to $m-th$ vehicle determined by its state of charge $\left(SoC_{m,l}\right)$, $CS_{m,l}$ is the rate of charge for $m-th$ vehicle.

The load profile of EVs is attained by anticipated LPMF as depicted in Figure 14.8. The optimized pattern of EVs charging and actual EVs load demand of a day is depicted in Figure 14.9. As observed, charging period is scheduled during

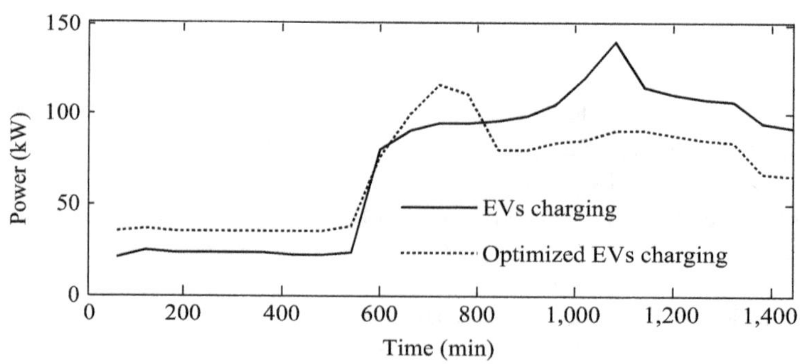

Figure 14.9 Peak load shifting optimization of EV charging

the valley of base load demand. In other words, EVs charging is shifted during the period when the load demand is low.

14.5.2 EV charging optimization according to real-time price

The consumers get the real-time electricity price before its use due to smart meter. Therefore, a high revenue can be saved on charging the EVs during low price hours. It also suggests to avoid EVs charging during peak hour and so reducing the demand stress on the network. The RTP curve is shown in Figure 14.10. Therefore, the objective function is expressed as

$$O_{f,r} = \sum_{m=1}^{m=N} \sum_{t=X_m}^{t=X_m+T_{Cm,r}} P[t] \tag{14.20}$$

$$T_{C_{m,r}} = \left(1 - SoC_{m,r}\right)/CS_{m,r} \tag{14.21}$$

With constraints as

$$X_m \geq A_{t_m}$$

$$0.2 \leq SoC_{m,r} \leq 1$$

where $O_{f,r}$ is the minimum cost.

The aim is to attain minimum cost along with the successful charging of maximum EVs. The optimized EV charging as according to RTP is shown in Figure 14.10. As observed, optimized load demand avoids the peak period of RTP by shifting to later hours.

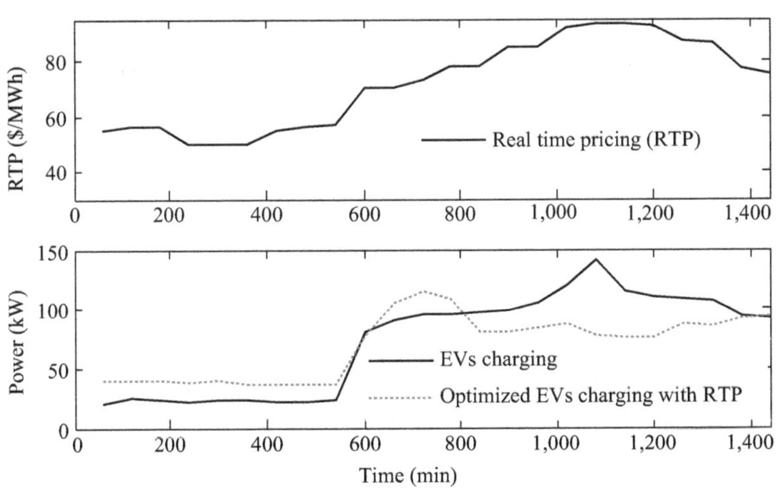

Figure 14.10 EV charging according to real-time pricing

14.5.3 EV charging optimization according to wind power and RTP

In distribution network having wind penetration, charging strategy may be better quantified based on the availability of wind power. The uncertainty associated with wind generation varies widely. There are many approaches to forecast the wind power output. If the forecasted load curve could be applied in optimization, this would aid to achieve better advantage of wind power. The variability in EVs as charging loads can be controlled to accommodate the availability of wind power output. Consider $W[t]$ and $P[t]$ as the wind power output and the RTP, respectively. It is clear that RTP curve and wind power output curve are not on the same scale; however, each one must be controlled thus equal prominence is given in optimization by taking two constants:

$$K_1 + K_2 = 1, \quad \text{i.e. } K_1 = K_2 = 0.5$$

Therefore the objective function is expressed as

$$O_{f,w} = K_1 \sum_{m=1}^{m=N} \sum_{t=X_m}^{t=X_m+T_{cm,w}} P[t] + K_2 \sum_{m=1}^{m=N} \sum_{t=X_m}^{t=X_m+T_{cm,w}} W[t] \tag{14.22}$$

$$T_{c_m,w} = \left(1 - SoC_{m,w}\right)/CS_{m,w} \tag{14.23}$$

With constraints as
$X_m \geq At_m$; $0.2 \leq SoC_{m,w} \leq 1$

$$P[t] = \frac{P[t] - Min\{P[t]\}}{Max\{P[t]\} - Min\{P[t]\}} \tag{14.24}$$

$$W[t] = \frac{W[t] - Min\{W[t]\}}{Max\{W[t]\} - Min\{W[t]\}} \tag{14.25}$$

where $O_{f,w}$ is the optimized load for EVs charging according to wind power output and the RTP data.

Figure 14.11 illustrates the optimized EV charging achieved with optimization of RTP and available wind power output. As observed, more numbers of EVs are charged when the wind output is high.

14.5.4 Charging/discharging shifting

The electricity cost and the net demand are correlated to each other [27]. The higher net demand will result into higher price. The charging/discharging shifting refers using EVs to charge energy (G2V) at the time of low demand (low price) and discharge (V2G) when demand is high (high price) in order to achieve economic benefit. An optimal EVs charging/discharging shifting is obtained by identifying the factors; 24-h net load demand and marginal cost (MC) of the particular feeder in study. The EV operates as G2V and V2G in real time; however, its charging/discharging periods are scheduled according to the peak times of

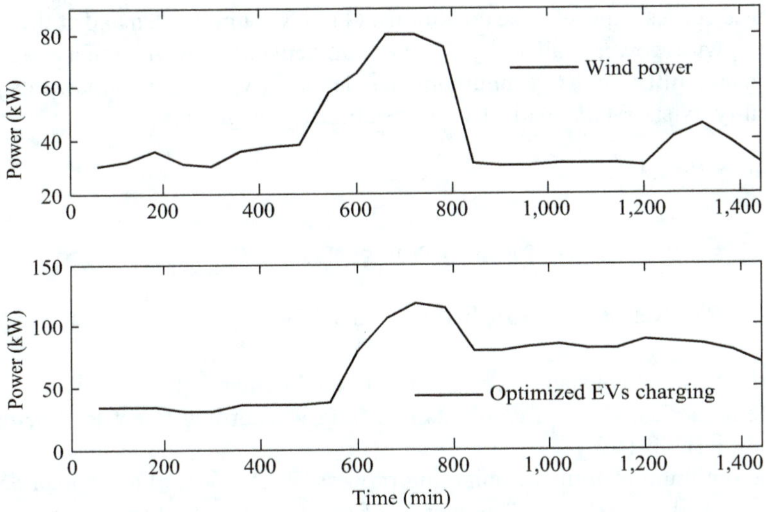

Figure 14.11 Optimized EV charging with wind and real-time pricing

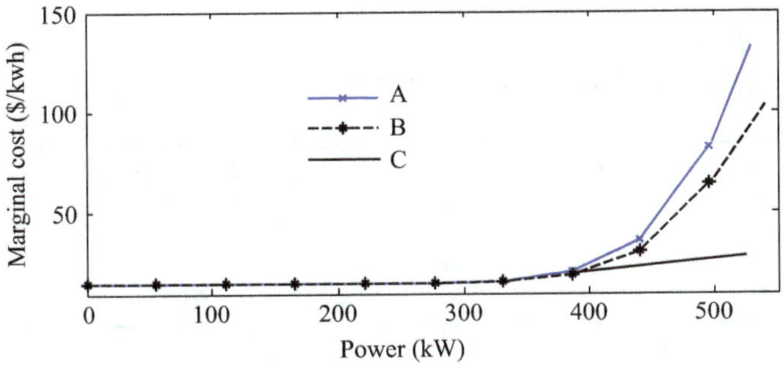

Figure 14.12 Best fit curves for MC vs. net demand

predicted 24-h load curve (Figure 14.3). The performance of EV as energy storage is measured as a function of the differential cost of energy (DCE), expressing the variance between the costs of energy acquired to charge the EV and the cost of energy supplied to grid.

The net demand varies due to variation in wind power availability; thus, three case studies have been performed considering high wind power, average wind power, and low wind power. Figure 14.12 represents the best fit curves from actual net demand varying in range of 280 to 600 kW [28] of a region. The three curves define the MC of electric energy, ($/kWh) against the net demand (kW) drawn by a feeder.

These curves express three conditions: (1) maximum net demand at the time of minimum wind power availability, (2) medium demand at the time of average wind power availability, and (3) minimum net demand when maximum wind power availability exists. Mathematically, the relationship is given as

$$C_w = \alpha + \beta P^{\gamma} \tag{14.26}$$

C_w = Marginal cost of electric energy (\$/kWh)

P = Net demand supplied to feeder (Kw)

$\alpha, \beta,$ *and* γ are the constant according to net load demand

The numerical value of constants $\alpha, \beta,$ *and* γ according to net load demand is depicted in Table 14.3 [27].

The optimum charging/discharging process of EVs according to load demand is depicted in Figure 14.13. The control of the charging/discharging procedure is indicated by the two areas; A_1 and A_2 that denote the charging/discharging energy.

During the charging process, the load demand increases with the energy transfer to the EV (W_{EVC}) specified by the area A_1 between T_o and T_e The charging plan is such that the area $A_1 = W_{EVC}$ occupies a section nearby to P_{min} in order to

Table 14.3 Marginal cost parameters [27]

Curve	α (\$/kWh)	β (\$/(kW$^{\gamma+1}$)h)	γ
A	15	1.43×10^{-3}	1.79
B	15	2.64×10^{-3}	1.65
C	15	4.47×10^{-4}	1.50

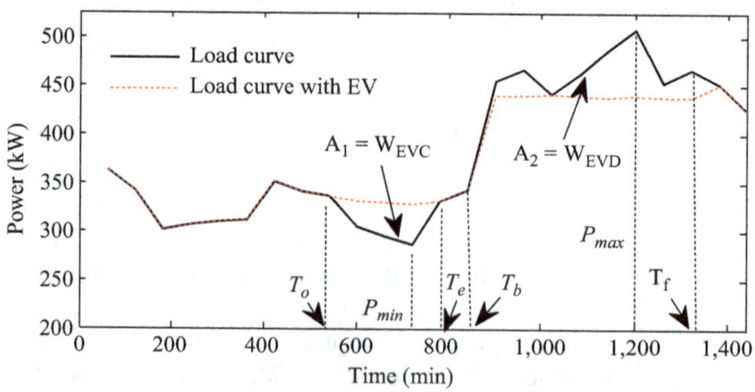

Figure 14.13 Optimal charging/discharging methodology

reduce the cost of energy purchased. The area A_2 represents the discharging energy of EV (W_{EVD}) between T_b and T_f. The discharge area, $A_2 = W_{EVD} = n \times W_{EVC}$ is nearby to P_{max} to maximize the cost of energy supplied; n represents the round trip efficiency of EVs battery.

The profit in charging/discharging shifting is estimated using the differential benefit (*DCE*), formulated as

$$DCE = C_C - C_D \tag{14.27}$$

where

C_c = extra cost of charging energy
C_D = compact cost of the energy distributed to feeders loads by the EV discharging energy that shifts the peak power.

Where,

$$C_C = \int_{T_o}^{Te} P(T_o)C(P(T_o))dT - \int_{T_o}^{T_e} P(T)C(P(T))dT \tag{14.28}$$

$$C_D = \int_{T_b}^{Tf} P(T)C(P(T))dT - \int_{T_b}^{T_f} P(T_b)C(P(T_b))dT \tag{14.29}$$

The procedure follows in determining the time period corresponding to relative minimum charging cost and maximum discharging cost of energy. Minimum MC befalls at P_{min} and maximal MC befalls at P_{max}. From these initial points the charging and discharging times are represented in Figure 14.12, subject to the constraints of EVs energy capacity W_{EVC}, $A_2 = n \times A_1$, $P(T_o) = P(T_e)$ and $P(T_b) = P(T_f)$.

An optimum time period leads to maximize the economic benefit with unlimited storage capacity of EVs. Thus, increase in the total charging time of storage capacity, allowing additional storage, will have DCE to reach peak at certain time, and then decrease thereafter. Figure 14.14 illustrates the variation in differential benefit as a function discharging time, T_B. The EV is assumed to get fully charged during the low demand and discharge during the peak demand.

The variation of differential benefit suggests that for a specified 24-h load demand and storage round trip efficiency, there is a storage size, i.e., number of EVs, characterized by a T_B, that provides the highest differential benefit (DCE). The DCE increases as the MC define a more extreme deviation of cost with net demand. For instance, with round trip efficiency, $n = 70\%$ and at $T_B = 5$ h, the differential benefit is $37/day for maximum wind penetration (minimum load demand) as depicted by denotation 3 and $64/day for minimum wind penetration (maximum load demand), depicted by denotation 1 in Figure 14.14.

Further, as observed by curve with denotation 1 in Figure 14.14, it is clear that DCE reaches maximum with lower discharging time period ($T_B = 3.3$ h) when the efficiency is low as compared to higher efficiency, i.e., $n = 85\%$. In other words, EV storage capacity undergoes into saturation earlier when the efficiency is lower.

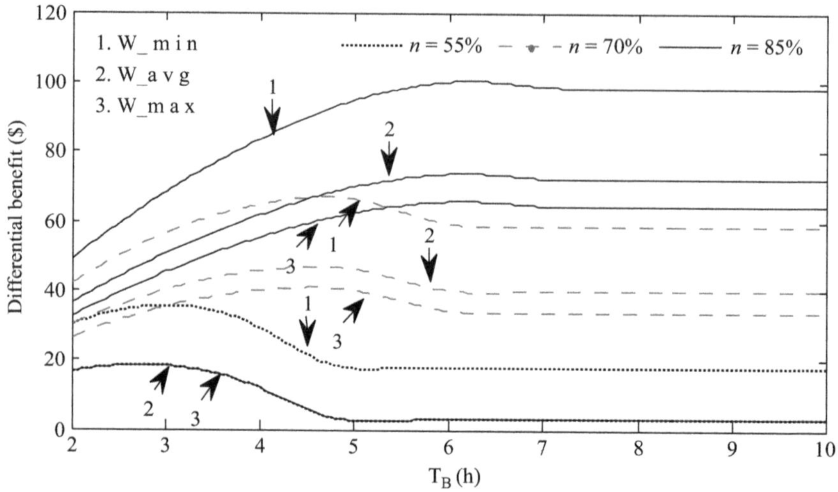

Figure 14.14 Differential benefit variation against discharging time period

14.6 Conclusion

The work presented a scheme for optimum location and charging strategy of EVs in a distribution network with renewable energy penetration. The steps included determining the voltage sensitivity indexes so that loadability violations do not take place with allocation of EVs load. Stochastic load profile of EVs is modeled by considering the arrival and departure time of EVs. This followed the optimum charging considering the RTP and wind power output. An optimization-based approach aids the owners of EVs to get revenue by saving electricity bills, as it shifts the charging from high price hours to low price hours. Additionally, shifting of charging period from peak demand period to low demand period reduces the demand stress on the network. The charging/discharging shifting demonstrated a large amount of differential benefit with the effects of EV storage characteristics. A large number of EVs as storage is cost-effective under the maximum load demand (P_{max}) and minimum load demand (P_{min}) with an extreme variation of MC.

References

[1] Gang Li Xuefei Zheng. "Thermal energy storage system integration forms for a sustainable future," *Renewable and Sustainable Energy Reviews*, 2016, 62, 736–757.

[2] Amini, M. H., and Arif Islam. "Allocation of electric vehicles' parking lots in distribution network," *Innovative Smart Grid Technologies Conference (ISGT), 2014 IEEE PES*. IEEE, 2014.

[3] Milos Pantos. "Stochastic optimal charging of electric-drive vehicles with renewable energy," *Energy*, 2011, 36, no. 11, 6567–6576.

[4] Amini, M. Hadi, Mohsen Parsa Moghaddam, and Orkun Karabasoglu. "Simultaneous allocation of electric vehicles' parking lots and distributed renewable resources in smart power distribution networks," *Sustainable Cities and Society*, 2017, 28, 332–342.

[5] Denholm, Paul, and Robert M. Margolis. "Evaluating the limits of solar photovoltaic (PV) in electric power systems utilizing energy storage and other enabling technologies," *Energy Policy*, 2007, 35.9, 4424–4433.

[6] Amini, M. H., and M. Parsa Moghaddam. "Probabilistic modelling of electric vehicles' parking lots charging demand," *Electrical Engineering (ICEE), 2013 21st Iranian Conference on. IEEE*, 2013.

[7] S. Deilami, A. S. Masoum, P. S. Moses, and M. A. S. Masoum. "Real-time coordination of plug-in electric vehicle charging in smart grids to minimize power losses and improve voltage profile," *IEEE Trans. Smart Grid*, 2011, 2, no. 3, 456–467.

[8] JuhaKiviluoma, and Peter Meibom. "Influence of wind power, plug-in electric vehicles, and heat storages on power system," *Energy*, 2010, 35, no. 3, 1244–1255.

[9] Yugong Luo, Tao Zhu, Shuang Wan, Shuwei Zhang, and Keqiang L. "Optimal charging scheduling for large-scale EV (electric vehicle) deployment based on the interaction of the smart-grid and intelligent-transport systems," *Energy*, 2016, 97, 359–368.

[10] Kempton, Willett, and Steven E. Letendre. "Electric vehicles as a new power source for electric utilities," *Transportation Research Part D: Transport and Environment 2.3,* 1997, 157–175.

[11] Amini, M. Hadi, Amin Kargarian, and Orkun Karabasoglu. "ARIMA-based decoupled time series forecasting of electric vehicle charging demand for stochastic power system operation," *Electric Power Systems Research,* 2016, 140, 378–390.

[12] João Soares, Nuno Borges, Mohammad Ali Fotouhi Ghazvini, Zita Vale, and P. B. de Moura Oliveira. "Scenario generation for electric vehicles' uncertain behavior in a smart city environment," *Energy*, 2016, 111, 664–675.

[13] L. F. Montoya, Q. Fu, A. Nasiri, V. Bhavaraju, and D. Yu. "Novel method to determine the optimal energy storage location in a microgrid and address power quality and stability," May 2014.

[14] N. Neyestani, M. Yazdani Damavandi, M. Shafie-Khah, J. Contreras, and J. Catalão. "Allocation of plug-in vehicles' parking lots in distribution systems considering network-constrained objectives," *IEEE Trans. Power Syst*, 2014, PP, no. 99, 1–14, 2014.

[15] SretenDavidov, and Miloš Pantoš. "Planning of electric vehicle infrastructure based on charging reliability and quality of service," *Energy*, 2017, *118*, 1156–1167.

[16] Sajad Tabatabaee, Seyed Saeedallah Mortazavi, and Taher Niknam. "Stochastic scheduling of local distribution systems considering high penetration of plug-in electric vehicles and renewable energy sources," *Energy*, 2017, 121, 480–490.

[17] Ioannis Zenginis, John S. Vardakas, Nizar Zorba, and Christos V. Verikoukis. "Analysis and quality of service evaluation of a fast charging station for electric vehicles," *Energy*, 2016, 112, 669–678.

[18] Y. Zheng, Z. Y. Dong, Y. Xu, K. Meng, J. H. Zhao, and J. Qiu. "Electric vehicle battery charging – Swap stations in distribution systems – Comparison study and optimal planning," *IEEE Trans. Power System.*, 2014, 29, no. 1, 221–229.

[19] I.J. Fernández, C.F. Calvillo, A. Sánchez-Miralles, and J. Boal. "Capacity fade and aging models for electric batteries and optimal charging strategy for electric vehicles," *Energy*, 2013, 60, 35–43.

[20] S. Sachan and N. Kishor. "Optimal location for centralized charging of electric vehicle in distribution network," *2016 18th Mediterranean Electro technical Conference (MELECON)*, Lemesos, Cyprus, 2016, pp. 1–6.

[21] IEEE 34 Node Test Feeder—IEEE Distribution System Analysis Subcommittee [Online]. Available: http://www.ewh.ieee.org/soc/pes/dsacom/testfeeders.html

[22] National Household Travel Survey, 2009 [Online]. Available: http://nhts.ornl.gov

[23] S. Shafiee, M. Firuzabad, and M. Rastegar. "Investigating the impacts of plug-in hybrid electric vehicles on power distribution systems," *IEEE Trans. Smart Grid*, 2013, 4, no. 3, 1351–1360.

[24] BunyaminYagcitekin, and Mehmet Uzunoglu. "A double-layer smart charging strategy of electric vehicles taking routing and charge scheduling into account," *Applied Energy*, 2016, 167, 407–419.

[25] J. Tan, and L. Wang. "Integration of plug-in hybrid electric vehicles into residential distribution grid based on two layer intelligent optimization," *IEEE Tran. Smart Grid*, 2014, 5, no. 4, 1774–1784.

[26] Olle Suudstrom, and Carl Binding. "Flexible charging optimization for EV considering distribution grid constraints," *IEEE Tran. Smart Grid*, 2012, 3, no. 1, 26–37.

[27] ISO-USA. HourlyZonalInformation [Online]. Available: http://www.isone.com/markets/hstdata/znl_info/hourly/index.html

[28] ComEd. Comed Real Time Saving, 2012 [Online]. Available: https://www.thewattspot.com/rtsavings.php

Chapter 15

Optimal coordination of vehicle-to-grid batteries and renewable generators in a distribution system

Lu Wang[1,2]*, Suleiman Sharkh*[1] *and Andy Chipperfield*[1]

The previous chapter proposed a dispatch strategy for electric vehicles (EVs) only. However, without renewable energy sources (RES), the CO_2 emissions cannot be effectively reduced. Therefore, RES should be integrated into the power grid and EVs should be at least partially charged from these sources. The increasing penetration of EVs and RESs will challenge the distribution network, since it has constrained capacity and most EVs and distributed renewable generators are directly connected to it. However, appropriate dispatch of EVs via vehicle-to-grid operation in coordination with the distributed renewable generations can provide support for the grid, reduce the reliance on traditional fossil-fuel generators and also benefit EV users. This chapter develops a novel agent-based coordinated dispatch strategy for EVs and distributed renewable generators, taking into account both grid's and EV users' concerns and their priorities. This optimal dispatch problem is formulated as a distributed multi-objective constraint optimization problem utilizing analytic hierarchy process and is solved using an A*-based algorithm. The proposed coordinated dispatch strategy is tested on a modified UK Generic Distribution System (UKGDS). The simulation results are presented to demonstrate the feasibility and stability of this dispatch strategy.

15.1 Introduction

To reap the desired environmental benefits, renewable generators (RGs) should provide at least a part of electric vehicles' (EVs') charging energy [1]. Therefore, a novel approach is required, so that EVs can be dispatched in coordination with renewable power flow.

In this work, a novel agent coordinated dispatch strategy for EVs and RGs is developed for real-time application, which aims at satisfying the concerns and requirements of both EV users and the grid, including (1) cost, as saving charging

[1]Faculty of Engineering and Physical Sciences, University of Southampton, Southampton, UK
[2]Global Energy Interconnection Development and Cooperation Organization, Beijing, China

cost is a very common request of EV users, which are taken into account in many publications [2–5]; (2) sufficient state of charge (SOC) for the next journey, which is important because mobility is the basic function of an EV [2,5,6–8]; (3) improved utilization of renewable energy (RE), because it is crucial for carbon emission reduction, as discussed earlier and in [6] and [9–11] and (4) load levelling, which is one of the main grid operational support that EV batteries can provide and is also discussed in many papers [5,12–14].

In this strategy, each node in the network is represented by a software agent, which is only aware of the elements that are locally connected to it and manages the dispatch of EVs and RGs connected to it, based on information received from the agents of other nodes that are directly linked to it, so that the stability of the network is ensured and all the objectives of dispatch are best achieved.

Accordingly, the dispatch problem is formulated as a distributed multi-objective constraint optimization problem (DMOCOP) and then solved using an A*-based algorithm to derive an optimal set of dispatch actions for EVs and RGs within a distribution network. DMOCOP is developed from the distributed constraint optimization problem (DCOP) that was proposed in [15], using an analytic hierarchy process (AHP) [16] to take several different objectives of dispatch, as discussed earlier, into account at the same time. Moreover, the inherent uncertainty of EV travel patterns and renewable power generation is modelled and simulated using Gaussian copulas, so that the correlation between each pair of random variables is taken into account to better reflect the characteristics of the real case. The proposed A*-based optimal dispatch algorithm is tested on a radial distribution network, a modified UK Generic Distribution System (UKGDS), for its stability, feasibility and efficacy at satisfying the requirements of both EV users and the grid. In practice, the particle swarm optimization (PSO) is supposed to be in charge of this optimal dispatch problem.

The remainder of this chapter is organized as follows. A general model of the electricity network that is used for optimal dispatch is described in Section 15.2, while in Section 15.3, a DMOCOP is formulated for the optimal coordination of EVs and RGs. Following Section 15.4, where the proposed A*-based algorithm is presented in detail, its application to solving the DMOCOP while considering the uncertainties of EV travel patterns and RG outputs are demonstrated in Section 15.5. The simulation results are also presented and discussed in this section to verify its feasibility and efficacy. Finally, the conclusions of the work are presented in Section 15.6.

This work has been published in *Energy*, vol. 113, pp. 1250–1264, 2016 [17] and in the *International Journal of Electrical Power & Energy Systems*, vol. 98, pp. 474–487, 2018 [18].

15.2 General description of the electricity network and agents

Figure 15.1 shows a radial distribution network, which is derived from a UKGDS [19]. It includes RGs and EVs. Node v_0 is the slack bus, which is connected to the rest of the power grid.

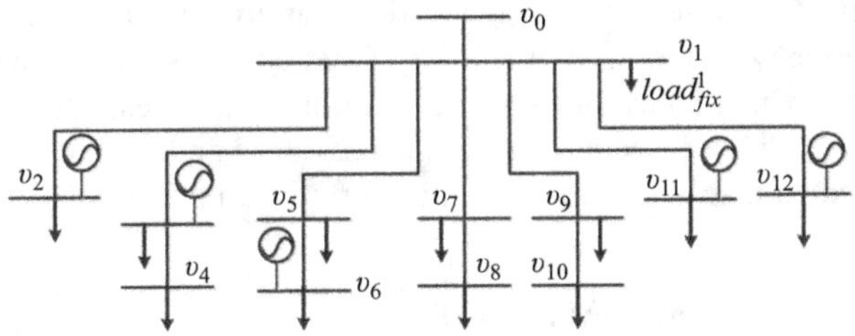

Figure 15.1 Diagram of a modified generic radial distribution network

In more detail, the network contains a set of RGs, which are denoted by $G = \{g_1, \ldots, g_n\}$ with each generator g_i producing a certain amount of renewable power $p_i \in RG_i$ MW. $RG_i = \{0, \ldots, P_i^{max}\}$, where RG_i is discretized into 1-MW steps to simplify the computational burden and $P_i^{max} \in \mathbb{Z}^+$ is the maximum power output of RG g_i rounded to the nearest whole MW at a particular moment in time; P_i^{max} will vary depending on the environmental conditions (e.g. cloud cover or wind speed variation). Let $p = \{p_1, \ldots, p_n\}$ denote a set of power output variables for the RGs in G.

Moreover, the network provides the capability of connecting a great number of EVs, each of which can be either charged or discharged. EV represents a set of EVs, where $EV = \{ev_1, \ldots, ev_m\}$. Each EV has seven choices of dispatch mode when it is parked and connected to the grid: charge (+) or discharge (–) at high (3), medium (2) or low-level (1) current, OR idle (0). Thus, each EV ev_i is dispatched in a certain mode $\delta_i \in DM_i = \{-3, -2, -1, 0, 1, 2, 3\}$, where the sign indicates the dispatch action (i.e. charge/discharge) and the absolute value represents the dispatch current level. Let $\mathbf{d} = \{\delta_1, \ldots, \delta_m\}$ denote a set of dispatch modes for EVs in EV.

$V = \{v_1, \ldots, v_k\}$ denotes the set of nodes within the network. A node v_i exchanges power with other nodes and contains a combination of fixed loads, EVs and RGs. $chi(v_i)$ denotes a set of children nodes of v_i, while $adj(v_i)$ represents a set of adjacent nodes of v_i, that is, nodes that are directly connected to v_i via distribution cables, including its children nodes and its parent node. $load^i_{fix}$ represents the fixed load at v_i, while $EV(v_i)$ and $G(v_i)$ denote the sets of EVs and RGs connected at v_i, respectively. Note that $EV(v_i) = \varnothing$ and $G(v_i) = \varnothing$, respectively, mean that v_i contains no EV and no RG.

The set of distribution cables in the network denoted by T and t_{ij} refers to the distribution cable between nodes i and j. The power flow along the cable t_{ij} is denoted by f_{ij}, and cannot exceed the thermal capacity of the cable, C_{ij}. Distribution losses have not been considered in this work as is often done in many studies [20,21]. The effect of distribution loss will be the subject of future investigations.

The network is controlled by agents in a decentralized way. Each node v_i is represented by an agent, which is aware of the output power domains of the local RGs, $G(v_i)$, and all possible dispatch modes of the connected EVs, $EV(v_i)$, and has control over the dispatch of renewable power output and EVs. Each agent has a utility function, carries out a part of the computation required to achieve optimal and stable operation of the network (i.e. utility computation) and communicates the computation results with its adjacent agents that map to the adjacent nodes $adj(v_i)$ of its designated node v_i. In this work, the utility function maps to the penalty cost at the designated node, which takes into account multiple objectives, using the AHP, and measures how well these objectives are achieved. Based on the above definitions, this optimal coordinated dispatch problem can be formulated as a DMOCOP.

15.3 Formulation of DMOCOP

The DMOCOP extends the DCOP procedure described in [15] to a multi-objective optimization problem. It contains three main elements as follows:

1. Variables: a set of h variables $X = \{x_1, \ldots, x_h\}$. In this work, x_i can be an RG's power output or an EV's battery dispatch mode. Thus, in this work $X = \{p, \mathbf{d}\}$.
2. Domains: a set of finite domains $D = \{d_1, \ldots, d_h\}$, which include all possible values of variables X. In this work, d_i can be represented as follows:

$$d_i = \begin{cases} RG_i & \text{when } x_i \text{ is a renewable generator} \\ DM_i & \text{when } x_i \text{ is an EV battery} \end{cases} \tag{15.1}$$

3. Utilities: a set of k utilities $U = \{U_1, \ldots, U_k\}$, each of which corresponds to an agent. In this work, U_i maps to the penalty cost at agent i, that is, how unsatisfactory a combination of dispatch actions is in terms of the objectives. They are formed by using the AHP, as discussed in the following section.

15.3.1 Objectives

The objectives include saving cost while leaving sufficient charge in an EV's battery, network load levelling and reduction of wasted RE.

15.3.1.1 Objective 1 (*RE*): Reduce wasted RE

Assume that the weather forecast is accurate enough, and it is predicted that an RG's power output $\widetilde{P_{RG}}$ is available during the following time interval of 30 min. The penalty cost of wasting renewable power C_{RG} is measured as follows:

$$C_{RG} = (\widetilde{P_{RG}} - P_{RG})/\widetilde{P_{RG}}, \tag{15.2}$$

where P_{RG} is the actual amount of power that is injected by an RG into the network.

15.3.1.2 Objective 2 *BS*: Sufficient EV battery SOC

Assuming that EVs' travel patterns are available (either directly entered by the user or estimated based on information in their diary or passed travel history), then the expected SOC at the end of the current time interval SOC_p is estimated to be

$$SOC_p = \begin{cases} \begin{matrix} SOC \\ \dfrac{SOC_{pf} - SOC}{T_p} t + SOC \\ SOC_{pf} \end{matrix} \end{cases} \tag{15.3}$$

If EV has no travel plan during the next 2 h;

If EV has travel plan within 2 h and current SOC is not enough for its next journey ($t = 1$ time interval);

If EV has travel plan within 2 h and has enough SOC for its next journey.

When the EV is expected to travel within the next 2 h, a linear interpolation is used to determine the value of SOC_p as illustrated in Figure 15.2. T_p is the available preparation time before the next journey in multiples of dispatch action time interval, that is, multiples of 30 min. SOC_{pf} is the desired SOC of the EV battery before the next journey, which may be lower than the current SOC value, in which case the EV will be able to participate in vehicle to grid (V2G) operations.

The aim is to keep the SOC at about 0.5. The penalty cost of insufficient SOC, C_{SOC}, is measured as follows:

1. If the EV has no travel plan during the next 2 h and the battery is more than half full at the end of this time interval OR if the EV will be used in 2 h and battery's SOC is not less than SOC_p at the end of the time interval, then

$$C_{SOC} = 0. \tag{15.4}$$

2. If the EV has no travel plan during the next 2 h and the battery is less than half full at the end of this time interval, then

$$C_{SOC} = \frac{0.5 - SOC_n}{0.5 - S_{min}}, \tag{15.5}$$

where SOC_n is the battery's SOC after a time interval of charging/discharging. In this work, the lower limit of SOC S_{min} is selected to be 0.4 to provide a good margin above the absolute minimum of 0.3.

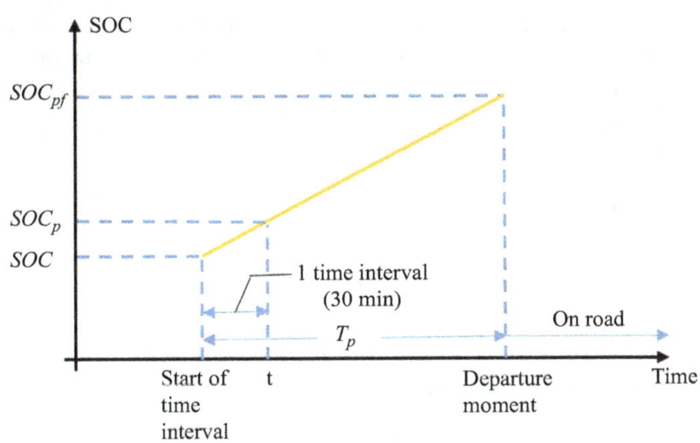

Figure 15.2 Illustration of the calculation of SOC_p when EV has travel plan within 2 h and current SOC is not adequate

3. If the EV has travel plans within 2 h and the battery's SOC_n is lower than SOC_p, then

$$C_{SOC} = \frac{SOC_p - SOC_n}{SOC_p - S_{min}}. \tag{15.6}$$

15.3.1.3 Objective 3 (*CC*): Save charging cost to EV users

The penalty cost is sure to be low when an EV is charged at a low electricity buying price while discharged at a high selling price. And the penalty cost C_{ep} could vary depending on how fast the EV is charged/discharged, that is, the charging/discharging current.

1. Charge at high current (i.e. 30 A):

$$C_{ep} = \begin{cases} \dfrac{bp - bp_{min}}{hbp - bp_{min}} & bp < hbp \\ 1 & bp \geqslant hbp \end{cases}. \tag{15.7}$$

2. Charge at mid-level current (i.e. 10 A):

$$C_{ep} = \begin{cases} \dfrac{|bp - (bp_{min} + hbp)/2|}{(hbp - bp_{min})/2} & bp < hbp \\ 1 & bp \geqslant hbp \end{cases}. \tag{15.8}$$

3. Charge at low current (i.e. 2 A):

$$C_{ep} = \begin{cases} \dfrac{hbp - bp}{hbp - bp_{min}} & bp < hbp \\ 0 & bp \geqslant hbp \end{cases}, \tag{15.9}$$

where bp is the electricity buying price. bp_{max} and bp_{min} are, respectively, the top and bottom buying prices, while hbp is the high buying price threshold defined as $hbp = 0.9 * (bp_{max} - bp_{min}) + bp_{min}$. Equations (15.7)–(15.9) give progressively lower penalties for relatively high electricity buying prices as the charging current reduces.

Similarly, the penalty cost for selling is defined as follows:

4. Discharge at high current (i.e. 30 A):

$$C_{ep} = \begin{cases} \dfrac{sp_{max} - sp}{sp_{max} - lsp} & sp > lsp \\ 1 & sp \leqslant lsp \end{cases}. \tag{15.10}$$

5. Discharge at mid-level current (i.e. 10 A):

$$C_{ep} = \begin{cases} \dfrac{|sp - (sp_{max} + lsp)/2|}{(sp_{max} - lsp)/2} & sp > lsp \\ 1 & sp \leqslant lsp \end{cases}. \tag{15.11}$$

6. Discharge at low current (i.e. 2 A):

$$C_{ep} = \begin{cases} \dfrac{sp - lsp}{sp_{max} - lsp} & sp > lsp \\ 0 & sp \leqslant lsp \end{cases}. \tag{15.12}$$

7. No dispatch (i.e. 0 A):

$$C_{ep} = \begin{cases} 0 & sp \leqslant lsp \& bp \geqslant hbp \\ 1 & otherwise \end{cases}, \tag{15.13}$$

where sp is the electricity selling price. sp_{max} and sp_{min} are, respectively, the top and bottom selling prices, while lsp is the low selling price threshold defined as $lsp = 0.1 * (sp_{max} - sp_{min}) + sp_{min}$. These price data can be derived from historical data [22].

15.3.1.4 Objective 4 (*LL*): Load levelling in the distribution network

The daily fixed (non-controllable) load demand is assumed to be available for both peak and off-peak periods. With the integration of EVs and RGs, the peak load could be pulled up further and might result in some spikes, or high power could be generated in the network and transferred to the rest of the grid through the slack bus during the off-peak period. Therefore, one of the objectives of coordinated dispatch of EVs and RGs is peak shaving and valley filling, that is, load levelling. The penalty cost of failing to level the load, C_{ll}, is evaluated as follows:

$$C_{ll} = \begin{cases} 0 & load_l \leqslant load_{tot} \leqslant load_h \\ \dfrac{load_{tot} - load_h}{load_{max} - load_h} & load_h < load_{tot} < load_{max} \\ \dfrac{load_l - load_{tot}}{load_l - load_{min}} & load_{min} < load_{tot} < load_l \\ \dfrac{load_{min}}{load_{tot}} & 0 < load_{tot} \leqslant load_{min} \\ \dfrac{load_{tot}}{load_{max}} & load_{tot} \geqslant load_{max} \end{cases}, \tag{15.14}$$

where $load_{min}$ and $load_{max}$ are, respectively, the minimum and maximum fixed load demand during a day. $load_h$ and $load_l$ are the high and low fixed load thresholds, respectively, and they are determined as

$$load_h = (load_{max} + load_{ave})/2, \tag{15.15}$$

$$load_l = (load_{min} + load_{ave})/2, \tag{15.16}$$

where $load_{ave}$ is the average of daily fixed load demand. $load_{tot}$ is the total load at a node, which can be calculated as

$$load_{tot} = load_{fix} + load_{ev}^{c} - load_{ev}^{d} - P_{RG}, \tag{15.17}$$

where $load_{fix}$ is the fixed load at a node, which is not controllable. $load_{ev}^{c}$ and $load_{ev}^{d}$ are the EVs' charging load and the EVs' discharging power, respectively. P_{RG} is the RG's output power.

15.3.2 The AHP

In order to jointly consider these objectives and thus the corresponding penalty costs in the optimization process, they should be weighted depending on their relative importance in determining how EV batteries and RGs should be dispatched in coordination.

Two different hierarchy models are, respectively, built up for the agents that only involve EV battery variables and those that have both EV battery and RG variables, as shown in Figure 15.3.

The priorities of these agents' objectives are determined by forming a pairwise comparison matrix for each of the AHP hierarchy models shown in Figure 15.3; the scale numbers in the matrices are determined by the aggregator in practice based on experience and common sense. The matrices PC_a (Figure 15.3(a)) and PC_b (Figure 15.3(b)) are as follows:

$$PC_a = \begin{array}{c} \\ RE \\ BS \\ CC \\ LL \end{array} \begin{array}{c} \begin{array}{cccc} RE & BS & CC & LL \end{array} \\ \begin{pmatrix} 1 & \frac{1}{2} & 2 & 2 \\ 2 & 1 & 2 & 2 \\ \frac{1}{2} & \frac{1}{2} & 1 & 1 \\ \frac{1}{2} & \frac{1}{2} & 1 & 1 \end{pmatrix} \end{array}, \tag{15.18}$$

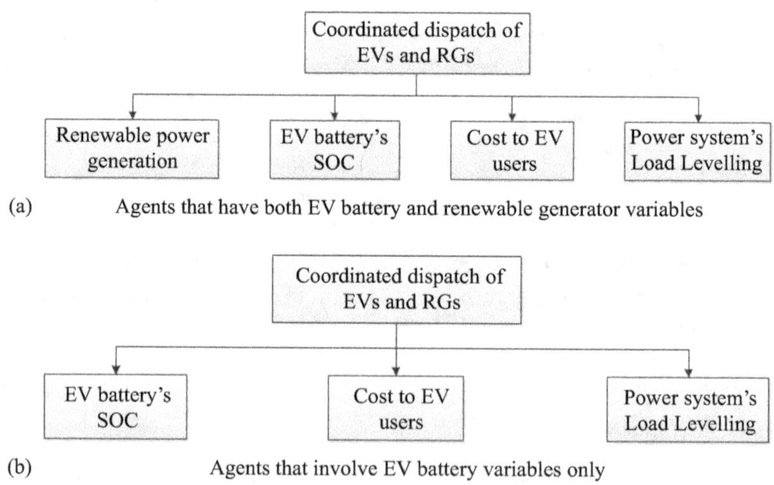

Figure 15.3 AHP hierarchy models of the utilities of two types of agents for the coordinated dispatch of EVs and RGs. (a) Agents that have both EV battery and RG variables. (b) Agents that involve EV battery variables only

$$PC_b = \begin{array}{c} \\ BS \\ CC \\ LL \end{array} \begin{array}{c} \begin{array}{ccc} BS & CC & LL \end{array} \\ \begin{pmatrix} 1 & 2 & 2 \\ \frac{1}{2} & 1 & 1 \\ \frac{1}{2} & 1 & 1 \end{pmatrix} \end{array} \tag{15.19}$$

By calculating the principle eigenvectors of PC_a and PC_b, the priorities of RE, BS, CC and LL are found to be 27.81%, 39.52%, 16.34% and 16.34%, respectively, for AHP model (a), and the priorities of BS, CC and LL are, respectively, 50%, 25% and 25% for AHP model (b).

The utilities U of agents that involve both EV battery and RG variables are then calculated as follows:

$$\begin{aligned} U &= 27.81\% \times C_{RG} + 39.52\% \times C_{SOC} \\ &\quad + 16.34\% \times C_{ep} + 16.34\% \times C_{LL}, \end{aligned} \tag{15.20}$$

while the utilities U of agents that only involve EV battery variables are calculated as follows:

$$U = 50\% \times C_{SOC} + 25\% \times C_{ep} + 25\% \times C_{LL}. \tag{15.21}$$

15.3.3 Constraints

For the stability of the electricity network and the better performance of EVs and RGs, several constraints are applied. The goal of agents is to find an assignment X^* for the variables in X (i.e. a combination of dispatch actions of EVs and RGs) that minimizes the sum of penalty costs (i.e. the sum of utilities):

$$\arg\min_{X^*} \sum_{i=0}^{k} U_i, \tag{15.22}$$

subject to the following constraints:

Constraint 1: The sum of power flow into a node v_i should be equal to the sum of power flow out:

$$\sum_{j \in adj(v_i)} f_{ij} + load_{fix} + load_{ev}^{c} - load_{ev}^{d} - P_{RG} = 0, \tag{15.23}$$

where $adj(v_i)$ is the set of nodes that are connected to the node v_i. f_{ij} is the power flow from node i to j, and $f_{ij} = -f_{ji}$.

Constraint 2: The power flow along a distribution cable should not exceed its capacity:

$$|f_{ij}| \leqslant C_{ij}, \tag{15.24}$$

where C_{ij} is the thermal capacity of the distribution cable between nodes v_i and v_j.

Constraint 3: The SOC of EV batteries should be within the range from 0 to 1:

$$0 \leqslant SOC \leqslant 1. \tag{15.25}$$

Constraint 4: When the EV is not going to be used within the next 2 h and the battery is currently less than half full of electricity, the EV has to be charged:

$$SOC_n \geqslant SOC, \quad if \;\; SOC < 0.5 \tag{15.26}$$

Constraint 5: When the EV has a travel plan within the next 2 h and the battery's SOC is currently less than the desired at the end of the time interval, the EV has to be charged:

$$SOC_n \geqslant SOC, \quad if \;\; SOC < SOC_p. \tag{15.27}$$

15.4 A* optimal dispatch procedure

In order to solve the DCOP defined earlier, A* is used to find the optimal assignment of the control variables X^* in the network.

A* is a well-known optimal path search procedure, discovering the shortest route between a starting point and a destination. A simple example is as follows.

Figure 15.4 describes a route map from city A to city D (numbers present the distance between two cities), and the objective is to find the shortest path from A to D. Starting from city A, two paths can be expanded: $A - B$ (A to B) and $A - C$ (A to C) with path lengths of 10 and 5, respectively. Thus, two states can be formed as $\{A - B, 10\}$ and $\{A - C, 5\}$, which comprise a state queue, Q, in which the states are lined up in an ascending order of path length:

$$Q = [\{A - C, 5\}, \{A - B, 10\}]. \tag{15.28}$$

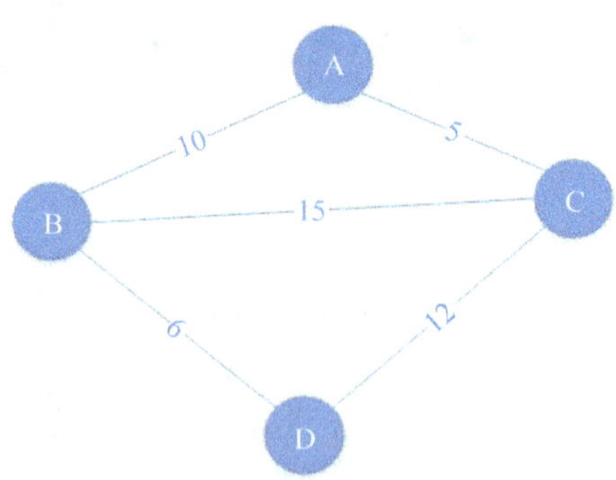

Figure 15.4 A route map

Only the first state of the queue is extended: $\{A-C-B, 20\}$ and $\{A-C-D, 17\}$. These two new states replace the old state $\{A-C, 5\}$ in the queue and the queue is sorted again with the shortest patch in the front. The state queue Q now becomes:

$$Q = [\{A-B, 10\}, \{A-C-D, 17\}, \{A-C-B, 20\}] \tag{15.29}$$

Again, the first state of the queue is extended as $\{A-B-C, 25\}$ and $\{A-B-D, 16\}$, which replaces the old state in the queue and the states in the queue are reordered. The new state queue Q becomes:

$$Q = [\{A-B-D, 16\}, \{A-C-D, 17\}, \{A-C-B, 20\}, \{A-B-C, 25\}] \tag{15.30}$$

Now, the first state, which has the shortest length, terminates at the destination D. The A* search procedure ends, and the optimal path is $A-B-D$ with a length of 16.

The procedure is also illustrated graphically in Figure 15.5. We refer the interested reader to [23] for a detailed description of the A* search procedure.

In the remainder of this section, we describe how A* is applied to our specific case of optimal dispatch of RGs and EVs in a distribution network. To facilitate the explanation, a 10-node electricity network presented in Figure 15.6 is used as an example.

This problem can be interpreted by comparison with the optimal path search problem described earlier. As an electricity network has more than one leaf node, this problem is like an optimal path search problem with multiple starting points. The path length in the path search problem maps to the accumulated utilities in this optimal dispatch problem, as they are, respectively, the objective function of each problem that needs to be minimized. The path maps to the power flow from a node

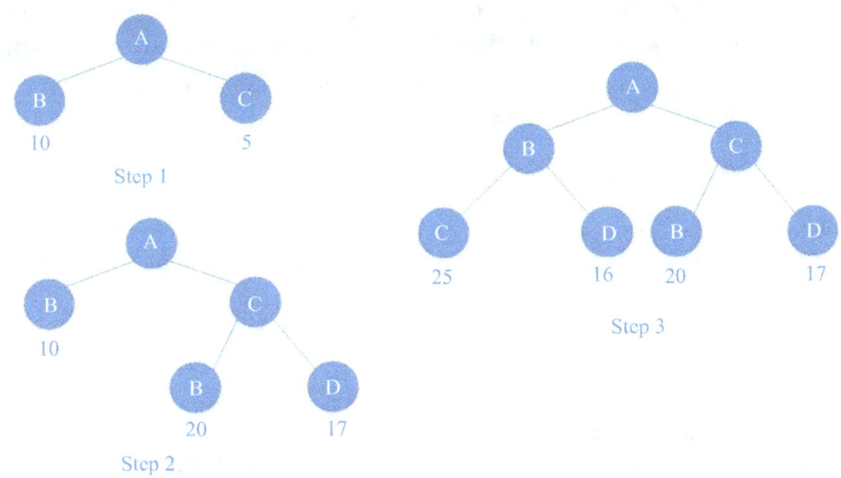

*Figure 15.5 Optimal path searching using A**

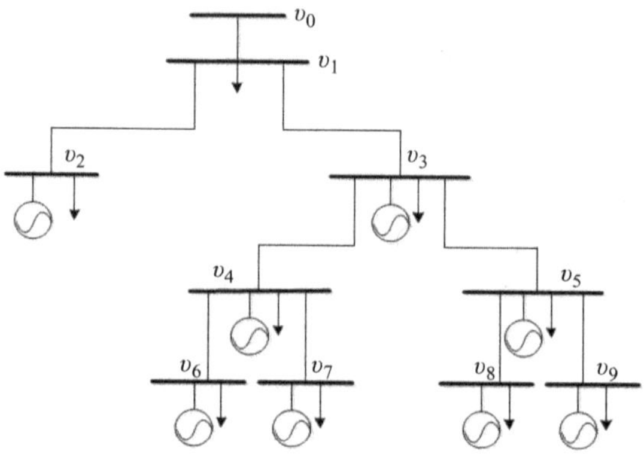

Figure 15.6 Diagram of a 10-node radial distribution network

to its parent node. Both path and power flow are extendible and their extensions are dependent on the current states: given power flow from a node v_i to its parent node $\widehat{v_i}$, the power flow from node $\widehat{v_i}$ to $\widehat{v_i}$'s parent node can be calculated.

The A*-based optimal dispatch of a distribution network can be applied as follows.

Each node is managed by an agent, as discussed earlier. The computational results of the agents are stored and managed in a central memory for the future usage in the A* search procedure. Figure 15.7 illustrates ą thorough procedure of A*-based dispatch. The communication of central memory and agents is illustrated in Figure 15.8.

The A* terminology used in these figures is as follows:

- State and state queue: A state S_j in a state queue Q_i consists of two elements ï¼š the power flow value $f_{i\hat{i}}$ from the node v_i to its parent node $\widehat{v_i}$ and the corresponding accumulated utility $U_{\sum}(f_{i\hat{i}})$ that are resulted from a certain given assignment of control variables $X_{v_i}(S_j)$ at this node v_i and $X_{v_c}(S_j)$ at its children nodes $chi(v_i)$. Therefore,

$$S_j = \{f_{i\hat{i}}, U_{\sum}(f_{i\hat{i}})\}, \tag{15.31}$$

where $|f_{i\hat{i}}| \leqslant C_{i\hat{i}}$ for the stability of power system. Each state S_j maps to a *Pstate* which records the values of control variables at v_i and its children nodes that results in the accumulated utility described by the function $U_{\sum}$:

$$Pstate(S_j) = \{S_j, X(S_j)\}, \tag{15.32}$$

where $X(S_j) = \{X_{v_i}(S_j), X_{v_c}(S_j) | v_c \in chi(v_i)\}$.

A state queue Q_i consists of M states that are generated from the corresponding agents and lined up in an ascending order of the utility value $U_{\sum}$:

$$Q_i = [S_1, S_2 \ldots, S_M] \tag{15.33}$$

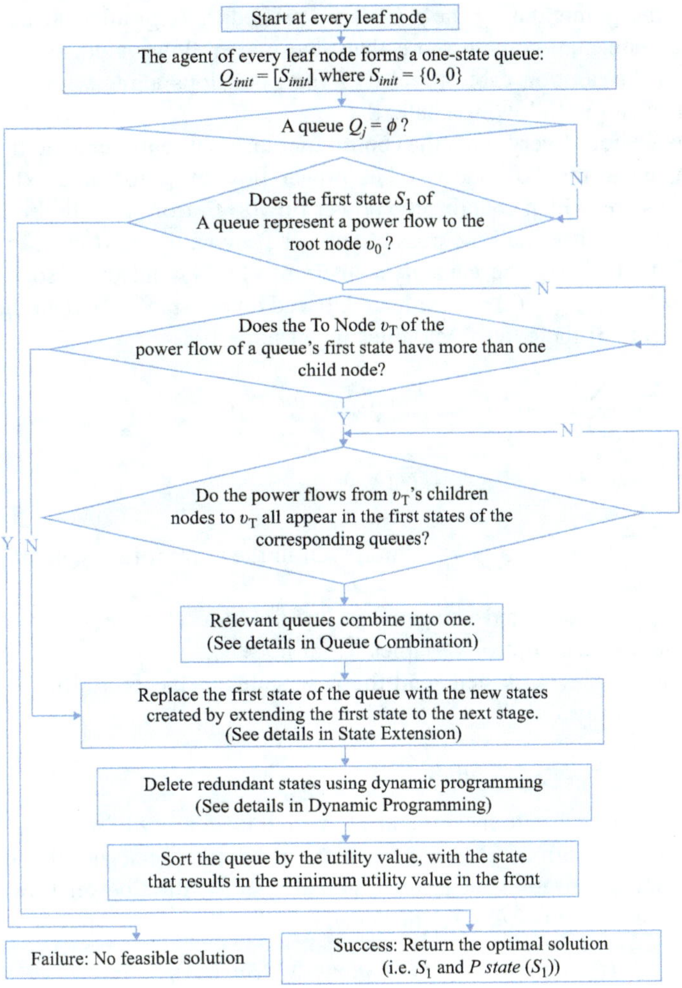

Figure 15.7 The A-based dispatch algorithm*

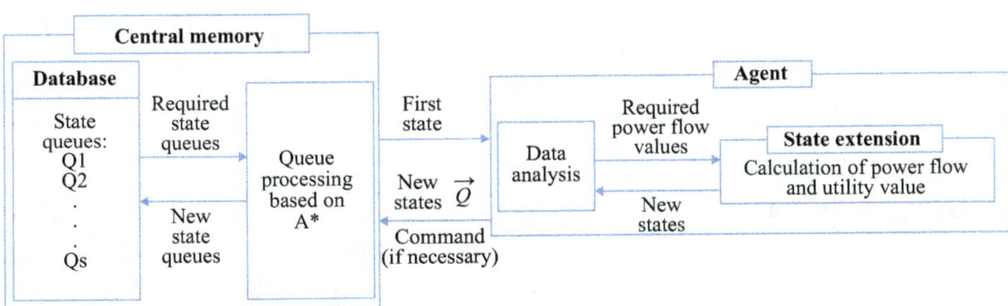

Figure 15.8 A general graph of data communication within A-based optimal dispatch*

At some points during the A* search procedure (e.g. after queue combination), a state may contain more than one power flow value, associated with different distribution cables, and its *Pstate* develops along with it, as will be explained in the following context.

- State extension: Every time, the central memory will only send the first state S_1 of a queue to the To node v_i of its power flow(s) f_{ci} for state extension. To extend a state, which contains the power flow(s) f_{ci} from v_i's child(ren) node(s) $v_c \in chi(v_i)$ to itself, all the possible power flow values $f_{i\hat{i}}$ from v_i to its parent node $\hat{v}_i$ that satisfy the capacity constraint (15.24) and the associated accumulated utility $U_{\sum}(f_{i\hat{i}})$ are evaluated, based on the state S_1 and all the possible assignments of the control variables X_{v_i} at node v_i:

$$f_{i\hat{i}} = \sum_{v_c \in chi(v_i)} f_{ci} + p^i_{con} - load^i_{con} - load^i_{fix}, \tag{15.34}$$

$$U_{\sum}(f_{i\hat{i}}) = U_{\sum current} + U_i(X_{v_i}), \tag{15.35}$$

where $\sum_{v_c \in chi(v_i)} f_{ci}$ and $U_{\sum current}$ are given in the current first state of the queue S_1. $U_i(X_{v_i})$ is calculated using the utility function at node v_i given a certain assignment of the control variables X_{v_i} at node v_i.

The agent at v_i will then send the new states to the central memory in the form of an array $\vec{Q}$:

$$\vec{Q} = [Sp_1, Sp_2 \ldots, Sp_m], \tag{15.36}$$

where $\vec{Q}$ is a new state message array, which is an array of *mSp* messages. An Sp_j message is formed of two elements: a new state *Snew* and the assignment of the control variables X_{v_i} at node v_i that results in the power flow of $f_{i\hat{i}}$ and the associated utility $U_{\sum}(f_{i\hat{i}})$ of *Snew*, as follows:

$$Sp_j = \{f_{i\hat{i}}, U_{\sum}(f_{i\hat{i}}), X_{v_i}\} = \{Snew_j, X_{v_i}(Snew_j)\}. \tag{15.37}$$

Once the $\vec{Q}$ is received at the central memory, the original first state S_1 of the queue is replaced by the new states that are created from it, that is, $Snew_1, Snew_2, \ldots, Snew_m$ in $\vec{Q}$, and the corresponding $Pstate(Snew_j)$ is developed from the $Pstate(S_1)$ of S_1:

$$Pstate(Snew_j) = \{Snew_j, X(Snew_j)\}, \tag{15.38}$$

where $X(Snew_j) = \{X_{v_i}(Snew_j), X(S_1)\}$.

- Dynamic programming: In the A* procedure, according to dynamic programming, if there are more than one states representing the same power flow, only the state with the minimum accumulated utility value remains while the others are redundant and deleted.
- Queue combination: When the procedure proceeds to a point where the first state of a queue represents a power flow to a node v_i that has more than one

child node, the extension of this state requires the state(s) from other queue(s) that contain(s) the power flow(s) from v_i's other child(ren) node(s) to v_i. Therefore, those involved queues need to be combined, but only after their first states all represents a power flow to the same node v_i. Combining state queues $Q_1, Q_2, \ldots, Q_s$ is to perform Cartesian product over those queues to form a new state queue Q_{comb}:

$$Q_{comb} = \{Q_1 \times Q_2 \times \ldots \times Q_s\}, \tag{15.39}$$

- with the state within becoming:

$$S_{comb} = \{(f_{vq_1\widehat{vq_1}}, f_{vq_2\widehat{vq_2}} \ldots f_{vq_s\widehat{vq_s}}), \sum_{x=1}^{s} U_{\sum}(f_{vq_x\widehat{vq_x}})\}, \tag{15.40}$$

- where $f_{vq_x\widehat{vq_x}}$ is a power flow value presented by a state S_j^x in the queue $Q_x \in \{Q_1, Q_2, \ldots, Q_s\}$. And the $Pstate(S_{comb})$ of S_{comb} becomes:

$$Pstate(S_{comb}) = \{S_{comb}, X(S_{comb})\}, \tag{15.41}$$

- where $X(S_{comb}) = \{X(S_\alpha^1), X(S_\beta^2), \ldots, X(S_\gamma^s)\}$.

As shown in Figure 15.7, the A* procedure starts from every leaf node with an initial state $S_{init} = \{0, 0\}$. At a leaf node v_i, the agent extends S_{init} using (15.34) and (15.35) to form new states. Figure 15.9 illustrates the state extension at a leaf node v_6 in Figure 15.6 for an example. These new states, together with the corresponding control variables' values, are sent to the central memory in the form of $\vec{Q}$ and stored in the database for the future usage in the A* procedure, as demonstrated in Figure 15.8. After $\vec{Q}$ is received at the central memory, the corresponding first state of the queue is replaced with the new states in the received $\vec{Q}$. The redundant states are deleted according to dynamic programming. The queue is then sorted in an ascending order of the accumulated utility values, as presented in Figure 15.7. Thus the new first state of the queue will have the minimum accumulated utility value so far and will be sent to the To Node of the state's power flow (e.g. node v_j of f_{ij}) for its agent to extend, as presented in Figure 15.8. Therefore, every time, in every queue the state with the least accumulated utility value will be extended to a set of new states, which will be sent back to the central memory and with the original state queue they will form into a new state queue according to the A* procedure.

Queue combination will be conducted using (15.39) when the procedure reaches a stage where a queue's first state represents a power flow $f_{i\hat{i}}$ to a node $\widehat{v_i}$

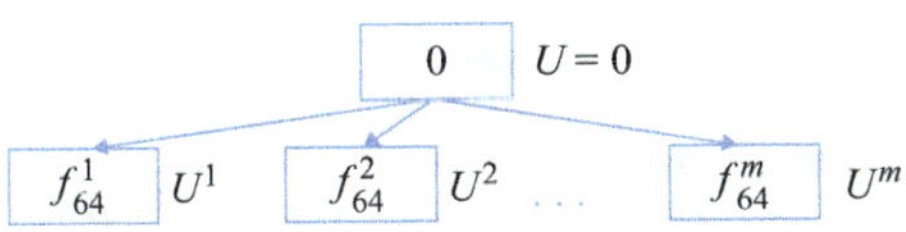

Figure 15.9 An illustration of the A procedure at a leaf node v_6 in the distribution network shown in Figure 15.6*

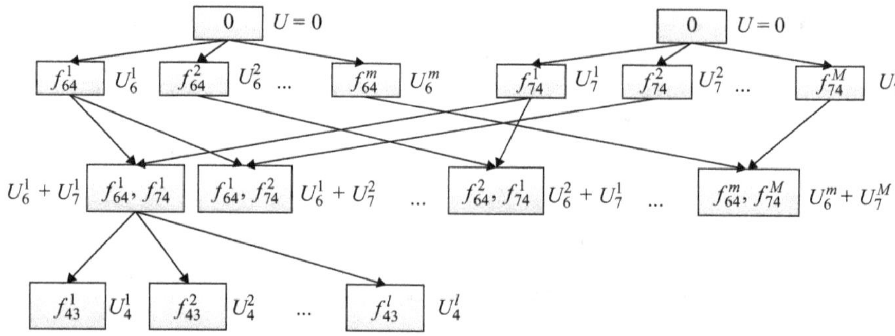

Figure 15.10 An illustration of state queues' combination and extension of the A procedure at a node v_4 in the distribution network shown in Figure 15.6*

that has more than one child node. The agent of node $\widehat{v}_i$ will wait till it receives from central memory the associated combined queue's first state $S_{1,comb}$ which contains all the power flow values from its children nodes $\{f_{c\widehat{i}} | c \in chi(\widehat{v}_i)\}$. It will then extend the received $S_{1,comb}$ to a set of new states using (15.34) and (15.35). The combined queue will then be updated with the new states, as stated in Figures 15.7 and 15.8. Taking node v_4 in Figure 15.6 as an example, Figure 15.10 illustrates how the state queues for nodes v_6 and v_7 combine for the agent at node v_4 to extend their first states and create new states.

The aforementioned A* search procedure will iterate until terminating at the root node of power network (i.e. node v_0 in Figure 15.6), at which stage the first state S_1 of a queue is the optimal solution that optimizes the total utility of the network and thus best achieves the objective of dispatch. The optimal assignment of all control variables X^* within the power network are recorded in the corresponding $Pstate(S_1)$. The optimal dispatch solution is then returned and success is announced, as stated in Figure 15.7. However, if the procedure ends up with an empty queue, no feasible solution can be found for this optimal dispatch problem and thus failure is announced.

15.5 The application of A* search to optimal decentralized coordination of EVs and RGs in a distribution network

With the definition of control variables X and the optimal dispatch problem DMOCOP, the A* algorithm is customized for this application of coordinated dispatch of RGs and EVs.

15.5.1 Stochastic modelling of uncertainties

In order to test the A* optimal dispatch algorithm on the EVs and RGs within a distribution network, the UKGDS in Figure 15.1 and information on EV travel patterns and RE sources (only wind power are incorporated in this case) are required. However, due to the information's intrinsic uncertainties, stochastic modelling of EV travel patterns and wind power were modelled using Copula [24,25].

15.5.1.1 Copula

In most practical cases, random variables are not completely dependent – they could have complicated relationships with each other, and these variables could come from different marginal distributions individually. Therefore, copulae, which represent the dependence structure between variables, are used to generate data from multivariate distributions in order to reflect the traits of input data [24].

Copulae are basically multivariate distribution functions whose 1-D probability distributions are uniform in the interval [0, 1], which join univariate marginal distributions to form multivariate distribution functions [26]. Specifically, according to Sklar's theorem [27], the continuous random variables $X_1, \ldots, X_k$ with cumulative distribution functions (CDFs) $F_{X_1}, \ldots, F_{X_k}$, are joined by copula C if their joint distribution can be written as

$$F_{X_1,\ldots,X_k}(X_1, \ldots, X_k) = C(F_{X_1}(x_1), \ldots, F_{X_k}(x_k)). \tag{15.42}$$

As CDFs F_{X_i} transform random variables X_i into uniform variables U_i, $F_{X_i}(x_i) = u_i$ for $i = 1, \ldots, k$ where $u_1 \ldots u_k$ are, respectively, realizations of uniform variables $U_1, \ldots, U_k$. Therefore, (15.42) can be rewritten as

$$C(u_1, \ldots, u_k) = F_{X_1,\ldots,X_k}(F_{X_1}^{-1}(u_1), \ldots, F_{X_k}^{-1}(u_k)). \tag{15.43}$$

Gaussian copulae are a frequently utilized family of copulae, which are constructed from the multivariate Gaussian distribution, which can be defined as

$$C_\rho(u_1, \ldots, u_k) = \Phi_\rho(\Phi^{-1}(u_1), \ldots, \Phi^{-1}(u_k)), \tag{15.44}$$

where Φ denotes the standard normal CDF and ρ is the linear correlation matrix between the standard normal variables $\Phi^{-1}(u_1), \ldots, \Phi^{-1}(u_k)$. The details are as follows:

Step 1: Calculate the rank correlation, such as Spearman's ρ (ρ_s), between random variables $X_1, \ldots, X_k$. Then calculate the linear correlation ρ between the corresponding normal variables $Y_{X_1} = \Phi^{-1}(F_{X_1}(X_1)), \ldots, Y_{X_k} = \Phi^{-1}(F_{X_k}(X_k))$, using

$$\rho = 2\sin\left(\frac{\pi}{6}\rho_s\right), \tag{15.45}$$

where $F_{X_1}, \ldots, F_{X_k}$ are, respectively, CDFs of continuous random variables $X_1, \ldots, X_k$. Φ^{-1} is the inverse standard normal CDF.

Step 2: Simulate $y_{x_1}, \ldots, y_{x_k}$ from multivariate standard normal distribution with correlation ρ (e.g. use mvnrnd command in MATLAB®).

Step 3: Transform the simulated values $y_{x_1}, \ldots, y_{x_k}$ back to the original domain via standard normal CDF Φ and inverse CDF of each random variable: $x_1 = F_{X_1}^{-1}(\Phi(y_{x_1})), \ldots, x_k = F_{X_k}^{-1}(\Phi(y_{x_k}))$.

15.5.1.2 Modelling of EV travel patterns and on-road energy consumption

In this case, we focus on domestic dispatch of EV batteries, that is, charging or discharging of EVs when parked at home. Therefore, three key parameters of EV driving activities are needed by the proposed optimal dispatch algorithm: time of departure from home, travel distance and time of arrival at home. In order to model the stochastic driving behaviours of EV users, the methodology presented in [24] is utilized, which leveraged Copula to simulate the values of these parameters considering their dependence on each other.

As introduced in [24], 75% of the users had a single home-to-home (h2h) trip in a day while 21% take double h2h trips and the rest 4% take more than two h2h trips. Therefore, the modelling focuses mainly on the single and double h2h trips as they account for most of the daily driving activities.

Modelling EVs' travel patterns requires capturing the stochastic behaviour of the three parameters mentioned earlier. That, in terms of single h2h trip, incorporates three random variables representing it as [24]: departure time T_d^s, arrival time T_a^s and travel distance D^s of the h2h trip, while six random variables are needed to model double h2h trips: departure time, arrival time and travel distance of the first h2h trip (T_d^{d1}, T_a^{d1}, D^{d1}) and the second h2h trip (T_d^{d2}, T_a^{d2}, D^{d2}). Dependence between each pair of random variables is an essential element in modelling together with the marginal distribution of each random variable. Gaussian copula, which is described earlier, is used in this stochastic modelling.

The historical data used and presented in [24] are directly leveraged to simulate the stochastic EV driving activities in this work. The marginal distributions of random variables of single and double h2h trips are shown in Figures 15.11–15.13.

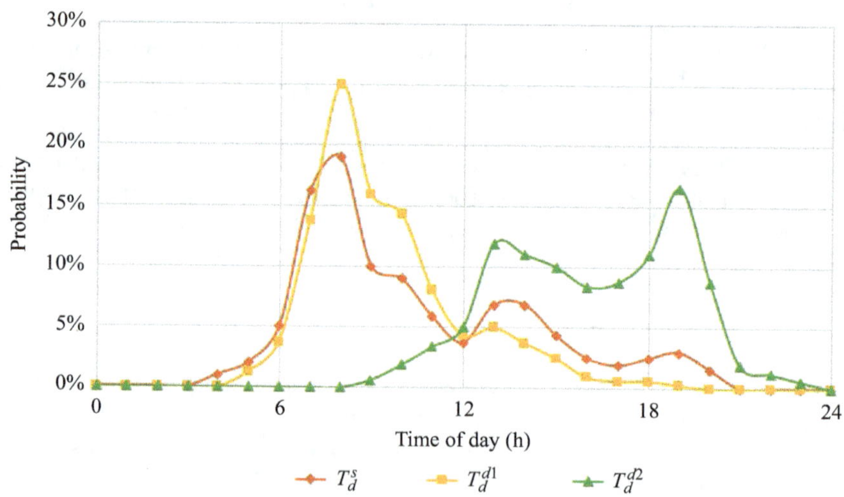

Figure 15.11 Marginal distributions of departure time of single and double h2h trips

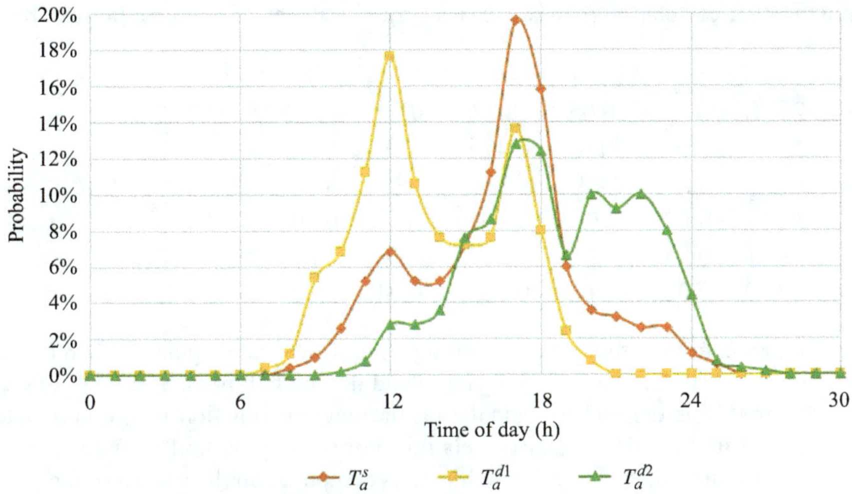

Figure 15.12 Marginal distributions of arrival time of single and double h2h trips

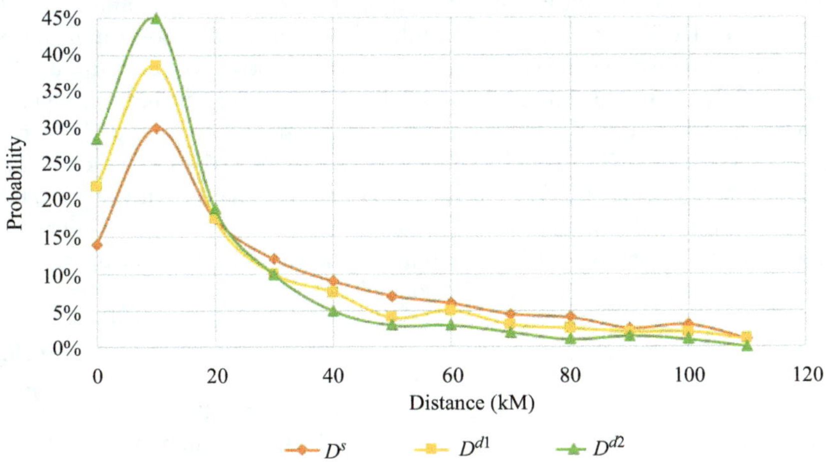

Figure 15.13 Marginal distributions of travelled distance of single and double h2h trips

The rank correlation matrix (Spearman's ρ) of the three random variables representing single h2h trips, ρ_s^s, is obtained from [24] as follows:

$$\rho_s^s = \begin{array}{c} \\ T_d^s \\ T_a^s \\ D^s \end{array} \begin{array}{c} \begin{array}{ccc} T_d^s & T_a^s & D^s \end{array} \\ \begin{pmatrix} 1 & 0.13 & -0.35 \\ 0.13 & 1 & 0.29 \\ -0.35 & 0.29 & 1 \end{pmatrix} \end{array}, \tag{15.46}$$

while the rank correlation matrix of six random variables of double h2h trips, ρ_s^d, is

$$\rho_s^d = \begin{array}{c} \\ T_d^{d^1} \\ T_a^{d^1} \\ D^{d^1} \\ T_d^{d2} \\ T_a^{d^2} \\ D^{d^2} \end{array} \begin{array}{c} \begin{array}{cccccc} T_d^{d^1} & T_a^{d^1} & D^{d^1} & T_d^{d2} & T_a^{d^2} & D^{d^2} \end{array} \\ \begin{pmatrix} 1 & -0.06 & -0.37 & -0.02 & -0.03 & -0.02 \\ -0.06 & 1 & 0.42 & 0.70 & 0.57 & -0.06 \\ -0.37 & 0.42 & 1 & 0.3 & 0.24 & 0.17 \\ -0.02 & 0.70 & 0.30 & 1 & 0.80 & -0.04 \\ -0.03 & 0.57 & 0.24 & 0.80 & 1 & 0.20 \\ -0.02 & -0.06 & 0.17 & -0.04 & 0.20 & 1 \end{pmatrix} \end{array}, \tag{15.47}$$

The procedure of stochastic modelling using Gaussian Copulas is then carried out to simulate random variables of single and double h2h trips. The CDF of each random variable is derived by calculating the integral function of the associated marginal distribution. The linear correlation ρ of underlying multivariate standard normal distribution of each type of h2h trips is computed on the corresponding rank correlation (ρ_s^s, ρ_s^d) using the 1-1 mapping formula between ρ and ρ_s described in (15.45).

As the simulated EV driving activities need to be viable, the departure time must be earlier than arrival time of an h2h trip: $T_d^s < T_a^s$, $T_d^{d1} < T_a^{d1}$ and $T_d^{d2} < T_a^{d2}$. If the simulated values of random variables do not meet this constraint, the stochastic simulation using Gaussian Copulas described above will be rerun until this viability constraint is satisfied and the simulated values are ensured to be viable.

In order to compute the energy consumed on road, the value of EV driving velocity is essential and needs to be simulated. Due to the lack of historical velocity data, and thus its actual marginal distribution and correlation with other random variables, Monte Carlo simulation is adopted to derive the stochastic values of EV driving velocity, where a normal distribution is assumed with a mean value μ of 40 mph and a standard deviation σ formulated as

$$\sigma = 0.5 \times \min(v_{max} - \mu, \ \mu - v_{min}), \tag{15.48}$$

where v_{max} is set to be 70 mph, which, generally speaking, is the maximum velocity a vehicle can drive in the urban area. $v_{min} = \frac{D}{T_a - T_d}$ is the minimum average velocity an EV must reach in order to travel a distance of D (the travelled distance of a h2h trip, which can be D^s, D^{d1} or D^{d2}) within the time interval of $T_a - T_d$ (the duration of an h2h trip, i.e. the difference between the departure and arrival time of a h2h trip, which can be $T_a^s - T_d^s$, $T_a^{d1} - T_d^{d1}$ or $T_a^{d2} - T_d^{d2}$). If the simulated velocity is outside the range $[v_{min}, v_{max}]$, the above Monte Carlo simulation is rerun until the simulated velocity is within that reasonable range.

Therefore, the energy consumed on an h2h trip is computed as follows.

It is assumed that EV driving velocity is constant on road, therefore the propulsion force is equal to the sum of rolling friction and air resistance:

$$F = C_r mg + \frac{1}{2}\rho_a A_f C_D v^2 \tag{15.49}$$

where rolling resistance coefficient $C_r = 0.015$, mass of the vehicle $m = 1150kg$, air density $\rho_a = 1.29kg/m^3$, area of vehicle's face $A_f = 1.5^2m^2$ and air resistance coefficient $C_D = 0.3$. Hence, the power of propulsion is

$$P = Fv = C_r mgv + \frac{1}{2}\rho_a A_f C_D v^3. \tag{15.50}$$

The energy of propulsion E_w is thus the integral of P:

$$E_w = \int P dt = C_r mg \int v dt + \frac{1}{2}\rho_a A_f C_D \int v^3 dt = D(A + Bv^2), \tag{15.51}$$

where $A = C_r mg$, $B = \frac{1}{2}\rho_a A_f C_D$ and D is the travelled distance of an h2h trip. Therefore, the electric energy consumed E_e is

$$E_e = E_w/\eta, \tag{15.52}$$

where the efficiency of the electrical engine $\eta = 70\%$ [28]. With the on-road electric energy consumption calculated, the SOC of an EV battery after an h2h trip can be estimated.

15.5.1.3 Wind power modelling

In this work, distributed wind power generators are considered only and are spread in the distribution network under investigation. As these wind turbines are located in proximity, the wind speeds at these turbine sites have strong correlation with each other. Due to the lack of historical data, the correlation matrix calculated in [25] is used in this work:

$$\rho_{wind} = \begin{array}{c} \\ S_w^1 \\ S_w^2 \\ S_w^3 \\ S_w^4 \\ S_w^5 \end{array} \begin{array}{c} \begin{array}{ccccc} S_w^1 & S_w^2 & S_w^3 & S_w^4 & S_w^5 \end{array} \\ \begin{pmatrix} 1 & 0.82 & 0.85 & 0.74 & 0.78 \\ 0.82 & 1 & 0.83 & 0.74 & 0.82 \\ 0.85 & 0.83 & 1 & 0.81 & 0.74 \\ 0.74 & 0.74 & 0.81 & 1 & 0.65 \\ 0.78 & 0.82 & 0.74 & 0.65 & 1 \end{pmatrix} \end{array}, \tag{15.53}$$

The marginal distributions of wind speeds at different sites are derived from [29], as shown in Figures 15.14 and 15.15.

The copula-based stochastic modelling procedure discussed earlier is then applied to simulate random wind speeds at these different sites.

Provided the simulated wind speed, the wind power generated at a wind turbine generator (WTG) confronting the wind with given speed is calculated according to Figure 15.16*, where the cut-in, nominal and cut-out wind speeds are 3.5, 11.2 and 25 m/s and its nominal power is 2 MW.

*The figure is from [25]. In the real case, when wind speed $\leqslant$ nominal speed, WTG power output $\propto$ wind speed^3, but for the simplification of modelling, we use a first approximation here, i.e., WTG power output $\propto$ wind speed.

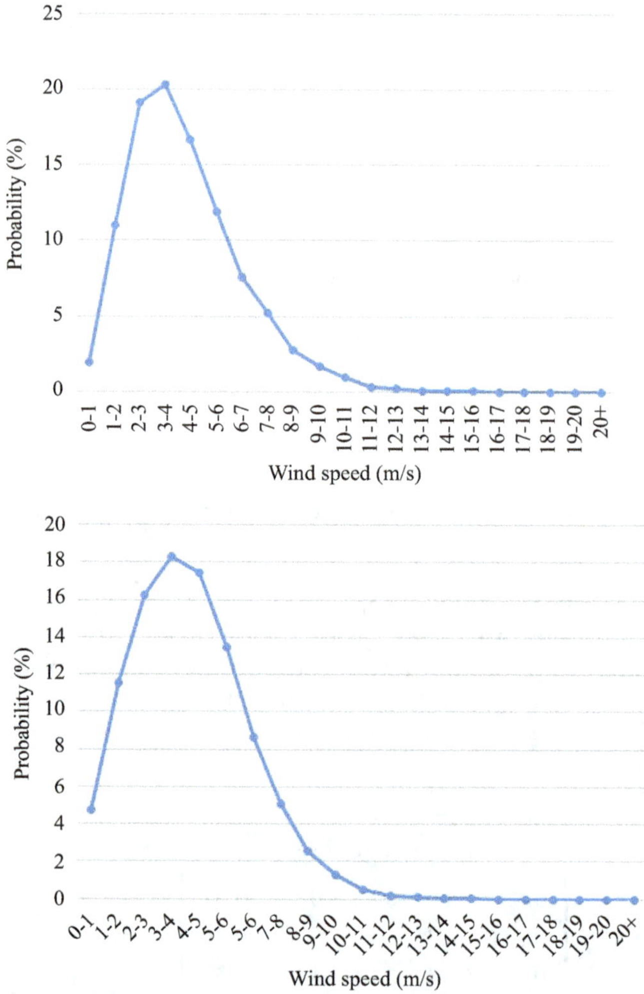

Figure 15.14 Marginal distributions of wind speed at sites 1 (top) and 2 (bottom)

Due to lack of information about autocorrelation of wind speed in time series, the simulated wind speed data are arranged manually in temporal dimension only, so that the variation of wind speed is reasonable and more realistic and the correlations between different WTG sites remain untouched. However, in the future, autocorrelation should also be considered in the stochastic modelling of wind speed in order to better reflect the reality. Figure 15.17 shows the sum of simulated wind power generated at five WTG sites in the network.

15.5.2 Simulation results

The proposed A*-based optimal dispatch algorithm is tested in a radial distribution network; a modified UKGDS shown in Figure 15.18. In this distribution network,

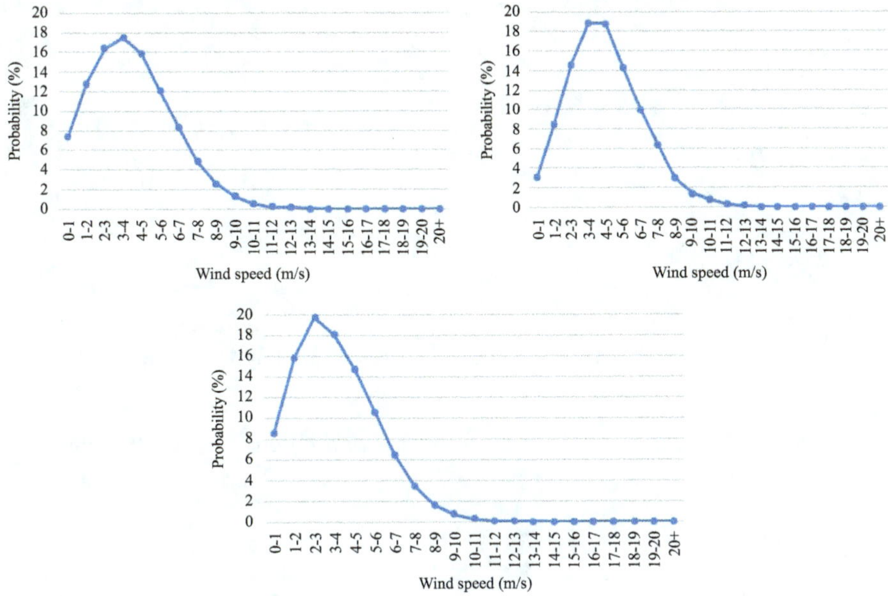

Figure 15.15 Marginal distributions of wind speed at sites 3 (top-left), 4 (top-right) and 5 (middle)

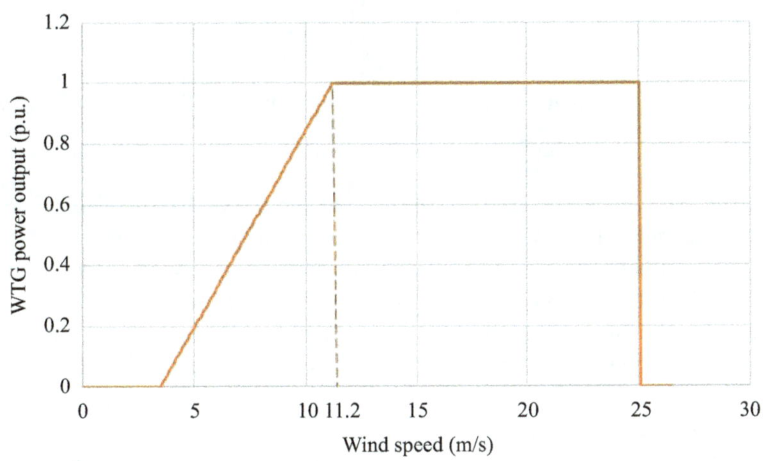

Figure 15.16 WTG wind speed/power characteristics

five RG sites are located at nodes v_2, v_3, v_6, v_{11} and v_{12}, respectively, each of which site has three wind turbines, thus a total nominal wind power of 6 MW can be generated at each site. A total number of 33,000 EVs spread in the network and are capable of connecting to the grid via nodes $v_2, v_3, \ldots, v_{12}$, each of which can

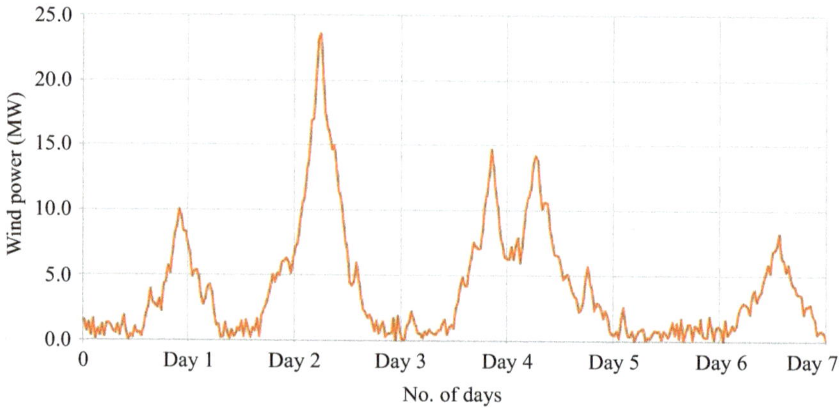

Figure 15.17 Simulated wind power generated in the power network during a week

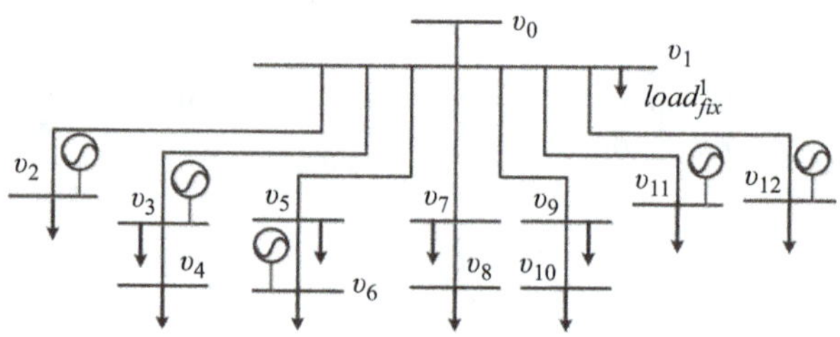

Figure 15.18 Diagram of a modified generic radial distribution network

provide synchronous V2G/G2V services for at most 3,000 EVs. The random EV travel patterns and wind power are simulated using the stochastic models discussed earlier. The results of the simulated EV travel patterns and wind power are used in the simulation test of proposed A*-based dispatch algorithm.

The thermal capacity of distribution cables are derived from the UKGDS model [19], as shown in Tables 15.1 and 15.2. The parameters of EV batteries are obtained from [3]. The simulation is implemented on a laptop with a Dual Core 2 GHz CPU and 8 GB RAM using MATLAB.

The total load demand and system selling/buying prices of electricity during a day were taken from [22]. The total load demand is scaled down so that the peak demand during a day is 350 MW, which approximately represents the daily load demand in a typical UK regional distribution network, as shown in Figure 15.19.

Due to scarcity of information on the payments by/to EV users when the EVs are charged/discharged, the adjusted system selling price (ASSP) and the adjusted system buying price (ASBP) in Figure 8.7 (in Chapter 8) are, respectively, used as the EV charging and discharging prices.

Table 15.1 Thermal capacity of distribution cables

Distribution cable		Thermal capacity (MVA)
From Node	To Node	
v_0	v_1	500
v_1	v_2	52
v_1	v_3	154
v_1	v_5	102
v_1	v_7	76
v_1	v_9	76
v_1	v_{11}	54
v_1	v_{12}	154
v_3	v_4	81
v_5	v_6	110
v_7	v_8	46
v_9	v_{10}	51

Table 15.2 Fixed load at each node of the distribution network

Node	Fixed active load (MW)	Fixed reactive load (MVAR)
v_1	38.48	11.28
v_2	22.98	9.11
v_3	56.05	3.63
v_4	36.63	9.77
v_5	25.32	5.55
v_6	55.54	6.19
v_7	17.70	0
v_8	15.73	2.28
v_9	13.14	0
v_{10}	20.93	3.70
v_{11}	99.05	21.09
v_{12}	87.07	16.58

The time step is set to be 30 min. The dispatch actions are determined by the proposed A*-based optimal dispatch strategy at the beginning of every time interval and lasts for the entire time interval of 30 min, as mentioned earlier.

In order to test the potential economic benefits that participating in the proposed optimal coordinated scheme can bring to EV users, the net costs of EVs are calculated and presented in Table 15.3. An average cost of £0.45 per vehicle per day is expected for those who adopt the proposed A*-based dispatch strategy, in comparison with an average cost of £0.73 for an uncontrolled way of treating EV batteries, which indicates an approximate amount of £103 annually for each EV user. Furthermore, the simulation results in Table 15.4 show that the proposed

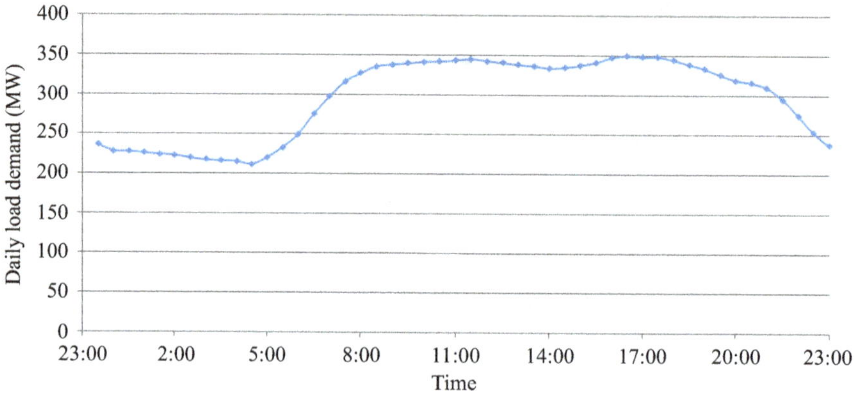

Figure 15.19 Total fixed load demand during a day

Table 15.3 Daily costs of EVs calculated from simulations starting with different initial SOCs

	Total daily costs to EV users in a network (£)
Average	14773.5
Standard deviation	1184.6

Table 15.4 Minimum and maximum SOCs of EVs during a day

Node	Minimum SOC	Maximum SOC	Node	Minimum SOC	Maximum SOC
v_2	0.41	1.00	v_3	0.65	1.00
v_4	0.50	1.00	v_5	0.49	1.00
v_6	0.49	1.00	v_7	0.64	1.00
v_8	0.49	1.00	v_9	0.51	1.00
v_{10}	0.57	1.00	v_{11}	0.34	0.99
v_{12}	0.50	1.00			

dispatch strategy is safe in the sense that it prevents EVs from overcharging their batteries (the stored energy never exceeds 100% of the battery's capacity) and also from running out of energy during the day with, at least 34% of battery's available capacity remaining.

As for the RGs, the simulation results show that overall 99.5% of wind power is utilized to support network load demand and/or EV charging demand, as shown in Table 15.5. The difference between wind power utilization rates at different WTG sites is because the wind speed, the battery conditions of locally parked EVs and the local fixed load demand varies from site to site.

Table 15.5 Utilization of wind power generated at WTG sites of the distribution network during a week

Site no.	**1**	**2**	**3**	**4**	**5**
Utilization of generated wind power	99.6%	100%	98.1%	100%	100%

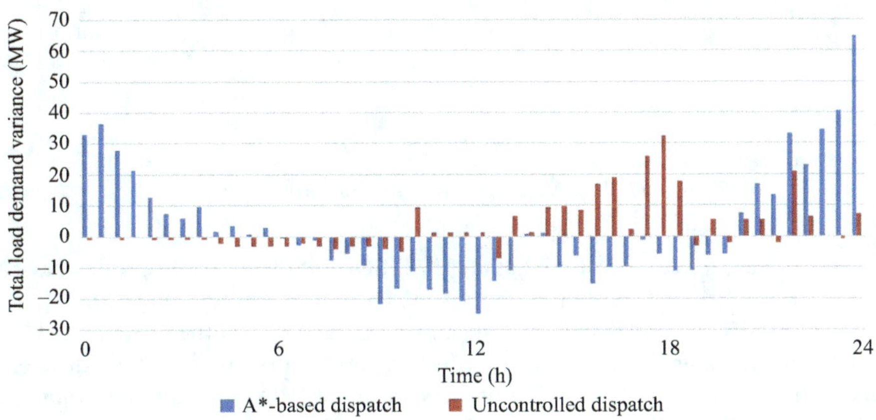

Figure 15.20 Network load variance caused by the integration of RGs and EVs

In terms of the network load variance due to the integration of RGs and EVs, the simulation results are demonstrated in Figure 15.20, where the comparison between the proposed optimal dispatch and uncontrolled charging is made and the load variance value $\Delta load$ is calculated as

$$\Delta load = load(with\ \ EVs\ \ \&\ \ RGs) - load(without\ \ EVs\ \ \&\ \ RGs). \qquad (15.54)$$

Therefore, the positive value implies valley filling while negative value peak shaving. From Figure 15.20, it is obvious to find that A*-based optimal dispatch strategy works well in levelling the network's load demand (peak shaving with EV load = −5.7 MW and RG output = 3.7 MW on average, and valley filling with EV load = 22.5MW and RG output = 4.1 MW on average), while uncontrolled charging of EVs pushes the peak of network's load even higher.

Furthermore, in order to test the stability of the proposed A*-based optimal dispatch strategy, the simulation for 7 incessant days was run and the results are demonstrated in Figure 15.21. It is clear that these 7 daily load curves are mostly overlapped with each other, indicating the stability of the proposed dispatch strategy. This figure also shows that load levelling is generally achieved on the daily basis without increasing the peak load demand of the network.

Further testing the stability of the proposed dispatch algorithm is done by examining the distribution the EV batteries' SOCs at the start of each day.

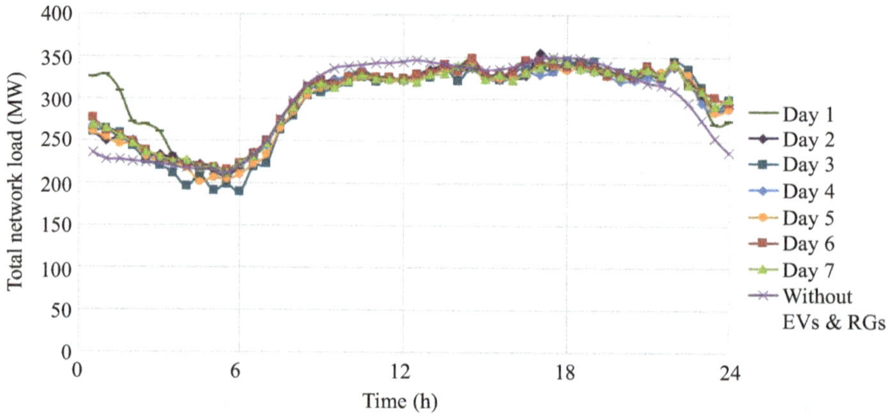

Figure 15.21 Network load demand during a week

In Figure 15.22, it is easy to find that the EVs' SOCs at the end of every day have a similar probability distribution, with most of EVs' SOCs in the range [0.9,1] and the smallest SOC no less than 0.5. If the initial SOCs of EVs at the beginning of the simulation is assigned based on a different normal distribution with different means, the distribution of EV SOCs at the start of the following days are evaluated and shown in Table 15.6. It is clear from this table that starting from different assignments of initial SOCs, the EVs' SOCs at the outset of a day always stabilize to a distribution that has a mean of 0.88 and a standard deviation of 0.15 on average, with very slight variation. Again, the stability of the proposed algorithm is verified.

Table 15.6 Key factors of EVs' SOC distribution at the end of a day tested with different mean values of the random assignments of the initial SOCs

Initial SOC assignment	**Mean of SOCs**		**Standard deviation of SOCs**	
	Mean	**Standard deviation**	**Mean**	**Standard deviation**
$\mu = \mathbf{0.5}, \sigma = \mathbf{0.1}$	0.88	0.01	0.15	0.01
$\mu = \mathbf{0.6}, \sigma = \mathbf{0.1}$	0.88	0.01	0.15	0.01
$\mu = \mathbf{0.7}, \sigma = \mathbf{0.1}$	0.88	0.01	0.15	0.01

15.6 Complexity discussion

In this section, the computational and communication complexities of A*-based algorithm are discussed, with the pseudo code presented below.

The number of states it iterates grows linearly with the number of possible assignments of X, as can be seen from line 16 of the pseudo code. If the control

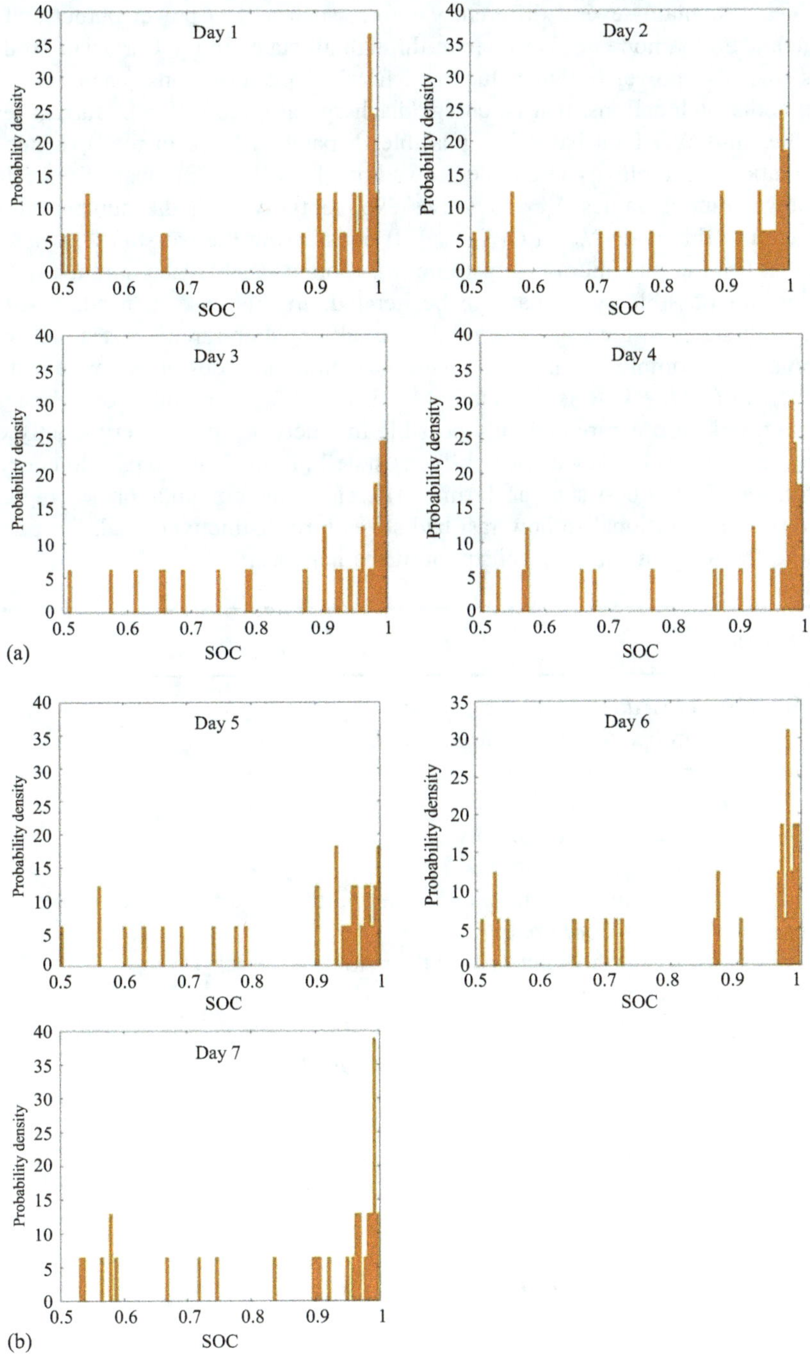

Figure 15.22 SOC distributions at the end of every day

variables X contain several different variables such as RG power output and EV dispatch action, a node needs to iterate through all states in the Cartesian products of its own RG power output values and EVs' dispatch actions (each EV has 7 possible dispatch actions, that is, charge/discharge at three different current levels and idle, thus N_{ev} EVs have $7^{N_{ev}}$ possible dispatch actions in total), hence the computational complexity at a node grows linearly with the number of discretized RG power output values N_p and grows exponentially with the number of EVs connected to this node N_{ev} in $O(N_p 7^{N_{ev}})$. If considering the worst case, where the state that has the maximum utility value is the only feasible state and the optimal solution, all possible states have to be iterated. In this case, it needs to iterate through all states in the Cartesian product of all of its children's states, hence the computational complexity at a node with N_{chi} children nodes grows exponentially with N_{chi} in $O(M^{N_{chi}})$. M is the number of different states a child node generates. However, this case is rare and only possible in a network that is very strictly constrained and severely loaded. Generally, as stated in line 7 of the pseudo code, the algorithm picks the best state each time. Therefore, the computational complexity is inversely proportional to how fast bad states turn distinctively bad, but proportional to the level of capacity constraint of the network.

A* algorithm

---

1. Initialize *closedlist* := *the empty set*;
2. Initialize an *openlist* for each *leafnode*;
3. $openlist_i := \{(leafnode_i, 0)\}$;
4. $U(leafnode_i, 0) := 0$;
5. while any *openlist* is not empty
6. *for each* non-empty *openlist*
7. *currentnode* = the node in *openlist* having the lowest utility value U;
8. *if currentnode* = *root node*
9. {We have found the solution and thus stop the search;}
10. *if* N_{chi} (currentnode) > 1
11.   {N_{chi} *openlists that have the same currentnode are combined*;
12.   *Remove* $(currentnode, \mathbf{f}_{chi \to current})$ *from the* $openlist_{comb}$; }
13.   *else*
14.   {*Remove* $(currentnode, f_{chi \to current})$ *from the openlist*; }
15.   *Generate Nextnode* := $par(currentnode)$;
16.   *for each possible assignment of* X *at currentnode*
17. $P_X := powercal(X)$;
18. $U_X := utilitycal(X)$;
19. $f_{current \to next} := sum(\mathbf{f}_{chi \to current}) + P_X$;
20.   *if* $f_{current \to next} > C_{current \to next}$
21.   {*continues*; }
22. $UtilityValue := U(currentnode, \mathbf{f}_{chi \to current}) + U_X$;
23.   *if* $(Nextnode, f_{current \to next})$ *is in the openlist*

```
24.         {if U(Nextnode, f_current→next) > UtilityValue
25.           {U(Nextnode, f_current→next) := UtilityValue;
26.      Pstate(Nextnode, f_current→next) := {X, Pstate(currentnode, f_chi→current)}; }
27.           else
28.             {continue; }}
29.           if (Nextnode, f_current→next) is in the closedlist
30.           {if U(Nextnode, f_current→next) > UtilityValue
31.           {U(Nextnode, f_current→next) := UtilityValue;
32.      Pstate(Nextnode, f_current→next) := {X, Pstate(currentnode, f_chi→current)}; }
33.           else
34.             {continue; }}
35.      Add (Next node, f_current→next) to the corresponding openlist;
36.    end
37.      Add (current node, f_chi→current) to the closedlist;
38.  end
39. end
```

As the states generated by the A* algorithm will be sent to the central memory for comparisons and for future usage (lines 22–35), the communication complexity grows linearly with the computational complexity. Moreover, the total size of messages sent by the A* algorithm equals the sum of the size of messages that each node creates and sends, hence increases linearly with the size of the network in $O(N_v)$. N_v is the number of nodes in a network.

In contrast, the computational complexity of a centralized algorithm, such as simplex algorithm, which does not explicitly consider the topology of the network, grows exponentially with the size of the network, as discussed in [15]. Therefore, it will quickly become infeasible for a centralized algorithm to solve an optimal dispatch problem when the network expands or integrate with more EVs and RGs.

15.7 Conclusion

In this chapter, an A*-based optimal dispatch algorithm is proposed and discussed thoroughly. The feasibility and efficacy of the A* algorithm are verified by the simulation results of its application to the coordinated dispatch of EVs and RGs, where the uncertainties of RG power outputs and EVs' driving activities are also modelled and simulated. Its stability and feasibility in dispatching the RGs and EVs to best achieve the objectives demonstrate its robustness to overcome the uncertainties of RGs ad EVs and ensure a stable and good performance of the power network. It costs EV users £0.45 per vehicle per day averagely, which saves EV users around £103 per year per vehicle compared with uncontrolled charging of EVs, while ensuring all driving activities complete with at least 34% of battery's capacity remaining. Moreover, overall more than 99% of generated wind power is absorbed and utilized by the network to support various load demands. Different local network conditions and wind speeds result in the discrepant wind power utilization rates among different WTG sites. The proposed optimal coordination of EVs and RGs proves to work well and effectively in load levelling without intensifying the peak loading of a constrained distribution network. Tests in terms of total load demand curves of 7 successive days and distributions of EV SOCs at the start of each day all verify the stability of the proposed A*-based optimal dispatch strategy, as they all follow a stable and repeatable pattern, respectively, since day 2. (The results for day 1 is mostly dominated by transient responses of network.)

In the future, other published algorithms will be tested in the same scenario for further comparison and investigation, such as the centralized optimal dispatch model proposed in [30] that minimize load variance, the hierarchical control approach proposed in [31] that minimizes operating cost, the fuzzy logic controller proposed in [32] that improves the voltage profile of a network, the two-stage stochastic programming approach proposed in [33] that optimizes pricing and maximizes the energy aggregator's profits and so forth. By examining these existing approaches in the optimal coordination of EVs and RGs, different characteristics of these approaches can be better understood, such as computational and communication complexity and impacts on the power network and EVs. Therefore, better decision can be made in terms of which approach will be applied in reality.

References

[1] N. Masuch, J. Keiser, M. Lutzenberger, and S. Albayrak, "Wind power-aware vehicle-to-grid algorithms for sustainable EV energy management systems," in *2012 IEEE International Electric Vehicle Conference (IEVC)*, March 2012, pp. 1–7.

[2] L. Yang, J. Zhang, and H. Poor, "Risk-aware day-ahead scheduling and real-time dispatch for electric vehicle charging," *IEEE Transactions on Smart Grid*, vol. 5, no. 2, pp. 693–702, 2014.

[3] Y. Ma, T. Houghton, A. Cruden, and D. Infield, "Modeling the benefits of vehicle-to-grid technology to a power system," *IEEE Transactions on Power Systems*, vol. 27, no. 2, pp. 1012–1020, 2012.

[4] O. Erdinc, N. Paterakis, T. Mendes, A. Bakirtzis, and J. Catalao, "Smart household operation considering bi-directional EV and ess utilization by real-time pricing-based dr," *IEEE Transactions on Smart Grid*, vol. 6, no. 3, pp. 1281–1291, 2015.

[5] L. Hua, J. Wang, and C. Zhou, "Adaptive electric vehicle charging coordination on distribution network," *IEEE Transactions on Smart Grid*, vol. 5, no. 6, pp. 2666–2675, 2014.

[6] Z. Li, Q. Guo, H. Sun, Y. Wang, and S. Xin, "Emission-concerned wind-EV coordination on the transmission grid side with network constraints: Concept and case study," *IEEE Transactions on Smart Grid*, vol. 4, no. 3, pp. 1692–1704, 2013.

[7] E. B. Iversen, J. M. Morales, and H. Madsen, "Optimal charging of an electric vehicle using a markov decision process," *Applied Energy*, vol. 123, no. 0, pp. 1–12, 2014.

[8] T. Zhang, W. Chen, Z. Han, and Z. Cao, "Charging scheduling of electric vehicles with local renewable energy under uncertain electric vehicle arrival and grid power price," *IEEE Transactions on Vehicular Technology*, vol. 63, no. 6, pp. 2600–2612, 2014.

[9] A. Verma, R. Raj, M. Kumar, S. Ghandehariun, and A. Kumar, "Assessment of renewable energy technologies for charging electric vehicles in Canada," *Energy*, vol. 86, pp. 548 – 559, 2015.

[10] C. Li, C. Ahn, H. Peng, and J. Sun, "Synergistic control of plug-in vehicle charging and wind power scheduling," *IEEE Transactions on Power Systems*, vol. 28, no. 2, pp. 1113–1121, 2013.

[11] D. Dallinger, S. Gerda, and M. Wietschel, "Integration of intermittent renewable power supply using grid-connected vehicles – a 2030 case study for California and Germany," *Applied Energy*, vol. 104, no. 0, pp. 666–682, 2013.

[12] H. Fathabadi, "Utilization of electric vehicles and renewable energy sources used as distributed generators for improving characteristics of electric power distribution systems," *Energy*, vol. 90, Part 1, pp. 1100–1110, 2015.

[13] M. Singh, P. Kumar, and I. Kar, "Implementation of vehicle to grid infrastructure using fuzzy logic controller," *IEEE Transactions on Smart Grid*, vol. 3, no. 1, pp. 565–577, 2012.

[14] E. L. Karfopoulos and N. D. Hatziargyriou, "Distributed coordination of electric vehicles providing V2G services," *IEEE Transactions on Power Systems*, vol. 31, no. 1, pp. 329–338, 2016.

[15] S. Miller, S. D. Ramchurn, and A. Rogers, "Optimal decentralised dispatch of embedded generation in the smart grid," in *Proceedings of 11th International Conference on Autonomous Agents and Multi-Agent Systems (AAMAS)*, Spain, June 2012, pp. 281–288.

[16] T. L. Saaty, *The Analytic Hierarchy. Process*. 1em plus 0.5em minus 0.4em New York: McGraw Hill, 1980.

[17] L. Wang, S. Sharkh, and A. Chipperfield, "Optimal coordination of vehicle-to-grid batteries and renewable generators in a distribution system," *Energy*, vol. 113, pp. 1250 – 1264, 2016.

[18] L. Wang, S. Sharkh, and A. Chipperfield, "Optimal decentralized coordination of electric vehicles and renewable generators in a distribution network using a* search," *International Journal of Electrical Power & Energy Systems*, vol. 98, pp. 474–487, 2018.

[19] "UK generic distribution network," [Online], Oct 2013, Available at: http://www.sedg.ac.uk/ukgds.htm

[20] E. Sortomme, M. M. Hindi, S. D. J. MacPherson, and S. S. Venkata, "Coordinated charging of plug-in hybrid electric vehicles to minimize distribution system losses," *IEEE Transactions on Smart Grid*, vol. 2, no. 1, pp. 198–205, 2011.

[21] K. Clement-Nyns, E. Haesen, and J. Driesen, "The impact of charging plug-in hybrid electric vehicles on a residential distribution grid," *IEEE Transactions on Power Systems*, vol. 25, no. 1, pp. 371–380, 2010.

[22] "UK national grid," [Online], Jan 2014, Available at: http://www.bmreports.com/bwx_reporting.htm

[23] P. H. Winston, *Artificial Intelligence*. 1em plus 0.5em minus 0.4em USA: Addison Wesley, 1992.

[24] A. Lojowska, D. Kurowicka, G. Papaefthymiou, and L. van der Sluis, "Stochastic modeling of power demand due to evs using copula," *IEEE Transactions on Power Systems*, vol. 27, no. 4, pp. 1960–1968, 2012.

[25] G. Papaefthymiou and D. Kurowicka, "Using copulas for modeling stochastic dependence in power system uncertainty analysis," *IEEE Transactions on Power Systems*, vol. 24, no. 1, pp. 40–49, 2009.

[26] R.B.Nelsen, An Introduction to Copulas. *1em plus 0.5em minus 0.4em New York*: Springer, 1999.

[27] A. Sklar, "Fonctions de répartition à n dimensions et leurs marges," *Publications de l'Institut de Statistique de L'Université de Paris*, vol. 8, pp. 229–231, 1959.

[28] M. Markowitz, "Wells to wheels: electric car efficiency," [Online], Feb 2013, Available at: https://matter2energy.wordpress.com/2013/02/22/wells-to-wheels-electric-car-efficiency/

[29] "Wind speed distribution for different sites," [Online], 2016, Available at: http://www.rensmart.com/Weather/Metar

[30] X. Jiang, J. Wang, Y. Han, and Q. Zhao, "Coordination dispatch of electric vehicles charging/discharging and renewable energy resources power in microgrid," *Procedia Computer Science*, vol. 107, pp. 157–163, 2017.

[31] S. Gao, K. Chau, C. Liu, D. Wu, and C. Chan, "Integrated energy management of plug-in electric vehicles in power grid with renewables," *IEEE Transactions on Vehicular Technology*, vol. 63, no. 7, pp. 3019–3027, 2014.

[32] S. Faddel, A. A. Mohamed, and O. A. Mohammed, "Fuzzy logic-based autonomous controller for electric vehicles charging under different conditions in residential distribution systems," *Electric Power Systems Research*, vol. 148, pp. 48–58, 2017.

[33] J. Soares, M. A. F. Ghazvini, N. Borges, and Z. Vale, "Dynamic electricity pricing for electric vehicles using stochastic programming," *Energy*, vol. 122, pp. 111–127, 2017.

Index

www.ingramcontent.com/pod-product-compliance
Lightning Source LLC
LaVergne TN
LVHW011828200926
848484LV00002B/335
* 9 7 8 1 7 8 5 6 1 7 6 2 1 *